MATLAB®
Simulations for Radar Systems Design

Bassem R. Mahafza, Ph.D.
Decibel Research, Inc.
Huntsville, Alabama

Atef Z. Elsherbeni
Professor
Electrical Engineering Department
The University of Mississippi
Oxford, Mississippi

CHAPMAN & HALL/CRC

A CRC Press Company
Boca Raton   London   New York   Washington, D.C.

Library of Congress Cataloging-in-Publication Data

Mahafza, Bassem R.
  MATLAB simulations for radar systems design / Bassem R. Mahafza, Atef Z. Elsherbeni
       p. cm.
  Includes bibliographical references and index.
    ISBN 1-58488-392-8 (alk. paper)
    1. Radar–Computer simulation. 2. Radar–Equipment and
supplies–Design and construction–Data processing. 3. MATLAB. I.
Elsherbeni, Atef Z. II. Title
TK6585.M34 2003
621.3848′01′13—dc22   2003065397

This book contains information obtained from authentic and highly regarded sources. Reprinted material is quoted with permission, and sources are indicated. A wide variety of references are listed. Reasonable efforts have been made to publish reliable data and information, but the author and the publisher cannot assume responsibility for the validity of all materials or for the consequences of their use.

Neither this book nor any part may be reproduced or transmitted in any form or by any means, electronic or mechanical, including photocopying, microfilming, and recording, or by any information storage or retrieval system, without prior permission in writing from the publisher.

The consent of CRC Press LLC does not extend to copying for general distribution, for promotion, for creating new works, or for resale. Specific permission must be obtained in writing from CRC Press LLC for such copying.

Direct all inquiries to CRC Press LLC, 2000 N.W. Corporate Blvd., Boca Raton, Florida 33431.

Trademark Notice: Product or corporate names may be trademarks or registered trademarks, and are used only for identification and explanation, without intent to infringe.

Visit the CRC Press Web site at www.crcpress.com

© 2004 by Chapman & Hall/CRC CRC Press LLC

No claim to original U.S. Government works
International Standard Book Number 1-58488-392-8
Library of Congress Catalog Number 2003065397
Printed in the United States of America 1 2 3 4 5 6 7 8 9 0
Printed on acid-free paper

To: My wife and four sons;

Wayne and Shirley;

and

in the memory of my parents

Bassem R. Mahafza

To: My wife and children;

my mother;

and

in the memory of my father

Atef Z. Elsherbeni

Preface

The emphasis of *"MATLAB Simulations for Radar Systems Design"* is on radar systems design. However, a strong presentation of the theory is provided so that the reader will be equipped with the necessary background to perform radar systems analysis. The organization of this book is intended to teach a conceptual design process of radars and related trade-off analysis and calculations. It is intended to serve as an engineering reference for radar engineers working in the field of radar systems. The MATLAB®[1] code provided in this book is designed to provide the user with hands-on experience in radar systems, analysis and design.

A radar design case study is introduced in Chapter 1 and carried throughout the text, where the authors' view of how to design this radar is detailed and analyzed. Trade off analyses and calculations are performed. Additionally, several mini design case studies are scattered throughout the book.

"MATLAB Simulations for Radar Systems Design" is divided into two parts: Part I provides a comprehensive description of radar systems, analyses and design. A design case study, which is carried throughout the text, is introduced in Chapter 1. In each chapter the authors' view of how to design the case-study radar is presented based on the theory covered up to that point in the book. As the material coverage progresses through the book, and new theory is discussed, the design case-study requirements are changed and/or updated, and of course the design level of complexity is also increased. This design process is supported by a comprehensive set of MATLAB 6 simulations developed for this purpose. This part will serve as a valuable tool to students and radar engineers in helping them understand radar systems, design process. This includes 1) learning how to go about selecting different radar parameters to meet the design requirements; 2) performing detailed trade-off analysis in the context of radar sizing, modes of operations, frequency selection, waveforms and signal processing; 3) establishing and developing loss and error budgets associated with the design; and 4) generating an in-depth understanding of radar operations and design philosophy. Additionally, Part I includes several mini design case studies pertinent to different chapters in order to help enhance understanding of radar design in the context of the material presented in different chapters.

Part II includes few chapters that cover specialized radar topics, some of which is authored and/or coauthored by other experts in the field. The material

---

1. MATLAB is a registered trademark of the The MathWorks, Inc. For product information, please contact: The MathWorks, Inc., 3 Apple Hill Drive, Natick, MA 01760-2098 USA. Web: *www.mathworks.com*.

included in Part II is intended to further enhance the understanding of radar system analysis by providing detailed and comprehensive coverage of these radar related topics. For this purpose, MATLAB 6 code has also been developed and made available.

All MATLAB programs and functions provided in this book can be downloaded from the CRC Press Web site (*www.crcpress.com*). For this purpose, follow this procedure: 1) from your Web browser type *"http://www.crcpress.com"*, 2) click on *"Electronic Products"*, 3) click on *"Download & Updates"*, and finally 4) follow instructions of how to download a certain set of code off that Web page. Furthermore, this MATLAB code can also be downloaded from The MathWorks Web site by following these steps: 1) from your Web browser type: *"http://mathworks.com/matlabcentral/fileexchange/"*, 2) place the curser on *"Companion Software for Books"* and click on *"Communications"*. The MATLAB functions and programs developed in this book include all forms of the radar equation: pulse compression, stretch processing, matched filter, probability of detection calculations with all Swerling models, High Range Resolution (HRR), stepped frequency waveform analysis, ghk tracking filter, Kalman filter, phased array antennas, clutter calculations, radar ambiguity functions, ECM, chaff, and many more.

Chapter 1 describes the most common terms used in radar systems, such as range, range resolution, and Doppler frequency. This chapter develops the radar range equation. Finally, a radar design case study entitled *"MyRadar Design Case Study"* is introduced. Chapter 2 is intended to provide an overview of the radar probability of detection calculations and related topics. Detection of fluctuating targets including Swerling I, II, III, and IV models is presented and analyzed. Coherent and non-coherent integration are also introduced. Cumulative probability of detection analysis is in this chapter. Visit 2 of the design case study *"MyRadar"* is introduced.

Chapter 3 reviews radar waveforms, including CW, pulsed, and LFM. High Range Resolution (HRR) waveforms and stepped frequency waveforms are also analyzed. The concept of the Matched Filter (MF) is introduced and analyzed. Chapter 4 presents in detail the principles associated with the radar ambiguity function. This includes the ambiguity function for single pulse, Linear Frequency Modulated pulses, train of unmodulated pulses, Barker codes, and PRN codes. Pulse compression is introduced in Chapter 5. Both the MF and the stretch processors are analyzed.

Chapter 6 contains treatment of the concepts of clutter. This includes both surface and volume clutter. Chapter 7 presents clutter mitigation using Moving Target Indicator (MTI). Delay line cancelers implementation to mitigate the effects of clutter is analyzed.

Chapter 8 presents detailed analysis of Phased Arrays. Linear arrays are investigated and detailed and MATLAB code is developed to calculate and plot the associated array patterns. Planar arrays, with various grid configurations, are also presented.

Chapter 9 discusses target tracking radar systems. The first part of this chapter covers the subject of single target tracking. Topics such as sequential lobing, conical scan, monopulse, and range tracking are discussed in detail. The second part of this chapter introduces multiple target tracking techniques. Fixed gain tracking filters such as the $\alpha\beta$ and the $\alpha\beta\gamma$ filters are presented in detail. The concept of the Kalman filter is introduced. Special cases of the Kalman filter are analyzed in depth.

Chapter 10 is coauthored with Mr. J. Michael Madewell from the US Army Space and Missile Defense Command, in Huntsville, Alabama. This chapter presents an overview of Electronic Counter Measures (ECM) techniques. Topics such as self screening and stand off jammers are presented. Radar chaff is also analyzed and a chaff mitigation technique for Ballistic Missile Defense (BMD) is introduced.

Chapter 11 is concerned with the Radar Cross Section (RCS). RCS dependency on aspect angle, frequency, and polarization is discussed. The target scattering matrix is developed. RCS formulas for many simple objects are presented. Complex object RCS is discussed, and target fluctuation models are introduced. Chapter 12 is coauthored with Dr. Brian Smith from the US Army Aviation and Missile Command (AMCOM), Redstone Arsenal in Alabama. This chapter presents the topic of Tactical Synthetic Aperture Radar (SAR). The topics of this chapter include: SAR signal processing, SAR design considerations, and the SAR radar equation. Finally Chapter 13 presents an overview of signal processing.

Using the material presented in this book and the MATLAB code designed by the authors by any entity or person is strictly at will. The authors and the publisher are neither liable nor responsible for any material or non-material losses, loss of wages, personal or property damages of any kind, or for any other type of damages of any and all types that may be incurred by using this book.

Bassem R. Mahafza
Huntsville, Alabama
July, 2003

Atef Z. Elsherbeni
Oxford, Mississippi
July, 2003

Acknowledgment

The authors first would like to thank God for giving us the endurance and perseverance to complete this work. Many thanks are due to our families who have given up and sacrificed many hours in order to allow us to complete this book. The authors would like to also thank all of our colleagues and friends for their support during the preparation of this book. Special thanks are due to Brian Smith, James Michael Madewell, Patrick Barker, David Hall, Mohamed Al-Sharkawy, and Matthew Inman who have coauthored and/or reviewed some of the material in this reference book.

Table of Contents

Preface
Acknowledgment

PART I

Chapter 1

Introduction to Radar Basics **1**

1.1. Radar Classifications **1**
1.2. Range **3**
1.3. Range Resolution **5**
1.4. Doppler Frequency **7**
1.5. The Radar Equation **13**
    1.5.1. Radar Reference Range **19**
1.6. Search (Surveillance) **20**
    1.6.1. Mini Design Case Study 1.1 **25**
1.7. Pulse Integration **27**
    1.7.1. Coherent Integration **29**
    1.7.2. Non-Coherent Integration **30**
    1.7.3. Detection Range with Pulse Integration **30**
    1.7.4. Mini Design Case Study 1.2 **31**
1.8. Radar Losses **35**
    1.8.1. Transmit and Receive Losses **36**
    1.8.2. Antenna Pattern Loss and Scan Loss **36**
    1.8.3. Atmospheric Loss **37**
    1.8.4. Collapsing Loss **37**
    1.8.5. Processing Losses **38**
    1.8.6. Other Losses **41**
1.9. *"MyRadar"* Design Case Study - Visit 1 **41**

  1.9.1 Authors and Publisher Disclaimer **41**
  1.9.2. Problem Statement **42**
  1.9.3. A Design **42**
  1.9.4. A Design Alternative **46**
 1.10. MATLAB Program and Function Listings **48**
  Listing 1.1. Function *"radar_eq.m"* **48**
  Listing 1.2. Program *"fig1_12.m"* **49**
  Listing 1.3. Program *"fig1_13.m"* **50**
  Listing 1.4. Program *"ref_snr.m"* **50**
  Listing 1.5. Function *"power_aperture.m"* **51**
  Listing 1.6. Program *"fig1_16.m"* **51**
  Listing 1.7. Program *"casestudy1_1.m"* **52**
  Listing 1.8. Program *"fig1_19.m"* **54**
  Listing 1.9. Program *"fig1_21.m"* **55**
  Listing 1.10. Function *"pulse_integration.m"* **55**
  Listing 1.11. Program *"myradarvisit1_1.m"* **55**
  Listing 1.12. Program *"fig1_27.m"* **56**

Appendix 1A

Pulsed Radar **59**

 1A.1. Introduction **59**
 1A.2. Range and Doppler Ambiguities **60**
 1A.3. Resolving Range Ambiguity **61**
 1A.4. Resolving Doppler Ambiguity **64**

Appendix 1B

Noise Figure **69**

Chapter 2

Radar Detection **75**

 2.1. Detection in the Presence of Noise **75**
 2.2. Probability of False Alarm **79**
 2.3. Probability of Detection **81**
 2.4. Pulse Integration **83**
  2.4.1. Coherent Integration **83**
  2.4.2. Non-Coherent Integration **84**
  2.4.3. Mini Design Case Study 2.1 **88**
 2.5. Detection of Fluctuating Targets **90**
  2.5.1. Threshold Selection **92**

2.6. Probability of Detection Calculation **95**
  2.6.1. Detection of Swerling V Targets **96**
  2.6.2. Detection of Swerling I Targets **97**
  2.6.3. Detection of Swerling II Targets **100**
  2.6.4. Detection of Swerling III Targets **100**
  2.6.5. Detection of Swerling IV Targets **102**
2.7. The Radar Equation Revisited **104**
2.8. Cumulative Probability of Detection **106**
  2.8.1. Mini Design Case Study 2.2 **107**
2.9. Constant False Alarm Rate (CFAR) **109**
  2.9.1. Cell-Averaging CFAR (Single Pulse) **110**
  2.9.2. Cell-Averaging CFAR with Non-Coherent Integration **111**
2.10. *"MyRadar"* Design Case Study - Visit 2 **113**
  2.10.1. Problem Statement **113**
  2.10.2. A Design **113**
    2.10.2.1. Single Pulse (per Frame) Design Option **114**
    2.10.2.2. Non-Coherent Integration Design Option **116**
2.11. MATLAB Program and Function Listings **117**
  Listing 2.1. Program *"fig2_2.m"* **118**
  Listing 2.2. Function *"que_func.m"* **119**
  Listing 2.3. Program *"fig2_3.m"* **119**
  Listing 2.4. Function *"marcumsq.m"* **119**
  Listing 2.5. Program *"prob_snr1.m"* **120**
  Listing 2.6. Program *"fig2_6a.m"* **121**
  Listing 2.7. Function *"improv_fac.m"* **121**
  Listing 2.8 Program *"fig2_6b.m"* **122**
  Listing 2.9. Function *"incomplete_gamma.m"* **122**
  Listing 2.10. Function *"factor.m"* **124**
  Listing 2.11. Program *"fig2_7.m"* **124**
  Listing 2.12. Function *"threshold.m"* **124**
  Listing 2.13. Program *"fig2_8.m"* **125**
  Listing 2.14. Function *"pd_swerling5.m"* **125**
  Listing 2.15. Program *"fig2_9.m"* **126**
  Listing 2.16. Function *"pd_swerling1.m"* **127**
  Listing 2.17. Program *"fig2_10.m"* **128**
  Listing 2.18. Program *"fig2_11ab.m"* **128**
  Listing 2.19. Function *"pd_swerling2.m"* **129**
  Listing 2.20. Program *"fig2_12.m"* **130**
  Listing 2.21. Function *"pd_swerling3.m"* **130**
  Listing 2.22. Program *"fig2_13.m"* **131**
  Listing 2.23 Function *"pd_swerling4.m"* **132**
  Listing 2.24. Program *"fig2_14.m"* **133**

Listing 2.25. Function *"fluct_loss.m"* **134**
Listing 2.26. Program *"fig2_16.m"* **136**
Listing 2.27. Program *"myradar_visit2_1.m"* **136**
Listing 2.28. Program *"myradar_visit2_2.m"* **137**
Listing 2.29. Program *"fig2_21.m"* **139**

Chapter 3

Radar Waveforms **141**

3.1. Low Pass, Band Pass Signals and Quadrature Components **141**
3.2. The Analytic Signal **143**
3.3. CW and Pulsed Waveforms **144**
3.4. Linear Frequency Modulation Waveforms **148**
3.5. High Range Resolution **153**
3.6. Stepped Frequency Waveforms **154**
  3.6.1. Range Resolution and Range Ambiguity in SWF **157**
  3.6.2. Effect of Target Velocity **162**
3.7. The Matched Filter **165**
3.8. The Replica **169**
3.9. Matched Filter Response to LFM Waveforms **170**
3.10. Waveform Resolution and Ambiguity **172**
  3.10.1. Range Resolution **172**
  3.10.2. Doppler Resolution **174**
  3.10.3. Combined Range and Doppler Resolution **176**
3.11. *"Myradar"* Design Case Study - Visit 3 **177**
  3.11.1. Problem Statement **177**
  3.11.2. A Design **177**
3.12. MATLAB Program and Function Listings **182**
  Listing 3.1. Program *"fig3_7.m"* **182**
  Listing 3.2. Program *"fig3_8.m"* **182**
  Listing 3.3. Function *"hrr_profile.m"* **183**
  Listing 3.4. Program *"fig3_17.m"* **185**

Chapter 4

The Radar Ambiguity Function **187**

4.1. Introduction **187**
4.2. Examples of the Ambiguity Function **188**
  4.2.1. Single Pulse Ambiguity Function **188**
  4.2.2. LFM Ambiguity Function **193**

4.2.3. Coherent Pulse Train Ambiguity Function **198**
4.3. Ambiguity Diagram Contours **204**
4.4. Digital Coded Waveforms **206**
   4.4.1. Frequency Coding (Costas Codes) **206**
   4.4.2. Binary Phase Codes **209**
   4.4.3. Pseudo-Random (PRN) Codes **215**
4.5. *"MyRadar"* Design Case Study - Visit 4 **223**
   4.5.1. Problem Statement **223**
   4.5.2. A Design **223**
4.6. MATLAB Program and Function Listings **224**
   Listing 4.1. Function *"single_pulse_ambg.m"* **224**
   Listing 4.2. Program *"fig4_2.m"* **225**
   Listing 4.3. Program *"fig4_4.m"* **226**
   Listing 4.4. Function *"lfm_ambg.m"* **226**
   Listing 4.5. Program *"fig4_5.m"* **227**
   Listing 4.6. Program *"fig4_6.m"* **227**
   Listing 4.7. Function *"train_ambg.m"* **228**
   Listing 4.8. Program *"fig4_8.m"* **229**
   Listing 4.9. Function *"barker_ambg.m"* **229**
   Listing 4.10. Function *"prn_ambg.m"* **231**
   Listing 4.11. Program *"myradar_visit4.m"* **232**

Chapter 5

Pulse Compression **235**

5.1. Time-Bandwidth Product **235**
5.2. Radar Equation with Pulse Compression **236**
5.3. LFM Pulse Compression **237**
   5.3.1. Correlation Processor **240**
   5.3.2. Stretch Processor **247**
   5.3.3. Distortion Due to Target Velocity **254**
5.4. *"MyRadar"* Design Case Study - Visit 5 **257**
   5.4.1. Problem Statement **257**
   5.4.2. A Design **257**
5.5. MATLAB Program and Function Listings **262**
   Listing 5.1. Program *"fig5_3.m"* **262**
   Listing 5.2. Function *"matched_filter.m"* **262**
   Listing 5.3. Function *"power_integer_2.m"* **264**
   Listing 5.4. Function *"stretch.m"* **264**
   Listing 5.5. Program *"fig5_14.m"* **265**

Chapter 6

Surface and Volume Clutter 267

6.1. Clutter Definition **267**
6.2. Surface Clutter **268**
  6.2.1. Radar Equation for Area Clutter - Airborne Radar **270**
  6.6.2. Radar Equation for Area Clutter - Ground Based Radar **272**
6.3. Volume Clutter **280**
  6.3.1. Radar Equation for Volume Clutter **282**
6.4. Clutter Statistical Models **283**
6.5. *"MyRadar"* Design Case Study - Visit 6 **284**
  6.5.1. Problem Statement **284**
  6.5.2. A Design **284**
6.6. MATLAB Program and Function Listings **288**
  Listing 6.1. Function *"clutter_rcs.m"* **288**
  Listing 6.2. Program *"myradar_visit6.m"* **290**

Chapter 7

Moving Target Indicator (MTI) and Clutter Mitigation 293

7.1. Clutter Spectrum **293**
7.2. Moving Target Indicator (MTI) **294**
7.3. Single Delay Line Canceler **296**
7.4. Double Delay Line Canceler **298**
7.5. Delay Lines with Feedback (Recursive Filters) **300**
7.6. PRF Staggering **302**
7.7. MTI Improvement Factor **303**
  7.7.1. Two-Pulse MTI Case **307**
  7.7.2. The General Case **309**
7.8. *"MyRadar"* Design Case Study - Visit 7 **309**
  7.8.1. Problem Statement **309**
  7.8.2. A Design **310**
7.9. MATLAB Program and Function Listings **313**
  Listing 7.1. Function *"single_canceler.m"* **313**
  Listing 7.2. Function *"double_canceler.m"* **313**
  Listing 7.3. Program *"fig7_9.m"* **314**
  Listing 7.4. Program *"fig7_10.m"* **314**
  Listing 7.5. Program *"fig7_11.m"* **315**
  Listing 7.4. Program *"myradar_visit7.m"* **315**

Chapter 8

Phased Arrays 319

8.1. Directivity, Power Gain, and Effective Aperture **319**
8.2. Near and Far Fields **321**
8.3. General Arrays **322**
8.4. Linear Arrays **325**
  8.4.1. Array Tapering **330**
  8.4.2. Computation of the Radiation Pattern via the DFT **333**
8.5. Planar Arrays **341**
8.6. Array Scan Loss **375**
8.7. *"MyRadar"* Design Case Study - Visit 8 **378**
  8.7.1. Problem Statement **378**
  8.7.2. A Design **378**
8.8. MATLAB Program and Function Listings **380**
  Listing 8.1. Program *"fig8_5.m"* **381**
  Listing 8.2. Program *"fig8_7.m"* **382**
  Listing 8.3. Function *"linear_array.m"* **382**
  Listing 8.4. Program *"circular_array.m"* **384**
  Listing 8.5. Function *"rect_array.m"* **391**
  Listing 8.6. Function *"circ_array.m"* **395**
  Listing 8.7. Function *"rec_to_circ.m"* **399**
  Listing 8.8. Program *"fig8_53.m"* **400**

Chapter 9

Target Tracking 401

Single Target Tracking
9.1. Angle Tracking **401**
  9.1.1. Sequential Lobing **402**
  9.1.2. Conical Scan **403**
9.2. Amplitude Comparison Monopulse **407**
9.3. Phase Comparison Monopulse **416**
9.4. Range Tracking **418**

Multiple Target Tracking
9.5. Track-While-Scan (TWS) **420**
9.6. State Variable Representation of an LTI System **422**
9.7. The LTI System of Interest **426**
9.8. Fixed-Gain Tracking Filters **427**
  9.8.1. The $\alpha\beta$ Filter **430**
  9.8.2. The $\alpha\beta\gamma$ Filter **434**

9.9. The Kalman Filter **445**
    9.9.1. The Singer $\alpha\beta\gamma$-Kalman Filter **447**
    9.9.2. Relationship between Kalman and $\alpha\beta\gamma$ Filters **450**
9.10. *"MyRadar"* Design Case Study - Visit 9 **454**
    9.10.1. Problem Statement **454**
    9.10.2. A Design **454**
9.11. MATLAB Program and Function Listings **462**
    Listing 9.1. Function *"mono_pulse.m"* **462**
    Listing 9.2. Function *"ghk_tracker.m"* **463**
    Listing 9.3. Program *"fig9_21.m"* **464**
    Listing 9.4. Function *"kalman_filter.m"* **465**
    Listing 9.5. Program *"fig9_28.m"* **466**
    Listing 9.6. Function *"maketraj.m"* **467**
    Listing 9.7. Function *"addnoise.m"* **468**
    Listing 9.8. Function *"kalfilt.m"* **469**

PART II

Chapter 10

Electronic Countermeasures (ECM) **471**

10.1. Introduction **471**
10.2. Jammers **472**
    10.2.1. Self-Screening Jammers (SSJ) **473**
    10.2.2. Stand-Off Jammers (SOJ) **480**
10.3. Range Reduction Factor **482**
10.4. Chaff **485**
    10.4.1. Multiple MTI Chaff Mitigation Technique **486**
10.5. MATLAB Program and Function Listings **493**
    Listing 10.1. Function *"ssj_req.m"* **493**
    Listing 10.2. Function *"sir.m"* **495**
    Listing 10.3. Function *"burn_thru.m"* **495**
    Listing 10.4. Function *"soj_req.m"* **496**
    Listing 10.5. Function *"range_red_factor.m"* **497**
    Listing 10.6. Program *"fig8_10.m"* **498**

Chapter 11

Radar Cross Section (RCS) **501**

11.1. RCS Definition **501**
11.2. RCS Prediction Methods **503**
11.3. Dependency on Aspect Angle and Frequency **504**

11.4. RCS Dependency on Polarization **508**
   11.4.1. Polarization **510**
   11.4.2. Target Scattering Matrix **515**
11.5 RCS of Simple Objects **517**
   11.5.1. Sphere **518**
   11.5.2. Ellipsoid **520**
   11.5.3. Circular Flat Plate **522**
   11.5.4. Truncated Cone (Frustum) **524**
   11.5.5. Cylinder **528**
   11.5.6. Rectangular Flat Plate **531**
   11.5.7. Triangular Flat Plate **534**
11.6. Scattering From a Dielectric-Capped Wedge **536**
   11.6.1. Far Scattered Field **541**
   11.6.2. Plane Wave Excitation **542**
   11.6.3. Special Cases **542**
11.7. RCS of Complex Objects **557**
11.8. RCS Fluctuations and Statistical Models **558**
   11.8.1. RCS Statistical Models - Scintillation Models **559**
11.9. MATLAB Program and Function Listings **560**
   Listing 11.1. Function *"rcs_aspect.m"* **561**
   Listing 11.2. Function *"rcs_frequency.m"* **561**
   Listing 11.3. Program *"example11_1.m"* **562**
   Listing 11.4. Program *"rcs_sphere.m"* **564**
   Listing 11.5. Function *"rcs_ellipsoid.m"* **565**
   Listing 11.6. Program *"fig11_18a.m"* **566**
   Listing 11.7. Function *"rcs_circ_plate.m"* **567**
   Listing 11.8. Function *"rcs_frustum.m"* **568**
   Listing 11.9. Function *"rcs_cylinder.m"* **570**
   Listing 11.10. Function *"rcs_rect_plate.m"* **572**
   Listing 11.11. Function *"rcs_isosceles.m"* **573**
   Listing 11.12. Program *"Capped_WedgeTM.m"* **574**
   Listing 11.13. Function *"DielCappedWedgeTM Fields_LS.m"* **580**
   Listing 11.14. Function *"DielCappedWedgeTMFields_PW.m"* **581**
   Listing 11.15. Function *"polardb.m"* **582**
   Listing 11.16. Function *"dbesselj.m"* **586**
   Listing 11.17. Function *"dbessely.m"* **586**
   Listing 11.18. Function *"dbesselh.m"* **586**
   Listing 11.19. Program *"rcs_cylinder_complex.m"* **586**
   Listing 11.20. Program *"Swerling_models.m"* **587**

Chapter 12

High Resolution Tactical Synthetic Aperture Radar (TSAR) 589

12.1. Introduction **589**
12.2. Side Looking SAR Geometry **590**
12.3. SAR Design Considerations **592**
12.4. SAR Radar Equation **599**
12.5. SAR Signal Processing **600**
12.6. Side Looking SAR Doppler Processing **601**
12.7. SAR Imaging Using Doppler Processing **606**
12.8. Range Walk **606**
12.9. A Three-Dimensional SAR Imaging Technique **608**
   12.9.1. Background **608**
   12.9.2. DFTSQM Operation and Signal Processing **609**
   12.9.3. Geometry for DFTSQM SAR Imaging **612**
   12.9.4. Slant Range Equation **614**
   12.9.5. Signal Synthesis **616**
   12.9.6. Electronic Processing **618**
   12.9.7. Derivation of Eq. (12.71) **619**
   12.9.8. Non-Zero Taylor Series Coefficients for the kth Range Cell **621**
12.10. MATLAB Programs and Functions **623**
   Listing 12.1. Program *"fig12_12-13.m"* **623**

Chapter 13

Signal Processing 625

13.1. Signal and System Classifications **625**
13.2. The Fourier Transform **627**
13.3. The Fourier Series **629**
13.4. Convolution and Correlation Integrals **631**
13.5. Energy and Power Spectrum Densities **632**
13.6. Random Variables **635**
13.7. Multivariate Gaussian Distribution **638**
13.8. Random Processes **641**
13.9. Sampling Theorem **642**
13.10. The Z-Transform **644**
13.11. The Discrete Fourier Transform **648**
13.12. Discrete Power Spectrum **648**
13.13. Windowing Techniques **650**
13.14. MATLAB Programs **654**
   Listing 13.1. Program *"figs13.m"* **654**

Appendix 13A

    Fourier Transform Table **657**

Appendix 13B

    Some Common Probability Densities **659**

Appendix 13C

    Z - Transform Table **661**

Chapter 14

    MATLAB Program and Function Name List **663**

***Bibliography* 671**
***Index* 677**

Chapter 1
Introduction to Radar Basics

1.1. Radar Classifications

The word radar is an abbreviation for RAdio Detection And Ranging. In general, radar systems use modulated waveforms and directive antennas to transmit electromagnetic energy into a specific volume in space to search for targets. Objects (targets) within a search volume will reflect portions of this energy (radar returns or echoes) back to the radar. These echoes are then processed by the radar receiver to extract target information such as range, velocity, angular position, and other target identifying characteristics.

Radars can be classified as ground based, airborne, spaceborne, or ship based radar systems. They can also be classified into numerous categories based on the specific radar characteristics, such as the frequency band, antenna type, and waveforms utilized. Another classification is concerned with the mission and/or the functionality of the radar. This includes: weather, acquisition and search, tracking, track-while-scan, fire control, early warning, over the horizon, terrain following, and terrain avoidance radars. Phased array radars utilize phased array antennas, and are often called multifunction (multimode) radars. A phased array is a composite antenna formed from two or more basic radiators. Array antennas synthesize narrow directive beams that may be steered mechanically or electronically. Electronic steering is achieved by controlling the phase of the electric current feeding the array elements, and thus the name phased array is adopted.

Radars are most often classified by the types of waveforms they use, or by their operating frequency. Considering the waveforms first, radars can be Continuous Wave (CW) or Pulsed Radars (PR).[1] CW radars are those that continuously emit electromagnetic energy, and use separate transmit and receive antennas. Unmodulated CW radars can accurately measure target radial velocity (Doppler shift) and angular position. Target range information cannot be extracted without utilizing some form of modulation. The primary use of unmodulated CW radars is in target velocity search and track, and in missile guidance. Pulsed radars use a train of pulsed waveforms (mainly with modulation). In this category, radar systems can be classified on the basis of the Pulse Repetition Frequency (PRF) as low PRF, medium PRF, and high PRF radars. Low PRF radars are primarily used for ranging where target velocity (Doppler shift) is not of interest. High PRF radars are mainly used to measure target velocity. Continuous wave as well as pulsed radars can measure both target range and radial velocity by utilizing different modulation schemes.

Table 1.1 has the radar classifications based on the operating frequency.

TABLE 1.1. Radar frequency bands.

Letter designation	Frequency (GHz)	New band designation (GHz)
HF	0.003 - 0.03	A
VHF	0.03 - 0.3	A<0.25; B>0.25
UHF	0.3 - 1.0	B<0.5; C>0.5
L-band	1.0 - 2.0	D
S-band	2.0 - 4.0	E<3.0; F>3.0
C-band	4.0 - 8.0	G<6.0; H>6.0
X-band	8.0 - 12.5	I<10.0; J>10.0
Ku-band	12.5 - 18.0	J
K-band	18.0 - 26.5	J<20.0; K>20.0
Ka-band	26.5 - 40.0	K
MMW	Normally >34.0	L<60.0; M>60.0

High Frequency (HF) radars utilize the electromagnetic waves' reflection off the ionosphere to detect targets beyond the horizon. Very High Frequency (VHF) and Ultra High Frequency (UHF) bands are used for very long range Early Warning Radars (EWR). Because of the very large wavelength and the sensitivity requirements for very long range measurements, large apertures are needed in such radar systems.

---

1. See Appendix 1A.

Radars in the L-band are primarily ground based and ship based systems that are used in long range military and air traffic control search operations. Most ground and ship based medium range radars operate in the S-band. Most weather detection radar systems are C-band radars. Medium range search and fire control military radars and metric instrumentation radars are also C-band.

The X-band is used for radar systems where the size of the antenna constitutes a physical limitation; this includes most military multimode airborne radars. Radar systems that require fine target detection capabilities and yet cannot tolerate the atmospheric attenuation of higher frequency bands may also be X-band. The higher frequency bands (Ku, K, and Ka) suffer severe weather and atmospheric attenuation. Therefore, radars utilizing these frequency bands are limited to short range applications, such as police traffic radar, short range terrain avoidance, and terrain following radar. Milli-Meter Wave (MMW) radars are mainly limited to very short range Radio Frequency (RF) seekers and experimental radar systems.

1.2. Range

Figure 1.1 shows a simplified pulsed radar block diagram. The time control box generates the synchronization timing signals required throughout the system. A modulated signal is generated and sent to the antenna by the modulator/transmitter block. Switching the antenna between the transmitting and receiving modes is controlled by the duplexer. The duplexer allows one antenna to be used to both transmit and receive. During transmission it directs the radar electromagnetic energy towards the antenna. Alternatively, on reception, it directs the received radar echoes to the receiver. The receiver amplifies the radar returns and prepares them for signal processing. Extraction of target information is performed by the signal processor block. The target's range, R, is computed by measuring the time delay, Δt, it takes a pulse to travel the two-way path between the radar and the target. Since electromagnetic waves travel at the speed of light, $c = 3 \times 10^8 m/\text{sec}$, then

$$R = \frac{c\Delta t}{2} \tag{1.1}$$

where R is in meters and Δt is in seconds. The factor of $\frac{1}{2}$ is needed to account for the two-way time delay.

In general, a pulsed radar transmits and receives a train of pulses, as illustrated by Fig. 1.2. The Inter Pulse Period (IPP) is T, and the pulsewidth is τ. The IPP is often referred to as the Pulse Repetition Interval (PRI). The inverse of the PRI is the PRF, which is denoted by f_r,

$$f_r = \frac{1}{PRI} = \frac{1}{T} \tag{1.2}$$

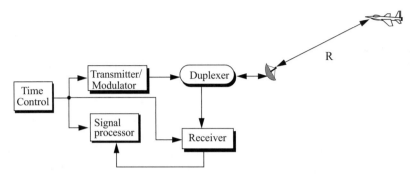

Figure 1.1. A simplified pulsed radar block diagram.

During each PRI the radar radiates energy only for τ seconds and listens for target returns for the rest of the PRI. The radar transmitting duty cycle (factor) d_t is defined as the ratio $d_t = \tau/T$. The radar average transmitted power is

$$P_{av} = P_t \times d_t, \tag{1.3}$$

where P_t denotes the radar peak transmitted power. The pulse energy is $E_p = P_t \tau = P_{av} T = P_{av}/f_r$.

The range corresponding to the two-way time delay T is known as the radar unambiguous range, R_u. Consider the case shown in Fig. 1.3. Echo 1 represents the radar return from a target at range $R_1 = c\Delta t/2$ due to pulse 1. Echo 2 could be interpreted as the return from the same target due to pulse 2, or it may be the return from a faraway target at range R_2 due to pulse 1 again. In this case,

$$R_2 = \frac{c\Delta t}{2} \quad or \quad R_2 = \frac{c(T + \Delta t)}{2} \tag{1.4}$$

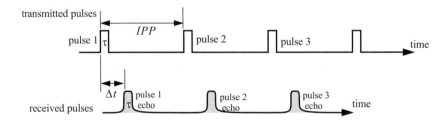

Figure 1.2. Train of transmitted and received pulses.

Range Resolution

Clearly, range ambiguity is associated with echo 2. Therefore, once a pulse is transmitted the radar must wait a sufficient length of time so that returns from targets at maximum range are back before the next pulse is emitted. It follows that the maximum unambiguous range must correspond to half of the PRI,

$$R_u = c\frac{T}{2} = \frac{c}{2f_r} \tag{1.5}$$

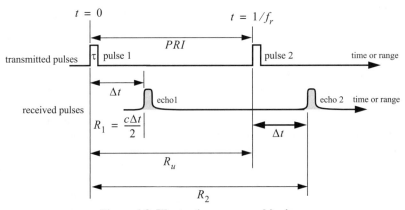

Figure 1.3. Illustrating range ambiguity.

1.3. Range Resolution

Range resolution, denoted as ΔR, is a radar metric that describes its ability to detect targets in close proximity to each other as distinct objects. Radar systems are normally designed to operate between a minimum range R_{min}, and maximum range R_{max}. The distance between R_{min} and R_{max} is divided into M range bins (gates), each of width ΔR,

$$M = (R_{max} - R_{min})/\Delta R \tag{1.6}$$

Targets separated by at least ΔR will be completely resolved in range. Targets within the same range bin can be resolved in cross range (azimuth) utilizing signal processing techniques. Consider two targets located at ranges R_1 and R_2, corresponding to time delays t_1 and t_2, respectively. Denote the difference between those two ranges as ΔR:

$$\Delta R = R_2 - R_1 = c\frac{(t_2 - t_1)}{2} = c\frac{\delta t}{2} \tag{1.7}$$

Now, try to answer the following question: What is the minimum δt such that target 1 at R_1 and target 2 at R_2 will appear completely resolved in range (different range bins)? In other words, what is the minimum ΔR?

First, assume that the two targets are separated by $c\tau/4$, where τ is the pulsewidth. In this case, when the pulse trailing edge strikes target 2 the leading edge would have traveled backwards a distance $c\tau$, and the returned pulse would be composed of returns from both targets (i.e., unresolved return), as shown in Fig. 1.4a. However, if the two targets are at least $c\tau/2$ apart, then as the pulse trailing edge strikes the first target the leading edge will start to return from target 2, and two distinct returned pulses will be produced, as illustrated by Fig. 1.4b. Thus, ΔR should be greater or equal to $c\tau/2$. And since the radar bandwidth B is equal to $1/\tau$, then

$$\Delta R = \frac{c\tau}{2} = \frac{c}{2B} \tag{1.8}$$

In general, radar users and designers alike seek to minimize ΔR in order to enhance the radar performance. As suggested by Eq. (1.8), in order to achieve fine range resolution one must minimize the pulsewidth. However, this will reduce the average transmitted power and increase the operating bandwidth. Achieving fine range resolution while maintaining adequate average transmitted power can be accomplished by using pulse compression techniques.

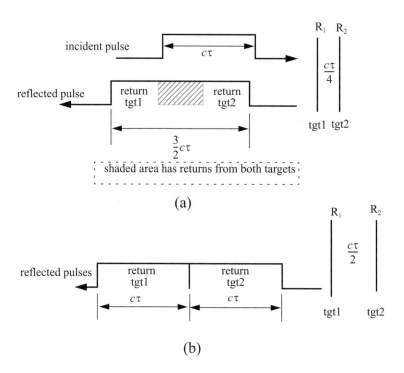

Figure 1.4. (a) Two unresolved targets. (b) Two resolved targets.

1.4. Doppler Frequency

Radars use Doppler frequency to extract target radial velocity (range rate), as well as to distinguish between moving and stationary targets or objects such as clutter. The Doppler phenomenon describes the shift in the center frequency of an incident waveform due to the target motion with respect to the source of radiation. Depending on the direction of the target's motion, this frequency shift may be positive or negative. A waveform incident on a target has equiphase wavefronts separated by λ, the wavelength. A closing target will cause the reflected equiphase wavefronts to get closer to each other (smaller wavelength). Alternatively, an opening or receding target (moving away from the radar) will cause the reflected equiphase wavefronts to expand (larger wavelength), as illustrated in Fig. 1.5.

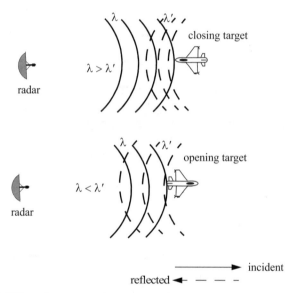

Figure 1.5. Effect of target motion on the reflected equiphase waveforms.

Consider a pulse of width τ (seconds) incident on a target which is moving towards the radar at velocity v, as shown in Fig. 1.6. Define d as the distance (in meters) that the target moves into the pulse during the interval Δt,

$$d = v \Delta t \quad (1.9)$$

where Δt is equal to the time between the pulse leading edge striking the target and the trailing edge striking the target. Since the pulse is moving at the speed of light and the trailing edge has moved distance $c\tau - d$, then

$$c\tau = c \Delta t + v \Delta t \quad (1.10)$$

and

8   MATLAB Simulations for Radar Systems Design

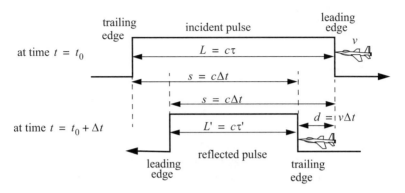

Figure 1.6. Illustrating the impact of target velocity on a single pulse.

$$c\tau' = c\Delta t - v\Delta t \tag{1.11}$$

Dividing Eq. (1.11) by Eq. (1.10) yields,

$$\frac{c\tau'}{c\tau} = \frac{c\Delta t - v\Delta t}{c\Delta t + v\Delta t} \tag{1.12}$$

which after canceling the terms c and Δt from the left and right side of Eq. (1.12) respectively, one establishes the relationship between the incident and reflected pulses widths as

$$\tau' = \frac{c-v}{c+v}\tau \tag{1.13}$$

In practice, the factor $(c-v)/(c+v)$ is often referred to as the time dilation factor. Notice that if $v = 0$, then $\tau' = \tau$. In a similar fashion, one can compute τ' for an opening target. In this case,

$$\tau' = \frac{v+c}{c-v}\tau \tag{1.14}$$

To derive an expression for Doppler frequency, consider the illustration shown in Fig. 1.7. It takes the leading edge of pulse 2 Δt seconds to travel a distance $(c/f_r)-d$ to strike the target. Over the same time interval, the leading edge of pulse 1 travels the same distance $c\Delta t$. More precisely,

$$d = v\Delta t \tag{1.15}$$

$$\frac{c}{f_r} - d = c\Delta t \tag{1.16}$$

solving for Δt yields

Doppler Frequency

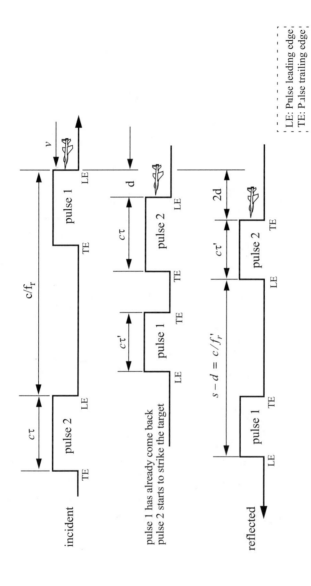

Figure 1.7. Illustration of target motion effects on the radar pulses.

$$\Delta t = \frac{c/f_r}{c+v} \qquad (1.17)$$

$$d = \frac{cv/f_r}{c+v} \qquad (1.18)$$

The reflected pulse spacing is now $s-d$ and the new PRF is f_r', where

$$s - d = \frac{c}{f_r'} = c\Delta t - \frac{cv/f_r}{c+v} \qquad (1.19)$$

It follows that the new PRF is related to the original PRF by

$$f_r' = \frac{c+v}{c-v} f_r \qquad (1.20)$$

However, since the number of cycles does not change, the frequency of the reflected signal will go up by the same factor. Denoting the new frequency by f_0', it follows

$$f_0' = \frac{c+v}{c-v} f_0 \qquad (1.21)$$

where f_0 is the carrier frequency of the incident signal. The Doppler frequency f_d is defined as the difference $f_0' - f_0$. More precisely,

$$f_d = f_0' - f_0 = \frac{c+v}{c-v} f_0 - f_0 = \frac{2v}{c-v} f_0 \qquad (1.22)$$

but since $v \ll c$ and $c = \lambda f_0$, then

$$f_d \approx \frac{2v}{c} f_0 = \frac{2v}{\lambda} \qquad (1.23)$$

Eq. (1.23) indicates that the Doppler shift is proportional to the target velocity, and, thus, one can extract f_d from range rate and vice versa.

The result in Eq. (1.23) can also be derived using the following approach: Fig. 1.8 shows a closing target with velocity v. Let R_0 refer to the range at time t_0 (time reference); then the range to the target at any time t is

$$R(t) = R_0 - v(t - t_0) \qquad (1.24)$$

The signal received by the radar is then given by

$$x_r(t) = x(t - \psi(t)) \qquad (1.25)$$

where $x(t)$ is the transmitted signal, and

$$\psi(t) = \frac{2}{c}(R_0 - vt + vt_0) \qquad (1.26)$$

Doppler Frequency

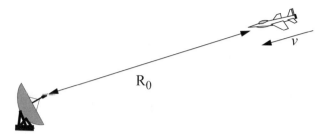

Figure 1.8. Closing target with velocity v.

Substituting Eq. (1.26) into Eq. (1.25) and collecting terms yield

$$x_r(t) = x\left(\left(1 + \frac{2v}{c}\right)t - \psi_0\right) \quad (1.27)$$

where the constant phase ψ_0 is

$$\psi_0 = \frac{2R_0}{c} + \frac{2v}{c} t_0 \quad (1.28)$$

Define the compression or scaling factor γ by

$$\gamma = 1 + \frac{2v}{c} \quad (1.29)$$

Note that for a receding target the scaling factor is $\gamma = 1 - (2v/c)$. Utilizing Eq. (1.29) we can rewrite Eq. (1.27) as

$$x_r(t) = x(\gamma t - \psi_0) \quad (1.30)$$

Eq. (1.30) is a time-compressed version of the return signal from a stationary target ($v = 0$). Hence, based on the scaling property of the Fourier transform, the spectrum of the received signal will be expanded in frequency to a factor of γ.

Consider the special case when

$$x(t) = y(t)\cos\omega_0 t \quad (1.31)$$

where ω_0 is the radar center frequency in radians per second. The received signal $x_r(t)$ is then given by

$$x_r(t) = y(\gamma t - \psi_0)\cos(\gamma\omega_0 t - \psi_0) \quad (1.32)$$

The Fourier transform of Eq. (1.32) is

$$X_r(\omega) = \frac{1}{2\gamma}\left(Y\left(\frac{\omega}{\gamma} - \omega_0\right) + Y\left(\frac{\omega}{\gamma} + \omega_0\right)\right) \quad (1.33)$$

where for simplicity the effects of the constant phase ψ_0 have been ignored in Eq. (1.33). Therefore, the bandpass spectrum of the received signal is now centered at $\gamma \omega_0$ instead of ω_0. The difference between the two values corresponds to the amount of Doppler shift incurred due to the target motion,

$$\omega_d = \omega_0 - \gamma \omega_0 \qquad (1.34)$$

ω_d is the Doppler frequency in radians per second. Substituting the value of γ in Eq. (1.34) and using $2\pi f = \omega$ yield

$$f_d = \frac{2v}{c} f_0 = \frac{2v}{\lambda} \qquad (1.35)$$

which is the same as Eq. (1.23). It can be shown that for a receding target the Doppler shift is $f_d = -2v/\lambda$. This is illustrated in Fig. 1.9.

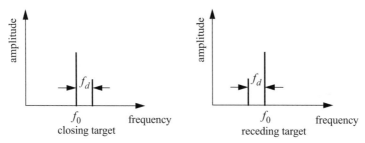

Figure 1.9. Spectra of received signal showing Doppler shift.

In both Eq. (1.35) and Eq. (1.23) the target radial velocity with respect to the radar is equal to v, but this is not always the case. In fact, the amount of Doppler frequency depends on the target velocity component in the direction of the radar (radial velocity). Fig. 1.10 shows three targets all having velocity v: target 1 has zero Doppler shift; target 2 has maximum Doppler frequency as defined in Eq. (1.35). The amount of Doppler frequency of target 3 is $f_d = 2v\cos\theta/\lambda$, where $v\cos\theta$ is the radial velocity; and θ is the total angle between the radar line of sight and the target.

Thus, a more general expression for f_d that accounts for the total angle between the radar and the target is

$$f_d = \frac{2v}{\lambda} \cos\theta \qquad (1.36)$$

and for an opening target

$$f_d = \frac{-2v}{\lambda} \cos\theta \qquad (1.37)$$

where $\cos\theta = \cos\theta_e \cos\theta_a$. The angles θ_e and θ_a are, respectively, the elevation and azimuth angles; see Fig. 1.11.

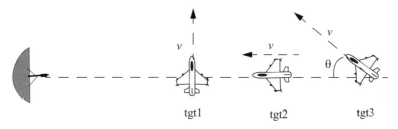

Figure 1.10. Target 1 generates zero Doppler. Target 2 generates maximum Doppler. Target 3 is in between.

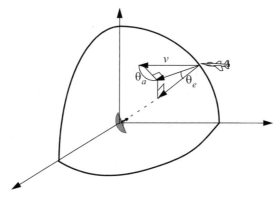

Figure 1.11. Radial velocity is proportional to the azimuth and elevation angles.

1.5. The Radar Equation

Consider a radar with an omni directional antenna (one that radiates energy equally in all directions). Since these kinds of antennas have a spherical radiation pattern, we can define the peak power density (power per unit area) at any point in space as

$$P_D = \frac{Peak\ transmitted\ power}{area\ of\ a\ sphere} \qquad \frac{watts}{m^2} \qquad (1.38)$$

The power density at range R away from the radar (assuming a lossless propagation medium) is

$$P_D = \frac{P_t}{4\pi R^2} \qquad (1.39)$$

where P_t is the peak transmitted power and $4\pi R^2$ is the surface area of a sphere of radius R. Radar systems utilize directional antennas in order to increase the power density in a certain direction. Directional antennas are usually characterized by the antenna gain G and the antenna effective aperture A_e. They are related by

$$G = \frac{4\pi A_e}{\lambda^2} \tag{1.40}$$

where λ is the wavelength. The relationship between the antenna's effective aperture A_e and the physical aperture A is

$$A_e = \rho A \tag{1.41}$$
$$0 \leq \rho \leq 1$$

ρ is referred to as the aperture efficiency, and good antennas require $\rho \to 1$. In this book we will assume, unless otherwise noted, that A and A_e are the same. We will also assume that antennas have the same gain in the transmitting and receiving modes. In practice, $\rho = 0.7$ is widely accepted.

The gain is also related to the antenna's azimuth and elevation beamwidths by

$$G = k\frac{4\pi}{\theta_e \theta_a} \tag{1.42}$$

where $k \leq 1$ and depends on the physical aperture shape; the angles θ_e and θ_a are the antenna's elevation and azimuth beamwidths, respectively, in radians. An excellent approximation of Eq. (1.42) introduced by Stutzman and reported by Skolnik is

$$G \approx \frac{26000}{\theta_e \theta_a} \tag{1.43}$$

where in this case the azimuth and elevation beamwidths are given in degrees.

The power density at a distance R away from a radar using a directive antenna of gain G is then given by

$$P_D = \frac{P_t G}{4\pi R^2} \tag{1.44}$$

When the radar radiated energy impinges on a target, the induced surface currents on that target radiate electromagnetic energy in all directions. The amount of the radiated energy is proportional to the target size, orientation, physical shape, and material, which are all lumped together in one target-specific parameter called the Radar Cross Section (RCS) denoted by σ.

The radar cross section is defined as the ratio of the power reflected back to the radar to the power density incident on the target,

The Radar Equation

$$\sigma = \frac{P_r}{P_D} \; m^2 \tag{1.45}$$

where P_r is the power reflected from the target. Thus, the total power delivered to the radar signal processor by the antenna is

$$P_{Dr} = \frac{P_t G \sigma}{(4\pi R^2)^2} A_e \tag{1.46}$$

Substituting the value of A_e from Eq. (1.40) into Eq. (1.46) yields

$$P_{Dr} = \frac{P_t G^2 \lambda^2 \sigma}{(4\pi)^3 R^4} \tag{1.47}$$

Let S_{min} denote the minimum detectable signal power. It follows that the maximum radar range R_{max} is

$$R_{max} = \left(\frac{P_t G^2 \lambda^2 \sigma}{(4\pi)^3 S_{min}} \right)^{1/4} \tag{1.48}$$

Eq. (1.48) suggests that in order to double the radar maximum range one must increase the peak transmitted power P_t sixteen times; or equivalently, one must increase the effective aperture four times.

In practical situations the returned signals received by the radar will be corrupted with noise, which introduces unwanted voltages at all radar frequencies. Noise is random in nature and can be described by its Power Spectral Density (PSD) function. The noise power N is a function of the radar operating bandwidth, B. More precisely

$$N = Noise \; PSD \times B \tag{1.49}$$

The input noise power to a lossless antenna is

$$N_i = kT_e B \tag{1.50}$$

where $k = 1.38 \times 10^{-23} \; joule/degree \; Kelvin$ is Boltzman's constant, and T_e is the effective noise temperature in degrees Kelvin. It is always desirable that the minimum detectable signal (S_{min}) be greater than the noise power. The fidelity of a radar receiver is normally described by a figure of merit called the noise figure F (see Appendix 1B for details). The noise figure is defined as

$$F = \frac{(SNR)_i}{(SNR)_o} = \frac{S_i/N_i}{S_o/N_o} \tag{1.51}$$

$(SNR)_i$ and $(SNR)_o$ are, respectively, the Signal to Noise Ratios (SNR) at the input and output of the receiver. S_i is the input signal power; N_i is the input noise power. S_o and N_o are, respectively, the output signal and noise power. Substituting Eq. (1.50) into Eq. (1.51) and rearranging terms yields

$$S_i = kT_e BF(SNR)_o \tag{1.52}$$

Thus, the minimum detectable signal power can be written as

$$S_{min} = kT_e BF(SNR)_{o_{min}} \tag{1.53}$$

The radar detection threshold is set equal to the minimum output SNR, $(SNR)_{o_{min}}$. Substituting Eq. (1.53) in Eq. (1.48) gives

$$R_{max} = \left(\frac{P_t G^2 \lambda^2 \sigma}{(4\pi)^3 kT_e BF(SNR)_{o_{min}}} \right)^{1/4} \tag{1.54}$$

or equivalently,

$$(SNR)_{o_{min}} = \frac{P_t G^2 \lambda^2 \sigma}{(4\pi)^3 kT_e BF R_{max}^4} \tag{1.55}$$

In general, radar losses denoted as L reduce the overall SNR, and hence

$$(SNR)_o = \frac{P_t G^2 \lambda^2 \sigma}{(4\pi)^3 kT_e BFLR^4} \tag{1.56}$$

Although it may take on many different forms, Eq. (1.56) is what is widely known as the Radar Equation. It is a common practice to perform calculations associated with the radar equation using decibel (dB) arithmetic. A review is presented in Appendix A.

MATLAB Function "radar_eq.m"

The function *"radar_eq.m"* implements Eq. (1.56); it is given in Listing 1.1 in Section 1.10. The syntax is as follows:

[snr] = radar_eq (pt, freq, g, sigma, te, b, nf, loss, range)

where

Symbol	Description	Units	Status
pt	peak power	Watts	input
freq	radar center frequency	Hz	input
g	antenna gain	dB	input
sigma	target cross section	m^2	input
te	effective noise temperature	Kelvin	input

The Radar Equation

Symbol	Description	Units	Status
b	bandwidth	Hz	input
nf	noise figure	dB	input
loss	radar losses	dB	input
range	target range (can be either a single value or a vector)	meters	input
snr	SNR (single value or a vector, depending on the input range)	dB	output

The function *"radar_eq.m"* is designed such that it can accept a single value for the input *"range"*, or a vector containing many range values. Figure 1.12 shows some typical plots generated using MATLAB program *"fig1_12.m"* which is listed in Listing 1.2 in Section 1.10. This program uses the function *"radar_eq.m"*, with the following default inputs: Peak power $P_t = 1.5 MW$, operating frequency $f_0 = 5.6 GHz$, antenna gain $G = 45 dB$, effective temperature $T_e = 290 K$, radar losses $L = 6 dB$, noise figure $F = 3 dB$. The radar bandwidth is $B = 5 MHz$. The radar minimum and maximum detection range are $R_{min} = 25 Km$ and $R_{max} = 165 Km$. Assume target cross section $\sigma = 0.1 m^2$.

Note that one can easily modify the MATLAB function *"radar_eq.m"* so that it solves Eq. (1.54) for the maximum detection range as a function of the minimum required SNR for a given set of radar parameters. Alternatively, the radar equation can be modified to compute the pulsewidth required to achieve a certain SNR for a given detection range. In this case the radar equation can be written as

$$\tau = \frac{(4\pi)^3 kT_e FLR^4 SNR}{P_t G^2 \lambda^2 \sigma} \tag{1.57}$$

Figure 1.13 shows an implementation of Eq. (1.57) for three different detection range values, using the radar parameters used in MATLAB program *"fig1_13.m"*. It is given in Listing 1.3 in Section 1.10.

When developing radar simulations, Eq. (1.57) can be very useful in the following sense. Radar systems often utilize a finite number of pulsewidths (waveforms) to accomplish all designated modes of operations. Some of these waveforms are used for search and detection, others may be used for tracking, while a limited number of wideband waveforms may be used for discrimination purposes. During the search mode of operation, for example, detection of a certain target with a specific RCS value is established based on a pre-determined probability of detection P_D. The probability of detection, P_D, is used to calculate the required detection SNR (this will be addressed in Chapter 2).

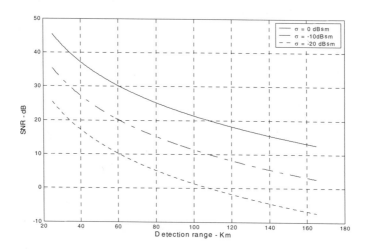

Figure 1.12a. SNR versus detection range for three different values of RCS.

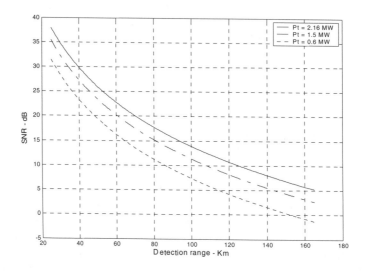

Figure 1.12b. SNR versus detection range for three different values of radar peak power.

The Radar Equation

Once the required SNR is computed, Eq. (1.57) can then be used to find the most suitable pulse (or waveform) that achieves the required SNR (or equivalently the required P_D). Often, it may be the case that none of the available radar waveforms may be able to guarantee the minimum required SNR for a particular RCS value at a particular detection range. In this case, the radar has to wait until the target is close enough in range to establish detection, otherwise pulse integration (coherent or non-coherent) can be used. Alternatively, cumulative probability of detection can be used. All these issues will be addressed in Chapter 2.

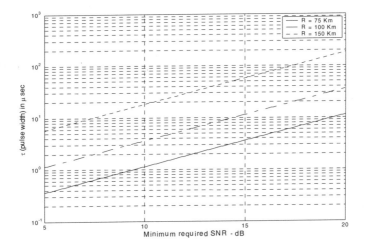

Figure 1.13. Pulsewidth versus required SNR for three different detection range values.

1.5.1. Radar Reference Range

Many radar design issues can be derived or computed based on the radar reference range R_{ref} which is often provided by the radar end user. It simply describes that range at which a certain SNR value, referred to as SNR_{ref}, has to be achieved using a specific reference pulsewidth τ_{ref} for a pre-determined target cross section, σ_{ref}. Radar reference range calculations assume that the target is on the line defined by the maximum antenna gain within a beam (broad side to the antenna). This is often referred to as the radar line of sight, as illustrated in Fig. 1.14.

The radar equation at the reference range is

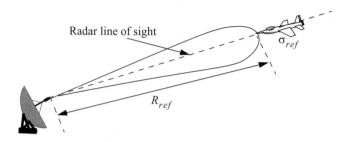

Figure 1.14. Definition of radar line of sight and radar reference range.

$$R_{ref} = \left(\frac{P_t G^2 \lambda^2 \sigma_{ref} \tau_{ref}}{(4\pi)^3 k T_e F L (SNR)_{ref}} \right)^{1/4} \quad (1.58)$$

The radar equation at any other detection range for any other combination of SNR, RCS, and pulsewidth can be given as

$$R = R_{ref} \left(\frac{\tau}{\tau_{ref}} \frac{\sigma}{\sigma_{ref}} \frac{SNR_{ref}}{SNR} \frac{1}{L_p} \right)^{1/4} \quad (1.59)$$

where the additional loss term L_p is introduced to account for the possibility that the non-reference target may not be on the radar line of sight, and to account for other losses associated with the specific scenario. Other forms of Eq. (1.59) can be in terms of the SNR. More precisely,

$$SNR = SNR_{ref} \frac{\tau}{\tau_{ref}} \frac{1}{L_p} \frac{\sigma}{\sigma_{ref}} \left(\frac{R_{ref}}{R} \right)^4 \quad (1.60)$$

As an example, consider the radar described in the previous section, in this case, define $\sigma_{ref} = 0.1 m^2$, $R_{ref} = 86 Km$, and $SNR_{ref} = 20 dB$. The reference pulsewidth is $\tau_{ref} = 0.1 \mu sec$. Using Eq. (1.60) we compute the SNR at $R = 120 Km$ for a target whose RCS is $\sigma = 0.2 m^2$. Assume that $L_p = 2dB$ to be equal to $(SNR)_{120Km} = 15.2dB$. For this purpose, the MATLAB program "ref_snr.m" has been developed; it is given in Listing 1.4 in Section 1.10.

1.6. Search (Surveillance)

The first task a certain radar system has to accomplish is to continuously scan a specified volume in space searching for targets of interest. Once detection is established, target information such as range, angular position, and possibly target velocity are extracted by the radar signal and data processors. Depending on the radar design and antenna, different search patterns can be

Search (Surveillance)

adopted. A two-dimensional (2-D) fan beam search pattern is shown in Fig.1.15a. In this case, the beamwidth is wide enough in elevation to cover the desired search volume along that coordinate; however, it has to be steered in azimuth. Figure 1.15b shows a stacked beam search pattern; here the beam has to be steered in azimuth and elevation. This latter kind of search pattern is normally employed by phased array radars.

Search volumes are normally specified by a search solid angle Ω in steradians. Define the radar search volume extent for both azimuth and elevation as Θ_A and Θ_E. Consequently, the search volume is computed as

$$\Omega = (\Theta_A \Theta_E)/(57.296)^2 \; steradians \tag{1.61}$$

where both Θ_A and Θ_E are given in degrees. The radar antenna $3dB$ beamwidth can be expressed in terms of its azimuth and elevation beamwidths θ_a and θ_e, respectively. It follows that the antenna solid angle coverage is $\theta_a \theta_e$ and, thus, the number of antenna beam positions n_B required to cover a solid angle Ω is

$$n_B = \frac{\Omega}{(\theta_a \theta_e)/(57.296)^2} \tag{1.62}$$

In order to develop the search radar equation, start with Eq. (1.56) which is repeated here, for convenience, as Eq. (1.63).

$$SNR = \frac{P_t G^2 \lambda^2 \sigma}{(4\pi)^3 k T_e BFLR^4} \tag{1.63}$$

Using the relations $\tau = 1/B$ and $P_t = P_{av} T/\tau$, where T is the PRI and τ is the pulsewidth, yields

$$SNR = \frac{T}{\tau} \frac{P_{av} G^2 \lambda^2 \sigma \tau}{(4\pi)^3 k T_e FLR^4} \tag{1.64}$$

(a)   (b)

Figure 1.15. (a) 2-D fan search pattern; (b) stacked search pattern.

Define the time it takes the radar to scan a volume defined by the solid angle Ω as the scan time T_{sc}. The time on target can then be expressed in terms of T_{sc} as

$$T_i = \frac{T_{sc}}{n_B} = \frac{T_{sc}}{\Omega}\theta_a\theta_e \tag{1.65}$$

Assume that during a single scan only one pulse per beam per PRI illuminates the target. It follows that $T_i = T$ and, thus, Eq. (1.64) can be written as

$$SNR = \frac{P_{av}G^2\lambda^2\sigma}{(4\pi)^3 kT_e FLR^4}\frac{T_{sc}}{\Omega}\theta_a\theta_e \tag{1.66}$$

Substituting Eqs. (1.40) and (1.42) into Eq. (1.66) and collecting terms yield the search radar equation (based on a single pulse per beam per PRI) as

$$SNR = \frac{P_{av}A_e\sigma}{4\pi kT_e FLR^4}\frac{T_{sc}}{\Omega} \tag{1.67}$$

The quantity $P_{av}A$ in Eq. (1.67) is known as the power aperture product. In practice, the power aperture product is widely used to categorize the radar's ability to fulfill its search mission. Normally, a power aperture product is computed to meet a predetermined SNR and radar cross section for a given search volume defined by Ω.

As a special case, assume a radar using a circular aperture (antenna) with diameter D. The 3-dB antenna beamwidth θ_{3dB} is

$$\theta_{3dB} \approx \frac{\lambda}{D} \tag{1.68}$$

and when aperture tapering is used, $\theta_{3dB} \approx 1.25\lambda/D$. Substituting Eq. (1.68) into Eq. (1.62) yields

$$n_B = \frac{D^2}{\lambda^2}\Omega \tag{1.69}$$

For this case, the scan time T_{sc} is related to the time-on-target by

$$T_i = \frac{T_{sc}}{n_B} = \frac{T_{sc}\lambda^2}{D^2\Omega} \tag{1.70}$$

Substitute Eq. (1.70) into Eq. (1.64) to get

$$SNR = \frac{P_{av}G^2\lambda^2\sigma}{(4\pi)^3 R^4 kT_e FL}\frac{T_{sc}\lambda^2}{D^2\Omega} \tag{1.71}$$

Search (Surveillance)

and by using Eq. (1.40) in Eq. (1.71) we can define the search radar equation for a circular aperture as

$$SNR = \frac{P_{av} A \sigma}{16 R^4 k T_e LF} \frac{T_{sc}}{\Omega} \qquad (1.72)$$

where the relation $A = \pi D^2 / 4$ (aperture area) is used.

MATLAB Function "power_aperture.m"

The function *"power_aperture.m"* implements the search radar equation given in Eq. (1.67); it is given in Listing 1.5 in Section 1.10. The syntax is as follows:

PAP = power_aperture (snr, tsc, sigma, range, te, nf, loss, az_angle, el_angle)

where

Symbol	Description	Units	Status
snr	sensitivity snr	dB	input
tsc	scan time	seconds	input
sigma	target cross section	m^2	input
range	target range (can be either single value or a vector)	meters	input
te	effective temperature	Kelvin	input
nf	noise figure	dB	input
loss	radar losses	dB	input
az_angle	search volume azimuth extent	degrees	input
el_angle	search volume elevation extent	degrees	input
PAP	power aperture product	dB	output

Plots of the power aperture product versus range and plots of the average power versus aperture area for three RCS choices are shown in Figure 1.16. MATLAB program *"fig1_16.m"* was used to produce these figures. It is given in Listing 1.6 in Section 1.10. In this case, the following radar parameters were used

σ	T_{sc}	$\theta_e = \theta_a$	R	T_e	$nf \times loss$	snr
$0.1\ m^2$	2.5 sec	2°	250 Km	900 K	13 dB	15 dB

24 *MATLAB Simulations for Radar Systems Design*

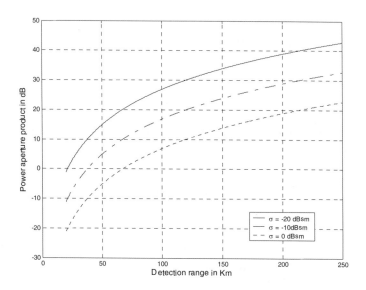

Figure 1.16a. Power aperture product versus detection range.

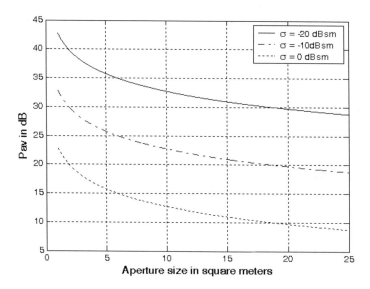

Figure 1.16b. Radar average power versus power aperture product.

Example:

Compute the power aperture product corresponding to the radar that has the following parameters: Scan time $T_{sc} = 2\sec$, Noise figure $F = 8dB$, losses $L = 6dB$, search volume $\Omega = 7.4$ steradians, range of interest is $R = 75Km$, and the required SNR is $20dB$. Assume that $T_e = 290 Kelvin$ and $\sigma = 3.162m^2$.

Solution:

Note that $\Omega = 7.4$ steradians corresponds to a search sector that is three fourths of a hemisphere. Thus, using Eq. (1.61) we conclude that $\theta_a = 180°$ and $\theta_e = 135°$. Using the MATLAB function "power_aperture.m" with the following syntax:

PAP = power_aperture(20, 2, 3.162, 75e3, 290, 8, 6, 180, 135)

we compute the power aperture product as 36.7 dB.

1.6.1. Mini Design Case Study 1.1

Problem Statement:

Design a ground based radar that is capable of detecting aircraft and missiles at 10 Km and 2 Km altitudes, respectively. The maximum detection range for either target type is 60 Km. Assume that an aircraft average RCS is 6 dBsm, and that a missile average RCS is -10 dBsm. The radar azimuth and elevation search extents are respectively $\Theta_A = 360°$ and $\Theta_E = 10°$. The required scan rate is 2 seconds and the range resolution is 150 meters. Assume a noise figure $F = 8$ dB, and total receiver noise $L = 10$ dB. Use a fan beam with azimuth beamwidth less than 3 degrees. The SNR is 15 dB.

A Design:

The range resolution requirement is $\Delta R = 150m$; thus by using Eq. (1.8) we calculate the required pulsewidth $\tau = 1\mu\sec$, or equivalently require the bandwidth $B = 1MHz$. The statement of the problem lends itself to radar sizing in terms of power aperture product. For this purpose, one must first compute the maximum search volume at the detection range that satisfies the design requirements. The radar search volume is

$$\Omega = \frac{\Theta_A \Theta_E}{(57.296)^2} = \frac{360 \times 10}{(57.296)^2} = 1.097 \text{ steradians} \quad (1.73)$$

At this point, the designer is ready to use the radar search equation (Eq. (1.67)) to compute the power aperture product. For this purpose, one can modify the MATLAB function "power_aperture.m" to compute and plot the power aperture product for both target types. To this end, the MATLAB program "casestudy1_1.m", which is given in Listing 1.7 in Section 1.10, was developed. Use the parameters in Table 1.2 as inputs for this program. Note that the selection of $T_e = 290 Kelvin$ is arbitrary.

TABLE 1.2: Input parameters to MATLAB program "*casestudy1_1.m*".

Symbol	Description	Units	Value
snr	sensitivity snr	dB	15
tsc	scan time	seconds	2
sigma_tgtm	missile radar cross section	dBsm	-10
sigma_tgta	aircraft radar cross section	dBsm	6
rangem	missile detection range	Km	60
rangea	aircraft detection range	Km	60
te	effective temperature	Kelvin	290
nf	noise figure	dB	8
loss	radar losses	dB	10
az_angle	search volume azimuth extent	degrees	360
el_angle	search volume elevation extent	degrees	10

Figure 1.17 shows a plot of the output produced by this program. The same program also calculates the corresponding power aperture product for both the missile and aircraft cases, which can also be read from the plot,

$$PAP_{missile} = 38.53 dB$$
$$PAP_{aircraft} = 22.53 dB$$
(1.74)

Choosing the more stressing case for the design baseline (i.e., select the power-aperture-product resulting from the missile analysis) yields

$$P_{av} \times A_e = 10^{3.853} = 7128.53 \Rightarrow A_e = \frac{7128.53}{P_{av}}$$
(1.75)

Choose $A_e = 1.75 m^2$ to calculate the average power as

$$P_{av} = \frac{7128.53}{1.75} = 4.073 KW$$
(1.76)

and assuming an aperture efficiency of $\rho = 0.8$ yields the physical aperture area. More precisely,

$$A = \frac{A_e}{\rho} = \frac{1.75}{0.8} = 2.1875 m^2$$
(1.77)

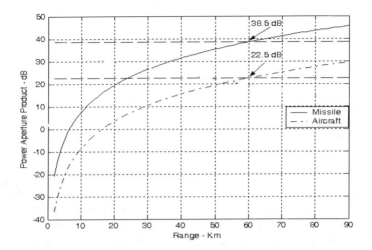

Figure 1.17. Power aperture product versus detection range for radar in mini design case study 1.1.

Use $f_0 = 2.0 GHz$ as the radar operating frequency. Then by using $A_e = 1.75 m^2$ we calculate using Eq. (1.40) $G = 29.9 dB$. Now one must determine the antenna azimuth beamwidth. Recall that the antenna gain is also related to the antenna 3-dB beamwidth by the relation

$$G = \frac{26000}{\theta_e \theta_a} \quad (1.78)$$

where (θ_a, θ_e) are the antenna 3-dB azimuth and elevation beamwidths, respectively. Assume a fan beam with $\theta_e = \Theta_E = 15°$. It follows that

$$\theta_a = \frac{26000}{\theta_e G} = \frac{26000}{10 \times 977.38} = 2.66° \Rightarrow \theta_a = 46.43 mrad \quad (1.79)$$

1.7. Pulse Integration

When a target is located within the radar beam during a single scan it may reflect several pulses. By adding the returns from all pulses returned by a given target during a single scan, the radar sensitivity (SNR) can be increased. The number of returned pulses depends on the antenna scan rate and the radar PRF. More precisely, the number of pulses returned from a given target is given by

$$n_P = \frac{\theta_a T_{sc} f_r}{2\pi} \quad (1.80)$$

where θ_a is the azimuth antenna beamwidth, T_{sc} is the scan time, and f_r is the radar PRF. The number of reflected pulses may also be expressed as

$$n_P = \frac{\theta_a f_r}{\dot{\theta}_{scan}} \qquad (1.81)$$

where $\dot{\theta}_{scan}$ is the antenna scan rate in degrees per second. Note that when using Eq. (1.80), θ_a is expressed in radians, while when using Eq. (1.81) it is expressed in degrees. As an example, consider a radar with an azimuth antenna beamwidth $\theta_a = 3°$, antenna scan rate $\dot{\theta}_{scan} = 45°/\sec$ (antenna scan time, $T_{sc} = 8 seconds$), and a PRF $f_r = 300 Hz$. Using either Eq.s (1.80) or (1.81) yields $n_P = 20$ pulses.

The process of adding radar returns from many pulses is called radar pulse integration. Pulse integration can be performed on the quadrature components prior to the envelope detector. This is called coherent integration or pre-detection integration. Coherent integration preserves the phase relationship between the received pulses. Thus a build up in the signal amplitude is achieved. Alternatively, pulse integration performed after the envelope detector (where the phase relation is destroyed) is called non-coherent or post-detection integration.

Radar designers should exercise caution when utilizing pulse integration for the following reasons. First, during a scan a given target will not always be located at the center of the radar beam (i.e., have maximum gain). In fact, during a scan a given target will first enter the antenna beam at the 3-dB point, reach maximum gain, and finally leave the beam at the 3-dB point again. Thus, the returns do not have the same amplitude even though the target RCS may be constant and all other factors which may introduce signal loss remain the same. This is illustrated in Fig. 1.18, and is normally referred to as antenna beam-shape loss.

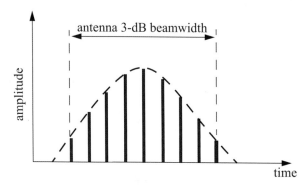

Figure 1.18. Pulse returns from a point target using a rotating (scanning) antenna

Other factors that may introduce further variation to the amplitude of the returned pulses include target RCS and propagation path fluctuations. Additionally, when the radar employs a very fast scan rate, an additional loss term is introduced due to the motion of the beam between transmission and reception. This is referred to as scan loss. A distinction should be made between scan loss due to a rotating antenna (which is described here) and the term scan loss that is normally associated with phased array antennas (which takes on a different meaning in that context). These topics will be discussed in more detail in other chapters.

Finally, since coherent integration utilizes the phase information from all integrated pulses, it is critical that any phase variation between all integrated pulses be known with a great level of confidence. Consequently, target dynamics (such as target range, range rate, tumble rate, RCS fluctuation, etc.) must be estimated or computed accurately so that coherent integration can be meaningful. In fact, if a radar coherently integrates pulses from targets without proper knowledge of the target dynamics it suffers a loss in SNR rather than the expected SNR build up. Knowledge of target dynamics is not as critical when employing non-coherent integration; nonetheless, target range rate must be estimated so that only the returns from a given target within a specific range bin are integrated. In other words, one must avoid range walk (i.e., avoid having a target cross between adjacent range bins during a single scan).

A comprehensive analysis of pulse integration should take into account issues such as the probability of detection P_D, probability of false alarm P_{fa}, the target statistical fluctuation model, and the noise or interference statistical models. These topics will be discussed in Chapter 2. However, in this section an overview of pulse integration is introduced in the context of radar measurements as it applies to the radar equation. The basic conclusions presented in this chapter concerning pulse integration will still be valid, in the general sense, when a more comprehensive analysis of pulse integration is presented; however, the exact implementation, the mathematical formulation, and /or the numerical values used will vary.

1.7.1. Coherent Integration

In coherent integration, when a perfect integrator is used (100% efficiency), to integrate n_P pulses the SNR is improved by the same factor. Otherwise, integration loss occurs, which is always the case for non-coherent integration. Coherent integration loss occurs when the integration process is not optimum. This could be due to target fluctuation, instability in the radar local oscillator, or propagation path changes.

Denote the single pulse SNR required to produce a given probability of detection as $(SNR)_1$. The SNR resulting from coherently integrating n_P pulses is then given by

$$(SNR)_{CI} = n_P(SNR)_1 \quad (1.82)$$

Coherent integration cannot be applied over a large number of pulses, particularly if the target RCS is varying rapidly. If the target radial velocity is known and no acceleration is assumed, the maximum coherent integration time is limited to

$$t_{CI} = \sqrt{\lambda/2a_r} \quad (1.83)$$

where λ is the radar wavelength and a_r is the target radial acceleration. Coherent integration time can be extended if the target radial acceleration can be compensated for by the radar.

1.7.2. Non-Coherent Integration

Non-coherent integration is often implemented after the envelope detector, also known as the quadratic detector. Non-coherent integration is less efficient than coherent integration. Actually, the non-coherent integration gain is always smaller than the number of non-coherently integrated pulses. This loss in integration is referred to as post detection or square law detector loss. Marcum and Swerling showed that this loss is somewhere between $\sqrt{n_P}$ and n_P. DiFranco and Rubin presented an approximation of this loss as

$$L_{NCI} = 10\log(\sqrt{n_P}) - 5.5 \ dB \quad (1.84)$$

Note that as n_P becomes very large, the integration loss approaches $\sqrt{n_P}$.

The subject of integration loss is treated in great levels of detail in the literature. Different authors use different approximations for the integration loss associated with non-coherent integration. However, all these different approximations yield very comparable results. Therefore, in the opinion of these authors the use of one formula or another to approximate integration loss becomes somewhat subjective. In this book, the integration loss approximation reported by Barton and used by Curry will be adopted. In this case, the non-coherent integration loss which can be used in the radar equation is

$$L_{NCI} = \frac{1 + (SNR)_1}{(SNR)_1} \quad (1.85)$$

It follows that the SNR when n_P pulses are integrated non-coherently is

$$(SNR)_{NCI} = \frac{n_P(SNR)_1}{L_{NCI}} = n_P(SNR)_1 \times \frac{(SNR)_1}{1 + (SNR)_1} \quad (1.86)$$

1.7.3. Detection Range with Pulse Integration

The process of determining the radar sensitivity or equivalently the maximum detection range when pulse integration is used is as follows: First, decide

Pulse Integration

whether to use coherent or non-coherent integration. Keep in mind the issues discussed in the beginning of this section when deciding whether to use coherent or non-coherent integration.

Second, determine the minimum required $(SNR)_{CI}$ or $(SNR)_{NCI}$ required for adequate detection and track. Typically, for ground based surveillance radars that can be on the order of 13 to 15 dB. The third step is to determine how many pulses should be integrated. The choice of n_P is affected by the radar scan rate, the radar PRF, the azimuth antenna beamwidth, and of course by the target dynamics (remember that range walk should be avoided or compensated for, so that proper integration is feasible). Once n_P and the required SNR are known one can compute the single pulse SNR (i.e., the reduction in SNR). For this purpose use Eq. (1.82) in the case of coherent integration. In the non-coherent integration case, Curry presents an attractive formula for this calculation, as follows

$$(SNR)_1 = \frac{(SNR)_{NCI}}{2n_P} + \sqrt{\frac{(SNR)^2_{NCI}}{4n_P^2} + \frac{(SNR)_{NCI}}{n_P}} \qquad (1.87)$$

Finally, use $(SNR)_1$ from Eq. (1.87) in the radar equation to calculate the radar detection range. Observe that due to the integration reduction in SNR the radar detection range is now larger than that for the single pulse when the same SNR value is used. This is illustrated using the following mini design case study.

1.7.4. Mini Design Case Study 1.2

Problem Statement:

A MMW radar has the following specifications: Center frequency $f = 94 GHz$, pulsewidth $\tau = 50 \times 10^{-9}$ sec, peak power $P_t = 4W$, azimuth coverage $\Delta \alpha = \pm 120°$, Pulse repetition frequency $PRF = 10KHz$, noise figure $F = 7dB$; antenna diameter $D = 12in$; antenna gain $G = 47dB$; radar cross section of target is $\sigma = 20m^2$; system losses $L = 10dB$; radar scan time $T_{sc} = 3$ sec. Calculate: The wavelength λ; range resolution ΔR; bandwidth B; antenna half power beamwidth; antenna scan rate; time on target. Compute the range that corresponds to 10 dB SNR. Plot the SNR as a function of range. Finally, compute the number of pulses on the target that can be used for integration and the corresponding new detection range when pulse integration is used, assuming that the SNR stays unchanged (i.e., the same as in the case of a single pulse). Assume $T_e = 290$ Kelvin.

A Design:

The wavelength λ is

$$\lambda = \frac{c}{f} = \frac{3 \times 10^8}{94 \times 10^9} = 0.00319 m$$

The range resolution ΔR is

$$\Delta R = \frac{c\tau}{2} = \frac{(3 \times 10^8)(50 \times 10^{-9})}{2} = 7.5 m$$

Radar operating bandwidth B is

$$B = \frac{1}{\tau} = \frac{1}{50 \times 10^{-9}} = 20 MHz$$

The antenna 3-dB beamwidth is

$$\theta_{3dB} = 1.25 \frac{\lambda}{D} = 0.7499°$$

Time on target is

$$T_i = \frac{\theta_{3dB}}{\dot{\theta}_{scan}} = \frac{0.7499°}{80°/\sec} = 9.38 m\sec$$

It follows that the number of pulses available for integration is calculated using Eq. (1.81),

$$n_P = \frac{\theta_{3dB}}{\dot{\theta}_{scan}} f_r = 9.38 \times 10^{-3} \times 10 \times 10^3 \Rightarrow 94 \ pulses$$

Coherent Integration case:

Using the radar equation given in Eq. (1.58) yields $R_{ref} = 2.245 Km$. The SNR improvement due to coherently integrating 94 pulses is 19.73dB. However, since it is requested that the SNR remains at 10dB, we can calculate the new detection range using Eq. (1.59) as

$$R_{CI}\big|_{n_P = 94} = 2.245 \times (94)^{1/4} = 6.99 Km$$

Using the MATLAB Function "radar_eq.m" with the following syntax

[snr] = radar_eq (4, 94e9, 47, 20, 290, 20e6, 7, 10, 6.99e3)

yields SNR = -9.68 dB. This means that using 94 pulses integrated coherently at 6.99 Km where each pulse has a SNR of -9.68 dB provides the same detection criteria as using a single pulse with SNR = 10dB at 2.245Km. This is illustrated in Fig. 1.19, using the MATLAB program "fig1_19.m", which is given in Listing 1.8 in Section 1.10. Figure 1.19 shows the improvement of the detection range if the SNR is kept constant before and after integration.

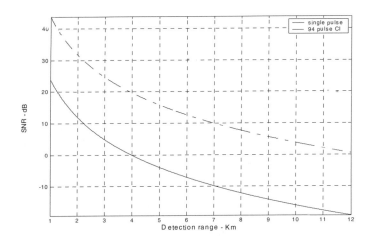

Figure 1.19. SNR versus detection range, using parameters from example.

Non-coherent Integration case:

Start with Eq. (1.87) with $(SNR)_{NCI} = 10dB$ *and* $n_P = 94$,

$$(SNR)_1 = \frac{10}{2 \times 94} + \sqrt{\frac{(10)^2}{4 \times 94^2} + \frac{10}{94}} = 0.38366 \Rightarrow -4.16dB$$

Therefore, the single pulse SNR when 94 pulses are integrated non-coherently is -4.16dB. You can verify this result by using Eq. (1.86). The integration loss L_{NCI} *is calculated using Eq. (1.85). It is*

$$L_{NCI} = \frac{1 + 0.38366}{0.38366} = 3.6065 \Rightarrow 5.571 dB$$

Therefore, the net non-coherent integration gain is

$$10 \times \log(94) - 5.571 = 14.16 dB \Rightarrow 26.06422$$

and, consequently, the maximum detection range is

$$R_{NCI}\big|_{n_P = 94} = 2.245 \times (26.06422)^{1/4} = 5.073 Km$$

This means that using 94 pulses integrated non-coherently at 5.073 Km where each pulse has SNR of -4.16dB provides the same detection criterion as using a single pulse with SNR = 10dB at 2.245Km. This is illustrated in Fig. 1.20, using the MATLAB program "fig1_19.m".

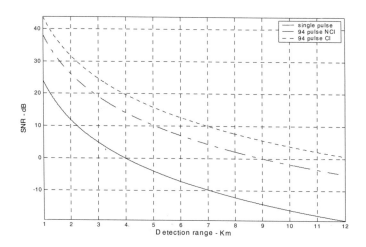

Figure 1.20. SNR versus detection range, for the same example.

MATLAB Function "pulse_integration.m"

Figure 1.21 shows the SNR gain versus the number of integrated pulses for both coherent and non-coherent integration. This figure corresponds to parameters from the previous example at $R = 5.01 Km$. Figure 1.22 shows the general case SNR improvement versus number of integrated pulses. Both figures were generated using MATLAB program *"fig1_21.m"* which is given in Listing 1.9 in Section 1.10. For this purpose the MATLAB function *"pulse_integration.m"* was developed. It is given in Listing 1.10 in Section 1.10. This function calculates the radar equation given in Eq. (1.56) with pulse integration. The syntax for MATLAB function *"pulse_integration.m"* is as follows

[snr] = pulse_integration (pt, freq, g, sigma, te, b, nf, loss, range, np, ci_nci)

where

Symbol	Description	Units	Status
pt	peak power	Watts	input
freq	radar center frequency	Hz	input
g	antenna gain	dB	input
sigma	target cross section	m^2	input
te	effective noise temperature	Kelvin	input
b	bandwidth	Hz	input

Radar Losses

Symbol	Description	Units	Status
nf	noise figure	dB	input
loss	radar losses	dB	input
range	target range (can be either a single value or a vector)	meters	input
np	number of integrated pulses	none	input
ci_nci	1 for CI; 2 for NCI	none	input
snr	SNR (single value or a vector, depending on the input range)	dB	output

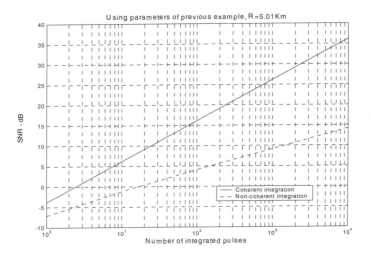

Figure 1.21. SNR improvement when integration is utilized.

1.8. Radar Losses

As indicated by the radar equation, the receiver SNR is inversely proportional to the radar losses. Hence, any increase in radar losses causes a drop in the SNR, thus decreasing the probability of detection, as it is a function of the SNR. Often, the principal difference between a good radar design and a poor radar design is the radar losses. Radar losses include ohmic (resistance) losses and statistical losses. In this section we will briefly summarize radar losses.

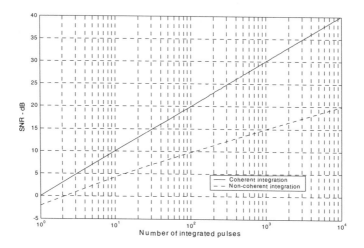

Figure 1.22. SNR improvement when integration is utilized.

1.8.1. Transmit and Receive Losses

Transmit and receive losses occur between the radar transmitter and antenna input port, and between the antenna output port and the receiver front end, respectively. Such losses are often called plumbing losses. Typically, plumbing losses are on the order of 1 to 2 dB.

1.8.2. Antenna Pattern Loss and Scan Loss

So far, when we used the radar equation we assumed maximum antenna gain. This is true only if the target is located along the antenna's boresight axis. However, as the radar scans across a target the antenna gain in the direction of the target is less than maximum, as defined by the antenna's radiation pattern. The loss in SNR due to not having maximum antenna gain on the target at all times is called the antenna pattern (shape) loss. Once an antenna has been selected for a given radar, the amount of antenna pattern loss can be mathematically computed.

For example, consider a $\sin x / x$ antenna radiation pattern as shown in Fig. 1.23. It follows that the average antenna gain over an angular region of $\pm \theta / 2$ about the boresight axis is

$$G_{av} \approx 1 - \left(\frac{\pi r}{\lambda}\right)^2 \frac{\theta^2}{36} \tag{1.88}$$

Radar Losses

where r is the aperture radius and λ is the wavelength. In practice, Gaussian antenna patterns are often adopted. In this case, if θ_{3dB} denotes the antenna 3dB beamwidth, then the antenna gain can be approximated by

$$G(\theta) = \exp\left(-\frac{2.776\theta^2}{\theta_{3dB}^2}\right) \qquad (1.89)$$

If the antenna scanning rate is so fast that the gain on receive is not the same as on transmit, additional scan loss has to be calculated and added to the beam shape loss. Scan loss can be computed in a similar fashion to beam shape loss. Phased array radars are often prime candidates for both beam shape and scan losses.

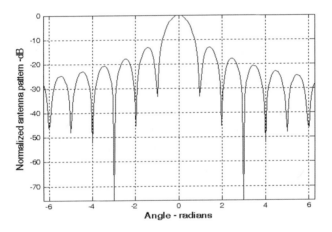

Figure 1.23. Normalized (sin x / x) antenna pattern.

1.8.3. Atmospheric Loss

Detailed discussion of atmospheric loss and propagation effects is in a later chapter. Atmospheric attenuation is a function of the radar operating frequency, target range, and elevation angle. Atmospheric attenuation can be as high as a few dB.

1.8.4. Collapsing Loss

When the number of integrated returned noise pulses is larger than the target returned pulses, a drop in the SNR occurs. This is called collapsing loss. The collapsing loss factor is defined as

$$\rho_c = \frac{n+m}{n} \tag{1.90}$$

where n is the number of pulses containing both signal and noise, while m is the number of pulses containing noise only. Radars detect targets in azimuth, range, and Doppler. When target returns are displayed in one coordinate, such as range, noise sources from azimuth cells adjacent to the actual target return converge in the target vicinity and cause a drop in the SNR. This is illustrated in Fig. 1.24.

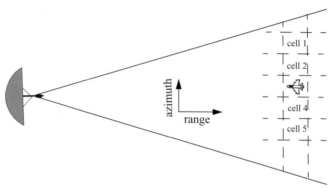

Figure 1.24. Illustration of collapsing loss. Noise sources in cells 1, 2, 4, and 5 converge to increase the noise level in cell 3.

1.8.5. Processing Losses

a. Detector Approximation:

The output voltage signal of a radar receiver that utilizes a linear detector is

$$v(t) = \sqrt{v_I^2(t) + v_Q^2(t)} \tag{1.91}$$

where (v_I, v_Q) are the in-phase and quadrature components. For a radar using a square law detector, we have $v^2(t) = v_I^2(t) + v_Q^2(t)$.

Since in real hardware the operations of squares and square roots are time consuming, many algorithms have been developed for detector approximation. This approximation results in a loss of the signal power, typically 0.5 to 1 dB.

b. Constant False Alarm Rate (CFAR) Loss:

In many cases the radar detection threshold is constantly adjusted as a function of the receiver noise level in order to maintain a constant false alarm rate. For this purpose, Constant False Alarm Rate (CFAR) processors are utilized in order to keep the number of false alarms under control in a changing and unknown background of interference. CFAR processing can cause a loss in the SNR level on the order of 1 dB.

Three different types of CFAR processors are primarily used. They are adaptive threshold CFAR, nonparametric CFAR, and nonlinear receiver techniques. Adaptive CFAR assumes that the interference distribution is known and approximates the unknown parameters associated with these distributions. Nonparametric CFAR processors tend to accommodate unknown interference distributions. Nonlinear receiver techniques attempt to normalize the root mean square amplitude of the interference.

c. Quantization Loss:

Finite word length (number of bits) and quantization noise cause an increase in the noise power density at the output of the Analog to Digital (A/D) converter. The A/D noise level is $q^2/12$, where q is the quantization level.

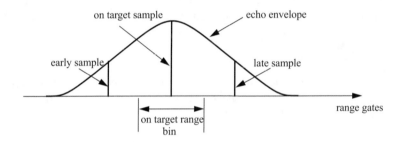

(a) Target on the center of a range gate.

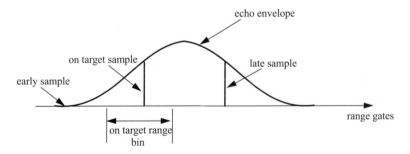

(b) Target on the boundary between two range gates.

Figure 1.25. Illustration of range gate straddling.

d. Range Gate Straddle:

The radar receiver is normally mechanized as a series of contiguous range gates (bins). Each range bin is implemented as an integrator matched to the transmitted pulsewidth. Since the radar receiver acts as a filter that smears (smooths), the received target echoes. The smoothed target return envelope is normally straddled to cover more than one range gate.

Typically, three gates are affected; they are called the early, on, and late gates. If a point target is located exactly at the center of a range gate, then the early and late samples are equal. However, as the target starts to move into the next gate, the late sample becomes larger while the early sample gets smaller. In any case, the amplitudes of all three samples should always roughly add up to the same value. Fig. 1.25 illustrates the concept of range straddling. The envelope of the smoothed target echo is likely to be Gaussian shaped. In practice, triangular shaped envelopes may be easier and faster to implement. Since the target is likely to fall anywhere between two adjacent range bins, a loss in the SNR occurs (per range gate). More specifically, a target's returned energy is split between three range bins. Typically, straddle loss of about 2 to 3 dB is not unusual.

Example:

Consider the smoothed target echo voltage shown below. Assume 1Ω resistance. Find the power loss due to range gate straddling over the interval $\{0, \tau\}$.

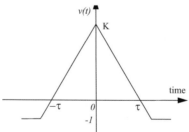

Solution:

The smoothed voltage can be written as

$$v(t) = \begin{cases} K + \left(\dfrac{K+1}{\tau}\right)t & ; \ t < 0 \\ K - \left(\dfrac{K+1}{\tau}\right)t & ; \ t \geq 0 \end{cases}$$

The power loss due to straddle over the interval $\{0, \tau\}$ is

$$L_s = \frac{v^2}{K^2} = 1 - 2\left(\frac{K+1}{K\tau}\right)t + \left(\frac{K+1}{K\tau}\right)^2 t^2$$

The average power loss is then

$$\bar{L}_s = \frac{2}{\tau}\int_0^{\tau/2}\left(1 - 2\left(\frac{K+1}{K\tau}\right)t + \left(\frac{K+1}{K\tau}\right)^2 t^2\right) dt$$

$$= 1 - \frac{K+1}{2K} + \frac{(K+1)^2}{12K^2}$$

and, for example, if $K = 15$, then $\bar{L}_s = 2.5 dB$.

e. Doppler Filter Straddle:

Doppler filter straddle is similar to range gate straddle. However, in this case the Doppler filter spectrum is spread (widened) due to weighting functions. Weighting functions are normally used to reduce the sidelobe levels. Since the target Doppler frequency can fall anywhere between two Doppler filters, signal loss occurs.

1.8.6. Other Losses

Other losses may include equipment losses due to aging radar hardware, matched filter loss, and antenna efficiency loss. Tracking radars suffer from crossover (squint) loss.

1.9. *"MyRadar" Design Case Study - Visit 1*

In this section, a design case study, referred to as *"MyRadar"* design case study, is introduced. For this purpose, only the theory introduced in this chapter is used to fulfill the design requirements. Note that since only a limited amount of information has been introduced in this chapter, the design process may seem illogical to some readers. However, as new material is introduced in subsequent chapters, the design requirements are updated and/or new design requirements are introduced based on the particular material of that chapter. Consequently, the design process will also be updated to accommodate the new theory and techniques learned in that chapter.

1.9.1. Authors and Publisher Disclaimer

The design case study *"MyRadar"* is a ground based air defense radar derived and based on Brookner's[1] open literature source. However, the design approach introduced in this book is based on the authors' point of view of how to design such radar. Thus, the design process takes on a different flavor than that introduced by Brookner. Additionally, any and all design alternatives presented in this book are based on and can be easily traced to open literature sources.

Furthermore, the design approach adopted in this book is based on modeling many of the radar system components with no regards to any hardware constraints nor to any practical limitations. The design presented in this book is intended to be tutorial and academic in nature and does not adhere to any other requirements. Finally, the MATLAB code presented in this book is intended to be illustrative and academic and is not designed nor intended for any other uses.

Using the material presented in this book and the MATLAB code designed by the authors of this book by any entity or person is strictly at will. The authors and the publisher are neither liable nor responsible for any material or non-material losses, loss of wages, personal or property damages of any kind, or for any other type of damages of any and all types that may be incurred by using this book.

1.9.2. Problem Statement

You are to design a ground based radar to fulfill the following mission: Search and Detection. The threat consists of aircraft with an average RCS of 6 dBsm ($\sigma_a = 4m^2$), and missiles with an average RCS of -3 dBsm ($\sigma_m = 0.5m^2$). The missile altitude is 2Km, and the aircraft altitude is about 7 Km. Assume a scanning radar with 360 degrees azimuth coverage. The scan rate is less than or equal to 1 revolution every 2 seconds. Assume L to X band. We need range resolution of 150 m. No angular resolution is specified at this time. Also assume that only one missile and one aircraft constitute the whole threat. Assume a noise figure $F = 6$ dB, and total receiver loss $L = 8$ dB. For now use a fan beam with azimuth beamwidth of less than 3 degrees. Assume that 13 dB SNR is a reasonable detection threshold. Finally, assume flat earth.

1.9.3. A Design

The desired range resolution is $\Delta R = 150m$. Thus, using Eq. (1.8) one calculates the required pulsewidth as $\tau = 1\mu sec$, or equivalently the required bandwidth $B = 1MHz$. At this point a few preliminary decisions must be made. This includes the selection of the radar operating frequency, the aperture size, and the single pulse peak power.

---

1. Brookner, Eli, Editor, *Practical Phased Array Antenna Systems*, Artech House, 1991, Chapter 7.

The choice of an operating frequency that can fulfill the design requirements is driven by many factors, such as aperture size, antenna gain, clutter, atmospheric attenuation, and the maximum peak power, to name a few. In this design, an operating frequency $f = 3GHz$ is selected. This choice is somewhat arbitrary at this point; however, as we proceed with the design process this choice will be better clarified.

Second, the transportability (mobility) of the radar drives the designer in the direction of a smaller aperture type. A good choice would be less than 5 meters squared. For now choose $A_e = 2.25m^2$. The last issue that one must consider is the energy required per pulse. Note that this design approach assumes that the minimum detection SNR (13 dB) requirement is based on pulse integration. This condition is true because the target is illuminated with several pulses during a single scan, provided that the antenna azimuth beamwidth and the PRF choice satisfy Eq. (1.81).

The single pulse energy is $E = P_t \tau$. Typically, a given radar must be designed such that it has a handful of pulsewidths (waveforms) to choose from. Different waveforms (pulsewidths) are used for definite modes of operations (search, track, etc.). However, for now only a single pulse which satisfies the range resolution requirement is considered. To calculate the minimum single pulse energy required for proper detection, use Eq. (1.57). More precisely,

$$E = P_t \tau = \frac{(4\pi)^3 k T_e F L R^4 SNR_1}{G^2 \lambda^2 \sigma} \quad (1.92)$$

All parameters in Eq. (1.92) are known, except for the antenna gain, the detection range, and the single pulse SNR. The antenna gain is calculated from

$$\left(G = \frac{4\pi A_e}{\lambda^2} = \frac{4\pi \times 2.25}{(0.1)^2} = 2827.4 \right) \Rightarrow G = 34.5 dB \quad (1.93)$$

where the relation ($\lambda = c/f$) was used.

In order to estimate the detection range, consider the following argument. Since an aircraft has a larger RCS than a missile, one would expect an aircraft to be detected at a much longer range than that of a missile. This is depicted in Fig. 1.26, where R_a refers to the aircraft detection range and R_m denotes the missile detection range. As illustrated in this figure, the minimum search elevation angle θ_1 is driven by the missile detection range, assuming that the missiles are detected, with the proper SNR, as soon as they enter the radar beam. Alternatively, the maximum search elevation angle θ_2 is driven the aircraft's position along with the range that corresponds to the defense's last chance to intercept the threat (both aircraft and missile). This range is often called "keep-out minimum range" and is denoted by R_{min}. In this design approach,

$R_{min} = 30Km$ is selected. In practice, the keep-out minimum range is normally specified by the user as a design requirement.

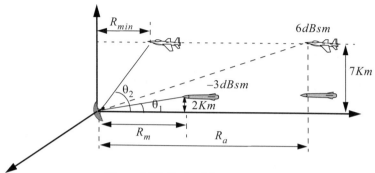

Figure 1.26. Radar / threat geometry.

The determination of R_a and R_m is dictated by how fast can a defense interceptor reach the keep-out minimum range and kill the threat. For example, assume that the threatening aircraft velocity is $400m/s$ and the threatening missile velocity is $150m/s$. Alternatively, assume that an interceptor average velocity is $250m/s$. It follows that, the interceptor time of flight, based on $R_{min} = 30Km$, is

$$T_{interceptor} = \frac{30 \times 10^3}{250} = 120 \sec \qquad (1.94)$$

Therefore, an aircraft and a missile must be detected by the radar at

$$R_a = 30Km + 120 \times 400 = 78Km$$
$$R_m = 30Km + 120 \times 150 = 48Km \qquad (1.95)$$

Note that these values should be used only as a guide. The actual detection range must also include a few more kilometers, in order to allow the defense better reaction time. In this design, choose $R_m = 55Km$; and $R_a = 90Km$. Therefore, the maximum PRF that guarantees an unambiguous range of at least 90Km is calculated from Eq. (1.5). More precisely,

$$f_r \leq \frac{c}{2R_u} = \frac{3 \times 10^8}{2 \times 90 \times 10^3} = 1.67KHz \qquad (1.96)$$

Since there are no angular resolution requirements imposed on the design at this point, then Eq. (1.96) is the only criterion that will be used to determine the radar operating PRF. Select,

$$f_r = 1000Hz \qquad (1.97)$$

The minimum and maximum elevation angles are, respectively, calculated as

$$\theta_1 = \operatorname{atan}\left(\frac{2}{55}\right) = 2.08° \tag{1.98}$$

$$\theta_2 = \operatorname{atan}\left(\frac{7}{30}\right) = 13.13° \tag{1.99}$$

These angles are then used to compute the elevation search extent (remember that the azimuth search extent is equal to 360°). More precisely, the search volume Ω (in steradians) is given by

$$\Omega = \frac{\theta_2 - \theta_1}{(57.296)^2} \times 360 \tag{1.100}$$

Consequently, the search volume is

$$\Omega = 360 \times \frac{\theta_2 - \theta_1}{(57.296)^2} = 360 \times \frac{13.13 - 2.08}{(57.296)^2} = 1.212 \; steradians \tag{1.101}$$

The desired antenna must have a fan beam; thus using a parabolic rectangular antenna will meet the design requirements. Select $A_e = 2.25 m^2$; the corresponding antenna 3-dB elevation and azimuth beamwidths are denoted as θ_e, θ_a, respectively. Select

$$\theta_e = \theta_2 - \theta_1 = 13.13 - 2.08 = 11.05° \tag{1.102}$$

The azimuth 3-dB antenna beamwidth is calculated using Eq. (1.42) as

$$\theta_a = \frac{4\pi}{G\theta_e} = \frac{4 \times \pi \times 180^2}{2827.4 \times \pi^2 \times 11} = 1.33° \tag{1.103}$$

It follows that the number of pulses that strikes a target during a single scan is calculated using Eq. (1.81) as

$$n_p \leq \frac{\theta_a f_r}{\dot{\theta}_{scan}} = \frac{1.33 \times 1000}{180} = 7.39 \Rightarrow n_p = 7 \tag{1.104}$$

The design approach presented in this book will only assume non-coherent integration (the reader is advised to re-calculate all results by assuming coherent integration, instead). The design requirement mandates a 13 dB SNR for detection. By using Eq. (1.87) one calculates the required single pulse SNR,

$$(SNR)_1 = \frac{10^{1.3}}{2 \times 7} + \sqrt{\frac{(10^{1.3})^2}{4 \times 7^2} + \frac{10^{1.3}}{7}} = 3.635 \Rightarrow (SNR)_1 = 5.6 dB \tag{1.105}$$

Furthermore the non-coherent integration loss associated with this case is computed from Eq. (1.85),

$$L_{NCI} = \frac{1 + 3.635}{3.635} = 1.27 \Rightarrow L_{NCI} = 1.056 dB \qquad (1.106)$$

It follows that the corresponding **single pulse** energy for the missile and the aircraft cases are respectively given by

$$E_m = \frac{(4\pi)^3 k T_e F L R_m^4 (SNR)_1}{G^2 \lambda^2 \sigma_m} \Rightarrow$$

$$E_m = \frac{(4\pi)^3 (1.38 \times 10^{-23})(290)(10^{0.8})(10^{0.6})(55 \times 10^3)^4 10^{0.56}}{(2827.4)^2 (0.1)^2 (0.5)} = 0.1658 \; Joules$$

(1.107)

$$E_a = \frac{(4\pi)^3 k T_e F L R_a^4 (SNR)_1}{G^2 \lambda^2 \sigma_a} \Rightarrow$$

$$E_a = \frac{(4\pi)^3 (1.38 \times 10^{-23})(290)(10^{0.8})(10^{0.6})(90 \times 10^3)^4 10^{0.56}}{(2827.4)^2 (0.1)^2 (4)} = 0.1487 \; Joules$$

(1.108)

Hence, the peak power that satisfies the single pulse detection requirement for both target types is

$$P_t = \frac{E}{\tau} = \frac{0.1658}{1 \times 10^{-6}} = 165.8 KW \qquad (1.109)$$

The radar equation with pulse integration is

$$SNR = \frac{P_t^1 G^2 \lambda^2 \sigma}{(4\pi)^3 k T_e B F L R^4} \frac{n_p}{L_{NCI}} \qquad (1.110)$$

Figure 1.27 shows the SNR versus detection range for both target-types with and without integration. To reproduce this figure use MATLAB program "*fig1_27.m*" which is given in Listing 1.12 in Section 1.10.

1.9.4. A Design Alternative

One could have elected not to reduce the single pulse peak power, but rather keep the single pulse peak power as computed in Eq. (1.109) and increase the radar detection range. For example, integrating 7 pulses coherently would improve the radar detection range by a factor of

$$R_{imp} = (7)^{0.25} = 1.63 \qquad (1.111)$$

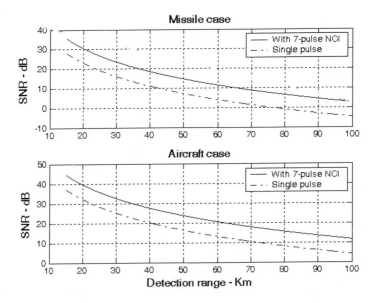

Figure 1.27. SNR versus detection range for both target types with and without pulse integration.

It follows that the new missile and aircraft detection ranges are

$$R_a = 78 \times 1.63 = 126.9 Km$$
$$R_m = 48 \times 1.63 = 78.08 Km$$
(1.112)

Note that extending the minimum detection range for a missile to $R_m = 78Km$ would increase the size of the extent of the elevation search volume. More precisely,

$$\theta_1 = \operatorname{atan}\left(\frac{2}{78}\right) = 1.47°$$
(1.113)

It follows that the search volume Ω (in steradians) is now

$$\Omega = 360 \times \frac{\theta_2 - \theta_1}{(57.296)^2} = 360 \times \frac{13.13 - 1.47}{(57.296)^2} = 1.279 \ steradians$$
(1.114)

Alternatively, integrating 7 pulses non-coherently with $(SNR)_{NCI} = 13dB$ yields

$$(SNR)_1 = 5.6 dB$$
(1.115)

and the integration loss is

$$L_{NCI} = 1.057 dB \tag{1.116}$$

Then, the net non-coherent integration gain is

$$NCI_{gain} = 10 \times \log(7) - 1.057 = 7.394 dB \Rightarrow NCI_{gain} = 5.488 \tag{1.117}$$

Thus, the radar detection range is now improved due to a 7-pulse non-coherent integration to

$$R_a = 78 \times (5.488)^{0.25} = 119.38 Km$$
$$R_m = 48 \times (5.488)^{0.25} = 73.467 Km \tag{1.118}$$

Again, the extent of the elevation search volume is changed to

$$\theta_1 = \operatorname{atan}\left(\frac{2}{73.467}\right) = 1.56° \tag{1.119}$$

It follows that the search volume Ω (in steradians) is now

$$\Omega = 360 \times \frac{\theta_2 - \theta_1}{(57.296)^2} = 360 \times \frac{13.13 - 1.56}{(57.296)^2} = 1.269 \; steradians \tag{1.120}$$

1.10. MATLAB Program and Function Listings

This section presents listings for all MATLAB functions and programs used in this chapter. Users are encouraged to vary the input parameters and rerun these programs in order to enhance their understanding of the theory presented in the text. All selected parameters and variables follow the same nomenclature used in the text; thus, understanding the structure and hierarchy of the presented code should be an easy task once the user has read the chapter.

Note that all MATLAB programs and functions developed in this book can be downloaded from CRC Press Web Site "www.crcpress.com". Additionally, all MATLAB code developed for this book was developed using MATLAB 6.5 Release 13 for Microsoft Windows.

Listing 1.1. MATLAB Function "radar_eq.m"

```
function [snr] = radar_eq(pt, freq, g, sigma, te, b, nf, loss, range)
% This program implements Eq. (1.56)
c = 3.0e+8; % speed of light
lambda = c / freq; % wavelength
p_peak = 10*log10(pt); % convert peak power to dB
lambda_sqdb = 10*log10(lambda^2); % compute wavelength square in dB
sigmadb = 10*log10(sigma); % convert sigma to dB
four_pi_cub = 10*log10((4.0 * pi)^3); % (4pi)^3 in dB
```

```
k_db = 10*log10(1.38e-23); % Boltzman's constant in dB
te_db = 10*log10(te); % noise temp. in dB
b_db = 10*log10(b); % bandwidth in dB
range_pwr4_db = 10*log10(range.^4); % vector of target range^4 in dB
% Implement Equation (1.56)
num = p_peak + g + g + lambda_sqdb + sigmadb;
den = four_pi_cub + k_db + te_db + b_db + nf + loss + range_pwr4_db;
snr = num - den;
return
```

Listing 1.2. MATLAB Program "fig1_12.m"

```
% Use this program to reproduce Fig. 1.12 of text.
close all
clear all
pt = 1.5e+6; % peak power in Watts
freq = 5.6e+9; % radar operating frequency in Hz
g = 45.0; % antenna gain in dB
sigma = 0.1; % radar cross section in m squared
te = 290.0; % effective noise temperature in Kelvins
b = 5.0e+6; % radar operating bandwidth in Hz
nf = 3.0; %noise figure in dB
loss = 6.0; % radar losses in dB
range = linspace(25e3,165e3,1000); % traget range 25 -165 Km, 1000 points
snr1 = radar_eq(pt, freq, g, sigma, te, b, nf, loss, range);
snr2 = radar_eq(pt, freq, g, sigma/10, te, b, nf, loss, range);
snr3 = radar_eq(pt, freq, g, sigma*10, te, b, nf, loss, range);
% plot SNR versus range
figure(1)
rangekm = range ./ 1000;
plot(rangekm,snr3,'k',rangekm,snr1,'k -.',rangekm,snr2,'k:')
grid
legend('\sigma = 0 dBsm','\sigma = -10dBsm','\sigma = -20 dBsm')
xlabel ('Detection range - Km');
ylabel ('SNR - dB');
snr1 = radar_eq(pt, freq, g, sigma, te, b, nf, loss, range);
snr2 = radar_eq(pt*.4, freq, g, sigma, te, b, nf, loss, range);
snr3 = radar_eq(pt*1.8, freq, g, sigma, te, b, nf, loss, range);
figure (2)
plot(rangekm,snr3,'k',rangekm,snr1,'k -.',rangekm,snr2,'k:')
grid
legend('Pt = 2.16 MW','Pt = 1.5 MW','Pt = 0.6 MW')
xlabel ('Detection range - Km');
ylabel ('SNR - dB');
```

Listing 1.3. MATLAB Program "fig1_13.m"

```
% Use this program to reproduce Fig. 1.13 of text.
close all
clear all
pt = 1.e+6; % peak power in Watts
freq = 5.6e+9; % radar operating frequency in Hz
g = 40.0; % antenna gain in dB
sigma = 0.1; % radar cross section in m squared
te =300.0; % effective noise temperature in Kelvins
nf = 5.0; %noise figure in dB
loss = 6.0; % radar losses in dB
range = [75e3,100e3,150e3]; % three range values
snr_db = linspace(5,20,200); % SNR values from 5 dB to 20 dB 200 points
snr = 10.^(0.1.*snr_db); % convert snr into base 10
gain = 10^(0.1*g); %convert antenna gain into base 10
loss = 10^(0.1*loss); % convert losses into base 10
F = 10^(0.1*nf); % convert noise figure into base 10
lambda = 3.e8 / freq; % compute wavelength
% Implement Eq.(1.57)
den = pt * gain * gain * sigma * lambda^2;
num1 = (4*pi)^3 * 1.38e-23 * te * F * loss * range(1)^4 .* snr;
num2 = (4*pi)^3 * 1.38e-23 * te * F * loss * range(2)^4 .* snr;
num3 = (4*pi)^3 * 1.38e-23 * te * F * loss * range(3)^4 .* snr;
tau1 = num1 ./ den ;
tau2 = num2 ./ den;
tau3 = num3 ./ den;
% plot tau versus snr
figure(1)
semilogy(snr_db,1e6*tau1,'k',snr_db,1e6*tau2,'k -.',snr_db,1e6*tau3,'k:')
grid
legend('R = 75 Km','R = 100 Km','R = 150 Km')
xlabel ('Minimum required SNR - dB');
ylabel ('\tau (pulsewidth) in \mu sec');
```

Listing 1.4. MATLAB Program "ref_snr.m"

```
% This program implements Eq. (1.60)
clear all
close all
Rref = 86e3; % ref. range
tau_ref = .1e-6; % ref. pulsewidth
SNRref = 20.; % Ref SNR in dB
snrref = 10^(SNRref/10);
```

```
Sigmaref = 0.1; % ref RCS in m^2
Lossp = 2; % processing loss in dB
lossp = 10^(Lossp/10);
% Enter desired value
tau = tau_ref;
R = 120e3;
rangeratio = (Rref / R)^4;
Sigma = 0.2;
% Implement Eq. (1.60)
snr = snrref * (tau / tau_ref) * (1. / lossp) * ...
   (Sigma / Sigmaref) * rangeratio;
snr = 10*log10(snr)
```

Listing 1.5. MATLAB Function "power_aperture.m"

```
function PAP = ...
              power_aperture(snr,tsc,sigma,range,te,nf,loss,az_angle,el_
              angle)
% This program implements Eq. (1.67)
Tsc = 10*log10(tsc); % convert Tsc into dB
Sigma = 10*log10(sigma); % convert sigma to dB
four_pi = 10*log10(4.0 * pi); % (4pi) in dB
k_db = 10*log10(1.38e-23); % Boltzman's constant in dB
Te = 10*log10(te); % noise temp. in dB
range_pwr4_db = 10*log10(range.^4); % target range^4 in dB
omega = az_angle * el_angle / (57.296)^2; % compute search volume in stera-
              dians
Omega = 10*log10(omega) % search volume in dB
% implement Eq. (1.67)
PAP = snr + four_pi + k_db + Te + nf + loss + range_pwr4_db + Omega ...
   - Sigma - Tsc;
return
```

Listing 1.6. MATLAB Program "fig1_16.m"

```
% Use this program to reproduce Fig. 1.16 of text.
close all
clear all
tsc = 2.5; % Scan time is 2.5 seconds
sigma = 0.1; % radar cross section in m squared
te = 900.0; % effective noise temperature in Kelvins
snr = 15; % desired SNR in dB
nf = 6.0; %noise figure in dB
loss = 7.0; % radar losses in dB
```

```
az_angle = 2; % search volume azimuth extent in degrees
el_angle = 2; % search volume elevation extent in degrees
range = linspace(20e3,250e3,1000); % range to target 20 Km 250 Km, 1000
                points
pap1 = power_aperture(snr,tsc,sigma/10,range,te,nf,loss,az_angle,el_angle);
pap2 = power_aperture(snr,tsc,sigma,range,te,nf,loss,az_angle,el_angle);
pap3 = power_aperture(snr,tsc,sigma*10,range,te,nf,loss,az_angle,el_angle);
% plot power aperture product versus range
% figure 1.16a
figure(1)
rangekm  = range ./ 1000;
plot(rangekm,pap1,'k',rangekm,pap2,'k -.',rangekm,pap3,'k:')
grid
legend('\sigma = -20 dBsm','\sigma = -10dBsm','\sigma = 0 dBsm')
xlabel ('Detection range in Km');
ylabel ('Power aperture product in dB');
% generate Figure 1.16b
lambda = 0.03; % wavelength in meters
G = 45; % antenna gain in dB
ae = linspace(1,25,1000);% aperture size 1 to 25 meter squared, 1000 points
Ae = 10*log10(ae);
range = 250e3; % range of interest is 250 Km
pap1 = power_aperture(snr,tsc,sigma/10,range,te,nf,loss,az_angle,el_angle);
pap2 = power_aperture(snr,tsc,sigma,range,te,nf,loss,az_angle,el_angle);
pap3 = power_aperture(snr,tsc,sigma*10,range,te,nf,loss,az_angle,el_angle);
Pav1 = pap1 - Ae;
Pav2 = pap2 - Ae;
Pav3 = pap3 - Ae;
figure(2)
plot(ae,Pav1,'k',ae,Pav2,'k -.',ae,Pav3,'k:')
grid
xlabel('Aperture size in square meters')
ylabel('Pav in dB')
legend('\sigma = -20 dBsm','\sigma = -10dBsm','\sigma = 0 dBsm')
```

Listing 1.7. MATLAB Program "casestudy1_1.m"

```
% This program is used to generate Fig. 1.17
% It implements the search radar equation defined in Eq. 1.67
clear all
close all
snr = 15.0;        % Sensitivity SNR in dB
tsc = 2.;          % Antenna scan time in seconds
sigma_tgtm = -10;  % Missile RCS in dBsm
```

```
sigma_tgta = 6;     % Aircraft RCS in dBsm
range = 60.0; % Sensitivity range in Km,
te = 290.0;         % Effective noise temperature in Kelvins
nf = 8;             % Noise figure in dB
loss = 10.0;        % Radar losses in dB
az_angle = 360.0;   % Search volume azimuth extent in degrees
el_angle = 10.0;    % Search volume elevation extent in degrees
c = 3.0e+8;         % Speed of light
% Compute Omega in steradians
omega = (az_angle / 57.296) * (el_angle /57.296);
omega_db = 10.0*log10(omega); % Convert Omega to dBs
k_db = 10.*log10(1.38e-23);
te_db = 10*log10(te);
tsc_db = 10*log10(tsc);
factor = 10*log10(4*pi);
rangemdb = 10*log10(range * 1000.);
rangeadb = 10*log10(range * 1000.);
PAP_Missile = snr - sigma_tgtm - tsc_db + factor + 4.0 * rangemdb + ...
   k_db + te_db + nf + loss + omega_db
PAP_Aircraft = snr - sigma_tgta - tsc_db + factor + 4.0 * rangeadb + ...
   k_db + te_db + nf + loss + omega_db
index = 0;
% vary range from 2Km to 90 Km
for rangevar = 2 : 1 : 90
   index = index + 1;
   rangedb = 10*log10(rangevar * 1000.0);
   papm(index) = snr - sigma_tgtm - tsc_db + factor + 4.0 * rangedb + ...
      k_db + te_db + nf + loss + omega_db;
   missile_PAP(index) = PAP_Missile;
   aircraft_PAP(index) = PAP_Aircraft;
   papa(index) = snr - sigma_tgta - tsc_db + factor + 4.0 * rangedb + ...
      k_db + te_db + nf + loss +omega_db;
end
var = 2 : 1 : 90;
figure (1)
plot (var,papm,'k',var,papa,'k-.')
legend ('Missile','Aircraft')
xlabel ('Range - Km');
ylabel ('Power Aperture Product - dB');
hold on
plot(var,missile_PAP,'k:',var,aircraft_PAP,'k:')
grid
hold off
```

Listing 1.8. MATLAB Program "fig1_19.m"

```
% Use this program to reproduce Fig. 1.19 and Fig. 1.20 of text.
close all
clear all
pt = 4; % peak power in Watts
freq = 94e+9; % radar operating frequency in Hz
g = 47.0; % antenna gain in dB
sigma = 20; % radar cross section in m squared
te = 293.0; % effective noise temperature in Kelvins
b = 20e+6; % radar operating bandwidth in Hz
nf = 7.0; %noise figure in dB
loss = 10.0; % radar losses in dB
range = linspace(1.e3,12e3,1000); % range to target from 1. Km 12 Km, 1000
                points
snr1 = radar_eq(pt, freq, g, sigma, te, b, nf, loss, range);
Rnewci = (94^0.25) .* range;
snrCI = snr1 + 10*log10(94); % 94 pulse coherent integration
% plot SNR versus range
figure(1)
rangekm = range ./ 1000;
plot(rangekm,snr1,'k',Rnewci./1000,snr1,'k -.')
axis([1 12 -20 45])
grid
legend('single pulse','94 pulse CI')
xlabel ('Detection range - Km');
ylabel ('SNR - dB');
% Generate Figure 1.20
snr_b10 = 10.^(snr1./10);
SNR_1 = snr_b10./(2*94) + sqrt(((snr_b10.^2) ./ (4*94*94)) + (snr_b10 ./
                94)); % Equation 1.80 of text
LNCI = (1+SNR_1) ./ SNR_1; % Equation 1.78 of text
NCIgain = 10*log10(94) - 10*log10(LNCI);
Rnewnci = ((10.^(0.1*NCIgain)).^0.25) .* range;
snrnci = snr1 + NCIgain;
figure (2)
plot(rangekm,snr1,'k',Rnewnci./1000,snr1,'k -.', Rnewci./1000,snr1,'k:')
axis([1 12 -20 45])
grid
legend('single pulse','94 pulse NCI','94 pulse CI')
xlabel ('Detection range - Km');
ylabel ('SNR - dB');
```

Listing 1.9. MATLAB Program "fig1_21.m"

```
%use this figure to generate Fig. 1.21 of text
clear all
close all
np = linspace(1,10000,1000);
snrci = pulse_integration(4,94.e9,47,20,290,20e6,7,10,5.01e3,np,1);
snrnci = pulse_integration(4,94.e9,47,20,290,20e6,7,10,5,01e3,np,2);
semilogx(np,snrci,'k',np,snrnci,'k:')
legend('Coherent integration','Non-coherent integration')
grid
xlabel ('Number of integrated pulses');
ylabel ('SNR - dB');
```

Listing 1.10. MATLAB Function "pulse_integration.m"

```
function [snrout] = pulse_integration(pt, freq, g, sigma, te, b, nf, loss,
                range,np,ci_nci)
 snr1 = radar_eq(pt, freq, g, sigma, te, b, nf, loss, range) % single pulse SNR
if (ci_nci == 1) % coherent integration
   snrout = snr1 + 10*log10(np);
else % non-coherent integration
   if (ci_nci == 2)
      snr_nci = 10.^(snr1./10);
      val1 = (snr_nci.^2) ./ (4.*np.*np);
      val2 = snr_nci ./ np;
      val3 = snr_nci ./ (2.*np);
      SNR_1 = val3 + sqrt(val1 + val2); % Equation 1.87 of text
      LNCI = (1+SNR_1) ./ SNR_1; % Equation 1.85 of text
      snrout = snr1 + 10*log10(np) - 10*log10(LNCI);
   end
end
return
```

Listing 1.11. MATLAB Program "myradarvisit1_1.m"

```
close all
clear all
pt = 724.2e+3; % peak power in Watts
freq = 3e+9; % radar operating frequency in Hz
g = 37.0; % antenna gain in dB
sigmam = 0.5; % missile RCS in m squared
sigmaa = 4.0; % aircraft RCS in m squared
te = 290.0; % effective noise temperature in Kelvins
b = 1.0e+6; % radar operating bandwidth in Hz
```

```
nf = 6.0; %noise figure in dB
loss = 8.0; % radar losses in dB
range = linspace(5e3,125e3,1000); % range to target from 25 Km 165 Km,
                1000 points
snr1 = radar_eq(pt, freq, g, sigmam, te, b, nf, loss, range);
snr2 = radar_eq(pt, freq, g, sigmaa, te, b, nf, loss, range);
% plot SNR versus range
figure(1)
rangekm = range ./ 1000;
plot(rangekm,snr1,'k',rangekm,snr2,'k:')
grid
legend('Misssile','Aircraft')
xlabel ('Detection range - Km');
ylabel ('SNR - dB');
```

Listing 1.12. MATLAB Program "fig1_27.m"

```
% Use this program to reproduce Fig. 1.27 of text.
close all
clear all
np = 7;
pt = 165.8e3; % peak power in Watts
freq = 3e+9; % radar operating frequency in Hz
g = 34.5139; % antenna gain in dB
sigmam = 0.5; % missile RCS m squared
sigmaa = 4; % aircraft RCS m squared
te = 290.0; % effective noise temperature in Kelvins
b = 1.0e+6; % radar operating bandwidth in Hz
nf = 6.0; %noise figure in dB
loss = 8.0; % radar losses in dB
% compute the single pulse SNR when 7-pulse NCI is used
SNR_1 = (10^1.3)/(2*7) + sqrt((((10^1.3)^2) / (4*7*7)) + ((10^1.3) / 7));
% compute the integration loss
LNCI = 10*log10((1+SNR_1)/SNR_1);
loss_total = loss + LNCI;
range = linspace(15e3,100e3,1000); % range to target from 15 to 100 Km,
                1000 points
% modify pt by np*pt to account for pulse integration
snrmnci = radar_eq(np*pt, freq, g, sigmam, te, b, nf, loss_total, range);
snrm = radar_eq(pt, freq, g, sigmam, te, b, nf, loss, range);
snranci = radar_eq(np*pt, freq, g, sigmaa, te, b, nf, loss_total, range);
snra = radar_eq(pt, freq, g, sigmaa, te, b, nf, loss, range);
% plot SNR versus range
rangekm = range ./ 1000;
```

```
figure(1)
subplot(2,1,1)
plot(rangekm,snrmnci,'k',rangekm,snrm,'k -.')
grid
legend('With 7-pulse NCI','Single pulse')
ylabel ('SNR - dB');
title('Missile case')
subplot(2,1,2)
plot(rangekm,snranci,'k',rangekm,snra,'k -.')
grid
legend('With 7-pulse NCI','Single pulse')
ylabel ('SNR - dB');
title('Aircraft case')
xlabel('Detection range - Km')
```

Appendix 1A — *Pulsed Radar*

1A.1. Introduction

Pulsed radars transmit and receive a train of modulated pulses. Range is extracted from the two-way time delay between a transmitted and received pulse. Doppler measurements can be made in two ways. If accurate range measurements are available between consecutive pulses, then Doppler frequency can be extracted from the range rate $\dot{R} = \Delta R / \Delta t$. This approach works fine as long as the range is not changing drastically over the interval Δt. Otherwise, pulsed radars utilize a Doppler filter bank.

Pulsed radar waveforms can be completely defined by the following: (1) carrier frequency which may vary depending on the design requirements and radar mission; (2) pulsewidth, which is closely related to the bandwidth and defines the range resolution; (3) modulation; and finally (4) the pulse repetition frequency. Different modulation techniques are usually utilized to enhance the radar performance, or to add more capabilities to the radar that otherwise would not have been possible. The PRF must be chosen to avoid Doppler and range ambiguities as well as maximize the average transmitted power.

Radar systems employ low, medium, and high PRF schemes. Low PRF waveforms can provide accurate, long, unambiguous range measurements, but exert severe Doppler ambiguities. Medium PRF waveforms must resolve both range and Doppler ambiguities; however, they provide adequate average transmitted power as compared to low PRFs. High PRF waveforms can provide superior average transmitted power and excellent clutter rejection capabilities. Alternatively, high PRF waveforms are extremely ambiguous in range. Radar systems utilizing high PRFs are often called Pulsed Doppler Radars (PDR). Range and Doppler ambiguities for different PRFs are in Table 1A.1.

TABLE 1A.1. PRF ambiguities.

PRF	Range Ambiguous	Doppler Ambiguous
Low PRF	No	Yes
Medium PRF	Yes	Yes
High PRF	Yes	No

Radars can utilize constant and varying (agile) PRFs. For example, Moving Target Indicator (MTI) radars use PRF agility to avoid blind speeds. This kind of agility is known as PRF staggering. PRF agility is also used to avoid range and Doppler ambiguities, as will be explained in the next three sections. Additionally, PRF agility is also used to prevent jammers from locking onto the radar's PRF. These two latter forms of PRF agility are sometimes referred to as PRF jitter.

Fig. 1A.1 shows a simplified pulsed radar block diagram. The range gates can be implemented as filters that open and close at time intervals that correspond to the detection range. The width of such an interval corresponds to the desired range resolution. The radar receiver is often implemented as a series of contiguous (in time) range gates, where the width of each gate is matched to the radar pulsewidth. The NBF bank is normally implemented using an FFT, where bandwidth of the individual filters corresponds to the FFT frequency resolution.

1A.2. Range and Doppler Ambiguities

As explained earlier, a pulsed radar can be range ambiguous if a second pulse is transmitted prior to the return of the first pulse. In general, the radar PRF is chosen such that the unambiguous range is large enough to meet the radar's operational requirements. Therefore, long-range search (surveillance) radars would require relatively low PRFs.

The line spectrum of a train of pulses has $\sin x/x$ envelope, and the line spectra are separated by the PRF, f_r, as illustrated in Fig. 1A.2. The Doppler filter bank is capable of resolving target Doppler as long as the anticipated Doppler shift is less than one half the bandwidth of the individual filters (i.e., one half the width of an FFT bin). Thus, pulsed radars are designed such that

$$f_r = 2f_{dmax} = \frac{2v_{rmax}}{\lambda} \quad (1A.1)$$

where f_{dmax} is the maximum anticipated target Doppler frequency, v_{rmax} is the maximum anticipated target radial velocity, and λ is the radar wavelength.

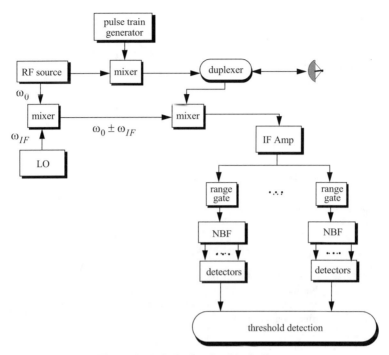

Figure 1A.1. Pulsed radar block diagram.

If the Doppler frequency of the target is high enough to make an adjacent spectral line move inside the Doppler band of interest, the radar can be Doppler ambiguous. Therefore, in order to avoid Doppler ambiguities, radar systems require high PRF rates when detecting high speed targets. When a long-range radar is required to detect a high speed target, it may not be possible to be both range and Doppler unambiguous. This problem can be resolved by using multiple PRFs. Multiple PRF schemes can be incorporated sequentially within each dwell interval (scan or integration frame) or the radar can use a single PRF in one scan and resolve ambiguity in the next. The latter technique, however, may have problems due to changing target dynamics from one scan to the next.

1A.3. Resolving Range Ambiguity

Consider a radar that uses two PRFs, f_{r1} and f_{r2}, on transmit to resolve range ambiguity, as shown in Fig. 1A.3. Denote R_{u1} and R_{u2} as the unambiguous ranges for the two PRFs, respectively. Normally, these unambiguous ranges are relatively small and are short of the desired radar unambiguous range R_u (where $R_u \gg R_{u1}, R_{u2}$). Denote the radar desired PRF that corresponds to R_u as f_{rd}.

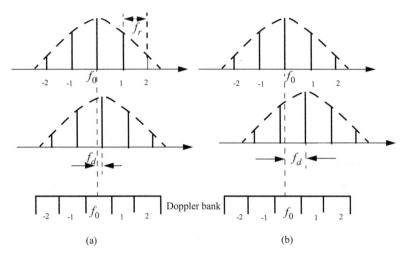

Figure 1A.2. Spectra of transmitted and received waveforms, and Doppler bank. (a) Doppler is resolved. (b) Spectral lines have moved into the next Doppler filter. This results in an ambiguous Doppler measurement.

We choose f_{r1} and f_{r2} such that they are relatively prime with respect to one another. One choice is to select $f_{r1} = Nf_{rd}$ and $f_{r2} = (N+1)f_{rd}$ for some integer N. Within one period of the desired PRI ($T_d = 1/f_{rd}$) the two PRFs f_{r1} and f_{r2} coincide only at one location, which is the true unambiguous target position. The time delay T_d establishes the desired unambiguous range. The time delays t_1 and t_2 correspond to the time between the transmit of a pulse on each PRF and receipt of a target return due to the same pulse.

Let M_1 be the number of PRF1 intervals between transmit of a pulse and receipt of the true target return. The quantity M_2 is similar to M_1 except it is for PRF2. It follows that, over the interval 0 to T_d, the only possible results are $M_1 = M_2 = M$ or $M_1 + 1 = M_2$. The radar needs only to measure t_1 and t_2. First, consider the case when $t_1 < t_2$. In this case,

$$t_1 + \frac{M}{f_{r1}} = t_2 + \frac{M}{f_{r2}} \qquad (1A.2)$$

for which we get

$$M = \frac{t_2 - t_1}{T_1 - T_2} \qquad (1A.3)$$

where $T_1 = 1/f_{r1}$ and $T_2 = 1/f_{r2}$. It follows that the round trip time to the true target location is

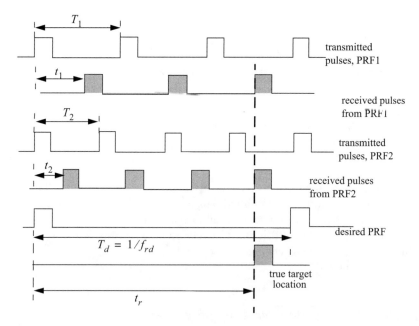

Figure 1A.3. Resolving range ambiguity.

$$t_r = MT_1 + t_1$$
$$t_r = MT_2 + t_2 \tag{1A.4}$$

and the true target range is

$$R = ct_r/2 \tag{1A.5}$$

Now if $t_1 > t_2$, then

$$t_1 + \frac{M}{f_{r1}} = t_2 + \frac{M+1}{f_{r2}} \tag{1A.6}$$

Solving for M we get

$$M = \frac{(t_2 - t_1) + T_2}{T_1 - T_2} \tag{1A.7}$$

and the round-trip time to the true target location is

$$t_{r1} = MT_1 + t_1 \tag{1A.8}$$

and in this case, the true target range is

$$R = \frac{ct_{r1}}{2} \tag{1A.9}$$

Finally, if $t_1 = t_2$, then the target is in the first ambiguity. It follows that

$$t_{r2} = t_1 = t_2 \tag{1A.10}$$

and

$$R = ct_{r2}/2 \tag{1A.11}$$

Since a pulse cannot be received while the following pulse is being transmitted, these times correspond to blind ranges. This problem can be resolved by using a third PRF. In this case, once an integer N is selected, then in order to guarantee that the three PRFs are relatively prime with respect to one another. In this case, one may choose $f_{r1} = N(N+1)f_{rd}$, $f_{r2} = N(N+2)f_{rd}$, and $f_{r3} = (N+1)(N+2)f_{rd}$.

1A.4. Resolving Doppler Ambiguity

The Doppler ambiguity problem is analogous to that of range ambiguity. Therefore, the same methodology can be used to resolve Doppler ambiguity. In this case, we measure the Doppler frequencies f_{d1} and f_{d2} instead of t_1 and t_2.

If $f_{d1} > f_{d2}$, then we have

$$M = \frac{(f_{d2} - f_{d1}) + f_{r2}}{f_{r1} - f_{r2}} \tag{1A.12}$$

And if $f_{d1} < f_{d2}$,

$$M = \frac{f_{d2} - f_{d1}}{f_{r1} - f_{r2}} \tag{1A.13}$$

and the true Doppler is

$$\begin{aligned} f_d &= Mf_{r1} + f_{d1} \\ f_d &= Mf_{r2} + f_{d2} \end{aligned} \tag{1A.14}$$

Finally, if $f_{d1} = f_{d2}$, then

$$f_d = f_{d1} = f_{d2} \tag{1A.15}$$

Again, blind Doppler can occur, which can be resolved using a third PRF.

Resolving Doppler Ambiguity

Example:

A certain radar uses two PRFs to resolve range ambiguities. The desired unambiguous range is $R_u = 100 Km$. Choose $N = 59$. Compute f_{r1}, f_{r2}, R_{u1}, and R_{u2}.

Solution:

First let us compute the desired PRF, f_{rd}

$$f_{rd} = \frac{c}{2R_u} = \frac{3 \times 10^8}{200 \times 10^3} = 1.5 KHz$$

It follows that

$$f_{r1} = Nf_{rd} = (59)(1500) = 88.5 KHz$$

$$f_{r2} = (N+1)f_{rd} = (59+1)(1500) = 90 KHz$$

$$R_{u1} = \frac{c}{2f_{r1}} = \frac{3 \times 10^8}{2 \times 88.5 \times 10^3} = 1.695 Km$$

$$R_{u2} = \frac{c}{2f_{r2}} = \frac{3 \times 10^8}{2 \times 90 \times 10^3} = 1.667 Km.$$

Example:

Consider a radar with three PRFs; $f_{r1} = 15 KHz$, $f_{r2} = 18 KHz$, and $f_{r3} = 21 KHz$. Assume $f_0 = 9 GHz$. Calculate the frequency position of each PRF for a target whose velocity is $550 m/s$. Calculate f_d (Doppler frequency) for another target appearing at $8 KHz$, $2 KHz$, and $17 KHz$ for each PRF.

Solution:

The Doppler frequency is

$$f_d = 2\frac{vf_0}{c} = \frac{2 \times 550 \times 9 \times 10^9}{3 \times 10^8} = 33 KHz$$

Then by using Eq. (1A.14) $n_i f_{ri} + f_{di} = f_d$ where $i = 1, 2, 3$, we can write

$$n_1 f_{r1} + f_{d1} = 15n_1 + f_{d1} = 33$$

$$n_2 f_{r2} + f_{d2} = 18n_2 + f_{d2} = 33$$

$$n_3 f_{r3} + f_{d3} = 21n_3 + f_{d3} = 33$$

We will show here how to compute n_1, and leave the computations of n_2 and n_3 to the reader. First, if we choose $n_1 = 0$, that means $f_{d1} = 33 KHz$, which cannot be true since f_{d1} cannot be greater than f_{r1}. Choosing $n_1 = 1$ is also *invalid since* $f_{d1} = 18KHz$ *cannot be true either. Finally, if we choose* $n_1 = 2$ *we get* $f_{d1} = 3KHz$, *which is an acceptable value. It follows that the minimum* n_1, n_2, n_3 *that may satisfy the above three relations are* $n_1 = 2$, $n_2 = 1$, *and* $n_3 = 1$. *Thus, the apparent Doppler frequencies are* $f_{d1} = 3KHz$, $f_{d2} = 15KHz$, *and* $f_{d3} = 12KHz$.

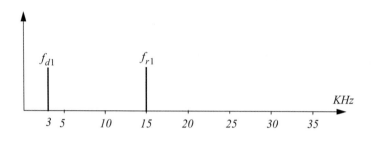

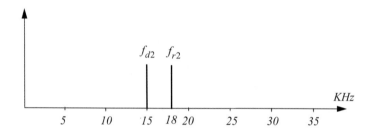

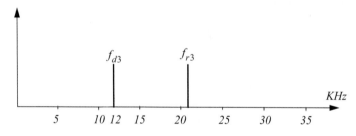

Now for the second part of the problem. Again by using Eq. (1A.14) we have

$$n_1 f_{r1} + f_{d1} = f_d = 15n_1 + 8$$

$$n_2 f_{r2} + f_{d2} = f_d = 18n_2 + 2$$

$$n_3 f_{r3} + f_{d3} = f_d = 21n_3 + 17$$

We can now solve for the smallest integers n_1, n_2, n_3 that satisfy the above three relations. See the table below.

n	0	1	2	3	4
f_d from f_{r1}	8	23	<u>38</u>	53	68
f_d from f_{r2}	2	20	<u>38</u>	56	
f_d from f_{r3}	17	<u>38</u>	39		

Thus, $n_1 = 2 = n_2$, and $n_3 = 1$, and the true target Doppler is $f_d = 38 KHz$. It follows that

$$v_r = 38000 \times \frac{0.0333}{2} = 632.7 \frac{m}{\text{sec}}$$

Appendix 1B   *Noise Figure*

1B.1. Noise Figure

Any signal other than the target returns in the radar receiver is considered to be noise. This includes interfering signals from outside the radar and thermal noise generated within the receiver itself. Thermal noise (thermal agitation of electrons) and shot noise (variation in carrier density of a semiconductor) are the two main internal noise sources within a radar receiver.

The power spectral density of thermal noise is given by

$$S_n(\omega) = \frac{|\omega|h}{\pi\left[\exp\left(\frac{|\omega|h}{2\pi kT}\right) - 1\right]} \tag{1B.1}$$

where $|\omega|$ is the absolute value of the frequency in radians per second, T is the temperature of the conducting medium in degrees Kelvin, k is Boltzman's constant, and h is Plank's constant ($h = 6.625 \times 10^{-34}$ joule seconds). When the condition $|\omega| \ll 2\pi kT/h$ is true, it can be shown that Eq. (1B.1) is approximated by

$$S_n(\omega) \approx 2kT \tag{1B.2}$$

This approximation is widely accepted, since, in practice, radar systems operate at frequencies less than 100 GHz; and, for example, if $T = 290K$, then $2\pi kT/h \approx 6000$ GHz.

The mean square noise voltage (noise power) generated across a 1 *ohm* resistance is then

$$\langle n^2 \rangle = \frac{1}{2\pi} \int_{-2\pi B}^{2\pi B} 2kT \, d\omega = 4kTB \qquad (1B.3)$$

where B is the system bandwidth in hertz.

Any electrical system containing thermal noise and having input resistance R_{in} can be replaced by an equivalent noiseless system with a series combination of a noise equivalent voltage source and a noiseless input resistor R_{in} added at its input. This is illustrated in Fig. 1B.1.

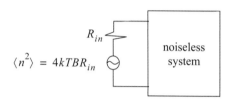

Figure 1B.1. Noiseless system with an input noise voltage source.

The amount of noise power that can physically be extracted from $\langle n^2 \rangle$ is one fourth the value computed in Eq. (1B.3). The proof is left as an exercise.

Consider a noisy system with power gain A_P, as shown in Fig. 1B.2. The noise figure is defined by

$$F_{dB} = 10 \, \log \frac{total \; noise \; power \; out}{noise \; power \; out \; due \; to \; R_{in} \; alone} \qquad (1B.4)$$

More precisely,

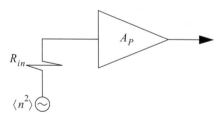

Figure 1B.2. Noisy amplifier replaced by its noiseless equivalent and an input voltage source in series with a resistor.

Noise Figure

$$F_{dB} = 10 \log \frac{N_o}{N_i A_p} \tag{1B.5}$$

where N_o and N_i are, respectively, the noise power at the output and input of the system.

If we define the input and output signal power by S_i and S_o, respectively, then the power gain is

$$A_p = \frac{S_o}{S_i} \tag{1B.6}$$

It follows that

$$F_{dB} = 10\log\left(\frac{S_i/N_i}{S_o/N_o}\right) = \left(\frac{S_i}{N_i}\right)_{dB} - \left(\frac{S_o}{N_o}\right)_{dB} \tag{1B.7}$$

where

$$\left(\frac{S_i}{N_i}\right)_{dB} > \left(\frac{S_o}{N_o}\right)_{dB} \tag{1B.8}$$

Thus, it can be said that the noise figure is the loss in the signal-to-noise ratio due to the added thermal noise of the amplifier $((SNR)_o = (SNR)_i - F \text{ in } dB)$.

We can also express the noise figure in terms of the system's effective temperature T_e. Consider the amplifier shown in Fig. 1B.2, and let its effective temperature be T_e. Assume the input noise temperature is T_o. Thus, the input noise power is

$$N_i = kT_o B \tag{1B.9}$$

and the output noise power is

$$N_o = kT_o B \, A_p + kT_e B \, A_p \tag{1B.10}$$

where the first term on the right-hand side of Eq. (1B.10) corresponds to the input noise, and the latter term is due to thermal noise generated inside the system. It follows that the noise figure can be expressed as

$$F = \frac{(SNR)_i}{(SNR)_o} = \frac{S_i}{kT_o B} kBA_p \frac{T_o + T_e}{S_o} = 1 + \frac{T_e}{T_o} \tag{1B.11}$$

Equivalently, we can write

$$T_e = (F-1)T_o \tag{1B.12}$$

Example:

An amplifier has a 4dB noise figure; the bandwidth is $B = 500$ KHz. Calculate the input signal power that yields a unity SNR at the output. Assume $T_o = 290$ degrees Kelvin and an input resistance of one ohm.

Solution:

The input noise power is

$$kT_oB = 1.38 \times 10^{-23} \times 290 \times 500 \times 10^3 = 2.0 \times 10^{-15} w$$

Assuming a voltage signal, then the input noise mean squared voltage is

$$\langle n_i^2 \rangle = kT_oB = 2.0 \times 10^{-15} \ v^2$$

$$F = 10^{0.4} = 2.51$$

From the noise figure definition we get

$$\frac{S_i}{N_i} = F\left(\frac{S_o}{N_o}\right) = F$$

and

$$\langle s_i^2 \rangle = F\langle n_i^2 \rangle = 2.51 \times 2.0 \times 10^{-15} = 5.02 \times 10^{-15} \ v^2$$

Finally,

$$\sqrt{\langle s_i^2 \rangle} = 70.852 nv$$

Consider a cascaded system as in Fig. 1B.3. Network 1 is defined by noise figure F_1, power gain G_1, bandwidth B, and temperature T_{e1}. Similarly, network 2 is defined by F_2, G_2, B, and T_{e2}. Assume the input noise has temperature T_0.

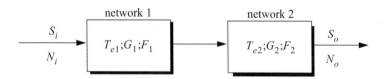

Figure 1B.3. Cascaded linear system.

Noise Figure

The output signal power is

$$S_o = S_i G_1 G_2 \tag{1B.13}$$

The input and output noise powers are, respectively, given by

$$N_i = kT_o B \tag{1B.14}$$

$$N_o = kT_0 B G_1 G_2 + kT_{e1} B G_1 G_2 + kT_{e2} B G_2 \tag{1B.15}$$

where the three terms on the right-hand side of Eq. (1B.15), respectively, correspond to the input noise power, thermal noise generated inside network 1, and thermal noise generated inside network 2.

Now if we use the relation $T_e = (F-1)T_0$ along with Eq. (1B.13) and Eq. (1B.14), we can express the overall output noise power as

$$N_o = F_1 N_i G_1 G_2 + (F_2 - 1) N_i G_2 \tag{1B.16}$$

It follows that the overall noise figure for the cascaded system is

$$F = \frac{(S_i/N_i)}{(S_o/N_o)} = F_1 + \frac{F_2 - 1}{G_1} \tag{1B.17}$$

In general, for an n-stage system we get

$$F = F_1 + \frac{F_2 - 1}{G_1} + \frac{F_3 - 1}{G_1 G_2} + \cdots + \frac{F_n - 1}{G_1 G_2 G_3 \cdots G_{n-1}} \tag{1B.18}$$

Also, the n-stage system effective temperatures can be computed as

$$T_e = T_{e1} + \frac{T_{e2}}{G_1} + \frac{T_{e3}}{G_1 G_2} + \cdots + \frac{T_{en}}{G_1 G_2 G_3 \cdots G_{n-1}} \tag{1B.19}$$

As suggested by Eq. (1B.18) and Eq. (1B.19), the overall noise figure is mainly dominated by the first stage. Thus, radar receivers employ low noise power amplifiers in the first stage in order to minimize the overall receiver noise figure. However, for radar systems that are built for low RCS operations every stage should be included in the analysis.

Example:

A radar receiver consists of an antenna with cable loss $L = 1dB = F_1$, an RF amplifier with $F_2 = 6dB$, and gain $G_2 = 20dB$, followed by a mixer whose noise figure is $F_3 = 10dB$ and conversion loss $L = 8dB$, and finally, an integrated circuit IF amplifier with $F_4 = 6dB$ and gain $G_4 = 60dB$. Find the overall noise figure.

Solution:

From Eq. (1B.18) we have

$$F = F_1 + \frac{F_2 - 1}{G_1} + \frac{F_3 - 1}{G_1 G_2} + \frac{F_4 - 1}{G_1 G_2 G_3}$$

G_1	G_2	G_3	G_4	F_1	F_2	F_3	F_4
$-1dB$	$20dB$	$-8dB$	$60dB$	$1dB$	$6dB$	$10dB$	$6dB$
0.7943	100	0.1585	10^6	1.2589	3.9811	10	3.9811

It follows that

$$F = 1.2589 + \frac{3.9811 - 1}{0.7943} + \frac{10 - 1}{100 \times 0.7943} + \frac{3.9811 - 1}{0.158 \times 100 \times 0.7943} = 5.3629$$

$$F = 10\log(5.3628) = 7.294 dB$$

Chapter 2 Radar Detection

2.1. Detection in the Presence of Noise

A simplified block diagram of a radar receiver that employs an envelope detector followed by a threshold decision is shown in Fig. 2.1. The input signal to the receiver is composed of the radar echo signal $s(t)$ and additive zero mean white Gaussian noise $n(t)$, with variance ψ^2. The input noise is assumed to be spatially incoherent and uncorrelated with the signal.

The output of the bandpass IF filter is the signal $v(t)$, which can be written as

$$v(t) = v_I(t)\cos\omega_0 t + v_Q(t)\sin\omega_0 t = r(t)\cos(\omega_0 t - \varphi(t))$$
$$v_I(t) = r(t)\cos\varphi(t) \qquad (2.1)$$
$$v_Q(t) = r(t)\sin\varphi(t)$$

where $\omega_0 = 2\pi f_0$ is the radar operating frequency, $r(t)$ is the envelope of $v(t)$, the phase is $\varphi(t) = \operatorname{atan}(v_Q/v_I)$, and the subscripts I, Q, respectively, refer to the in-phase and quadrature components.

A target is detected when $r(t)$ exceeds the threshold value V_T, where the decision hypotheses are

$$s(t) + n(t) > V_T \qquad Detection$$
$$n(t) > V_T \qquad False\ alarm$$

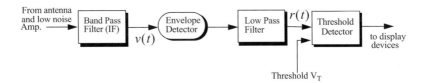

Figure 2.1. Simplified block diagram of an envelope detector and threshold receiver.

The case when the noise subtracts from the signal (while a target is present) to make $r(t)$ smaller than the threshold is called a miss. Radar designers seek to maximize the probability of detection for a given probability of false alarm.

The IF filter output is a complex random variable that is composed of either noise alone or noise plus target return signal (sine wave of amplitude A). The quadrature components corresponding to the first case are

$$v_I(t) = n_I(t) \qquad (2.2)$$
$$v_Q(t) = n_Q(t)$$

and for the second case,

$$v_I(t) = A + n_I(t) = r(t)\cos\varphi(t) \Rightarrow n_I(t) = r(t)\cos\varphi(t) - A$$
$$v_Q(t) = n_Q(t) = r(t)\sin\varphi(t) \qquad (2.3)$$

where the noise quadrature components $n_I(t)$ and $n_Q(t)$ are uncorrelated zero mean low pass Gaussian noise with equal variances, ψ^2. The joint Probability Density Function (*pdf*) of the two random variables $n_I; n_Q$ is

$$f(n_I, n_Q) = \frac{1}{2\pi\psi^2}\exp\left(-\frac{n_I^2 + n_Q^2}{2\psi^2}\right) \qquad (2.4)$$

$$= \frac{1}{2\pi\psi^2}\exp\left(-\frac{(r\cos\varphi - A)^2 + (r\sin\varphi)^2}{2\psi^2}\right)$$

The *pdfs* of the random variables $r(t)$ and $\varphi(t)$, respectively, represent the modulus and phase of $v(t)$. The joint *pdf* for the two random variables $r(t); \varphi(t)$ is given by

$$f(r, \varphi) = f(n_I, n_Q)|J| \qquad (2.5)$$

where $[J]$ is a matrix of derivatives defined by

$$[J] = \begin{bmatrix} \dfrac{\partial n_I}{\partial r} & \dfrac{\partial n_I}{\partial \varphi} \\ \dfrac{\partial n_Q}{\partial r} & \dfrac{\partial n_Q}{\partial \varphi} \end{bmatrix} = \begin{bmatrix} \cos\varphi & -r\sin\varphi \\ \sin\varphi & r\cos\varphi \end{bmatrix} \quad (2.6)$$

The determinant of the matrix of derivatives is called the Jacobian, and in this case it is equal to

$$|J| = r(t) \quad (2.7)$$

Substituting Eqs. (2.4) and (2.7) into Eq. (2.5) and collecting terms yield

$$f(r, \varphi) = \frac{r}{2\pi\psi^2} \exp\left(-\frac{r^2 + A^2}{2\psi^2}\right) \exp\left(\frac{rA\cos\varphi}{\psi^2}\right) \quad (2.8)$$

The *pdf* for r alone is obtained by integrating Eq. (2.8) over φ

$$f(r) = \int_0^{2\pi} f(r, \varphi) d\varphi = \frac{r}{\psi^2} \exp\left(-\frac{r^2 + A^2}{2\psi^2}\right) \frac{1}{2\pi} \int_0^{2\pi} \exp\left(\frac{rA\cos\varphi}{\psi^2}\right) d\varphi \quad (2.9)$$

where the integral inside Eq. (2.9) is known as the modified Bessel function of zero order,

$$I_0(\beta) = \frac{1}{2\pi} \int_0^{2\pi} e^{\beta\cos\theta} d\theta \quad (2.10)$$

Thus,

$$f(r) = \frac{r}{\psi^2} I_0\left(\frac{rA}{\psi^2}\right) \exp\left(-\frac{r^2 + A^2}{2\psi^2}\right) \quad (2.11)$$

which is the Rician probability density function. If $A/\psi^2 = 0$ (noise alone), then Eq. (2.11) becomes the Rayleigh probability density function

$$f(r) = \frac{r}{\psi^2} \exp\left(-\frac{r^2}{2\psi^2}\right) \quad (2.12)$$

Also, when (A/ψ^2) is very large, Eq. (2.11) becomes a Gaussian probability density function of mean A and variance ψ^2:

$$f(r) \approx \frac{1}{\sqrt{2\pi\psi^2}} \exp\left(-\frac{(r-A)^2}{2\psi^2}\right) \qquad (2.13)$$

Fig. 2.2 shows plots for the Rayleigh and Gaussian densities. For this purpose, use MATLAB program *"fig2_2.m"* given in Listing 2.1 in Section 2.11. This program uses MATLAB functions *"normpdf.m"* and *"raylpdf.m"*. Both functions are part of the MATLAB Statistics toolbox. Their associated syntax is as follows

normpdf(x,mu,sigma)

raylpdf(x,sigma)

"x" is the variable, *"mu"* is the mean, and *"sigma"* is the standard deviation.

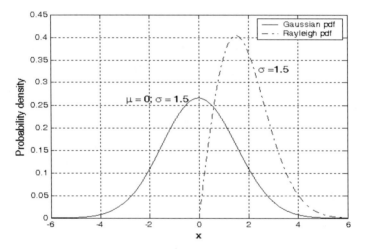

Figure 2.2. Gaussian and Rayleigh probability densities.

The density function for the random variable φ is obtained from

$$f(\varphi) = \int_0^r f(r, \varphi) \, dr \qquad (2.14)$$

While the detailed derivation is left as an exercise, the result of Eq. (2.14) is

$$f(\varphi) = \frac{1}{2\pi} \exp\left(\frac{-A^2}{2\psi^2}\right) + \frac{A\cos\varphi}{\sqrt{2\pi\psi^2}} \exp\left(\frac{-(A\sin\varphi)^2}{2\psi^2}\right) F\left(\frac{A\cos\varphi}{\psi}\right) \qquad (2.15)$$

where

$$F(x) = \int_{-\infty}^{x} \frac{1}{\sqrt{2\pi}} e^{-\zeta^2/2} d\zeta \qquad (2.16)$$

The function $F(x)$ can be found tabulated in most mathematical formula reference books. Note that for the case of noise alone ($A = 0$), Eq. (2.15) collapses to a uniform *pdf* over the interval $\{0, 2\pi\}$.

One excellent approximation for the function $F(x)$ is

$$F(x) = 1 - \left(\frac{1}{0.661x + 0.339\sqrt{x^2 + 5.51}}\right) \frac{1}{\sqrt{2\pi}} e^{-x^2/2} \qquad x \geq 0 \qquad (2.17)$$

and for negative values of x

$$F(-x) = 1 - F(x) \qquad (2.18)$$

MATLAB Function "que_func.m"

The function *"que_func.m"* computes $F(x)$ using Eqs. (2.17) and (2.18) and is given in Listing 2.2 in Section 2.11. The syntax is as follows:

$$fofx = que\_func\ (x)$$

2.2. Probability of False Alarm

The probability of false alarm P_{fa} is defined as the probability that a sample R of the signal $r(t)$ will exceed the threshold voltage V_T when noise alone is present in the radar,

$$P_{fa} = \int_{V_T}^{\infty} \frac{r}{\psi^2} \exp\left(-\frac{r^2}{2\psi^2}\right) dr = \exp\left(\frac{-V_T^2}{2\psi^2}\right) \qquad (2.19a)$$

$$V_T = \sqrt{2\psi^2 \ln\left(\frac{1}{P_{fa}}\right)} \qquad (2.19b)$$

Fig. 2.3 shows a plot of the normalized threshold versus the probability of false alarm. It is evident from this figure that P_{fa} is very sensitive to small changes in the threshold value. This figure can be reproduced using MATLAB program *"fig2_3.m"* given in Listing 2.3 in Section 2.11.

The false alarm time T_{fa} is related to the probability of false alarm by

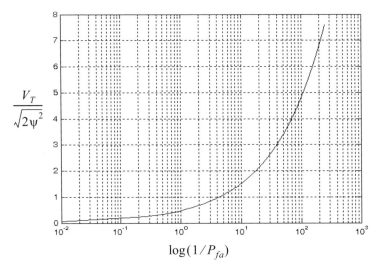

Figure 2.3. Normalized detection threshold versus probability of false alarm.

$$T_{fa} = t_{int}/P_{fa} \qquad (2.20)$$

where t_{int} represents the radar integration time, or the average time that the output of the envelope detector will pass the threshold voltage. Since the radar operating bandwidth B is the inverse of t_{int}, then by substituting Eq. (2.19) into Eq. (2.20) we can write T_{fa} as

$$T_{fa} = \frac{1}{B}\exp\left(\frac{V_T^2}{2\psi^2}\right) \qquad (2.21)$$

Minimizing T_{fa} means increasing the threshold value, and as a result the radar maximum detection range is decreased. Therefore, the choice of an acceptable value for T_{fa} becomes a compromise depending on the radar mode of operation.

Fehlner[1] defines the false alarm number as

$$n_{fa} = \frac{-\ln(2)}{\ln(1-P_{fa})} \approx \frac{\ln(2)}{P_{fa}} \qquad (2.22)$$

---

1. Fehlner, L. F., *Marcum's and Swerling's Data on Target Detection by a Pulsed Radar*, Johns Hopkins University, Applied Physics Lab. Rpt. # TG451, July 2, 1962, and Rpt. # TG451A, September 1964.

Probability of Detection

Other slightly different definitions for the false alarm number exist in the literature, causing a source of confusion for many non-expert readers. Other than the definition in Eq. (2.22), the most commonly used definition for the false alarm number is the one introduced by Marcum (1960). Marcum defines the false alarm number as the reciprocal of P_{fa}. In this text, the definition given in Eq. (2.22) is always assumed. Hence, a clear distinction is made between Marcum's definition of the false alarm number and the definition in Eq. (2.22).

2.3. Probability of Detection

The probability of detection P_D is the probability that a sample R of $r(t)$ will exceed the threshold voltage in the case of noise plus signal,

$$P_D = \int_{V_T}^{\infty} \frac{r}{\psi^2} I_0\left(\frac{rA}{\psi^2}\right) \exp\left(-\frac{r^2 + A^2}{2\psi^2}\right) dr \qquad (2.23)$$

If we assume that the radar signal is a sine waveform with amplitude A, then its power is $A^2/2$. Now, by using $SNR = A^2/2\psi^2$ (single-pulse SNR) and $(V_T^2/2\psi^2) = \ln(1/P_{fa})$, then Eq. (2.23) can be rewritten as

$$P_D = \int_{\sqrt{2\psi^2 \ln(1/P_{fa})}}^{\infty} \frac{r}{\psi^2} I_0\left(\frac{rA}{\psi^2}\right) \exp\left(-\frac{r^2 + A^2}{2\psi^2}\right) dr = \qquad (2.24)$$

$$Q\left[\sqrt{\frac{A^2}{\psi^2}}, \sqrt{2\ln\left(\frac{1}{P_{fa}}\right)}\right]$$

$$Q[\alpha, \beta] = \int_{\beta}^{\infty} \zeta I_0(\alpha\zeta) e^{-(\zeta^2 + \alpha^2)/2} \, d\zeta \qquad (2.25)$$

Q is called Marcum's Q-function. When P_{fa} is small and P_D is relatively large so that the threshold is also large, Eq. (2.24) can be approximated by

$$P_D \approx F\left(\frac{A}{\psi} - \sqrt{2\ln\left(\frac{1}{P_{fa}}\right)}\right) \qquad (2.26)$$

where $F(x)$ is given by Eq. (2.16). Many approximations for computing Eq. (2.24) can be found throughout the literature. One very accurate approximation presented by North (see bibliography) is given by

$$P_D \approx 0.5 \times erfc(\sqrt{-\ln P_{fa}} - \sqrt{SNR + 0.5}) \qquad (2.27)$$

where the complementary error function is

$$erfc(z) = 1 - \frac{2}{\sqrt{\pi}} \int_0^z e^{-v^2} dv \qquad (2.28)$$

MATLAB Function "marcumsq.m"

The integral given in Eq. (2.24) is complicated and can be computed using numerical integration techniques. Parl[1] developed an excellent algorithm to numerically compute this integral. It is summarized as follows:

$$Q[a, b] = \begin{cases} \frac{\alpha_n}{2\beta_n} \exp\left(\frac{(a-b)^2}{2}\right) & a < b \\ 1 - \left(\frac{\alpha_n}{2\beta_n}\right) \exp\left(\frac{(a-b)^2}{2}\right) & a \geq b \end{cases} \qquad (2.29)$$

$$\alpha_n = d_n + \frac{2n}{ab}\alpha_{n-1} + \alpha_{n-2} \qquad (2.30)$$

$$\beta_n = 1 + \frac{2n}{ab}\beta_{n-1} + \beta_{n-2} \qquad (2.31)$$

$$d_{n+1} = d_n d_1 \qquad (2.32)$$

$$\alpha_0 = \begin{cases} 1 & a < b \\ 0 & a \geq b \end{cases} \qquad (2.33)$$

$$d_1 = \begin{cases} a/b & a < b \\ b/a & a \geq b \end{cases} \qquad (2.34)$$

$\alpha_{-1} = 0.0$, $\beta_0 = 0.5$, and $\beta_{-1} = 0$. The recursive Eqs. (2.30) through (2.32) are computed continuously until $\beta_n > 10^p$ for values of $p \geq 3$. The accuracy of the algorithm is enhanced as the value of p is increased. The MATLAB function *"marcumsq.m"* given in Listing 2.4 in Section 2.11 implements Parl's algorithm to calculate the probability of detection defined in Eq. (2.24). The syntax is as follows:

Pd = marcumsq (alpha, beta)

where *alpha* and *beta* are from Eq. (2.25). Fig. 2.4 shows plots of the probability of detection, P_D, versus the single pulse SNR, with the P_{fa} as a parameter. This figure can be reproduced using the MATLAB program *"prob_snr1.m"* given in Listing 2.5 in Section 2.11.

---

1. Parl, S., A New Method of Calculating the Generalized Q Function, *IEEE Trans. Information Theory*, Vol. IT-26, No. 1, January 1980, pp. 121-124.

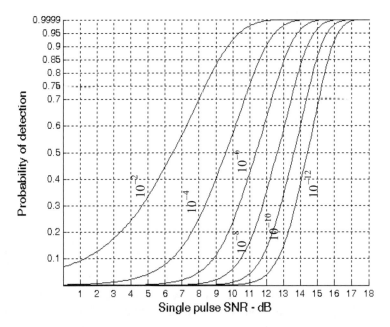

Figure 2.4. Probability of detection versus single pulse SNR, for several values of P_{fa}.

2.4. Pulse Integration

Pulse integration was discussed in Chapter 1 in the context of radar measurements. In this section a more comprehensive analysis of this topic is introduced in the context of radar detection. The overall principles and conclusions presented earlier will not change; however, the mathematical formulation and specific numerical values will change. Coherent integration preserves the phase relationship between the received pulses, thus achieving a build up in the signal amplitude. Alternatively, pulse integration performed after the envelope detector (where the phase relation is destroyed) is called non-coherent or post-detection integration.

2.4.1. Coherent Integration

In coherent integration, if a perfect integrator is used (100% efficiency), then integrating n_P pulses would improve the SNR by the same factor. Otherwise, integration loss occurs which is always the case for non-coherent integration. In order to demonstrate this signal buildup, consider the case where the radar return signal contains both signal plus additive noise. The mth pulse is

$$y_m(t) = s(t) + n_m(t) \tag{2.35}$$

where $s(t)$ is the radar return of interest and $n_m(t)$ is white uncorrelated additive noise signal. Coherent integration of n_P pulses yields

$$z(t) = \frac{1}{n_P}\sum_{m=1}^{n_P} y_m(t) = \sum_{m=1}^{n_P}\frac{1}{n_P}[s(t)+n_m(t)] = s(t) + \sum_{m=1}^{n_P}\frac{1}{n_P}n_m(t) \tag{2.36}$$

The total noise power in $z(t)$ is equal to the variance. More precisely,

$$\psi_{nz}^2 = E\left[\left(\sum_{m=1}^{n_P}\frac{1}{n_P}n_m(t)\right)\left(\sum_{l=1}^{n_P}\frac{1}{n_P}n_l(t)\right)^*\right] \tag{2.37}$$

where $E[\ ]$ is the expected value operator. It follows that

$$\psi_{nz}^2 = \frac{1}{n_P^2}\sum_{m,l=1}^{n_P} E[n_m(t)n_l^*(t)] = \frac{1}{n_P^2}\sum_{m,l=1}^{n_P}\psi_{ny}^2\delta_{ml} = \frac{1}{n_P}\psi_{ny}^2 \tag{2.38}$$

where ψ_{ny}^2 is the single pulse noise power and δ_{ml} is equal to zero for $m \neq l$ and unity for $m = l$. Observation of Eqs. (2.36) and (2.38) shows that the desired signal power after coherent integration is unchanged, while the noise power is reduced by the factor $1/n_P$. Thus, the SNR after coherent integration is improved by n_P.

Denote the single pulse SNR required to produce a given probability of detection as $(SNR)_1$. Also, denote $(SNR)_{n_P}$ as the SNR required to produce the same probability of detection when n_P pulses are integrated. It follows that

$$(SNR)_{n_P} = \frac{1}{n_P}(SNR)_1 \tag{2.39}$$

The requirements of knowing the exact phase of each transmitted pulse as well as maintaining coherency during propagation is very costly and challenging to achieve. Thus, radar systems would not utilize coherent integration during search mode, since target micro-dynamics may not be available.

2.4.2. Non-Coherent Integration

Non-coherent integration is often implemented after the envelope detector, also known as the quadratic detector. A block diagram of radar receiver utilizing a square law detector and non-coherent integration is illustrated in Fig. 2.5. In practice, the square law detector is normally used as an approximation to the optimum receiver.

Pulse Integration

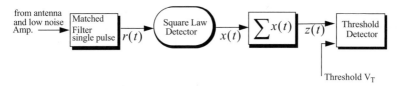

Figure 2.5. Simplified block diagram of a square law detector and non-coherent integration.

The *pdf* for the signal $r(t)$ was derived earlier and it is given in Eq. (2.11). Define a new dimensionless variable y as

$$y_n = \frac{r_n}{\psi} \tag{2.40}$$

and also define

$$\mathcal{R}_p = \frac{A^2}{\psi^2} = 2SNR \tag{2.41}$$

It follows that the *pdf* for the new variable is then given by

$$f(y_n) = f(r_n)\left|\frac{dr_n}{dy_n}\right| = y_n\, I_0(y_n\sqrt{\mathcal{R}_p})\, \exp\left(\frac{-(y_n^2 + \mathcal{R}_p)}{2}\right) \tag{2.42}$$

The output of a square law detector for the nth pulse is proportional to the square of its input, which, after the change of variable in Eq. (2.40), is proportional to y_n. Thus, it is convenient to define a new change variable,

$$x_n = \frac{1}{2}y_n^2 \tag{2.43}$$

The *pdf* for the variable at the output of the square law detector is given by

$$f(x_n) = f(y_n)\left|\frac{dy_n}{dx_n}\right| = \exp\left(-\left(x_n + \frac{\mathcal{R}_p}{2}\right)\right) I_0(\sqrt{2x_n\mathcal{R}_p}) \tag{2.44}$$

Non-coherent integration of n_p pulses is implemented as

$$z = \sum_{n=1}^{n_p} x_n \tag{2.45}$$

Since the random variables x_n are independent, the *pdf* for the variable z is

$$f(z) = f(x_1) \bullet f(x_2) \bullet \ldots \bullet f(x_{n_p}) \qquad (2.46)$$

The operator $\bullet$ symbolically indicates convolution. The characteristic functions for the individual *pdf*s can then be used to compute the joint *pdf* in Eq. (2.46). The details of this development are left as an exercise. The result is

$$f(z) = \left(\frac{2z}{n_p \Re_p}\right)^{(n_p-1)/2} \exp\left(-z - \frac{1}{2}n_p \Re_p\right) I_{n_p-1}(\sqrt{2 n_p z \Re_p}) \qquad (2.47)$$

I_{n_p-1} is the modified Bessel function of order $n_p - 1$. Therefore, the probability of detection is obtained by integrating $f(z)$ from the threshold value to infinity. Alternatively, the probability of false alarm is obtained by letting $\Re_p$ be zero and integrating the *pdf* from the threshold value to infinity. Closed form solutions to these integrals are not easily available. Therefore, numerical techniques are often utilized to generate tables for the probability of detection.

Improvement Factor and Integration Loss

Denote the SNR that is required to achieve a specific P_D given a particular P_{fa} when n_P pulses are integrated non-coherently by $(SNR)_{NCI}$. And thus, the single pulse SNR, $(SNR)_1$, is less than $(SNR)_{NCI}$. More precisely,

$$(SNR)_{NCI} = (SNR)_1 \times I(n_P) \qquad (2.48)$$

where $I(n_P)$ is called the integration improvement factor. An empirically derived expression for the improvement factor that is accurate within $0.8 dB$ is reported in Peebles[1] as

$$[I(n_P)]_{dB} = 6.79(1 + 0.235 P_D)\left(1 + \frac{\log(1/P_{fa})}{46.6}\right)\log(n_P) \qquad (2.49)$$

$$(1 - 0.140\log(n_P) + 0.018310(\log n_P)^2)$$

Fig. 2.6a shows plots of the integration improvement factor as a function of the number of integrated pulses with P_D and P_{fa} as parameters, using Eq. (2.49). This plot can be reproduced using the MATLAB program *"fig2_6a.m"* given in Listing 2.6 in Section 2.11. Note this program uses the MATLAB function *"improv_fac.m"*, which is given in Listing 2.7 in Section 2.11.

MATLAB Function "improv_fac.m"

The function *"improv_fac.m"* calculates the improvement factor using Eq. (2.49). It is given in Listing 2.7 in Section 2.11. The syntax is as follows:

[impr_of_np] = improv_fac (np, pfa, pd)

---

1. Peebles Jr., P. Z., *Radar Principles*, John Wiley & Sons, Inc., 1998.

Pulse Integration

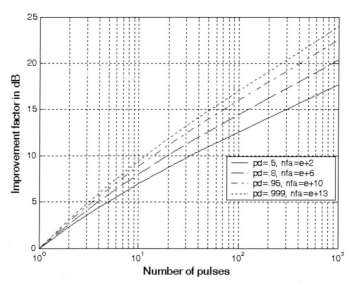

Figure 2.6a. Improvement factor versus number of non-coherently integrated pulses.

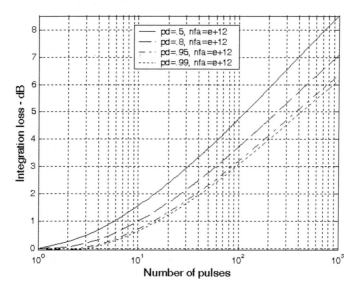

Figure 2.6b. Integration loss versus number of non-coherently integrated pulses.

where

Symbol	Description	Units	Status
np	number of integrated pulses	none	input
pfa	probability of false alarm	none	input
pd	probability of detection	none	input
impr_of_np	improvement factor	output	dB

The integration loss is defined as

$$L_{NCI} = n_P/I(n_P) \qquad (2.50)$$

Figure 2.6b shows a plot of the integration loss versus n_P. This figure can be reproduced using MATLAB program *"fig2_6b.m"* given in Listing 2.8 in Section 2.11. It follows that, when non-coherent integration is utilized, the corresponding SNR required to achieve a certain P_D given a specific P_{fa} is now given by

$$(SNR)_{NCI} = (n_P \times (SNR)_1)/L_{NCI} \qquad (2.51)$$

which is very similar to Eq. (1.86) derived in Chapter 1.

2.4.3. Mini Design Case Study 2.1

An L-band radar has the following specifications: operating frequency $f_0 = 1.5 GHz$, operating bandwidth $B = 2 MHz$, noise figure $F = 8 dB$, system losses $L = 4 dB$, time of false alarm $T_{fa} = 12$ minutes, detection range $R = 12 Km$, the minimum required SNR is $SNR = 13.85 dB$, antenna gain $G = 5000$, and target RCS $\sigma = 1 m^2$. (a) Determine the PRF f_r, the pulsewidth τ, the peak power P_t, the probability of false alarm P_{fa}, the corresponding P_D, and the minimum detectable signal level S_{min}. (b) How can you reduce the transmitter power to achieve the same performance when 10 pulses are integrated non-coherently? (c) If the radar operates at a shorter range in the single pulse mode, find the new probability of detection when the range decreases to $9 Km$.

A Solution

Assume that the maximum detection corresponds to the unambiguous range. From that the PRF is computed as

$$f_r = \frac{c}{2R_u} = \frac{3 \times 10^8}{2 \times 12000} = 12.5 KHz$$

The pulsewidth is proportional to the inverse of the bandwidth,

$$\tau = \frac{1}{B} = \frac{1}{2 \times 10^6} = 0.5\mu s$$

The probability of false alarm is

$$P_{fa} = \frac{1}{BT_{fa}} = \frac{1}{2 \times 10^6 \times 12 \times 60} = 6.94 \times 10^{-10}$$

It follows that by using MATLAB function "marcumsq.m" the probability of detection is calculated from

$$Q\left[\sqrt{\frac{A^2}{\psi^2}}, \sqrt{2\ln\left(\frac{1}{P_{fa}}\right)}\right]$$

with the following syntax

$$marcumsq(alpha, beta)$$

where

$$alpha = \sqrt{2} \times \sqrt{10^{13.85/10}} = 6.9665$$

$$beta = \sqrt{2\ln\left(\frac{1}{6.94 \times 10^{-10}}\right)} = 6.494$$

Remember that $(A^2/\psi^2) = 2SNR$. Thus, the detection probability is

$$P_D = marcumsq(6.9665, 6.944) = 0.508$$

Using the radar equation one can calculate the radar peak power. More precisely,

$$P_t = SNR \frac{(4\pi)^3 R^4 kT_0 BFL}{G^2 \lambda^2 \sigma} \Rightarrow$$

$$P_t = 10^{1.385} \frac{(4\pi)^3 \times 12000^4 \times 1.38 \times 10^{-23} \times 290 \times 2 \times 10^6 \times 6.309 \times 2.511}{5000^2 \times 0.2^2 \times 1}$$

$$= 126.61 \, Watts$$

And the minimum detectable signal is

$$S_{min} = \frac{P_t G^2 \lambda^2 \sigma}{(4\pi)^3 R^4 L} = \frac{126.61 \times 5000^2 \times 0.2^2 \times 1}{(4\pi)^3 \times 12000^4 \times 2.511} = 1.2254 \times 10^{-12} Volts$$

When 10 pulses are integrated non-coherently, the corresponding improvement factor is calculated from the MATLAB function "improv_fac.m" using the following syntax

improv_fac (10,1e-11,0.5)

which yields $I(10) = 6 \Rightarrow 7.78 dB$. Consequently, by keeping the probability of detection the same (with and without integration) the SNR can be reduced by a factor of almost 6 dB (13.85 - 7.78). The integration loss associated with a 10-pulse non-coherent integration is calculated from Eq. (2.50) as

$$L_{NCI} = \frac{n_P}{I(10)} = \frac{10}{6} = 1.67 \Rightarrow 2.2 dB$$

Thus the net single pulse SNR with 10-pulse non-coherent integration is

$$(SNR)_{NCI} = 13.85 - 7.78 + 2.2 = 8.27 dB.$$

Finally, the improvement in the SNR due to decreasing the detection range to 9 Km is

$$(SNR)_{9Km} = 10\log\left(\frac{12000}{9000}\right)^4 + 13.85 = 18.85 dB.$$

2.5. Detection of Fluctuating Targets

So far the probability of detection calculations assumed a constant target cross section (non-fluctuating target). This work was first analyzed by Marcum.[1] Swerling[2] extended Marcum's work to four distinct cases that account for variations in the target cross section. These cases have come to be known as Swerling models. They are: Swerling I, Swerling II, Swerling III, and Swerling IV. The constant RCS case analyzed by Marcum is widely known as Swerling 0 or equivalently Swerling V. Target fluctuation lowers the probability of detection, or equivalently reduces the SNR.

---

1. Marcum, J. I., *A Statistical Theory of Target Detection by Pulsed Radar*, IRE Transactions on Information Theory. Vol IT-6, pp 59-267. April 1960.
2. Swerling, P., *Probability of Detection for Fluctuating Targets*, IRE Transactions on Information Theory. Vol IT-6, pp 269-308. April 1960.

Detection of Fluctuating Targets

Swerling I targets have constant amplitude over one antenna scan; however, a Swerling I target amplitude varies independently from scan to scan according to a Chi-square probability density function with two degrees of freedom. The amplitude of Swerling II targets fluctuates independently from pulse to pulse according to a Chi-square probability density function with two degrees of freedom. Target fluctuation associated with a Swerling III model is similar to Swerling I, except in this case the target power fluctuates independently from pulse to pulse according to a Chi-square probability density function with four degrees of freedom. Finally, the fluctuation of Swerling IV targets is from pulse to pulse according to a Chi-square probability density function with four degrees of freedom. Swerling showed that the statistics associated with Swerling I and II models apply to targets consisting of many small scatterers of comparable RCS values, while the statistics associated with Swerling III and IV models apply to targets consisting of one large RCS scatterer and many small equal RCS scatterers. Non-coherent integration can be applied to all four Swerling models; however, coherent integration cannot be used when the target fluctuation is either Swerling II or Swerling IV. This is because the target amplitude decorrelates from pulse to pulse (fast fluctuation) for Swerling II and IV models, and thus phase coherency cannot be maintained.

The Chi-square *pdf* with $2N$ degrees of freedom can be written as

$$f(\sigma) = \frac{N}{(N-1)!\,\bar{\sigma}} \left(\frac{N\sigma}{\bar{\sigma}}\right)^{N-1} \exp\left(-\frac{N\sigma}{\bar{\sigma}}\right) \tag{2.52}$$

where $\bar{\sigma}$ is the average RCS value. Using this equation, the *pdf* associated with Swerling I and II targets can be obtained by letting $N = 1$, which yields a Rayleigh *pdf*. More precisely,

$$f(\sigma) = \frac{1}{\bar{\sigma}} \exp\left(-\frac{\sigma}{\bar{\sigma}}\right) \qquad \sigma \geq 0 \tag{2.53}$$

Letting $N = 2$ yields the *pdf* for Swerling III and IV type targets,

$$f(\sigma) = \frac{4\sigma}{\bar{\sigma}^2} \exp\left(-\frac{2\sigma}{\bar{\sigma}}\right) \qquad \sigma \geq 0 \tag{2.54}$$

The probability of detection for a fluctuating target is computed in a similar fashion to Eq. (2.23), except in this case $f(r)$ is replaced by the conditional *pdf* $f(r/\sigma)$. Performing the analysis for the general case (i.e., using Eq. (2.47)) yields

$$f(z/\sigma) = \left(\frac{2z}{n_p \sigma^2/\psi^2}\right)^{(n_p-1)/2} \exp\left(-z - \frac{1}{2}n_p\frac{\sigma^2}{\psi^2}\right) I_{n_p-1}\left(\sqrt{2n_p z \frac{\sigma^2}{\psi^2}}\right) \tag{2.55}$$

To obtain $f(z)$ use the relations

$$f(z, \sigma) = f(z/\sigma)f(\sigma) \tag{2.56}$$

$$f(z) = \int f(z, \sigma) d\sigma \tag{2.57}$$

Finally, using Eq. (2.56) in Eq. (2.57) produces

$$f(z) = \int f(z/\sigma)f(\sigma) d\sigma \tag{2.58}$$

where $f(z/\sigma)$ is defined in Eq. (2.55) and $f(\sigma)$ is in either Eq. (2.53) or (2.54). The probability of detection is obtained by integrating the *pdf* derived from Eq. (2.58) from the threshold value to infinity. Performing the integration in Eq. (2.58) leads to the incomplete Gamma function.

2.5.1. Threshold Selection

When only a single pulse is used, the detection threshold V_T is related to the probability of false alarm P_{fa} as defined in Eq. (2.19). DiFranco and Rubin[1] derived a general form relating the threshold and P_{fa} for any number of pulses when non-coherent integration is used. It is

$$P_{fa} = 1 - \Gamma_I\left(\frac{V_T}{\sqrt{n_P}}, n_P - 1\right) \tag{2.59}$$

where Γ_I is used to denote the incomplete Gamma function. It is given by

$$\Gamma_I\left(\frac{V_T}{\sqrt{n_P}}, n_P - 1\right) = \int_0^{V_T/\sqrt{n_P}} \frac{e^{-\gamma} \gamma^{(n_P-1)-1}}{((n_P-1)-1)!} d\gamma \tag{2.60}$$

Note that the limiting values for the incomplete Gamma function are

$$\Gamma_I(0, N) = 0 \qquad \Gamma_I(\infty, N) = 1 \tag{2.61}$$

For our purposes, the incomplete Gamma function can be approximated by

$$\Gamma_I\left(\frac{V_T}{\sqrt{n_P}}, n_P - 1\right) = 1 - \frac{V_T^{n_P-1} e^{-V_T}}{(n_P-1)!}\left[1 + \frac{n_P-1}{V_T} + \frac{(n_P-1)(n_P-2)}{V_T^2} + \ldots + \frac{(n_P-1)!}{V_T^{n_P-1}}\right] \tag{2.62}$$

---

1. DiFranco, J. V. and Rubin, W. L., *Radar Detection*, Artech House, 1980.

Detection of Fluctuating Targets

The threshold value V_T can then be approximated by the recursive formula used in the Newton-Raphson method. More precisely,

$$V_{T,m} = V_{T,m-1} - \frac{G(V_{T,m-1})}{G'(V_{T,m-1})} \quad ; \; m = 1, 2, 3, \ldots \quad (2.63)$$

The iteration is terminated when $|V_{T,m} - V_{T,m-1}| < V_{T,m-1}/10000.0$. The functions G and G' are

$$G(V_{T,m}) = (0.5)^{n_P/n_{fa}} - \Gamma_I(V_T, n_P) \quad (2.64)$$

$$G'(V_{T,m}) = -\frac{e^{-V_T} V_T^{n_P - 1}}{(n_P - 1)!} \quad (2.65)$$

The initial value for the recursion is

$$V_{T,0} = n_P - \sqrt{n_P} + 2.3 \sqrt{-\log P_{fa}} \left(\sqrt{-\log P_{fa}} + \sqrt{n_P} - 1 \right) \quad (2.66)$$

MATLAB Function "incomplete_gamma.m"

In general, the incomplete Gamma function for some integer N is

$$\Gamma_I(x, N) = \int_0^x \frac{e^{-v} v^{N-1}}{(N-1)!} dv \quad (2.67)$$

The function *"incomplete_gamma.m"* implements Eq. (2.67). It is given in Listing 2.9 in Section 2.11. Note that this function uses the MATLAB function *"factor.m"* which is given in Listing 2.10. The function *"factor.m"* calculates the factorial of an integer. Fig. 2.7 shows the incomplete Gamma function for $N = 1, 3, 6, 10$. This figure can be reproduced using the MATLAB program *"fig2_7.m"* given in Listing 2.11. The syntax for this function is as follows:

[value] = incomplete_gamma (x, N)

where

Symbol	Description	Units	Status
x	variable input to $\Gamma_I(x, N)$	units of x	input
N	variable input to $\Gamma_I(x, N)$	none / integer	input
value	$\Gamma_I(x, N)$	none	output

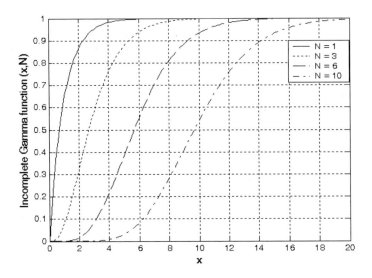

Figure 2.7. The incomplete Gamma function for four values of N.

MATLAB Function "threshold.m"

The function *"threshold.m"* calculates the threshold using the recursive formula used in the Newton-Raphson method. It is given in Listing 2.12 in Section 2.11. The syntax is as follows:

$$[pfa, vt] = threshold\ (nfa, np)$$

where

Symbol	Description	Units	Status
nfa	Marcum's false alarm number	none	input
np	number of integrated pulses	none	input
pfa	probability of false alarm	none	output
vt	threshold value	none	output

Fig. 2.8 shows plots of the threshold value versus the number of integrated pulses for several values of n_{fa}; remember that $P_{fa} \approx \ln(2)/n_{fa}$. This figure can be reproduced using MATLAB program *"fig2_8.m"* given in Listing 2.13. This program uses both *"threshold.m"* and *"incomplete_gamma"*.

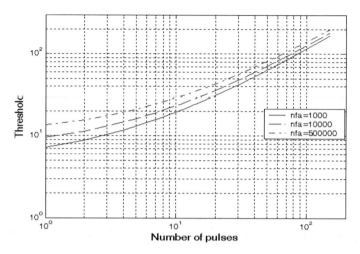

Figure 2.8. Threshold V_T versus n_p for several values of n_{fa}.

2.6. Probability of Detection Calculation

Marcum defined the probability of false alarm for the case when $n_p > 1$ as

$$P_{fa} \approx \ln(2)(n_p/n_{fa}) \tag{2.68}$$

The single pulse probability of detection for non-fluctuating targets is given in Eq. (2.24). When $n_p > 1$, the probability of detection is computed using the Gram-Charlier series. In this case, the probability of detection is

$$P_D \cong \frac{erfc(V/\sqrt{2})}{2} - \frac{e^{-V^2/2}}{\sqrt{2\pi}}[C_3(V^2-1) + C_4 V(3-V^2) \tag{2.69}$$

$$- C_6 V(V^4 - 10V^2 + 15)]$$

where the constants C_3, C_4, and C_6 are the Gram-Charlier series coefficients, and the variable V is

$$V = \frac{V_T - n_P(1+SNR)}{\varpi} \tag{2.70}$$

In general, values for C_3, C_4, C_6, and ϖ vary depending on the target fluctuation type.

2.6.1. Detection of Swerling V Targets

For Swerling V (Swerling 0) target fluctuations, the probability of detection is calculated using Eq. (2.69). In this case, the Gram-Charlier series coefficients are

$$C_3 = -\frac{SNR + 1/3}{\sqrt{n_p}(2SNR + 1)^{1.5}} \tag{2.71}$$

$$C_4 = \frac{SNR + 1/4}{n_p(2SNR + 1)^2} \tag{2.72}$$

$$C_6 = C_3^2/2 \tag{2.73}$$

$$\varpi = \sqrt{n_p(2SNR + 1)} \tag{2.74}$$

MATLAB Function "pd_swerling5.m"

The function *"pd_swerling5.m"* calculates the probability of detection for Swerling V targets. It is given in Listing 2.14. The syntax is as follows:

[pd] = pd_swerling5 (input1, indicator, np, snr)

where

Symbol	Description	Units	Status
input1	P_{fa} or n_{fa}	none	input
indicator	1 when input1 = P_{fa} 2 when input1 = n_{fa}	none	input
np	number of integrated pulses	none	input
snr	SNR	dB	input
pd	probability of detection	none	output

Fig. 2.9 shows a plot for the probability of detection versus SNR for cases $n_p = 1, 10$. This figure can be reproduced using the MATLAB program *"fig2_9.m"*. It is given in Listing 2.15 in Section 2.11.

Note that it requires less SNR, with ten pulses integrated non-coherently, to achieve the same probability of detection as in the case of a single pulse. Hence, for any given P_D the SNR improvement can be read from the plot. Equivalently, using the function *"improv_fac.m"* leads to about the same result. For example, when $P_D = 0.8$ the function *"improv_fac.m"* gives an SNR improvement factor of $I(10) \approx 8.55 dB$. Fig. 2.9 shows that the ten pulse SNR is about $6.03 dB$. Therefore, the single pulse SNR is about (from Eq. (2.49)) $14.5 dB$, which can be read from the figure.

Probability of Detection Calculation

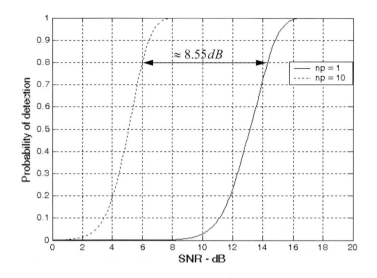

Figure 2.9. Probability of detection versus SNR, $P_{fa} = 10^{-9}$, and non-coherent integration.

2.6.2. Detection of Swerling I Targets

The exact formula for the probability of detection for Swerling I type targets was derived by Swerling. It is

$$P_D = e^{-V_T/(1+SNR)} \quad ; n_P = 1 \quad (2.75)$$

$$P_D = 1 - \Gamma_I(V_T, n_P - 1) + \left(1 + \frac{1}{n_P SNR}\right)^{n_p - 1} \Gamma_I\left(\frac{V_T}{1 + \frac{1}{n_P SNR}}, n_P - 1\right) \quad (2.76)$$

$$\times\, e^{-V_T/(1+n_P SNR)} \quad ; n_P > 1$$

MATLAB Function "pd_swerling1.m"

The function *"pd_swerling1.m"* calculates the probability of detection for Swerling I type targets. It is given in Listing 2.16 in Section 2.11. The syntax is as follows:

[pd] = pd_swerling1 (nfa, np, snr)

where

Symbol	Description	Units	Status
nfa	Marcum's false alarm number	none	input
np	number of integrated pulses	none	input
snr	SNR	dB	input
pd	probability of detection	none	output

Fig. 2.10 shows a plot of the probability of detection as a function of SNR for $n_p = 1$ and $P_{fa} = 10^{-9}$ for both Swerling I and V type fluctuations. Note that it requires more SNR, with fluctuation, to achieve the same P_D as in the case with no fluctuation. This figure can be reproduced using MATLAB program *"fig2_10.m"* given in Listing 2.17.

Fig. 2.11a shows a plot of the probability of detection versus SNR for $n_p = 1, 10, 50, 100$, where $P_{fa} = 10^{-8}$. Fig. 2.11b is similar to Fig. 2.11a; in this case $P_{fa} = 10^{-11}$. These figures can be reproduced using MATLAB program *"fig2_11ab.m"* given in Listing 2.18.

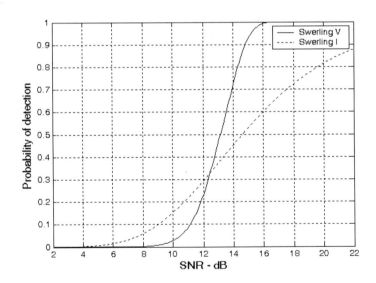

Figure 2.10. Probability of detection versus SNR, single pulse. $P_{fa} = 10^{-9}$.

Probability of Detection Calculation

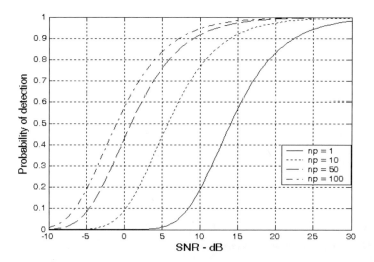

Figure 2.11a. Probability of detection versus SNR. Swerling I. $P_{fa} = 10^{-8}$.

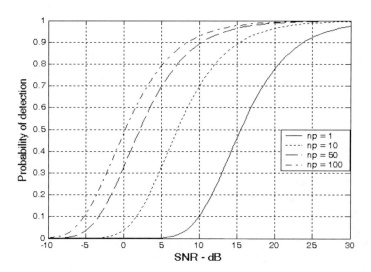

Figure 2.11b. Probability of detection versus SNR. Swerling I. $P_{fa} = 10^{-11}$.

2.6.3. Detection of Swerling II Targets

In the case of Swerling II targets, the probability of detection is given by

$$P_D = 1 - \Gamma_I\left(\frac{V_T}{(1+SNR)}, n_p\right) \quad ; \quad n_p \leq 50 \quad (2.77)$$

For the case when $n_p > 50$ Eq. (2.69) is used to compute the probability of detection. In this case,

$$C_3 = -\frac{1}{3\sqrt{n_p}} \quad , \quad C_6 = \frac{C_3^2}{2} \quad (2.78)$$

$$C_4 = \frac{1}{4n_p} \quad (2.79)$$

$$\varpi = \sqrt{n_p}\,(1+SNR) \quad (2.80)$$

MATLAB Function "pd_swerling2.m"

The function *"pd_swerling2.m"* calculates P_D for Swerling II type targets. It is given in Listing 2.19 in Section 2.11. The syntax is as follows:

[pd] = pd_swerling2 (nfa, np, snr)

where

Symbol	Description	Units	Status
nfa	Marcum's false alarm number	none	input
np	number of integrated pulses	none	input
snr	SNR	dB	input
pd	probability of detection	none	output

Fig. 2.12 shows a plot of the probability of detection as a function of SNR for $n_p = 1, 10, 50, 100$, where $P_{fa} = 10^{-10}$. This figure can be reproduced using MATLAB program *"fig2_12.m"* given in Listing 2.20.

2.6.4. Detection of Swerling III Targets

The exact formulas, developed by Marcum, for the probability of detection for Swerling III type targets when $n_p = 1, 2$ is

$$P_D = \exp\left(\frac{-V_T}{1+n_p SNR/2}\right)\left(1 + \frac{2}{n_p SNR}\right)^{n_p - 2} \times K_0 \quad (2.81)$$

$$K_0 = 1 + \frac{V_T}{1+n_p SNR/2} - \frac{2}{n_p SNR}(n_p - 2)$$

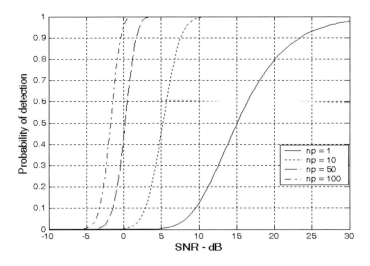

Figure 2.12. Probability of detection versus SNR. Swerling II. $P_{fa} = 10^{-10}$.

For $n_P > 2$ the expression is

$$P_D = \frac{V_T^{n_P-1} e^{-V_T}}{(1+n_P SNR/2)(n_P-2)!} + 1 - \Gamma_I(V_T, n_P-1) + K_0 \qquad (2.82)$$

$$\times \ \Gamma_I\left(\frac{V_T}{1+2/n_P SNR}, n_P-1\right)$$

MATLAB Function "pd_swerling3.m"

The function *"pd_swerling3.m"* calculates P_D for Swerling III type targets. It is given in Listing 2.21 in Section 2.11. The syntax is as follows:

$$[pd] = pd\_swerling3 \ (nfa, np, snr)$$

where

Symbol	Description	Units	Status
nfa	Marcum's false alarm number	none	input
np	number of integrated pulses	none	input
snr	SNR	dB	input
pd	probability of detection	none	output

Fig. 2.13 shows a plot of the probability of detection as a function of SNR for $n_p = 1, 10, 50, 100$, where $P_{fa} = 10^{-9}$. This figure can be reproduced using MATLAB program *"fig2_13.m"* given in Listing 2.22.

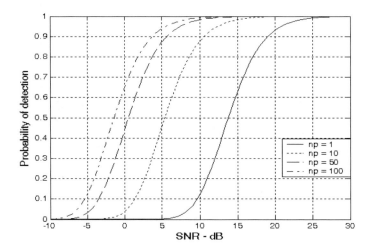

Figure 2.13. Probability of detection versus SNR. Swerling III. $P_{fa} = 10^{-9}$.

2.6.5. Detection of Swerling IV Targets

The expression for the probability of detection for Swerling IV targets for $n_p < 50$ is

$$P_D = 1 - \left[\gamma_0 + \left(\frac{SNR}{2}\right)n_p\gamma_1 + \left(\frac{SNR}{2}\right)^2 \frac{n_p(n_p-1)}{2!}\gamma_2 + \ldots + \left(\frac{SNR}{2}\right)^{n_p}\gamma_{n_p}\right]\left(1 + \frac{SNR}{2}\right)^{-n_p} \quad (2.83)$$

where

$$\gamma_i = \Gamma_I\left(\frac{V_T}{1+(SNR)/2}, n_p+i\right) \quad (2.84)$$

By using the recursive formula

$$\Gamma_I(x, i+1) = \Gamma_I(x, i) - \frac{x^i}{i! \exp(x)} \quad (2.85)$$

then only γ_0 needs to be calculated using Eq. (2.84) and the rest of γ_i are calculated from the following recursion:

Probability of Detection Calculation

$$\gamma_i = \gamma_{i-1} - A_i \quad ; \; i > 0 \tag{2.86}$$

$$A_i = \frac{V_T/(1+(SNR)/2)}{n_P + i - 1} A_{i-1} \quad ; \; i > 1 \tag{2.87}$$

$$A_1 = \frac{(V_T/(1+(SNR)/2))^{n_P}}{n_P! \exp(V_T/(1+(SNR)/2))} \tag{2.88}$$

$$\gamma_0 = \Gamma_I\left(\frac{V_T}{(1+(SNR)/2)}, n_P\right) \tag{2.89}$$

For the case when $n_P \geq 50$, the Gram-Charlier series and Eq. (2.69) can be used to calculate the probability of detection. In this case,

$$C_3 = \frac{1}{3\sqrt{n_P}} \frac{2\beta^3 - 1}{(2\beta^2 - 1)^{1.5}} \quad ; \; C_6 = \frac{C_3^2}{2} \tag{2.90}$$

$$C_4 = \frac{1}{4n_P} \frac{2\beta^4 - 1}{(2\beta^2 - 1)^2} \tag{2.91}$$

$$\varpi = \sqrt{n_P(2\beta^2 - 1)} \tag{2.92}$$

$$\beta = 1 + \frac{SNR}{2} \tag{2.93}$$

MATLAB Function "pd_swerling4.m"

The function *"pd_swerling4.m"* calculates P_D for Swerling IV type targets. It is given in Listing 2.23 in Section 2.11. The syntax is as follows:

[pd] = pd_swerling4 (nfa, np, snr)

where

Symbol	Description	Units	Status
nfa	Marcum's false alarm number	none	input
np	number of integrated pulses	none	input
snr	SNR	dB	input
pd	probability of detection	none	output

Figure 2.14 shows a plot of the probability of detection as a function of SNR for $n_P = 1, 10, 50, 100$, where $P_{fa} = 10^{-9}$. This figure can be reproduced using MATLAB program *"fig2_14.m"* given in Listing 2.24.

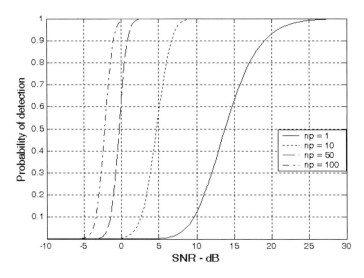

Figure 2.14. Probability of detection versus SNR. Swerling IV. $P_{fa} = 10^{-9}$.

2.7. The Radar Equation Revisited

The radar equation developed in Chapter 1 assumed a constant target RCS and did not account for integration loss. In this section, a more comprehensive form of the radar equation is introduced. In this case, the radar equation is given by

$$R^4 = \frac{P_{av} G_t G_r \lambda^2 \sigma I(n_p)}{(4\pi)^3 k T_e F \tau f_r L_t L_f (SNR)_1} \quad (2.94)$$

where $P_{av} = P_t \tau f_r$ is the average transmitted power, P_t is the peak transmitted power, τ is pulsewidth, f_r is PRF, G_t is transmitting antenna gain, G_r is receiving antenna gain, λ is wavelength, σ is target cross section, $I(n_p)$ is improvement factor, n_p is the number of integrated pulses, k is Boltzman's constant, T_e is effective noise temperature, F is the system noise figure, B is receiver bandwidth, L_t is total system losses including integration loss, L_f is loss due to target fluctuation, and $(SNR)_1$ is the minimum single pulse SNR required for detection.

The fluctuation loss, L_f, can be viewed as the amount of additional SNR required to compensate for the SNR loss due to target fluctuation, given a specific P_D value. This was demonstrated for a Swerling I fluctuation in Fig.

2.10. Kanter[1] developed an exact analysis for calculating the fluctuation loss. In this text the authors will take advantage of the computational power of MATLAB and the MATLAB functions developed for this text to numerically calculate the amount of fluctuation loss with an accuracy of $0.005 dB$ or better. For this purpose the MATLAB function *"fluct_loss.m"* was developed. It is given in Listing 2.25 in Section 2.11. Its syntax is as follows:

$$[Lf, Pd\_Sw5] = fluct\_loss(pd, pfa, np, sw\_case)$$

where

Symbol	Description	Units	Status
pd	desired probability of detection	none	input
pfa	probability of false alarm	none	input
np	number of pulses	none	input
sw_case	1, 2, 3, or 4 depending on the desired Swerling case	none	input
Lf	fluctuation loss	dB	output
Pd_Sw5	Probability of detection corresponding to a Swerling V case	none	output

For example, using the syntax

$$[Lf, Pd\_Sw5] = fluct\_loss(0.65, 1e-9, 10, 1)$$

will calculate the SNR corresponding to both Swerling V and Swerling I fluctuation when the desired probability of detection $P_D = 0.65$ and probability of false alarm $P_{fa} = 10^{-9}$ and 10 pulses of non-coherent integration. The following is a reprint of the output:

$$PD\_SW5 = 0.65096989459928$$
$$SNR\_SW5 = 5.52499999999990$$
$$PD\_SW1 = 0.65019653294095$$
$$SNR\_SW1 = 8.32999999999990$$
$$Lf = 2.80500000000000$$

Note that a negative value for L_f indicates a fluctuation SNR gain instead of loss. Finally, it must be noted that the function *"fluct_loss.m"* always assumes non-coherent integration. Fig. 2.15 shows a plot for the additional SNR (or fluctuation loss) required to achieve a certain probability of detection. This figure can be reproduced using MATLAB program *"fig2_16.m"* given in Listing 2.26 in Section 2.11.

---

1. Kanter, I., Exact Detection Probability for Partially Correlated Rayleigh Targets, IEEE Trans, AES-22, pp. 184-196, March 1986.

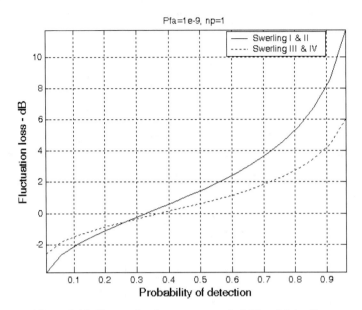

Figure 2.15. Fluctuation loss versus probability of detection.

2.8. Cumulative Probability of Detection

Denote the range at which the single pulse SNR is unity (0 dB) as R_0, and refer to it as the reference range. Then, for a specific radar, the single pulse SNR at R_0 is defined by the radar equation and is given by

$$(SNR)_{R_0} = \frac{P_t G^2 \lambda^2 \sigma}{(4\pi)^3 k T_0 BFLR_0^4} = 1 \tag{2.95}$$

The single pulse SNR at any range R is

$$SNR = \frac{P_t G^2 \lambda^2 \sigma}{(4\pi)^3 k T_0 BFLR^4} \tag{2.96}$$

Dividing Eq. (2.96) by Eq. (2.95) yields

$$\frac{SNR}{(SNR)_{R_0}} = \left(\frac{R_0}{R}\right)^4 \tag{2.97}$$

Therefore, if the range R_0 is known then the SNR at any other range R is

$$(SNR)_{dB} = 40\log\left(\frac{R_0}{R}\right) \tag{2.98}$$

Also, define the range R_{50} as the range at which $P_D = 0.5 = P_{50}$. Normally, the radar unambiguous range R_u is set equal to $2R_{50}$.

The cumulative probability of detection refers to detecting the target at least once by the time it is at range R. More precisely, consider a target closing on a scanning radar, where the target is illuminated only during a scan (frame). As the target gets closer to the radar, its probability of detection increases since the SNR is increased. Suppose that the probability of detection during the *nth* frame is P_{D_n}; then, the cumulative probability of detecting the target at least once during the *nth* frame (see Fig. 2.16) is given by

$$P_{C_n} = 1 - \prod_{i=1}^{n}(1 - P_{D_i}) \tag{2.99}$$

P_{D_1} is usually selected to be very small. Clearly, the probability of not detecting the target during the *nth* frame is $1 - P_{C_n}$. The probability of detection for the *ith* frame, P_{D_i}, is computed as discussed in the previous section.

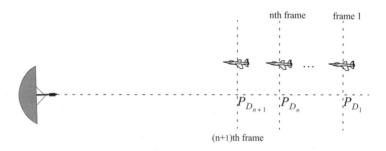

Figure 2.16. Detecting a target in many frames.

2.8.1. Mini Design Case Study 2.2

A radar detects a closing target at $R = 10Km$, with probability of detection P_D equal to 0.5. Assume $P_{fa} = 10^{-7}$. Compute and sketch the single look probability of detection as a function of normalized range (with respect to $R = 10Km$), over the interval $(2 - 20)Km$. If the range between two successive frames is $1Km$, what is the cumulative probability of detection at $R = 8Km$?

A Solution:

From the function "marcumsq.m" the SNR corresponding to $P_D = 0.5$ and $P_{fa} = 10^{-7}$ is approximately 12dB. By using a similar analysis to that which led to Eq. (2.98), we can express the SNR at any range R as

$$(SNR)_R = (SNR)_{10} + 40 \ \log\frac{10}{R} = 52 - 40 \ \log R$$

By using the function "marcumsq.m" we can construct the following table:

R Km	(SNR) dB	P_D
2	39.09	0.999
4	27.9	0.999
6	20.9	0.999
8	15.9	0.999
9	13.8	0.9
10	12.0	0.5
11	10.3	0.25
12	8.8	0.07
14	6.1	0.01
16	3.8	ε
20	0.01	ε

where ε is very small. A sketch of P_D versus normalized range is shown in Fig. 2.17.

The cumulative probability of detection is given in Eq. (2.95), where the probability of detection of the first frame is selected to be very small. Thus, we can arbitrarily choose frame 1 to be at $R = 16Km$. Note that selecting a different starting point for frame 1 would have a negligible effect on the cumulative probability (we only need P_{D_1} to be very small). Below is a range listing for frames 1 through 9, where frame 9 corresponds to $R = 8Km$. The cumulative

frame	1	2	3	4	5	6	7	8	9
range in Km	16	15	14	13	12	11	10	9	8

probability of detection at 8 Km is then

$$P_{C_9} = 1 - (1 - 0.999)(1 - 0.9)(1 - 0.5)(1 - 0.25)(1 - 0.07)$$
$$(1 - 0.01)(1 - \varepsilon)^2 \approx 0.9998$$

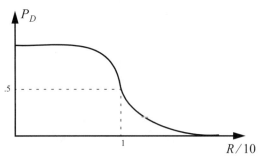

Figure 2.17. Cumulative probability of detection versus normalized range.

2.9. Constant False Alarm Rate (CFAR)

The detection threshold is computed so that the radar receiver maintains a constant pre-determined probability of false alarm. Eq. (2.19b) gives the relationship between the threshold value V_T and the probability of false alarm P_{fa}, and for convenience is repeated here as Eq. (2.100):

$$V_T = \sqrt{2\psi^2 \ln\left(\frac{1}{P_{fa}}\right)} \quad (2.100)$$

If the noise power ψ^2 is assumed to be constant, then a fixed threshold can satisfy Eq. (2.100). However, due to many reasons this condition is rarely true. Thus, in order to maintain a constant probability of false alarm the threshold value must be continuously updated based on the estimates of the noise variance. The process of continuously changing the threshold value to maintain a constant probability of false alarm is known as Constant False Alarm Rate (CFAR).

Three different types of CFAR processors are primarily used. They are adaptive threshold CFAR, nonparametric CFAR, and nonlinear receiver techniques. Adaptive CFAR assumes that the interference distribution is known and approximates the unknown parameters associated with these distributions. Nonparametric CFAR processors tend to accommodate unknown interference distributions. Nonlinear receiver techniques attempt to normalize the root mean square amplitude of the interference. In this book only analog Cell-Averaging CFAR (CA-CFAR) technique is examined. The analysis presented in this section closely follows Urkowitz[1].

---

1. Urkowitz, H., *Decision and Detection Theory*, unpublished lecture notes. Lockheed Martin Co., Moorestown, NJ.

2.9.1. Cell-Averaging CFAR (Single Pulse)

The CA-CFAR processor is shown in Fig. 2.18. Cell averaging is performed on a series of range and/or Doppler bins (cells). The echo return for each pulse is detected by a square law detector. In analog implementation these cells are obtained from a tapped delay line. The Cell Under Test (CUT) is the central cell. The immediate neighbors of the CUT are excluded from the averaging process due to a possible spillover from the CUT. The output of M reference cells ($M/2$ on each side of the CUT) is averaged. The threshold value is obtained by multiplying the averaged estimate from all reference cells by a constant K_0 (used for scaling). A detection is declared in the CUT if

$$Y_1 \geq K_0 Z \tag{2.101}$$

Cell-averaging CFAR assumes that the target of interest is in the CUT and all reference cells contain zero mean independent Gaussian noise of variance ψ^2. Therefore, the output of the reference cells, Z, represents a random variable with gamma probability density function (special case of the Chi-square) with $2M$ degrees of freedom. In this case, the gamma *pdf* is

$$f(z) = \frac{z^{(M/2)-1} e^{(-z/2\psi^2)}}{2^{M/2} \psi^M \Gamma(M/2)} \quad ; z > 0 \tag{2.102}$$

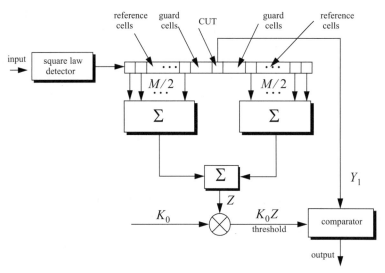

Figure 2.18. Conventional CA-CFAR.

The probability of false alarm corresponding to a fixed threshold was derived earlier. When CA-CFAR is implemented, then the probability of false alarm can be derived from the conditional false alarm probability, which is averaged over all possible values of the threshold in order to achieve an unconditional false alarm probability. The conditional probability of false alarm when $y = V_T$ can be written as

$$P_{fa}(V_T = y) = e^{-y/2\psi^2} \tag{2.103}$$

It follows that the unconditional probability of false alarm is

$$P_{fa} = \int_0^\infty P_{fa}(V_T = y) f(y) dy \tag{2.104}$$

where $f(y)$ is the *pdf* of the threshold, which except for the constant K_0 is the same as that defined in Eq. (2.102). Therefore,

$$f(y) = \frac{y^{M-1} e^{(-y/2K_0\psi^2)}}{(2K_0\psi^2)^M \Gamma(M)} \quad ; y \geq 0 \tag{2.105}$$

Performing the integration in Eq. (2.104) yields

$$P_{fa} = \frac{1}{(1+K_0)^M} \tag{2.106}$$

Observation of Eq. (2.106) shows that the probability of false alarm is now independent of the noise power, which is the objective of CFAR processing.

2.9.2. Cell-Averaging CFAR with Non-Coherent Integration

In practice, CFAR averaging is often implemented after non-coherent integration, as illustrated in Fig. 2.19. Now, the output of each reference cell is the sum of n_p squared envelopes. It follows that the total number of summed reference samples is Mn_p. The output Y_1 is also the sum of n_p squared envelopes. When noise alone is present in the CUT, Y_1 is a random variable whose *pdf* is a gamma distribution with $2n_p$ degrees of freedom. Additionally, the summed output of the reference cells is the sum of Mn_p squared envelopes. Thus, Z is also a random variable which has a gamma *pdf* with $2Mn_p$ degrees of freedom.

The probability of false alarm is then equal to the probability that the ratio Y_1/Z exceeds the threshold. More precisely,

$$P_{fa} = Prob\{Y_1/Z > K_1\} \tag{2.107}$$

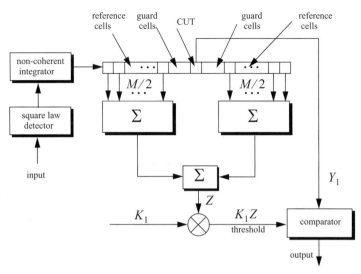

Figure 2.19. Conventional CA-CFAR with non-coherent integration.

Eq. (2.107) implies that one must first find the joint *pdf* for the ratio Y_1/Z. However, this can be avoided if P_{fa} is first computed for a fixed threshold value V_T, then averaged over all possible values of the threshold. Therefore, let the conditional probability of false alarm when $y = V_T$ be $P_{fa}(V_T = y)$. It follows that the unconditional false alarm probability is given by

$$P_{fa} = \int_0^\infty P_{fa}(V_T = y) f(y) dy \qquad (2.108)$$

where $f(y)$ is the *pdf* of the threshold. In view of this, the probability density function describing the random variable $K_1 Z$ is given by

$$f(y) = \frac{(y/K_1)^{Mn_p - 1} e^{(-y/2K_0\psi^2)}}{(2\psi^2)^{Mn_p} K_1 \, \Gamma(Mn_p)} \quad ; \; y \geq 0 \qquad (2.109)$$

It can be shown that in this case the probability of false alarm is independent of the noise power and is given by

$$P_{fa} = \frac{1}{(1 + K_1)^{Mn_p}} \sum_{k=0}^{n_p - 1} \frac{1}{k!} \frac{\Gamma(Mn_p + k)}{\Gamma(Mn_p)} \left(\frac{K_1}{1 + K_1}\right)^k \qquad (2.110)$$

which is identical to Eq. (2.106) when $K_1 = K_0$ and $n_P = 1$.

2.10. "MyRadar" Design Case Study - Visit 2[1]

2.10.1. Problem Statement

Modify the design introduced in Chapter 1 for the "MyRadar" design case study so that the effects of target RCS fluctuations are taken into account. For this purpose modify the design such that: The aircraft and missile target types follow Swerling I and Swerling III fluctuations, respectively. Also assume that a $P_D \geq 0.995$ is required at maximum range with $P_{fa} = 10^{-7}$ or better. You may use either non-coherent integration or cumulative probability of detection. Also, modify any other design parameters if needed.

2.10.2. A Design

The missile and the aircraft detection ranges were calculated in Chapter 1. They are $R_a = 90 Km$ for the aircraft and $R_m = 55 Km$ for the missile. First, determine the probability of detection for each target type with and without the 7-pulse non-coherent integration. For this purpose, use MATLAB program *"myradar_visit2_1.m"* given in Listing 2.27. This program first computes the improvement factor and the associated integration loss. Second it calculates the single pulse SNR. Finally it calculates the SNR when non-coherent integration is utilized. Executing this program yields:

SNR_single_pulse_missile = 5.5998 dB
SNR_7_pulse_NCI_missile = 11.7216 dB
SNR_single_pulse_aircraft = 6.0755 dB
SNR_7_pulse_NCI_aircrfat = 12.1973 dB

Using these values in functions *"pd_swerling1.m"* and *"pd_swerling3.m"* yields

Pd_single_pulse_missile = 0.013
Pd_7_pulse_NCI_missile= 0.9276
Pd_single_pulse_aircraft = 0.038
Pd_7_pulse_NCI_aircraft = 0.8273

Clearly in all four cases, there is not enough SNR to meet the design requirement of $P_D \geq 0.995$.

---

1. Please read disclaimer in Section 1.9.1.

Instead, resort to accomplishing the desired probability of detection by using cumulative probabilities. The single frame increment for the missile and aircraft cases are

$$R_{Missile} = scan\ rate \times v_m = 2 \times 150 = 300m \qquad (2.111)$$

$$R_{Aircraft} = scan\ rate \times v_a = 2 \times 400 = 800m \qquad (2.112)$$

2.10.2.1 Single Pulse (Per Frame) Design Option

As a first design option, consider the case where during each frame only a single pulse is used for detection (i.e., no integration). Consequently, if the single pulse detection does not achieve the desired probability of detection at 90 Km for the aircraft or at 55 Km for the missile, then non-coherent integration of a few pulses per frame can then be utilized. Keep in mind that only non-coherent integration can be used in the cases of Swerling type I and III fluctuations (see Section 2.4).

Assume that the first frame corresponding to detecting the aircraft is 106 Km. This assumption is arbitrary and it provides the designer with 21 frames. It follows that the first frame, when detecting the missile, is at 61 Km. Furthermore, assume that the SNR at $R = 90Km$ is $(SNR)_{aircraft} = 8.5dB$, for the aircraft case. And, for the missile case assume that at $R = 55Km$ the corresponding SNR is $(SNR)_{missile} = 9dB$. Note that these values are simply educated guesses, and the designer may be required to perform several iterations in order to accomplish the desired cumulative probability of detection, $P_D \geq 0.995$. In order to calculate the cumulative probability of detection at a certain range, the MATLAB program "myradar_visit2_2.m" was developed. This program is given in Listing 2.28 in Section 2.11.

Initialization of the program "myradar_visit2_2.m" includes entering the following inputs: The desired P_{fa}; the number of pulses to be used for non-coherent integration per frame; the range at which the desired cumulative operability of detection must be achieved; the frame size; and finally the target fluctuation type. For notational purposes, denote the range at which the desired cumulative probability of detection must be achieved as R_0. Then for each frame, the following list includes the outputs of this program: SNR, probability of detection, fluctuation loss, and cumulative probability of detection.

The logic used by this program for calculating the proper probability of detection at each frame and for computing the cumulative probability of detection is described as follows:

1. Initialize the program, by entering the desired input values. Assume Swerling V fluctuation and use Eq. (2.98) to calculate the frame-SNR, $(SNR)_i$.

1.1. For the *"MyRadar"* design case study, use $n_P = 1$, $R_0 = 90 Km$, and $(SNR_0)_{aircraft} = 8.5 dB$. Alternatively use $R_0 = 55 Km$ and $(SNR)_{missile} = 9 dB$ for the missile case. Note that the selected SNR values are best estimates or educated guesses, and it may require going through few iterations before finally selecting an acceptable set.

2. The program will then calculate the number of frames and their associated ranges. The program uses the function *"fluct_loss.m"* to calculate the Swerling V P_D at each frame and the additional SNR required to accomplish the same probability of detection when target fluctuation is included.

3. Depending on the fluctuation type, the program will then use the proper MATLAB function to calculate the probability of detection for each frame, P_{D_i}.

    3.1. For the *"MyRadar"* design case study, these functions are *"pd_swerling1.m"* and *"pd_swerling 3.m"*.

4. Finally, the program uses Eq. (2.99) to compute the cumulative probability of detection, P_{D_n}.

A Graphical User Interface (GUI) has been developed for this program; Fig. 2.20 shows its associated GUI workspace. To use this GUI, from the MATLAB command window type *"myradar_visit2_2_gui"*. Executing the program *"myradar_visit2_2.m"* using the input values stated above yields the following cumulative probabilities of detection for the aircraft and missile cases,

$$P_{DC_{Missile}} = 0.99872$$

$$P_{DC_{aircraft}} = 0.99687$$

These results clearly satisfy the design requirement of $P_D \geq 0.995$. However, one must re-validate the peak power requirement for the design. To do that, go back to Eq.s (1.107) and (1.108), and replace the SNR values used in Chapter 1 by the values adopted in this chapter (i.e., $(SNR_0)_{aircraft} = 8.5 dB$ and $(SNR)_{missile} = 9 dB$). It follows that the corresponding single pulse energy for the missile and the aircraft cases are respectively given by

$$E_m = 0.1658 \times \frac{10^{0.9}}{10^{0.56}} = 0.36273 Joules \qquad (2.113)$$

$$E_a = 0.1487 \times \frac{10^{0.85}}{10^{0.56}} = 0.28994 Joules \qquad (2.114)$$

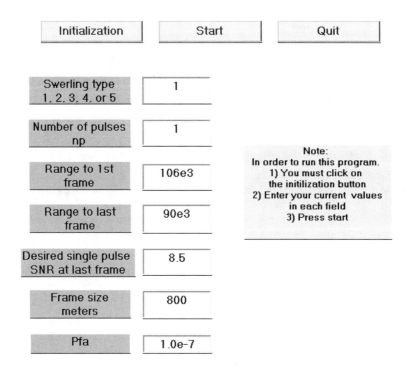

Figure 2.20. GUI workspace associated with program *"myradar_visit2_2_gui.m"*.

This indicates that the stressing single pulse peak power requirement (i.e., missile detection) exceeds $362KW$. This value for the single pulse peak power is high for a mobile ground based air defense radar and practical constraints would require using less peak power.

In order to bring the single pulse peak power requirement down, one can use non-coherent integration of a few pulses per frame prior to calculating the frame probability of detection. For this purpose, the program *"myradar_visit2_2.m"* can be used again. However, in this case $n_P > 1$. This is analyzed in the next section.

2.10.2.2. Non-Coherent Integration Design Option

The single frame probability of detection can be improved significantly when pulse integration is utilized. One may use coherent or non-coherent integration to improve the frame cumulative probability of detection. In this case, caution should be exercised since coherent integration would not be practical when the target fluctuation type is either Swerling I or Swerling III. Alternatively, using non-coherent integration will always reduce the minimum required SNR.

Rerun the MATLAB program "*myradar_visit2_2_gui*". Use $n_P = 4$ and use $SNR = 4dB$ (single pulse) for both the missile and aircraft single pulse SNR^1 at their respective reference ranges, $R_{0_{missile}} = 55Km$ and $R_{0_{aircraft}} = 90Km$. The resulting cumulative probabilities of detection are

$$P_{DC_{Missile}} = 0.99945$$

$$P_{DC_{aircraft}} = 0.99812$$

which are both within the desired design requirements. It follows that the corresponding minimum required single pulse energy for the missile and the aircraft cases are now given by

$$E_m = 0.1658 \times \frac{10^{0.4}}{10^{0.56}} = 0.1147 Joules \qquad (2.115)$$

$$E_a = 0.1487 \times \frac{10^{0.4}}{10^{0.56}} = 0.1029 Joules \qquad (2.116)$$

Thus, the minimum single pulse peak power (assuming the same pulsewidth as that given in Section1.9.2) is

$$P_t = \frac{0.1147}{1 \times 10^{-6}} = 114.7 KW \qquad (2.117)$$

Note that the peak power requirement will be significantly reduced while maintaining a very fine range resolution when pulse compression techniques are used. This will be discussed in a subsequent chapter.

Fig. 2.21 shows a plot of the SNR versus range for both target types. This plot assumes 4-pulse non-coherent integration. It can be reproduced using MATLAB program "*fig2_21.m*". It is given in Listing 2.29 in Section 2.11.

2.11. MATLAB Program and Function Listings

This section presents listings for all MATLAB programs/functions used in this chapter. The user is advised to rerun these programs with different input parameters.

---

1. Again these values are educated guesses. The designer my be required to go through a few iterations before arriving at an acceptable set of design parameters.

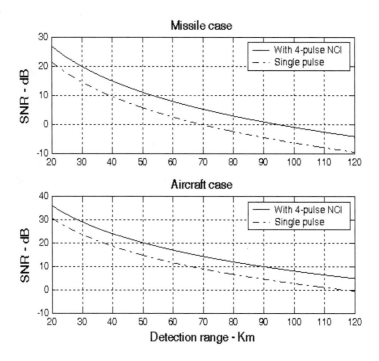

Figure 2.21. SNR versus detection range for both target types. The 4-pulse NCI curves correspond to 21 frame cumulative detection with the last frame at: 55 Km for the missile and 90 Km for the aircraft.

Listing 2.1. MATLAB Program "fig2_2.m"

```
% This program can be used to reproduce Figure 2.2 of the text
clear all
close all
xg = linspace(-6,6,1500); % random variable between -6 and 6
xr = linspace(0,6,1500); % random variable between 0 and 6
mu = 0; % zero mean Gaussian pdf mean
sigma = 1.5; % standard deviation (sqrt(variance))
ynorm = normpdf(xg,mu,sigma); % use MATLAB function normpdf
yray = raylpdf(xr,sigma); % use MATLAB function raylpdf
plot(xg,ynorm,'k',xr,yray,'k-.');
grid
legend('Gaussian pdf','Rayleigh pdf')
xlabel('x')
```

```
ylabel('Probability density')
gtext('\mu = 0; \sigma = 1.5')
gtext('\sigma =1.5')
```

Listing 2.2. MATLAB Function "que_func.m"

```
function fofx = que_func(x)
% This function computes the value of the Q-function
% listed in Eq.(2.16). It uses the approximation in Eqs. (2.17) and (2.18)
if (x >= 0)
  denom = 0.661 * x + 0.339 * sqrt(x^2 + 5.51);
  expo = exp(-x^2 /2.0);
  fofx = 1.0 - (1.0 / sqrt(2.0 * pi)) * (1.0 / denom) * expo;
else
  denom = 0.661 * x + 0.339 * sqrt(x^2 + 5.51);
  expo = exp(-x^2 /2.0);
  value = 1.0 - (1.0 / sqrt(2.0 * pi)) * (1.0 / denom) * expo;
  fofx = 1.0 - value;
end
```

Listing 2.3. MATLAB Program "fig2_3.m"

```
%This program generates Figure 2.3.
close all
clear all
logpfa = linspace(.01,250,1000);
var = 10.^(logpfa ./ 10.0);
vtnorm =  sqrt( log (var));
semilogx(logpfa, vtnorm,'k')
grid
```

Listing 2.4. MATLAB Function "marcumsq.m"

```
function Pd = marcumsq (a,b)
% This function uses Parl's method to compute PD
max_test_value = 5000.;
if (a < b)
  alphan0 = 1.0;
  dn = a / b;
else
  alphan0 = 0.;
  dn = b / a;
end
alphan_1 = 0.;
betan0 = 0.5;
```

```
betan_1 = 0.;
D1 = dn;
n = 0;
ratio = 2.0 / (a * b);
r1 = 0.0;
betan = 0.0;
alphan = 0.0;
while betan < 1000.,
  n = n + 1;
  alphan = dn + ratio * n * alphan0 + alphan;
  betan = 1.0 + ratio * n * betan0 + betan;
  alphan_1 = alphan0;
  alphan0 = alphan;
  betan_1 = betan0;
  betan0 = betan;
  dn = dn * D1;
end
PD = (alphan0 / (2.0 * betan0)) * exp( -(a-b)^2 / 2.0);
if ( a >= b)
  PD = 1.0 - PD;
end
return
```

Listing 2.5. MATLAB Program "prob_snr1.m"

```
% This program is used to produce Fig. 2.4
close all
clear all
for nfa = 2:2:12
  b = sqrt(-2.0 * log(10^(-nfa)));
  index = 0;
  hold on
  for snr = 0:.1:18
    index = index +1;
    a = sqrt(2.0 * 10^(.1*snr));
    pro(index) = marcumsq(a,b);
  end
  x = 0:.1:18;
  set(gca,'ytick',[.1 .2 .3 .4 .5 .6 .7 .75 .8 .85 .9 ...
      .95 .9999])
  set(gca,'xtick',[1 2 3 4 5 6 7 8 9 10 11 12 13 14 15 16 17 18])

  loglog(x, pro,'k');
end
```

```
hold off
xlabel ('Single pulse SNR - dB')
ylabel ('Probability of detection')
grid
```

Listing 2.6. MATLAB program "fig2_6a.m"

```
% This program is used to produce Fig. 2.6a
% It uses the function "improv_fac"
clear all
close all
pfa1 = 1.0e-2;
pfa2 = 1.0e-6;
pfa3 = 1.0e-10;
pfa4 = 1.0e-13;
pd1 = .5;
pd2 = .8;
pd3 = .95;
pd4 = .999;
index = 0;
for np = 1:1:1000
   index = index + 1;
   I1(index) = improv_fac (np, pfa1, pd1);
   I2(index) = improv_fac (np, pfa2, pd2);
   I3(index) = improv_fac (np, pfa3, pd3);
   I4(index) = improv_fac (np, pfa4, pd4);
end
np = 1:1:1000;
semilogx (np, I1, 'k', np, I2, 'k--', np, I3, 'k-.', np, I4, 'k:')
xlabel ('Number of pulses');
ylabel ('Improvement factor I - dB')
legend ('pd=.5, nfa=e+2','pd=.8, nfa=e+6','pd=.95, nfa=e+10','pd=.999, nfa=e+13');
grid
```

Listing 2.7. MATLAB Function "improv_fac.m"

```
function impr_of_np = improv_fac (np, pfa, pd)
% This function computes the non-coherent integration improvement
% factor using the empirical formula defined in Eq. (2.49)
fact1 = 1.0 + log10( 1.0 / pfa) / 46.6;
fact2 = 6.79 * (1.0 + 0.235 * pd);
fact3 = 1.0 - 0.14 * log10(np) + 0.0183 * (log10(np))^2;
```

impr_of_np = fact1 * fact2 * fact3 * log10(np);
return

Listing 2.8. MATLAB Program "fig2_6b.m"

```
% This program is used to produce Fig. 2.6b
% It uses the function "improv_fac".
clear all
close all
pfa1 = 1.0e-12;
pfa2 = 1.0e-12;
pfa3 = 1.0e-12;
pfa4 = 1.0e-12;
pd1 = .5;
pd2 = .8;
pd3 = .95;
pd4 = .99;
index = 0;
for np = 1:1:1000
   index = index+1;
   I1 = improv_fac (np, pfa1, pd1);
   i1 = 10.^(0.1*I1);
   L1(index) = -1*10*log10(i1 ./ np);
   I2 = improv_fac (np, pfa2, pd2);
   i2 = 10.^(0.1*I2);
   L2(index) = -1*10*log10(i2 ./ np);
   I3 = improv_fac (np, pfa3, pd3);
   i3 = 10.^(0.1*I3);
   L3(index) = -1*10*log10(i3 ./ np);
   I4 = improv_fac (np, pfa4, pd4);
   i4 = 10.^(0.1*I4);
   L4 (index) = -1*10*log10(i4 ./ np);
end
np = 1:1:1000;
semilogx (np, L1, 'k', np, L2, 'k--', np, L3, 'k-.', np, L4, 'k:')
axis tight
xlabel ('Number of pulses');
ylabel ('Integration loss - dB')
legend ('pd=.5, nfa=e+12','pd=.8, nfa=e+12','pd=.95, nfa=e+12','pd=.99, nfa=e+12');
grid
```

Listing 2.9. MATLAB Function "incomplete_gamma.m"

```
function [value] = incomplete_gamma ( vt, np)
% This function implements Eq. (2.67) to compute the Incomplete Gamma
Function
% This function needs "factor.m" to run
format long
eps = 1.000000001;
% Test to see if np = 1
if (np == 1)
  value1 = vt * exp(-vt);
  value = 1.0 - exp(-vt);
  return
end
sumold = 1.0;
sumnew = 1.0;
calc1 = 1.0;
calc2 = np;
xx = np * log(vt+0.0000000001) - vt - factor(calc2);
temp1 = exp(xx);
temp2 = np / (vt+0.0000000001);
diff = .0;
ratio = 1000.0;
if (vt >= np)
  while (ratio >= eps)
    diff = diff + 1.0;
    calc1 = calc1 * (calc2 - diff) / vt ;
    sumnew = sumold + calc1;
    ratio = sumnew / sumold;
    sumold = sumnew;
  end
  value = 1.0 - temp1 * sumnew * temp2;
  return
else
  diff = 0.;
  sumold = 1.;
  ratio = 1000.;
  calc1 = 1.;
  while(ratio >= eps)
    diff = diff + 1.0;
    calc1 = calc1 * vt / (calc2 + diff);
    sumnew = sumold + calc1;
    ratio = sumnew / sumold;
    sumold = sumnew;
```

124   *MATLAB Simulations for Radar Systems Design*

```
    end
  value = temp1 * sumnew;
end
```

Listing 2.10. MATLAB Function "factor.m"

```
function [val] = factor(n)
% Compute the factorial of n using logarithms to avoid overflow.
format long
n = n + 9.0;
n2 = n * n;
temp = (n-1) * log(n) - n + log(sqrt(2.0 * pi * n)) ...
    + ((1.0 - (1.0/30. + (1.0/105)/n2)/n2) / 12) / n;
val = temp - log((n-1)*(n-2)*(n-3)*(n-4)*(n-5)*(n-6) ...
    *(n-7)*(n-8));
return
```

Listing 2.11. MATLAB Program "fig2_7.m"

```
% This program can be used to reproduce Fig. 2.7
close all
clear all
format long
ii = 0;
for x = 0:.1:20
  ii = ii+1;
  val1(ii) = incomplete_gamma(x , 1);
  val2(ii) = incomplete_gamma(x , 3);
  val = incomplete_gamma(x , 6);
  val3(ii) = val;
  val = incomplete_gamma(x , 10);
  val4(ii) = val;
end
xx = 0:.1:20;
plot(xx,val1,'k',xx,val2,'k:',xx,val3,'k--',xx,val4,'k-.')
legend('N = 1','N = 3','N = 6','N = 10')
xlabel('x')
ylabel('Incomplete Gamma function (x,N)')
grid
```

Listing 2.12. MATLAB Function "threshold.m"

```
function [pfa, vt] = threshold (nfa, np)
% This function calculates the threshold value from nfa and np.
% The Newton-Raphson recursive formula is used (Eqs. (2-63) through (2-66))
```

```
% This function uses "incomplete_gamma.m".
delmax = .00001;
eps = 0.000000001;
delta =10000.;
pfa = np * log(2) / nfa;
sqrtpfa = sqrt(-log10(pfa));
sqrtnp = sqrt(np);
vt0 = np - sqrtnp + 2.3 * sqrtpfa * (sqrtpfa + sqrtnp - 1.0);
vt = vt0;
while (abs(delta) >= vt0)
   igf = incomplete_gamma(vt0,np);
   num = 0.5^(np/nfa) - igf;
   temp = (np-1) * log(vt0+eps) - vt0 - factor(np-1);
   deno = exp(temp);
   vt = vt0 + (num / (deno+eps));
   delta = abs(vt - vt0) * 10000.0;
   vt0 = vt;
end
```

Listing 2.13. MATLAB Program "fig2_8.m"

```
% Use this program to reproduce Fig. 2.8 of text
clear all
for n= 1: 1:150
   [pfa1 y1(n)] = threshold(1000,n);
   [pfa2 y3(n)] = threshold(10000,n);
   [pfa3 y4(n)] = threshold(500000,n);
end
n =1:1:150;
loglog(n,y1,'k',n,y3,'k--',n,y4,'k-.');
axis([0 200 1 300])
xlabel ('Number of pulses');
ylabel('Threshold')
legend('nfa=1000','nfa=10000','nfa=500000')
grid
```

Listing 2.14. MATLAB Function "pd_swerling5.m"

```
function pd = pd_swerling5 (input1, indicator, np, snrbar)
% This function is used to calculate the probability of
% for Swerling 5 or 0 targets for np>1.
if(np == 1)
   'Stop, np must be greater than 1'
   return
```

```
end
format long
snrbar = 10.0.^(snrbar./10.);
eps = 0.00000001;
delmax = .00001;
delta =10000.;
% Calculate the threshold Vt
if (indicator ~=1)
  nfa = input1;
  pfa = np * log(2) / nfa;
else
  pfa = input1;
  nfa = np * log(2) / pfa;
end
sqrtpfa = sqrt(-log10(pfa));
sqrtnp = sqrt(np);
vt0 = np - sqrtnp + 2.3 * sqrtpfa * (sqrtpfa + sqrtnp - 1.0);
vt = vt0;
while (abs(delta) >= vt0)
  igf = incomplete_gamma(vt0,np);
  num = 0.5^(np/nfa) - igf;
  temp = (np-1) * log(vt0+eps) - vt0 - factor(np-1);
  deno = exp(temp);
  vt = vt0 + (num / (deno+eps));
  delta = abs(vt - vt0) * 10000.0;
  vt0 = vt;
end
% Calculate the Gram-Charlier coefficients
temp1 = 2.0 .* snrbar + 1.0;
omegabar = sqrt(np .* temp1);
c3 = -(snrbar + 1.0 / 3.0) ./ (sqrt(np) .* temp1.^1.5);
c4 = (snrbar + 0.25) ./ (np .* temp1.^2.);
c6 = c3 .* c3 ./2.0;
V = (vt - np .* (1.0 + snrbar)) ./ omegabar;
Vsqr = V .*V;
val1 = exp(-Vsqr ./ 2.0) ./ sqrt( 2.0 * pi);
val2 = c3 .* (V.^2 -1.0) + c4 .* V .* (3.0 - V.^2) -...
    c6 .* V .* (V.^4 - 10. .* V.^2 + 15.0);
q = 0.5 .* erfc (V./sqrt(2.0));
pd =  q - val1 .* val2;
```

MATLAB Program and Function Listings   127

Listing 2.15. MATLAB Program "fig2_9.m"

```
% This program is used to produce Fig. 2.9
close all
clear all
pfa = 1e-9;
nfa = log(2) / pfa;
b = sqrt(-2.0 * log(pfa));
index = 0;
for snr = 0:.1:20
  index = index +1;
  a = sqrt(2.0 * 10^(.1*snr));
  pro(index) = marcumsq(a,b);
  prob205(index) = pd_swerling5 (pfa, 1, 10, snr);
end
x = 0:.1:20;
plot(x, pro,'k',x,prob205,'k:');
axis([0 20 0 1])
xlabel ('SNR - dB')
ylabel ('Probability of detection')
legend('np = 1','np = 10')
grid
```

Listing 2.16. MATLAB Function "pd_swerling1.m"

```
function pd = pd_swerling1 (nfa, np, snrbar)
% This function is used to calculate the probability of detection
% for Swerling 1 targets.
format long
snrbar = 10.0^(snrbar/10.);
eps = 0.00000001;
delmax = .00001;
delta =10000.;
% Calculate the threshold Vt
pfa =  np * log(2) / nfa;
sqrtpfa = sqrt(-log10(pfa));
sqrtnp = sqrt(np);
vt0 = np - sqrtnp + 2.3 * sqrtpfa * (sqrtpfa + sqrtnp - 1.0);
vt = vt0;
while (abs(delta) >= vt0)
  igf = incomplete_gamma(vt0,np);
  num = 0.5^(np/nfa) - igf;
  temp = (np-1) * log(vt0+eps) - vt0 - factor(np-1);
  deno = exp(temp);
```

```
    vt = vt0 + (num / (deno+eps));
    delta = abs(vt - vt0) * 10000.0;
    vt0 = vt;
end
if (np == 1)
    temp = -vt / (1.0 + snrbar);
    pd = exp(temp);
    return
end
    temp1 = 1.0 + np * snrbar;
    temp2 = 1.0 / (np *snrbar);
    temp = 1.0 + temp2;
    val1 = temp^(np-1.);
    igf1 = incomplete_gamma(vt,np-1);
    igf2 = incomplete_gamma(vt/temp,np-1);
    pd = 1.0 - igf1 + val1 * igf2 * exp(-vt/temp1);
```

Listing 2.17. MATLAB Program "fig2_10.m"

```
% This program is used to reproduce Fig. 2.10
close all
clear all
pfa = 1e-9;
nfa = log(2) / pfa;
b = sqrt(-2.0 * log(pfa));
index = 0;
for snr = 0:.1:22
    index = index +1;
    a = sqrt(2.0 * 10^(.1*snr));
    pro(index) = marcumsq(a,b);
    prob(index) = pd_swerling1 (nfa, 1, snr);
end
x = 0:.1:22;
plot(x, pro,'k',x,prob,'k:');
axis([2 22 0 1])
xlabel ('SNR - dB')
ylabel ('Probability of detection')
legend('Swerling V','Swerling I')
grid
```

Listing 2.18. MATLAB Program "fig2_11ab.m"

```
% This program is used to produce Fig. 2.11a&b
clear all
pfa = 1e-11;
nfa = log(2) / pfa;
index = 0;
for snr = -10:.5:30
   index = index +1;
   prob1(index) = pd_swerling1 (nfa, 1, snr);
   prob10(index) = pd_swerling1 (nfa, 10, snr);
   prob50(index) = pd_swerling1 (nfa, 50, snr);
   prob100(index) = pd_swerling1 (nfa, 100, snr);
end
x = -10:.5:30;
plot(x, prob1,'k',x,prob10,'k:',x,prob50,'k--', ...
   x, prob100,'k-.');
axis([-10 30 0 1])
xlabel ('SNR - dB')
ylabel ('Probability of detection')
legend('np = 1','np = 10','np = 50','np = 100')
grid
```

Listing 2.19. MATLAB Function "pd_swerling2.m"

```
function pd = pd_swerling2 (nfa, np, snrbar)
% This function is used to calculate the probability of detection
% for Swerling 2 targets.
format long
snrbar = 10.0^(snrbar/10.);
eps = 0.00000001;
delmax = .00001;
delta =10000.;
% Calculate the threshold Vt
pfa = np * log(2) / nfa;
sqrtpfa = sqrt(-log10(pfa));
sqrtnp = sqrt(np);
vt0 = np - sqrtnp + 2.3 * sqrtpfa * (sqrtpfa + sqrtnp - 1.0);
vt = vt0;
while (abs(delta) >= vt0)
   igf = incomplete_gamma(vt0,np);
   num = 0.5^(np/nfa) - igf;
   temp = (np-1) * log(vt0+eps) - vt0 - factor(np-1);
   deno = exp(temp);
   vt = vt0 + (num / (deno+eps));
```

```
    delta = abs(vt - vt0) * 10000.0;
    vt0 = vt;
end
if (np <= 50)
    temp = vt / (1.0 + snrbar);
    pd = 1.0 - incomplete_gamma(temp,np);
    return
else
    temp1 = snrbar + 1.0;
    omegabar = sqrt(np) * temp1;
    c3 = -1.0 / sqrt(9.0 * np);
    c4 = 0.25 / np;
    c6 = c3 * c3 /2.0;
    V = (vt - np * temp1) / omegabar;
    Vsqr = V *V;
    val1 = exp(-Vsqr / 2.0) / sqrt( 2.0 * pi);
    val2 = c3 * (V^2 -1.0) + c4 * V * (3.0 - V^2) - ...
        c6 * V * (V^4 - 10. * V^2 + 15.0);
    q = 0.5 * erfc (V/sqrt(2.0));
    pd = q - val1 * val2;
end
```

Listing 2.20. MATLAB Program "fig2_12.m"

```
% This program is used to produce Fig. 2.12
clear all
pfa = 1e-10;
nfa = log(2) / pfa;
index = 0;
for snr = -10:.5:30
    index = index +1;
    prob1(index) = pd_swerling2 (nfa, 1, snr);
    prob10(index) = pd_swerling2 (nfa, 10, snr);
    prob50(index) = pd_swerling2 (nfa, 50, snr);
    prob100(index) = pd_swerling2 (nfa, 100, snr);
end
x = -10:.5:30;
plot(x, prob1,'k',x,prob10,'k:',x,prob50,'k--', ...
    x, prob100,'k-.');
axis([-10 30 0 1])
xlabel ('SNR - dB')
ylabel ('Probability of detection')
legend('np = 1','np = 10','np = 50','np = 100')
grid
```

Listing 2.21. MATLAB Function "pd_swerling3.m"

```
function pd = pd_swerling3 (nfa, np, snrbar)
% This function is used to calculate the probability of detection
% for Swerling 3 targets.
format long
snrbar = 10.0^(snrbar/10.);
eps = 0.00000001;
delmax = .00001;
delta =10000.;
% Calculate the threshold Vt
pfa =  np * log(2) / nfa;
sqrtpfa = sqrt(-log10(pfa));
sqrtnp = sqrt(np);
vt0 = np - sqrtnp + 2.3 * sqrtpfa * (sqrtpfa + sqrtnp - 1.0);
vt = vt0;
while (abs(delta) >= vt0)
   igf = incomplete_gamma(vt0,np);
   num = 0.5^(np/nfa) - igf;
   temp = (np-1) * log(vt0+eps) - vt0 - factor(np-1);
   deno = exp(temp);
   vt = vt0 + (num / (deno+eps));
   delta = abs(vt - vt0) * 10000.0;
   vt0 = vt;
end
temp1 = vt / (1.0 + 0.5 * np *snrbar);
temp2 = 1.0 + 2.0 / (np * snrbar);
temp3 = 2.0 * (np - 2.0) / (np * snrbar);
ko = exp(-temp1) * temp2^(np-2.) * (1.0 + temp1 - temp3);
if (np <= 2)
   pd = ko;
   return
else
   temp4 = vt^(np-1.) * exp(-vt) / (temp1 * exp(factor(np-2.)));
   temp5 =  vt / (1.0 + 2.0 / (np *snrbar));
   pd = temp4 + 1.0 - incomplete_gamma(vt,np-1.) + ko * ...
      incomplete_gamma(temp5,np-1.);
end
```

Listing 2.22. MATLAB Program "fig2_13.m"

```
% This program is used to produce Fig. 2.13
clear all
pfa = 1e-9;
```

```
nfa = log(2) / pfa;
index = 0;
for snr = -10:.5:30
   index = index +1;
   prob1(index) =  pd_swerling3 (nfa, 1, snr);
   prob10(index) =  pd_swerling3 (nfa, 10, snr);
   prob50(index) =  pd_swerling3(nfa, 50, snr);
   prob100(index) =  pd_swerling3 (nfa, 100, snr);
end
x = -10:.5:30;
plot(x, prob1,'k',x,prob10,'k:',x,prob50,'k--', ...
   x, prob100,'k-.');
axis([-10 30 0 1])
xlabel ('SNR - dB')
ylabel ('Probability of detection')
legend('np = 1','np = 10','np = 50','np = 100')
grid
```

Listing 2.23. MATLAB Function "pd_swerling4.m"

```
function pd = pd_swerling4 (nfa, np, snrbar)
% This function is used to calculate the probability of detection
% for Swerling 4 targets.
format long
snrbar = 10.0^(snrbar/10.);
eps = 0.00000001;
delmax = .00001;
delta =10000.;
% Calculate the threshold Vt
pfa =  np * log(2) / nfa;
sqrtpfa = sqrt(-log10(pfa));
sqrtnp = sqrt(np);
vt0 = np - sqrtnp + 2.3 * sqrtpfa * (sqrtpfa + sqrtnp - 1.0);
vt = vt0;
while (abs(delta) >= vt0)
   igf = incomplete_gamma(vt0,np);
   num = 0.5^(np/nfa) - igf;
   temp = (np-1) * log(vt0+eps) - vt0 - factor(np-1);
   deno = exp(temp);
   vt = vt0 + (num / (deno+eps));
   delta = abs(vt - vt0) * 10000.0;
   vt0 = vt;
end
h8 = snrbar /2.0;
```

MATLAB Program and Function Listings 133

```
beta = 1.0 + h8;
beta2 = 2.0 * beta^2 - 1.0;
beta3 = 2.0 * beta^3;
if (np >= 50)
  temp1 = 2.0 * beta -1;
  omegabar = sqrt(np * temp1);
  c3 = (beta3 - 1.) / 3.0 / beta2 / omegabar;
  c4 = (beta3 * beta3 - 1.0) / 4. / np /beta2 /beta2;
  c6 = c3 * c3 /2.0;
  V = (vt - np * (1.0 + snrbar)) / omegabar;
  Vsqr = V *V;
  val1 = exp(-Vsqr / 2.0) / sqrt( 2.0 * pi);
  val2 = c3 * (V^2 -1.0) + c4 * V * (3.0 - V^2) - ...
    c6 * V * (V^4 - 10. * V^2 + 15.0);
  q = 0.5 * erfc (V/sqrt(2.0));
  pd = q - val1 * val2;
  return
else
  snr = 1.0;
  gamma0 = incomplete_gamma(vt/beta,np);
  a1 = (vt / beta)^np / (exp(factor(np)) * exp(vt/beta));
  sum = gamma0;
  for i = 1:1:np
    temp1 = 1;
    if (i == 1)
      ai = a1;
    else
      ai = (vt / beta) * a1 / (np + i -1);
    end
    a1 = ai;
    gammai = gamma0 - ai;
    gamma0 = gammai;
    a1 = ai;
    for ii = 1:1:i
      temp1 = temp1 * (np + 1 - ii);
    end
    term = (snrbar /2.0)^i * gammai * temp1 / exp(factor(i));
    sum = sum + term;
  end
  pd = 1.0 - sum / beta^np;
end
pd = max(pd,0.);
```

Listing 2.24. MATLAB Program "fig2_14.m"

```
% This program is used to produce Fig. 2.14
clear all
pfa = 1e-9;
nfa = log(2) / pfa;
index = 0;
for snr = -10:.5:30
   index = index +1;
   prob1(index) =  pd_swerling4 (nfa, 1, snr);
   prob10(index) =  pd_swerling4 (nfa, 10, snr);
   prob50(index) =  pd_swerling4(nfa, 50, snr);
   prob100(index) =  pd_swerling4 (nfa, 100, snr);
end
x = -10:.5:30;
plot(x, prob1,'k',x,prob10,'k:',x,prob50,'k--', ...
   x, prob100,'k-.');
axis([-10 30 0 1.1])
xlabel ('SNR - dB')
ylabel ('Probability of detection')
legend('np = 1','np = 10','np = 50','np = 100')
grid
axis tight
```

Listing 2.25. MATLAB Function "fluct_loss.m"

```
function [Lf,Pd_Sw5] = fluct_loss(pd, pfa, np, sw_case)
% This function calculates the SNR fluctuation loss for Swerling models
% A negative Lf value indicates SNR gain instead of loss
format long
% compute the false alarm number
nfa =  log(2) / pfa;
% *************** Swerling 5 case ****************
% check to make sure that np>1
if (np ==1)
   b = sqrt(-2.0 * log(pfa));
   Pd_Sw5 = 0.001;
   snr_inc = 0.1 - 0.005;
   while(Pd_Sw5 <= pd)
      snr_inc = snr_inc + 0.005;
      a = sqrt(2.0 * 10^(.1*snr_inc));
      Pd_Sw5 = marcumsq(a,b);
   end
   PD_SW5 = Pd_Sw5
```

```
      SNR_SW5 = snr_inc
else
   % np > 1 use MATLAB function pd_swerling5.m
   snr_inc = 0.1 - 0.005;
   Pd_Sw5 = 0.001;
   while(Pd_Sw5 <= pd)
      snr_inc = snr_inc + 0.005;
      Pd_Sw5 = pd_swerling5(pfa, 1, np, snr_inc);
   end
   PD_SW5 = Pd_Sw5
   SNR_SW5 = snr_inc
end
if sw_case == 5
   Lf = 0.
   return
end
% ************** End Swerling 5 case ***********
% ************** Swerling 1 case ***************
if (sw_case == 1)
   Pd_Sw1 = 0.001;
   snr_inc = 0.1 - 0.005;
   while(Pd_Sw1 <= pd)
      snr_inc = snr_inc + 0.005;
      Pd_Sw1 = pd_swerling1(nfa, np, snr_inc);
   end
   PD_SW1 = Pd_Sw1
   SNR_SW1 = snr_inc
   Lf = SNR_SW1 - SNR_SW5
end
% ************** End Swerling 1 case ***********
% ************** Swerling 2 case ***************
if (sw_case == 2)
   Pd_Sw2 = 0.001;
   snr_inc = 0.1 - 0.005;
   while(Pd_Sw2 <= pd)
      snr_inc = snr_inc + 0.005;
      Pd_Sw2 = pd_swerling2(nfa, np, snr_inc);
   end
   PD_SW2 = Pd_Sw2
   SNR_SW2 = snr_inc
   Lf = SNR_SW2 - SNR_SW5
end
% ************** End Swerling 2 case ***********
% ************** Swerling 3 case ***************
```

```
if (sw_case == 3)
   Pd_Sw3 = 0.001;
   snr_inc = 0.1 - 0.005;
   while(Pd_Sw3 <= pd)
      snr_inc = snr_inc + 0.005;
      Pd_Sw3 = pd_swerling3(nfa, np, snr_inc);
   end
   PD_SW3 = Pd_Sw3
   SNR_SW3 = snr_inc
   Lf = SNR_SW3 - SNR_SW5
end
% ************** End Swerling 3 case ************
% ************** Swerling 4 case ***************
if (sw_case == 4)
   Pd_Sw4 = 0.001;
   snr_inc = 0.1 - 0.005;
   while(Pd_Sw4 <= pd)
      snr_inc = snr_inc + 0.005;
      Pd_Sw4 = pd_swerling4(nfa, np, snr_inc);
   end
   PD_SW4 = Pd_Sw4
   SNR_SW4 = snr_inc
   Lf = SNR_SW4 - SNR_SW5
end
% ************** End Swerling 4 case ************
return
```

Listing 2.26. MATLAB Program "fig2_15.m"

```
% Use this program to reproduce Fig. 2.15 of text
clear all
close all
index = 0.;
for pd = 0.01:.05:1
   index = index + 1;
   [Lf,Pd_Sw5] = fluct_loss(pd, 1e-9,1,1);
   Lf1(index) = Lf;
   [Lf,Pd_Sw5] = fluct_loss(pd, 1e-9,1,4);
   Lf4(index) = Lf;
end
pd = 0.01:.05:1;
figure (2)
plot(pd, Lf1, 'k',pd, Lf4,'K:')
xlabel('Probability of detection')
```

```
ylabel('Fluctuation loss - dB')
legend('Swerling I & II','Swerling III & IV')
title('Pfa=1e-9, np=1')
grid
```

Listing 2.27. MATLAB Program "myradar_visit2_1.m"

```
% Myradar design case study visit 2_1
close all
clear all
pfa = 1e-7;
pd = 0.995;
np = 7;
pt = 165.8e3; % peak power in Watts
freq = 3e+9; % radar operating frequency in Hz
g = 34.5139; % antenna gain in dB
sigmam = 0.5; % missile RCS m squared
sigmaa = 4; % aircraft RCS m squared
te = 290.0; % effective noise temperature in Kelvins
b = 1.0e+6; % radar operating bandwidth in Hz
nf = 6.0; %noise figure in dB
loss = 8.0; % radar losses in dB
% compute the improvement factor due to 7-pulse non-coherent integration
Improv = improv_fac (np, pfa, pd);
% calculate the integration loss
lossnci = 10*log10(np) - Improv;
% calculate net gain in SNR due to integration
SNR_net = Improv - lossnci;
loss_total = loss + lossnci;
rangem = 55e3;
rangea = 90e3;
SNR_single_pulse_missile = radar_eq(pt, freq, g, sigmam, te, b, nf, loss, rangem)
SNR_7_pulse_NCI_missile = SNR_single_pulse_missile + SNR_net
SNR_single_pulse_aircraft = radar_eq(pt, freq, g, sigmaa, te, b, nf, loss, rangea)
SNR_7_pulse_NCI_aircraft = SNR_single_pulse_aircraft + SNR_net
```

Listing 2.28. MATLAB Program "myradar_visit2_2.m"

```
%clear all
% close all
% swid = 3;
```

```
% pfa = 1e-7;
% np = 1;
% R_1st_frame = 61e3; % Range for first frame
% R0 = 55e3; % range to last frame
% SNR0 = 9; % SNR at R0
% frame = 0.3e3; % frame size
nfa = log(2) / pfa;
range_frame = R_1st_frame:-frame:R0; % Range to each frame
% implement Eq. (2.98)
SNRi = SNR0 + 40 .* log10((R0 ./ range_frame));
% calculate the Swerling 5 Pd at each frame
b = sqrt(-2.0 * log(pfa));
if np ==1
   for frame = 1:1:size(SNRi,2)
      a = sqrt(2.0 * 10^(.1*SNRi(frame)));
      pd5(frame) = marcumsq(a,b);
   end
else
   [pd5] = pd_swerling5(pfa, 1, np, SNRi);
end
% compute additional SNR needed due to fluctuation
for frame = 1:1:size(SNRi,2)
   [Lf(frame),Pd_Sw5] = fluct_loss(pd5(frame), pfa, np, swid);
end
% adjust SNR at each frame
SNRi = SNRi - Lf;
% compute the frame Pd
for frame = 1:1:size(SNRi,2)
   if(swid==1)
      Pdi(frame) = pd_swerling1 (nfa, np, SNRi(frame));
   end
   if(swid==2)
      Pdi(frame) = pd_swerling2 (nfa, np, SNRi(frame));
   end
   if(swid==3)
      Pdi(frame) = pd_swerling3 (nfa, np, SNRi(frame));
   end
   if(swid==4)
      Pdi(frame) = pd_swerling4 (nfa, np, SNRi(frame));
   end
   if(swid==5)
      Pdi(frame) = pd5(frame);
   end
end
```

```
Pdc(1:size(SNRi,2)) = 0;
Pdc(1) = 1 - Pdi(1);
% compute the cumulative Pd
for frame = 2:1:size(SNRi,2)
    Pdc(frame) = (1-Pdi(frame)) * Pdc(frame-1);
end
PDC = 1 - Pdc(21)
```

Listing 2.29. MATLAB Program "fig2_21.m"

```
% Use this program to reproduce Fig. 2.20 of text.
close all
clear all
np = 4;
pfa = 1e-7;
pdm = 0.99945;
pda = 0.99812;
% calculate the improvement factor
Im = improv_fac(np,pfa, pdm);
Ia = improv_fac(np, pfa, pda);
% caculate the integration loss
Lm = 10*log10(np) - Im;
La = 10*log10(np) - Ia;
pt = 114.7e3; % peak power in Watts
freq = 3e+9; % radar operating frequency in Hz
g = 34.5139; % antenna gain in dB
sigmam = 0.5; % missile RCS m squared
sigmaa = 4; % aircraft RCS m squared
te = 290.0; % effective noise temperature in Kelvins
b = 1.0e+6; % radar operating bandwidth in Hz
nf = 6.0; % noise figure in dB
loss = 8.0; % radar losses in dB
losstm = loss + Lm; % total loss for missile
lossta = loss + La; % total loss for aircraft
range = linspace(20e3,120e3,1000); % range to target from 20 to 120 Km,
1000 points
% modify pt by np*pt to account for pulse integration
snrmnci = radar_eq(np*pt, freq, g, sigmam, te, b, nf, losstm, range);
snrm = radar_eq(pt, freq, g, sigmam, te, b, nf, loss, range);
snranci = radar_eq(np*pt, freq, g, sigmaa, te, b, nf, lossta, range);
snra = radar_eq(pt, freq, g, sigmaa, te, b, nf, loss, range);
% plot SNR versus range
rangekm = range ./ 1000;
figure(1)
```

```
subplot(2,1,1)
plot(rangekm,snrmnci,'k',rangekm,snrm,'k -.')
grid
legend('With 4-pulse NCI','Single pulse')
ylabel ('SNR - dB');
title('Missile case')
subplot(2,1,2)
plot(rangekm,snranci,'k',rangekm,snra,'k -.')
grid
legend('With 4-pulse NCI','Single pulse')
ylabel ('SNR - dB');
title('Aircraft case')
xlabel('Detection range - Km')
```

Chapter 3 Radar Waveforms

Choosing a particular waveform type and a signal processing technique in a radar system depends heavily on the radar's specific mission and role. The cost and complexity associated with a certain type of waveform hardware and software implementation constitute a major factor in the decision process. Radar systems can use Continuous Waveforms (CW) or pulsed waveforms with or without modulation. Modulation techniques can be either analog or digital. Range and Doppler resolutions are directly related to the specific waveform frequency characteristics. Thus, knowledge of the power spectrum density of a waveform is very critical. In general, signals or waveforms can be analyzed using time domain or frequency domain techniques. This chapter introduces many of the most commonly used radar waveforms. Relevant uses of a specific waveform will be addressed in the context of its time and frequency domain characteristics. In this book, the terms waveform and signal are used interchangeably to mean the same thing.

3.1. Low Pass, Band Pass Signals, and Quadrature Components

Signals that contain significant frequency composition at a low frequency band including DC are called Low Pass (LP) signals. Signals that have significant frequency composition around some frequency away from the origin are called Band Pass (BP) signals. A real BP signal $x(t)$ can be represented mathematically by

$$x(t) = r(t)\cos(2\pi f_0 t + \phi_x(t)) \tag{3.1}$$

where $r(t)$ is the amplitude modulation or envelope, $\phi_x(t)$ is the phase modulation, f_0 is the carrier frequency, and both $r(t)$ and $\phi_x(t)$ have frequency components significantly smaller than f_0. The frequency modulation is

$$f_m(t) = \frac{1}{2\pi} \frac{d}{dt} \phi_x(t) \tag{3.2}$$

and the instantaneous frequency is

$$f_i(t) = \frac{1}{2\pi} \frac{d}{dt} (2\pi f_0 t + \phi_x(t)) = f_0 + f_m(t) \tag{3.3}$$

If the signal bandwidth is B, and if f_0 is very large compared to B, the signal $x(t)$ is referred to as a narrow band pass signal.

Band pass signals can also be represented by two low pass signals known as the quadrature components; in this case Eq. (3.1) can be rewritten as

$$x(t) = x_I(t)\cos 2\pi f_0 t - x_Q(t)\sin 2\pi f_0 t \tag{3.4}$$

where $x_I(t)$ and $x_Q(t)$ are real LP signals referred to as the quadrature components and are given, respectively, by

$$\begin{aligned} x_I(t) &= r(t)\cos\phi_x(t) \\ x_Q(t) &= r(t)\sin\phi_x(t) \end{aligned} \tag{3.5}$$

Fig. 3.1 shows how the quadrature components are extracted.

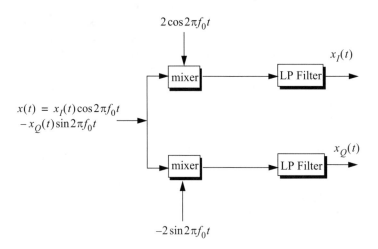

Figure 3.1. Extraction of quadrature components.

3.2. The Analytic Signal

The sinusoidal signal $x(t)$ defined in Eq. (3.1) can be written as the real part of the complex signal $\psi(t)$. More precisely,

$$x(t) = Re\{\psi(t)\} = Re\{r(t)e^{j\phi_x(t)}e^{j2\pi f_0 t}\} \qquad (3.6)$$

Define the "analytic signal" as

$$\psi(t) = v(t)e^{j2\pi f_0 t} \qquad (3.7)$$

where

$$v(t) = r(t)e^{j\phi_x(t)} \qquad (3.8)$$

and

$$\Psi(\omega) = \begin{pmatrix} 2X(\omega) & \omega \geq 0 \\ 0 & \omega < 0 \end{pmatrix} \qquad (3.9)$$

$\Psi(\omega)$ is the Fourier transform of $\psi(t)$ and $X(\omega)$ is the Fourier transform of $x(t)$. Eq. (3.9) can be written as

$$\Psi(\omega) = 2U(\omega)X(\omega) \qquad (3.10)$$

where $U(\omega)$ is the step function in the frequency domain. Thus, it can be shown that $\psi(t)$ is

$$\psi(t) = x(t) + j\tilde{x}(t) \qquad (3.11)$$

$\tilde{x}(t)$ is the Hilbert transform of $x(t)$.

Using Eqs. (3.6) and (3.11), one can then write (shown here without proof)

$$x(t) = u_{0I}(t)\cos\omega_0 t - u_{0Q}(t)\sin\omega_0 t \qquad (3.12)$$

which is similar to Eq. (3.4) with $\omega_0 = 2\pi f_0$.

Using Parseval's theorem it can be shown that the energy associated with the signal $x(t)$ is

$$E_x = \frac{1}{2}\int_{-\infty}^{\infty} x^2(t)\, dt = \frac{1}{2}\int_{-\infty}^{\infty} u^2(t)\, dt = \frac{1}{2}E_\psi \qquad (3.13)$$

3.3. CW and Pulsed Waveforms

The spectrum of a given signal describes the spread of its energy in the frequency domain. An energy signal (finite energy) can be characterized by its Energy Spectrum Density (ESD) function, while a power signal (finite power) is characterized by the Power Spectrum Density (PSD) function. The units of the ESD are Joules per Hertz and the PSD has units Watts per Hertz.

The signal bandwidth is the range of frequency over which the signal has a nonzero spectrum. In general, any signal can be defined using its duration (time domain) and bandwidth (frequency domain). A signal is said to be band-limited if it has finite bandwidth. Signals that have finite durations (time-limited) will have infinite bandwidths, while band-limited signals have infinite durations. The extreme case is a continuous sine wave, whose bandwidth is infinitesimal.

A time domain signal $f(t)$ has a Fourier Transform (FT) $F(\omega)$ given by

$$F(\omega) = \int_{-\infty}^{\infty} f(t) e^{-j\omega t} \, dt \tag{3.14}$$

where the Inverse Fourier Transform (IFT) is

$$f(t) = \frac{1}{2\pi} \int_{-\infty}^{\infty} F(\omega) e^{j\omega t} \, d\omega \tag{3.15}$$

The signal autocorrelation function $R_f(\tau)$ is

$$R_f(\tau) = \int_{-\infty}^{\infty} f^*(t) f(t+\tau) \, dt \tag{3.16}$$

The asterisk indicates the complex conjugate. The signal amplitude spectrum is $|F(\omega)|$. If $f(t)$ were an energy signal, then its ESD is $|F(\omega)|^2$; and if it were a power signal, then its PSD is $\bar{S}_f(\omega)$ which is the FT of the autocorrelation function

$$\bar{S}_f(\omega) = \int_{-\infty}^{\infty} \bar{R}_f(\tau) e^{-j\omega \tau} \, d\tau \tag{3.17}$$

First, consider a CW waveform given by

CW and Pulsed Waveforms

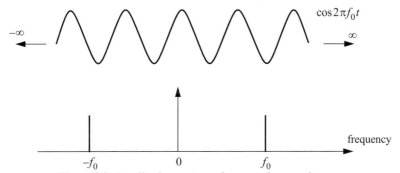

Figure 3.2. Amplitude spectrum for a continuous sine wave.

$$f_1(t) = A\cos\omega_0 t \tag{3.18}$$

The FT of $f_1(t)$ is

$$F_1(\omega) = A\pi[\delta(\omega - \omega_0) + \delta(\omega + \omega_0)] \tag{3.19}$$

where $\delta(\cdot)$ is the Dirac delta function, and $\omega_0 = 2\pi f_0$. As indicated by the amplitude spectrum shown in Fig. 3.2, the signal $f_1(t)$ has infinitesimal bandwidth, located at $\pm f_0$.

Next consider the time domain signal $f_2(t)$ given by

$$f_2(t) = A\operatorname{Rect}\left(\frac{t}{\tau}\right) = \begin{cases} A & -\frac{\tau}{2} \leq t \leq \frac{\tau}{2} \\ 0 & otherwise \end{cases} \tag{3.20}$$

It follows that the FT is

$$F_2(\omega) = A\tau\operatorname{Sinc}\left(\frac{\omega\tau}{2}\right) \tag{3.21}$$

where

$$\operatorname{Sinc}(x) = \frac{\sin(\pi x)}{\pi x} \tag{3.22}$$

The amplitude spectrum of $f_2(t)$ is shown in Fig. 3.3. In this case, the bandwidth is infinite. Since infinite bandwidths cannot be physically implemented, the signal bandwidth is approximated by $2\pi/\tau$ radians per second or $1/\tau$

Hertz. In practice, this approximation is widely accepted since it accounts for most of the signal energy.

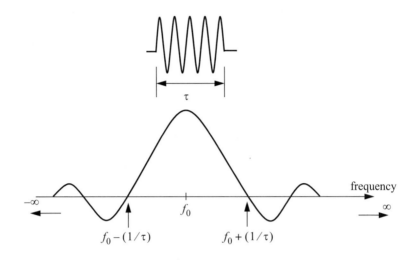

Figure 3.3. Amplitude spectrum for a single pulse, or a train of non-coherent pulses.

Now consider the coherent gated CW waveform $f_3(t)$ given by

$$f_3(t) = \sum_{n=-\infty}^{\infty} f_2(t - nT) \tag{3.23}$$

Clearly $f_3(t)$ is periodic, where T is the period (recall that $f_r = 1/T$ is the PRF). Using the complex exponential Fourier series we can rewrite $f_3(t)$ as

$$f_3(t) = \sum_{n=-\infty}^{\infty} F_n e^{\frac{j2\pi nt}{T}} \tag{3.24}$$

where the Fourier series coefficients F_n are given by

$$F_n = \frac{A\tau}{T} \operatorname{Sinc}\left(\frac{n\tau\pi}{T}\right) \tag{3.25}$$

It follows that the FT of $f_3(t)$ is

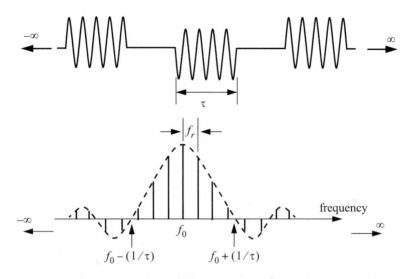

Figure 3.4. Amplitude spectrum for a coherent pulse train of infinite length.

$$F_3(\omega) = 2\pi \sum_{n=-\infty}^{\infty} F_n \delta(\omega - 2n\pi f_r) \quad (3.26)$$

The amplitude spectrum of $f_3(t)$ is shown in Fig. 3.4. In this case, the spectrum has a $\sin x/x$ envelope that corresponds to F_n. The spacing between the spectral lines is equal to the radar PRF, f_r.

Finally, define the function $f_4(t)$ as

$$f_4(t) = \sum_{n=0}^{N} f_2(t - nT) \quad (3.27)$$

Note that $f_4(t)$ is a limited duration of $f_3(t)$. The FT of $f_4(t)$ is

$$F_4(\omega) = AN\tau \left(\text{Sinc}\left(\omega \frac{NT}{2}\right) \bullet \sum_{n=-\infty}^{\infty} \text{Sinc}(n\pi\tau f_r)\delta(\omega - 2n\pi f_r) \right) \quad (3.28)$$

where the operator ($\bullet$) indicates convolution. The spectrum in this case is shown in Fig. 3.5. The envelope is still a $\sin x/x$ which corresponds to the pulsewidth. But the spectral lines are replaced by $\sin x/x$ spectra that correspond to the duration NT.

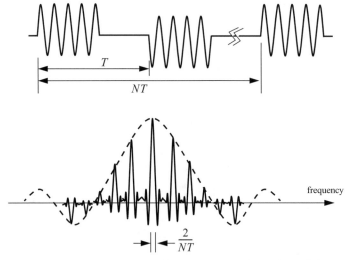

Figure 3.5. Amplitude spectrum for a coherent pulse train of finite length.

3.4. Linear Frequency Modulation Waveforms

Frequency or phase modulated waveforms can be used to achieve much wider operating bandwidths. Linear Frequency Modulation (LFM) is commonly used. In this case, the frequency is swept linearly across the pulsewidth, either upward (up-chirp) or downward (down-chirp). The matched filter bandwidth is proportional to the sweep bandwidth, and is independent of the pulsewidth. Fig. 3.6 shows a typical example of an LFM waveform. The pulsewidth is τ, and the bandwidth is B.

The LFM up-chirp instantaneous phase can be expressed by

$$\psi(t) = 2\pi\left(f_0 t + \frac{\mu}{2}t^2\right) \qquad -\frac{\tau}{2} \leq t \leq \frac{\tau}{2} \qquad (3.29)$$

where f_0 is the radar center frequency, and $\mu = (2\pi B)/\tau$ is the LFM coefficient. Thus, the instantaneous frequency is

$$f(t) = \frac{1}{2\pi}\frac{d}{dt}\psi(t) = f_0 + \mu t \qquad -\frac{\tau}{2} \leq t \leq \frac{\tau}{2} \qquad (3.30)$$

Similarly, the down-chirp instantaneous phase and frequency are given, respectively, by

Linear Frequency Modulation Waveforms

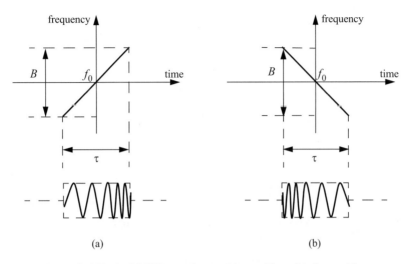

Figure 3.6. Typical LFM waveforms. (a) up-chirp; (b) down-chirp.

$$\psi(t) = 2\pi\left(f_0 t - \frac{\mu}{2}t^2\right) \qquad -\frac{\tau}{2} \leq t \leq \frac{\tau}{2} \qquad (3.31)$$

$$f(t) = \frac{1}{2\pi}\frac{d}{dt}\psi(t) = f_0 - \mu t \qquad -\frac{\tau}{2} \leq t \leq \frac{\tau}{2} \qquad (3.32)$$

A typical LFM waveform can be expressed by

$$s_1(t) = Rect\left(\frac{t}{\tau}\right) e^{j2\pi\left(f_0 t + \frac{\mu}{2}t^2\right)} \qquad (3.33)$$

where $Rect(t/\tau)$ denotes a rectangular pulse of width τ. Eq. (3.33) is then written as

$$s_1(t) = e^{j2\pi f_0 t} s(t) \qquad (3.34)$$

where

$$s(t) = Rect\left(\frac{t}{\tau}\right) e^{j\pi\mu t^2} \qquad (3.35)$$

is the complex envelope of $s_1(t)$.

The spectrum of the signal $s_1(t)$ is determined from its complex envelope $s(t)$. The complex exponential term in Eq. (3.34) introduces a frequency shift about the center frequency f_o. Taking the FT of $s(t)$ yields

$$S(\omega) = \int_{-\infty}^{\infty} Rect\left(\frac{t}{\tau}\right) e^{j\pi\mu t^2} e^{-j\omega t} dt = \int_{-\frac{\tau}{2}}^{\frac{\tau}{2}} \exp\left(\frac{j2\pi\mu t^2}{2}\right) e^{-j\omega t} dt \qquad (3.36)$$

Let $\mu' = 2\pi\mu = 2\pi B/\tau$, and perform the change of variable

$$x = \sqrt{\frac{\mu'}{\pi}}\left(t - \frac{\omega}{\mu'}\right) \quad ; \quad dx = \sqrt{\frac{\mu'}{\pi}} \, dt \qquad (3.37)$$

Thus, Eq. (3.36) can be written as

$$S(\omega) = \sqrt{\frac{\pi}{\mu'}} \, e^{-j\omega^2/2\mu'} \int_{-x_1}^{x_2} e^{j\pi x^2/2} dx \qquad (3.38)$$

$$S(\omega) = \sqrt{\frac{\pi}{\mu'}} \, e^{-j\omega^2/2\mu'} \left\{ \int_{0}^{x_2} e^{j\pi x^2/2} dx - \int_{0}^{-x_1} e^{j\pi x^2/2} dx \right\} \qquad (3.39)$$

where

$$x_1 = \sqrt{\frac{\mu'}{\pi}}\left(\frac{\tau}{2} + \frac{\omega}{\mu'}\right) = \sqrt{\frac{B\tau}{2}}\left(1 + \frac{f}{B/2}\right) \qquad (3.40)$$

$$x_2 = \sqrt{\frac{\mu'}{\pi}}\left(\frac{\tau}{2} - \frac{\omega}{\mu'}\right) = \sqrt{\frac{B\tau}{2}}\left(1 - \frac{f}{B/2}\right) \qquad (3.41)$$

The Fresnel integrals, denoted by $C(x)$ and $S(x)$, are defined by

$$C(x) = \int_{0}^{x} \cos\left(\frac{\pi\upsilon^2}{2}\right) d\upsilon \qquad (3.42)$$

$$S(x) = \int_{0}^{x} \sin\left(\frac{\pi\upsilon^2}{2}\right) d\upsilon \qquad (3.43)$$

Fresnel integrals are approximated by

$$C(x) \approx \frac{1}{2} + \frac{1}{\pi x}\sin\left(\frac{\pi}{2}x^2\right) \quad ; x \gg 1 \quad (3.44)$$

$$S(x) \approx \frac{1}{2} - \frac{1}{\pi x}\cos\left(\frac{\pi}{2}x^2\right) \quad ; x \gg 1 \quad (3.45)$$

Note that $C(-x) = -C(x)$ and $S(-x) = -S(x)$. Fig. 3.7 shows a plot of both $C(x)$ and $S(x)$ for $0 \le x \le 4.0$. This figure can be reproduced using MATLAB program "fig3_7.m" given in Listing 3.1 in Section 3.12.

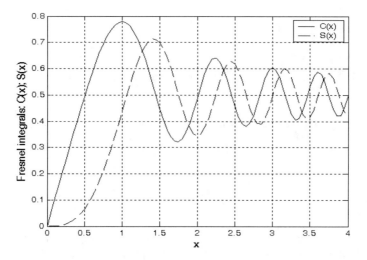

Figure 3.7. Fresnel integrals.

Using Eqs. (3.42) and (3.43) into (3.39) and performing the integration yield

$$S(\omega) = \tau\sqrt{\frac{1}{B\tau}}\, e^{-j\omega^2/(4\pi B)}\left\{\frac{[C(x_2) + C(x_1)] + j[S(x_2) + S(x_1)]}{\sqrt{2}}\right\} \quad (3.46)$$

Fig. 3.8 shows typical plots for the LFM real part, imaginary part, and amplitude spectrum. The square-like spectrum shown in Fig. 3.8c is widely known as the Fresnel spectrum. This figure can be reproduced using MATLAB program "fig3_8.m", given in Listing 3.2 in Section 3.12.

A MATLAB GUI (see Fig. 3.8d) was developed to input LFM data and display outputs as shown in Fig. 3.8. It is called "LFM_gui.m". Its inputs are the uncompressed pulsewidth and the chirp bandwidth.

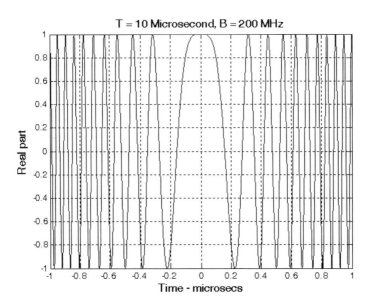

Figure 3.8a. Typical LFM waveform, real part.

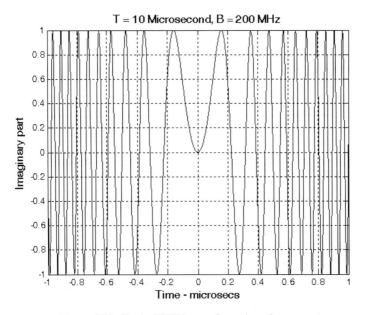

Figure 3.8b. Typical LFM waveform, imaginary part.

High Range Resolution 153

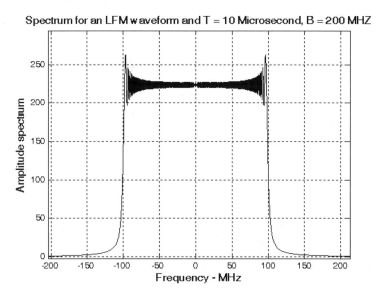

Figure 3.8c. Typical spectrum for an LFM waveform.

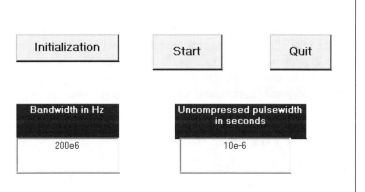

Figure 3.8d. GUI workspace *"LFM_gui.m"*.

3.5. High Range Resolution

An expression for range resolution ΔR in terms of the pulsewidth τ was derived in Chapter 1. When pulse compression is not used, the instantaneous bandwidth B of radar receiver is normally matched to the pulse bandwidth, and in most radar applications this is done by setting $B = 1/\tau$. Therefore, range resolution is given by

$$\Delta R = (c\tau)/2 = c/(2B) \tag{3.47}$$

Radar users and designers alike seek to accomplish High Range Resolution (HRR) by minimizing ΔR. However, as suggested by Eq. (3.47) in order to achieve HRR one must use very short pulses and consequently reduce the average transmitted power and impose severe operating bandwidth requirements. Achieving fine range resolution while maintaining adequate average transmitted power can be accomplished by using pulse compression techniques, which will be discussed in Chapter 5. By means of frequency or phase modulation, pulse compression allows us to achieve the average transmitted power of a relatively long pulse, while obtaining the range resolution corresponding to a very short pulse. As an example, consider an LFM waveform whose bandwidth is B and un-compressed pulsewidth (transmitted) is τ. After pulse compression the compressed pulsewidth is denoted by τ', where $\tau' \ll \tau$, and the HRR is

$$\Delta R = \frac{c\tau'}{2} \ll \frac{c\tau}{2} \tag{3.48}$$

Linear frequency modulation and Frequency-Modulated (FM) CW waveforms are commonly used to achieve HRR. High range resolution can also be synthesized using a class of waveforms known as the "Stepped Frequency Waveforms" (SFW). Stepped frequency waveforms require more complex hardware implementation as compared to LFM or FM-CW; however, the radar operating bandwidth requirements are less restrictive. This is true because the receiver instantaneous bandwidth is matched to the SFW sub-pulse bandwidth which is much smaller than the LFM or FM-CW bandwidth. A brief discussion of SFW waveforms is presented in the following section.

3.6. Stepped Frequency Waveforms

Stepped Frequency Waveforms (SFW) produce Synthetic HRR target profiles because the target range profile is computed by means of Inverse Discrete Fourier Transformation (IDFT) of frequency domain samples of the actual target range profile. The process of generating a synthetic HRR profile is described in Wehner.[1] It is summarized as follows:

1. A series of n narrow-band pulses are transmitted. The frequency from pulse to pulse is stepped by a fixed frequency step Δf. Each group of n pulses is referred to as a burst.

---

1. Wehner, D. R., *High Resolution Radar*, second edition, Artech House, 1993.

2. The received signal is sampled at a rate that coincides with the center of each pulse.
3. The quadrature components for each burst are collected and stored.
4. Spectral weighting (to reduce the range sidelobe levels) is applied to the quadrature components. Corrections for target velocity, phase, and amplitude variations are applied.
5. The IDFT of the weighted quadrature components of each burst is calculated to synthesize a range profile for that burst. The process is repeated for N bursts to obtain consecutive synthetic HRR profiles.

Fig. 3.9 shows a typical SFW burst. The Pulse Repetition Interval (PRI) is T, and the pulsewidth is τ'. Each pulse can have its own LFM, or other type of modulation; in this book LFM is assumed. The center frequency for the ith step is

$$f_i = f_0 + i\Delta f \quad ; \; i = 0, n-1 \tag{3.49}$$

Within a burst, the transmitted waveform for the ith step can be described as

$$s_i(t) = \begin{pmatrix} C_i \cos 2\pi f_i t + \theta_i & ; & iT \le t \le iT + \tau' \\ 0 & & elsewhere \end{pmatrix} \tag{3.50}$$

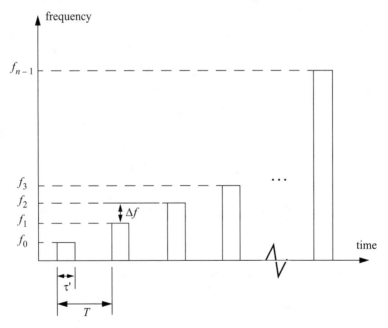

Figure 3.9. Stepped frequency waveform burst.

where θ_i are the relative phases and C_i are constants. The received signal from a target located at range R_0 at time $t = 0$ is then given by

$$s_{ri}(t) = C_i' \cos(2\pi f_i(t - \tau(t)) + \theta_i) \quad ; \quad iT + \tau(t) \leq t \leq iT + \tau' + \tau(t) \quad (3.51)$$

where C_i' are constant and the round trip delay $\tau(t)$ is given by

$$\tau(t) = \frac{R_0 - vt}{c/2} \quad (3.52)$$

c is the speed of light and v is the target radial velocity.

The received signal is down-converted to base-band in order to extract the quadrature components. More precisely, $s_{ri}(t)$ is mixed with the signal

$$y_i(t) = C \cos(2\pi f_i t + \theta_i) \quad ; \quad iT \leq t \leq iT + \tau' \quad (3.53)$$

After low pass filtering, the quadrature components are given by

$$\begin{pmatrix} x_I(t) \\ x_Q(t) \end{pmatrix} = \begin{pmatrix} A_i \cos \psi_i(t) \\ A_i \sin \psi_i(t) \end{pmatrix} \quad (3.54)$$

where A_i are constants, and

$$\psi_i(t) = -2\pi f_i \left(\frac{2R_0}{c} - \frac{2vt}{c} \right) \quad (3.55)$$

where now $f_i = \Delta f$. For each pulse, the quadrature components are then sampled at

$$t_i = iT + \frac{\tau_r}{2} + \frac{2R_0}{c} \quad (3.56)$$

τ_r is the time delay associated with the range that corresponds to the start of the range profile.

The quadrature components can then be expressed in complex form as

$$X_i = A_i e^{j\psi_i} \quad (3.57)$$

Eq. (3.57) represents samples of the target reflectivity, due to a single burst, in the frequency domain. This information can then be transformed into a series of range delay reflectivity (i.e., range profile) values by using the IDFT. It follows that

$$H_l = \frac{1}{n}\sum_{i=0}^{n-1} X_i \exp\left(j\frac{2\pi l i}{n}\right) \quad ; \; 0 \le l \le n-1 \quad (3.58)$$

Substituting Eqs. (3.57) and (3.55) into (3.58) and collecting terms yield

$$H_l = \frac{1}{n}\sum_{i=0}^{n-1} A_i \exp\left\{j\left(\frac{2\pi l i}{n} - 2\pi f_i\left(\frac{2R_0}{c} - \frac{2vt_i}{c}\right)\right)\right\} \quad (3.59)$$

By normalizing with respect to n and by assuming that $A_i = 1$ and that the target is stationary (i.e., $v = 0$), then Eq. (3.59) can be written as

$$H_l = \sum_{i=0}^{n-1} \exp\left\{j\left(\frac{2\pi l i}{n} - 2\pi f_i \frac{2R_0}{c}\right)\right\} \quad (3.60)$$

Using $f_i = i\Delta f$ inside Eq. (3.60) yields

$$H_l = \sum_{i=0}^{n-1} \exp\left\{j\frac{2\pi i}{n}\left(-\frac{2nR_0\Delta f}{c} + l\right)\right\} \quad (3.61)$$

which can be simplified to

$$H_l = \frac{\sin \pi \chi}{\sin \frac{\pi \chi}{n}} \exp\left(j\frac{n-1}{2}\frac{2\pi \chi}{n}\right) \quad (3.62)$$

where

$$\chi = \frac{-2nR_0\Delta f}{c} + l \quad (3.63)$$

Finally, the synthesized range profile is

$$|H_l| = \left|\frac{\sin \pi \chi}{\sin \frac{\pi \chi}{n}}\right| \quad (3.64)$$

3.6.1. Range Resolution and Range Ambiguity in SFW

As usual, range resolution is determined from the overall system bandwidth. Assuming a SFW with n steps, and step size Δf, then the corresponding range resolution is equal to

$$\Delta R = \frac{c}{2n\Delta f} \qquad (3.65)$$

Range ambiguity associated with a SFW can be determined by examining the phase term that corresponds to a point scatterer located at range R_0. More precisely,

$$\psi_i(t) = 2\pi f_i \frac{2R_0}{c} \qquad (3.66)$$

It follows that

$$\frac{\Delta\psi}{\Delta f} = \frac{4\pi(f_{i+1}-f_i)R_0}{(f_{i+1}-f_i)\,c} = \frac{4\pi R_0}{c} \qquad (3.67)$$

or equivalently,

$$R_0 = \frac{\Delta\psi}{\Delta f}\frac{c}{4\pi} \qquad (3.68)$$

It is clear from Eq. (3.68) that range ambiguity exists for $\Delta\psi = \Delta\psi + 2n\pi$. Therefore,

$$R_0 = \frac{\Delta\psi + 2n\pi}{\Delta f}\frac{c}{4\pi} = R_0 + n\left(\frac{c}{2\Delta f}\right) \qquad (3.69)$$

and the unambiguous range window is

$$R_u = \frac{c}{2\Delta f} \qquad (3.70)$$

Hence, a range profile synthesized using a particular SFW represents the relative range reflectivity for all scatterers within the unambiguous range window, with respect to the absolute range that corresponds to the burst time delay. Additionally, if a specific target extent is larger than R_u, then all scatterers falling outside the unambiguous range window will fold over and appear in the synthesized profile. This fold-over problem is identical to the spectral fold-over that occurs when using a Fast Fourier Transform (FFT) to resolve certain signal frequency contents. For example, consider an FFT with frequency resolution $\Delta f = 50Hz$, and size $NFFT = 64$. In this case, this FFT can resolve frequency tones between $-1600Hz$ and $1600Hz$. When this FFT is used to resolve the frequency content of a sine-wave tone equal to $1800Hz$, fold-over occurs and a spectral line at the fourth FFT bin (i.e., $200Hz$) appears. Therefore, in order to avoid fold-over in the synthesized range profile, the frequency step Δf must be

$$\Delta f \le c/2E \qquad (3.71)$$

where E is the target extent in meters.

Additionally, the pulsewidth must also be large enough to contain the whole target extent. Thus,

$$\Delta f \le 1/\tau' \qquad (3.72)$$

and, in practice,

$$\Delta f \le 1/2\tau' \qquad (3.73)$$

This is necessary in order to reduce the amount of contamination of the synthesized range profile caused by the clutter surrounding the target under consideration.

MATLAB Function "hrr_profile.m"

The function *"hrr_profile.m"* computes and plots the synthetic HRR profile for a specific SFW. It is given in Listing 3.3 in Section 3.12. This function utilizes an Inverse Fast Fourier Transform (IFFT) of a size equal to twice the number of steps. Hamming window of the same size is also assumed. The syntax is as follows:

[hl] = hrr_profile (nscat, scat_range, scat_rcs, n, deltaf, prf, v, r0, winid)

where

Symbol	Description	Units	Status
nscat	number of scatterers that make up the target	none	input
scat_range	vector containing range to individual scatterers	meters	input
scat_rcs	vector containing RCS of individual scatterers	meter square	input
n	number of steps	none	input
deltaf	frequency step	Hz	input
prf	PRF of SFW	Hz	input
v	target velocity	meter/second	input
r0	profile starting range	meters	input
winid	number>0 for Hamming window number < 0 for no window	none	input
hl	range profile	dB	output

For example, assume that the range profile starts at $R_0 = 900m$ and that

nscat	tau	n	deltaf	prf	v
3	100μsec	64	10MHz	10KHz	0.0

In this case,

$$\Delta R = \frac{3 \times 10^8}{2 \times 64 \times 10 \times 10^6} = 0.235m$$

$$R_u = \frac{3 \times 10^8}{2 \times 10 \times 10^6} = 15m$$

Thus, scatterers that are more than 0.235 meters apart will appear as distinct peaks in the synthesized range profile. Assume two cases; in the first case, *[scat_range] = [908, 910, 912] meters,* and in the second case, *[scat_range] = [908, 910, 910.2] meters.* In both cases, let *[scat_rcs] = [100, 10, 1] meters squared.*

Fig. 3.10 shows the synthesized range profiles generated using the function *"hrr_profile.m"* and the first case when the Hamming window is not used. Fig. 3.11 is similar to Fig. 3.10, except in this case the Hamming window is used. Fig. 3.12 shows the synthesized range profile that corresponds to the second case (Hamming window is used). Note that all three scatterers were resolved in Figs. 3.10 and 3.11; however, the last two scatterers are not resolved in Fig. 3.12, since they are separated by less than ΔR.

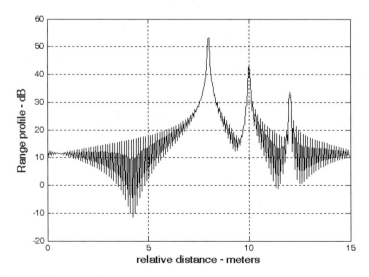

Figure 3.10. Synthetic range profile for three resolved scatterers. No window.

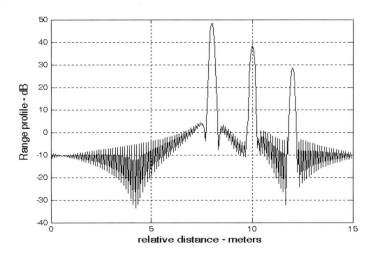

Figure 3.11. Synthetic range profile for three scatterers. Hamming window.

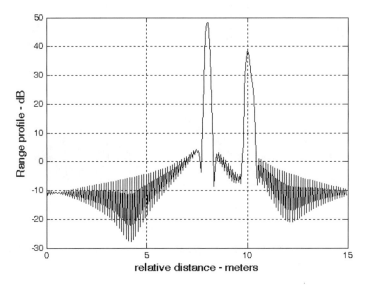

Figure 3.12. Synthetic range profile for three scatterers. Two are unresolved.

162   *MATLAB Simulations for Radar Systems Design*

Next, consider another case where *[scat_range] = [908, 912, 916] meters*. Fig. 3.13 shows the corresponding range profile. In this case, foldover occurs, and the last scatterer appears at the lower portion of the synthesized range profile. Also, consider the case where

$$[scat\_range] = [908, 910, 923] \text{ meters}$$

Fig. 3.14 shows the corresponding range profile. In this case, ambiguity is associated with the first and third scatterers since they are separated by $15m$. Both appear at the same range bin.

3.6.2. Effect of Target Velocity

The range profile defined in Eq. (3.64) is obtained by assuming that the target under examination is stationary. The effect of target velocity on the synthesized range profile can be determined by substituting Eqs. (3.55) and (3.56) into Eq. (3.58), which yields

$$H_l = \sum_{i=0}^{n-1} A_i \exp\left\{ j\frac{2\pi l i}{n} - j2\pi f_i \left[\frac{2R}{c} - \frac{2v}{c}\left(iT + \frac{\tau_1}{2} + \frac{2R}{c}\right) \right] \right\} \quad (3.74)$$

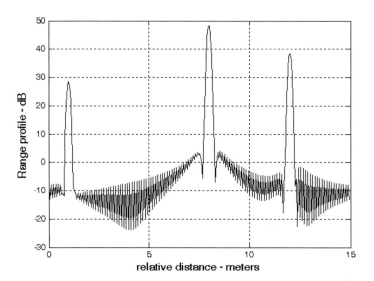

Figure 3.13. Synthetic range profile for three scatterers. Third scatterer folds over.

Stepped Frequency Waveforms

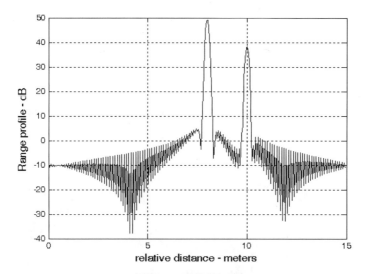

Figure 3.14. Synthetic range profile for three scatterers. The first and third scatterers appear in the same FFT bin.

The additional phase term present in Eq. (3.74) distorts the synthesized range profile. In order to illustrate this distortion, consider the SFW described in the previous section, and assume the three scatterers of the first case. Also, assume that $v = 100m/s$. Fig. 3.15 shows the synthesized range profile for this case. Comparisons of Figs. 3.11 and 3.15 clearly show the distortion effects caused by the uncompensated target velocity. Figure 3.16 is similar to Fig. 3.15 except in this case, $v = -100m/s$. Note in either case, the targets have moved from their expected positions (to the left or right) by $Disp = 2 \times n \times v/PRF$ (1.28 m).

This distortion can be eliminated by multiplying the complex received data at each pulse by the phase term

$$\Phi = \exp\left(-j2\pi f_i\left[\frac{2\underline{v}}{c}\left(iT + \frac{\tau_1}{2} + \frac{2\underline{R}}{c}\right)\right]\right) \tag{3.75}$$

$\underline{v}$ and $\underline{R}$ are, respectively, estimates of the target velocity and range. This process of modifying the phase of the quadrature components is often referred to as "phase rotation." In practice, when good estimates of $\underline{v}$ and $\underline{R}$ are not available, then the effects of target velocity are reduced by using frequency hopping between the consecutive pulses within the SFW. In this case, the frequency of each individual pulse is chosen according to a predetermined code. Waveforms of this type are often called Frequency Coded Waveforms (FCW). Costas waveforms or signals are a good example of this type of waveform.

164 *MATLAB Simulations for Radar Systems Design*

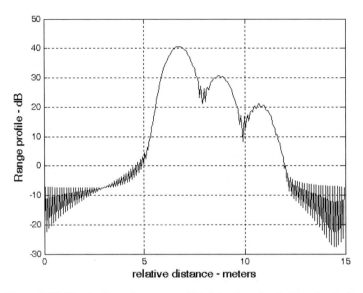

Figure 3.15. Illustration of range profile distortion due to target velocity.

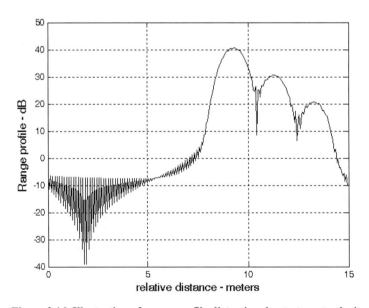

Figure 3.16. Illustration of range profile distortion due to target velocity.

Figure 3.17 shows a synthesized range profile for a moving target whose RCS is $\sigma = 10m^2$ and $v = 15m/s$. The initial target range is at $R = 912m$. All other parameters are as before. This figure can be reproduced using the MATLAB program *"fig3_17.m"* given in Listing 3.4 in Section 3.12.

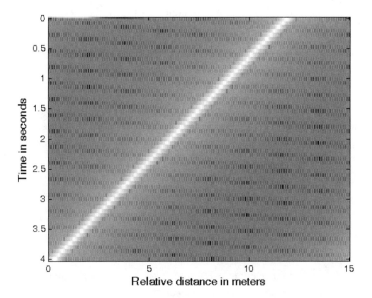

Figure 3.17. Synthesized range profile for a moving target (4 seconds long).

3.7. The Matched Filter

The most unique characteristic of the matched filter is that it produces the maximum achievable instantaneous SNR at its output when a signal plus additive white noise is present at the input. The noise does not need to be Gaussian. The peak instantaneous SNR at the receiver output can be achieved by matching the radar receiver transfer function to the received signal. We will show that the peak instantaneous signal power divided by the average noise power at the output of a matched filter is equal to twice the input signal energy divided by the input noise power, regardless of the waveform used by the radar. This is the reason why matched filters are often referred to as optimum filters in the SNR sense. Note that the peak power used in the derivation of the radar equation (SNR) represents the average signal power over the duration of the pulse, not the peak instantaneous signal power as in the case of a matched filter. In practice, it is sometimes difficult to achieve perfect matched filtering. In such cases, sub-optimum filters may be used. Due to this mismatch, degradation in the output SNR occurs.

Consider a radar system that uses a finite duration energy signal $s_i(t)$. Denote the pulsewidth as τ', and assume that a matched filter receiver is utilized. The main question that we need to answer is: What is the impulse, or frequency, response of the filter that maximizes the instantaneous SNR at the output of the receiver when a delayed version of the signal $s_i(t)$ plus additive white noise is at the input?

The matched filter input signal can then be represented by

$$x(t) = C\, s_i(t - t_1) + n_i(t) \tag{3.76}$$

where C is a constant, t_1 is an unknown time delay proportional to the target range, and $n_i(t)$ is input white noise. Since the input noise is white, its corresponding autocorrelation and Power Spectral Density (PSD) functions are given, respectively, by

$$\bar{R}_{n_i}(t) = \frac{N_0}{2}\delta(t) \tag{3.77}$$

$$\bar{S}_{n_i}(\omega) = \frac{N_0}{2} \tag{3.78}$$

where N_0 is a constant. Denote $s_o(t)$ and $n_o(t)$ as the signal and noise filter outputs, respectively. More precisely, we can define

$$y(t) = C\, s_o(t - t_1) + n_o(t) \tag{3.79}$$

where

$$s_o(t) = s_i(t) \bullet h(t) \tag{3.80}$$

$$n_o(t) = n_i(t) \bullet h(t) \tag{3.81}$$

The operator ($\bullet$) indicates convolution, and $h(t)$ is the filter impulse response (the filter is assumed to be linear time invariant).

Let $R_h(t)$ denote the filter autocorrelation function. It follows that the output noise autocorrelation and PSD functions are

$$\bar{R}_{n_o}(t) = \bar{R}_{n_i}(t) \bullet R_h(t) = \frac{N_0}{2}\delta(t) \bullet R_h(t) = \frac{N_0}{2} R_h(t) \tag{3.82}$$

$$\bar{S}_{n_o}(\omega) = \bar{S}_{n_i}(\omega)|H(\omega)|^2 = \frac{N_0}{2}|H(\omega)|^2 \tag{3.83}$$

The Matched Filter

where $H(\omega)$ is the Fourier transform for the filter impulse response, $h(t)$. The total average output noise power is equal to $\bar{R}_{n_o}(t)$ evaluated at $t = 0$. More precisely,

$$\bar{R}_{n_o}(0) = \frac{N_0}{2} \int_{-\infty}^{\infty} |h(u)|^2 du \tag{3.84}$$

The output signal power evaluated at time t is $|Cs_o(t-t_1)|^2$, and by using Eq. (3.80) we get

$$s_o(t-t_1) = \int_{-\infty}^{\infty} s_i(t-t_1-u) \, h(u) \, du \tag{3.85}$$

A general expression for the output SNR at time t can be written as

$$SNR(t) = \frac{|Cs_o(t-t_1)|^2}{\bar{R}_{n_o}(0)} \tag{3.86}$$

Substituting Eqs. (3.84) and (3.85) into Eq. (3.86) yields

$$SNR(t) = \frac{C^2 \left| \int_{-\infty}^{\infty} s_i(t-t_1-u) \, h(u) \, du \right|^2}{\frac{N_0}{2} \int_{-\infty}^{\infty} |h(u)|^2 du} \tag{3.87}$$

The Schwartz inequality states that

$$\left| \int_{-\infty}^{\infty} P(x)Q(x)dx \right|^2 \le \int_{-\infty}^{\infty} |P(x)|^2 dx \int_{-\infty}^{\infty} |Q(x)|^2 dx \tag{3.88}$$

where the equality applies only when $P = kQ^*$, where k is a constant and can be assumed to be unity. Then by applying Eq. (3.88) on the numerator of Eq. (3.87), we get

$$SNR(t) \le \frac{C^2 \int_{-\infty}^{\infty} |s_i(t-t_1-u)|^2 \, du \int_{-\infty}^{\infty} |h(u)|^2 \, du}{\frac{N_0}{2} \int_{-\infty}^{\infty} |h(u)|^2 du} = \frac{2C^2 \int_{-\infty}^{\infty} |s_i(t-t_1-u)|^2 \, du}{N_0} \tag{3.89}$$

Eq. (3.89) tells us that the peak instantaneous SNR occurs when equality is achieved (i.e., from Eq. (3.88) $h = ks_i^*$). More precisely, if we assume that equality occurs at $t = t_0$, and that $k = 1$, then

$$h(u) = s_i^*(t_0 - t_1 - u) \tag{3.90}$$

and the maximum instantaneous SNR is

$$SNR(t_0) = \frac{2C^2 \int_{-\infty}^{\infty} |s_i(t_0 - t_1 - u)|^2 \, du}{N_0} \tag{3.91}$$

Eq. (3.91) can be simplified using Parseval's theorem,

$$E = C^2 \int_{-\infty}^{\infty} |s_i(t_0 - t_1 - u)|^2 \, du \tag{3.92}$$

where E denotes the energy of the input signal; consequently we can write the output peak instantaneous SNR as

$$SNR(t_0) = \frac{2E}{N_0} \tag{3.93}$$

Thus, we can draw the conclusion that the peak instantaneous SNR depends only on the signal energy and input noise power, and is independent of the waveform utilized by the radar.

Finally, we can define the impulse response for the matched filter from Eq. (3.90). If we desire the peak to occur at $t_0 = t_1$, we get the non-causal matched filter impulse response,

$$h_{nc}(t) = s_i^*(-t) \tag{3.94}$$

Alternatively, the causal impulse response is

$$h_c(t) = s_i^*(\tau - t) \tag{3.95}$$

where, in this case, the peak occurs at $t_0 = t_1 + \tau$. It follows that the Fourier transforms of $h_{nc}(t)$ and $h_c(t)$ are given, respectively, by

$$H_{nc}(\omega) = S_i^*(\omega) \tag{3.96}$$

$$H_c(\omega) = S_i^*(\omega) e^{-j\omega\tau} \tag{3.97}$$

The Replica

where $S_i(\omega)$ is the Fourier transform of $s_i(t)$. Thus, the moduli of $H(\omega)$ and $S_i(\omega)$ are identical; however, the phase responses are opposite of each other.

Example:

Compute the maximum instantaneous SNR at the output of a linear filter whose impulse response is matched to the signal $x(t) = \exp(-t^2/2T)$.

Solution:

The signal energy is

$$E = \int_{-\infty}^{\infty} |x(t)|^2 dt = \int_{-\infty}^{\infty} e^{(-t^2)/T} dt = \sqrt{\pi T} \ Joules$$

It follows that the maximum instantaneous SNR is

$$SNR = \frac{\sqrt{\pi T}}{N_0/2}$$

where $N_0/2$ is the input noise power spectrum density.

3.8. The Replica

Again, consider a radar system that uses a finite duration energy signal $s_i(t)$, and assume that a matched filter receiver is utilized. The input signal is given in Eq. (3.76) and is repeated here as Eq. (3.98),

$$x(t) = C \ s_i(t - t_1) + n_i(t) \tag{3.98}$$

The matched filter output $y(t)$ can be expressed by the convolution integral between the filter's impulse response and $x(t)$,

$$y(t) = \int_{-\infty}^{\infty} x(u) h(t-u) du \tag{3.99}$$

Substituting Eq. (3.95) into Eq. (3.99) yields

$$y(t) = \int_{-\infty}^{\infty} x(u) s_i^*(\tau - t + u) du = \bar{R}_{xs_i}(t - \tau) \tag{3.100}$$

where $\bar{R}_{xs_i}(t - \tau)$ is a cross-correlation between $x(t)$ and $s_i(\tau - t)$. Therefore, the matched filter output can be computed from the cross-correlation between the radar received signal and a delayed replica of the transmitted waveform. If the input signal is the same as the transmitted signal, the output of the matched filter would be the autocorrelation function of the received (or transmitted) signal. In practice, replicas of the transmitted waveforms are normally computed and stored in memory for use by the radar signal processor when needed.

3.9. Matched Filter Response to LFM Waveforms

In order to develop a general expression for the matched filter output when an LFM waveform is utilized, we will consider the case when the radar is tracking a closing target with velocity v. The transmitted signal is

$$s_1(t) = Rect\left(\frac{t}{\tau}\right) e^{j2\pi\left(f_0 t + \frac{\mu}{2}t^2\right)} \tag{3.101}$$

The received signal is then given by

$$s_{r_1}(t) = s_1(t - \Delta(t)) \tag{3.102}$$

$$\Delta(t) = t_0 - \frac{2v}{c}(t - t_0) \tag{3.103}$$

where t_0 is the time corresponding to the target initial detection range, and c is the speed of light. Using Eq. (3.103) we can rewrite Eq. (3.102) as

$$s_{r_1}(t) = s_1\left(t - t_0 + \frac{2v}{c}(t - t_0)\right) = s_1(\gamma(t - t_0)) \tag{3.104}$$

and

$$\gamma = 1 + 2\frac{v}{c} \tag{3.105}$$

is the scaling coefficient. Substituting Eq. (3.101) into Eq. (3.104) yields

$$s_{r_1}(t) = Rect\left(\frac{\gamma(t - t_0)}{\tau'}\right) e^{j2\pi f_0 \gamma(t - t_0)} e^{j\pi\mu\gamma^2(t - t_0)^2} \tag{3.106}$$

which is the analytical signal representation for $s_{r_1}(t)$. The complex envelope of the signal $s_{r_1}(t)$ is obtained by multiplying Eq. (3.106) by $\exp(-j2\pi f_0 t)$. Denote the complex envelope by $s_r(t)$; then after some manipulation we get

$$s_r(t) = e^{-j2\pi f_0 t_0} Rect\left(\frac{\gamma(t - t_0)}{\tau'}\right) e^{j2\pi f_0(\gamma - 1)(t - t_0)} e^{j\pi\mu\gamma^2(t - t_0)^2} \tag{3.107}$$

The Doppler shift due to the target motion is

Matched Filter Response to LFM Waveforms

$$f_d = \frac{2v}{c} f_0 \tag{3.108}$$

and since $\gamma - 1 = 2v/c$, we get

$$f_d = (\gamma - 1) f_0 \tag{3.109}$$

Using the approximation $\gamma \approx 1$ and Eq. (3.109), Eq. (3.107) is rewritten as

$$s_r(t) \approx e^{j2\pi f_d(t-t_0)} s(t-t_0) \tag{3.110}$$

where

$$s(t-t_0) = e^{-j2\pi f_0 t} s_1(t-t_0) \tag{3.111}$$

$s_1(t)$ is given in Eq. (3.101). The matched filter response is given by the convolution integral

$$s_o(t) = \int_{-\infty}^{\infty} h(u) s_r(t-u) du \tag{3.112}$$

For a non-causal matched filter the impulse response $h(u)$ is equal to $s^*(-t)$; it follows that

$$s_o(t) = \int_{-\infty}^{\infty} s^*(-u) s_r(t-u) du \tag{3.113}$$

Substituting Eq. (3.111) into Eq. (3.113), and performing some algebraic manipulations, we get

$$s_o(t) = \int_{-\infty}^{\infty} s^*(u) e^{j2\pi f_d(t+u-t_0)} s(t+u-t_0) du \tag{3.114}$$

Finally, making the change of variable $t' = t + u$ yields

$$s_o(t) = \int_{-\infty}^{\infty} s^*(t'-t) s(t'-t_0) e^{j2\pi f_d(t'-t_0)} dt' \tag{3.115}$$

It is customary to set $t_0 = 0$. It follows that

$$s_o(t; f_d) = \int_{-\infty}^{\infty} s(t') s^*(t'-t) e^{j2\pi f_d t'} dt' \tag{3.116}$$

where we used the notation $s_o(t;f_d)$ to indicate that the output is a function of both time and Doppler frequency.

3.10. Waveform Resolution and Ambiguity

As indicated by Eq. (3.93) the radar sensitivity (in the case of white additive noise) depends only on the total energy of the received signal and is independent of the shape of the specific waveform. This leads us to ask the following question: If the radar sensitivity is independent of the waveform, then what is the best choice for a transmitted waveform? The answer depends on many factors; however, the most important consideration lies in the waveform's range and Doppler resolution characteristics.

As discussed in Chapter 1, range resolution implies separation between distinct targets in range. Alternatively, Doppler resolution implies separation between distinct targets in frequency. Thus, ambiguity and accuracy of this separation are closely associated terms.

3.10.1. Range Resolution

Consider radar returns from two stationary targets (zero Doppler) separated in range by distance ΔR. What is the smallest value of ΔR so that the returned signal is interpreted by the radar as two distinct targets? In order to answer this question, assume that the radar transmitted pulse is denoted by $s(t)$,

$$s(t) = A(t)\cos(2\pi f_0 t + \phi(t)) \tag{3.117}$$

where f_0 is the carrier frequency, $A(t)$ is the amplitude modulation, and $\phi(t)$ is the phase modulation. The signal $s(t)$ can then be expressed as the real part of the complex signal $\psi(t)$, where

$$\psi(t) = A(t)e^{j(\omega_0 t - \phi(t))} = u(t)e^{j\omega_0 t} \tag{3.118}$$

and

$$u(t) = A(t)e^{-j\phi(t)} \tag{3.119}$$

It follows that

$$s(t) = Re\{\psi(t)\} \tag{3.120}$$

The returns from both targets are respectively given by

$$s_{r1}(t) = \psi(t - \tau_0) \tag{3.121}$$

$$s_{r2}(t) = \psi(t - \tau_0 - \tau) \tag{3.122}$$

Waveform Resolution and Ambiguity

where τ is the difference in delay between the two returns. One can assume that the reference time is τ_0, and thus without any loss of generality one may set $\tau_0 = 0$. It follows that the two targets are distinguishable by how large or small the delay τ can be.

In order to measure the difference in range between the two targets consider the integral square error between $\psi(t)$ and $\psi(t-\tau)$. Denoting this error as ε_R^2, it follows that

$$\varepsilon_R^2 = \int_{-\infty}^{\infty} |\psi(t) - \psi(t-\tau)|^2 \, dt \tag{3.123}$$

Eq. (3.123) can be written as

$$\varepsilon_R^2 = \int_{-\infty}^{\infty} |\psi(t)|^2 \, dt + \int_{-\infty}^{\infty} |\psi(t-\tau)|^2 \, dt - \tag{3.124}$$

$$\int_{-\infty}^{\infty} \{(\psi(t)\psi^*(t-\tau) + \psi^*(t)\psi(t-\tau)) \, dt\}$$

Using Eq. (3.118) into Eq. (3.124) yields

$$\varepsilon_R^2 = 2\int_{-\infty}^{\infty} |u(t)|^2 \, dt - 2\mathrm{Re}\left\{\int_{-\infty}^{\infty} \psi^*(t)\psi(t-\tau) \, dt\right\} = \tag{3.125}$$

$$2\int_{-\infty}^{\infty} |u(t)|^2 \, dt - 2\mathrm{Re}\left\{e^{-j\omega_0\tau}\int_{-\infty}^{\infty} u^*(t)u(t-\tau) \, dt\right\}$$

The first term in the right hand side of Eq. (3.125) represents the signal energy, and is assumed to be constant. The second term is a varying function of τ with its fluctuation tied to the carrier frequency. The integral inside the right-most side of this equation is defined as the "range ambiguity function,"

$$\chi_R(\tau) = \int_{-\infty}^{\infty} u^*(t)u(t-\tau) \, dt \tag{3.126}$$

The maximum value of $\chi_R(\tau)$ is at $\tau = 0$. Target resolvability in range is measured by the squared magnitude $|\chi_R(\tau)|^2$. It follows that if $|\chi_R(\tau)| = \chi_R(0)$ for some nonzero value of τ, then the two targets are indistinguishable. Alternatively, if $|\chi_R(\tau)| \neq \chi_R(0)$ for some nonzero value of τ, then the two targets may be distinguishable (resolvable). As a consequence, the most desirable shape for $\chi_R(\tau)$ is a very sharp peak (thumb tack shape) centered at $\tau = 0$ and falling very quickly away from the peak.

The time delay resolution is

$$\Delta \tau = \frac{\int_{-\infty}^{\infty} |\chi_R(\tau)|^2 \, d\tau}{\chi_R^2(0)} \tag{3.127}$$

Using Parseval's theorem, Eq. (3.127) can be written as

$$\Delta \tau = 2\pi \frac{\int_{-\infty}^{\infty} |U(\omega)|^4 \, d\omega}{\left[\int_{-\infty}^{\infty} |U(\omega)|^2 \, d\omega\right]^2} \tag{3.128}$$

The minimum range resolution corresponding to $\Delta \tau$ is

$$\Delta R = c\Delta\tau/2 \tag{3.129}$$

However, since the signal effective bandwidth is

$$B = \frac{\left[\int_{-\infty}^{\infty} |U(\omega)|^2 \, d\omega\right]^2}{2\pi \int_{-\infty}^{\infty} |U(\omega)|^4 \, d\omega} \tag{3.130}$$

the range resolution is expressed as a function of the waveform's bandwidth as

$$\Delta R = c/(2B) \tag{3.131}$$

The comparison between Eqs. (3.116) and (3.126) indicates that the output of the matched filter and the range ambiguity function have the same envelope (in this case the Doppler shift f_d is set to zero). This indicates that the matched filter, in addition to providing the maximum instantaneous SNR at its output, also preserves the signal range resolution properties.

3.10.2. Doppler Resolution

It was shown in Chapter 1 that the Doppler shift corresponding to the target radial velocity is

$$f_d = \frac{2v}{\lambda} = \frac{2vf_0}{c} \tag{3.132}$$

where v is the target radial velocity, λ is the wavelength, f_0 is the frequency, and c is the speed of light.

Let

$$\Psi(f) = \int_{-\infty}^{\infty} \psi(t) e^{-j2\pi ft} \, dt \qquad (3.133)$$

Due to the Doppler shift associated with the target, the received signal spectrum will be shifted by f_d. In other words the received spectrum can be represented by $\Psi(f-f_d)$. In order to distinguish between the two targets located at the same range but having different velocities, one may use the integral square error. More precisely,

$$\varepsilon_f^2 = \int_{-\infty}^{\infty} |\Psi(f) - \Psi(f-f_d)|^2 \, df \qquad (3.134)$$

Using similar analysis as that which led to Eq. (3.125), one should minimize

$$2Re\left\{ \int_{-\infty}^{\infty} \Psi^*(f)\Psi(f-f_d) \, df \right\} \qquad (3.135)$$

By using the analytic signal in Eq. (3.118) it can be shown that

$$\Psi(f) = U(2\pi f - 2\pi f_0) \qquad (3.136)$$

Thus, Eq. (3.135) becomes

$$\int_{-\infty}^{\infty} U^*(2\pi f) U(2\pi f - 2\pi f_d) \, df = \int_{-\infty}^{\infty} U^*(2\pi f - 2\pi f_0) U(2\pi f - 2\pi f_0 - 2\pi f_d) \, df \qquad (3.137)$$

The complex frequency correlation function is then defined as

$$\chi_f(f_d) = \int_{-\infty}^{\infty} U^*(2\pi f) U(2\pi f - 2\pi f_d) \, df = \int_{-\infty}^{\infty} |u(t)|^2 e^{j2\pi f_d t} \, dt \qquad (3.138)$$

and the Doppler resolution constant Δf_d is

$$\Delta f_d = \frac{\int_{-\infty}^{\infty} |\chi_f(f_d)|^2 \, df_d}{\chi_f^2(0)} = \frac{\int_{-\infty}^{\infty} |u(t)|^4 \, dt}{\left[\int_{-\infty}^{\infty} |u(t)|^2 \, dt\right]^2} = \frac{1}{\tau'} \qquad (3.139)$$

where τ' is pulsewidth.

Finally, one can define the corresponding velocity resolution as

$$\Delta v = \frac{c \Delta f_d}{2 f_0} = \frac{c}{2 f_0 \tau'} \qquad (3.140)$$

Again observation of Eqs. (3.138) and (3.116) indicate that the output of the matched filter and the ambiguity function (when $\tau = 0$) are similar to each other. Consequently, one concludes that the matched filter preserves the waveform Doppler resolution properties as well.

3.10.3. Combined Range and Doppler Resolution

In this general case, one needs to use a two-dimensional function in the pair of variables (τ, f_d). For this purpose, assume that the complex envelope of the transmitted waveform is

$$\psi(t) = u(t) e^{j 2 \pi f_0 t} \qquad (3.141)$$

Then the delayed and Doppler-shifted signal is

$$\psi'(t - \tau) = u(t - \tau) e^{j 2 \pi (f_0 - f_d)(t - \tau)} \qquad (3.142)$$

Computing the integral square error between Eqs. (3.142) and (3.141) yields

$$\varepsilon^2 = \int_{-\infty}^{\infty} |\psi(t) - \psi'(t - \tau)|^2 \, dt = 2 \int_{-\infty}^{\infty} |\psi(t)|^2 dt - 2 \operatorname{Re}\left\{ \int_{-\infty}^{\infty} \psi^*(t) - \psi'(t - \tau) dt \right\} \qquad (3.143)$$

which can be written as

$$\varepsilon^2 = 2 \int_{-\infty}^{\infty} |u(t)|^2 \, dt - 2 \operatorname{Re}\left\{ e^{j 2 \pi (f_0 - f_d) \tau} \int_{-\infty}^{\infty} u(t) u^*(t - \tau) e^{j 2 \pi f_d t} dt \right\} \qquad (3.144)$$

Again, in order to maximize this squared error for $\tau \neq 0$ one must minimize the last term of Eq. (3.144).

Define the combined range and Doppler correlation function as

$$\chi(\tau, f_d) = \int_{-\infty}^{\infty} u(t) u^*(t - \tau) e^{j 2 \pi f_d t} dt \qquad (3.145)$$

In order to achieve the most range and Doppler resolution, the modulus square of this function must be minimized for $\tau \neq 0$ and $f_d \neq 0$. Note that the output of the matched filter in Eq. (3.116) is identical to that given in Eq. (3.145). This means that the output of the matched filter exhibits maximum instantaneous SNR as well as the most achievable range and Doppler resolutions.

3.11. "MyRadar" Design Case Study - Visit 3

3.11.1. Problem Statement

Assuming a matched filter receiver, select a set of waveforms that can meet the design requirements as stated in the previous two chapters. Assume linear frequency modulation. Do not use more than a total of 5 waveforms. Modify the design so that the range resolution $\Delta R = 30m$ during the search mode, and $\Delta R = 7.5m$ during tracking.

3.11.2. A Design

The major characteristics of radar waveforms include the waveform's energy, range resolution, and Doppler (or velocity) resolution. The pulse (waveform) energy is

$$E = P_t \tau \quad (3.146)$$

where P_t is the peak transmitted power and τ is the pulsewidth. Range resolution is defined in Eq. (3.131), while the velocity resolution is in Eq. (3.140).

Close attention should be paid to the selection process of the pulsewidth. In this design we will assume that the pulse energy is the same as that computed in Chapter 2. The radar operating bandwidth during search and track are calculated from Eq. (3.131) as

$$\begin{Bmatrix} B_{search} \\ B_{track} \end{Bmatrix} = \begin{Bmatrix} 3 \times 10^8/(2 \times 30) = 5 \ MHz \\ 3 \times 10^8/(2 \times 7.5) = 20 \ MHz \end{Bmatrix} \quad (3.147)$$

Since the design calls for a pulsed radar, then for each pulse transmitted (one PRI) the radar should not be allowed to receive any signal until that pulse has been completely transmitted. This limits the radar to a minimum operating range defined by

$$R_{min} = \frac{c\tau}{2} \quad (3.148)$$

In this design choose $R_{min} \geq 15 Km$. It follows that the minimum acceptable pulsewidth is $\tau_{max} \leq 100 \mu s$.

For this design select 5 waveforms, one for search and four for track. Typically search waveforms are longer than track waveforms; alternatively, tracking waveforms require wider bandwidths than search waveforms. However, in the context of range, more energy is required at longer ranges (for both track and search waveforms), since one would expect the SNR to get larger as range becomes smaller. This was depicted in the example shown in Fig. 1.13 in Chapter 1.

Assume that during search and initial detection the single pulse peak power is to be kept under 10 KW (i.e., $P_t \leq 20KW$). Then by using the single pulse energy calculated using Eq. (2.115) in Chapter 2, one can compute the minimum required pulsewidth as

$$\tau_{min} \geq \frac{0.1147}{20 \times 10^3} = 5.735 \mu s \qquad (3.149)$$

Choose $\tau_{search} = 20 \mu s$, with bandwidth $B = 5 MHz$ and use LFM modulation. Fig. 3.18 shows plots of the real part, imaginary part, and the spectrum of this search waveform. This figure was produced using the GUI workspace "LFM_gui.m". As far as the track waveforms, choose four waveforms of the same bandwidth ($B_{track} = 20 MHz$) and with the following pulsewidths.

TABLE 3.1. *"MyRadar"* **design case study track waveforms.**

Pulsewidth	Range window
$\tau_{t1} = 20 \mu s$	$R_{max} \rightarrow 0.75 R_{max}$
$\tau_{t2} = 17.5 \mu s$	$0.75 R_{max} \rightarrow 0.5 R_{max}$
$\tau_{t3} = 15 \mu s$	$0.5 R_{max} \rightarrow 0.25 R_{max}$
$\tau_{t4} = 12.5 \mu s$	$R \leq 0.25 R_{max}$

Note that R_{max} refers to the initial range at which track has been initiated. Fig. 3.19 is similar to Fig. 3.18 except it is for τ_{t3}.

For the waveform set selected in this design option, the radar duty cycle varies from 1.25% to 2.0%. Remember that the PRF was calculated in Chapter 1 as $f_r = 1 KHz$; thus the PRI is $T = 1 ms$.

At this point of the design, one must verify that the selected waveforms provide the radar with the desired SNR that meets or exceeds what was calculated in Chapter 2, and plotted in Fig. 2.21. In other words, one must now re-run these calculations and verify that the SNR has not been degraded. This task will be postponed until Chapter 5, where the radar equation with pulse compression is developed.

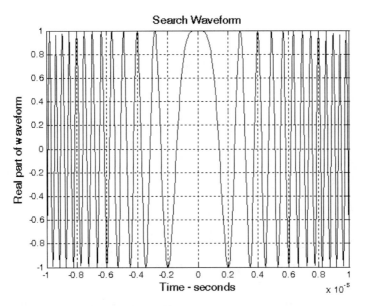

Figure 3.18a. Real part of search waveform.

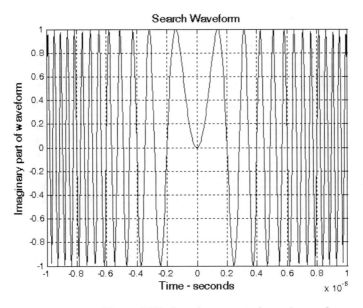

Figure 3.18b. Imaginary part of search waveform.

180   *MATLAB Simulations for Radar Systems Design*

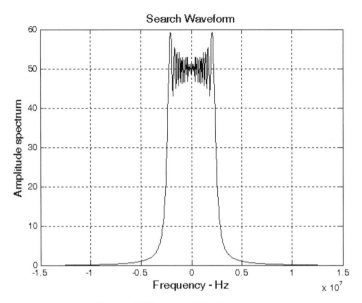

Figure 3.18c. Amplitude spectrum.

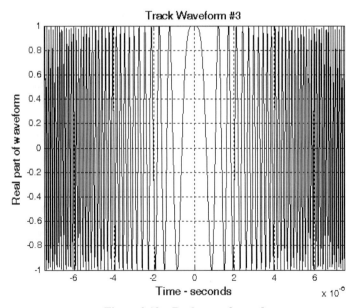

Figure 3.19a. Real part of waveform.

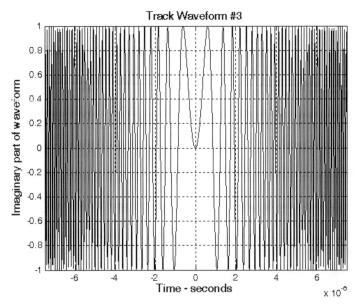

Figure 3.19b. Imaginary part of waveform.

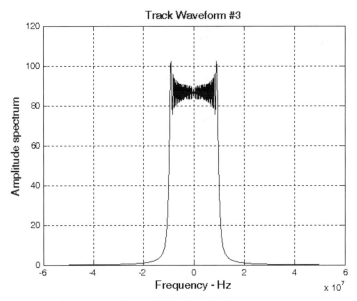

Figure 3.19c. Amplitude spectrum.

3.12. MATLAB Program and Function Listings

This section presents listings for all MATLAB programs/functions used in this chapter.

Listing 3.1. MATLAB Program "fig3_7.m"

```
% Use this program to reproduce Fig 3.7 from text
clear all
close all
n = 0;
for x = 0:.05:4
  n = n+1;
  sx(n) = quadl('fresnels',.0,x);
  cx(n) = quadl('fresnelc',.0,x);
end
plot(cx)
x=0:.05:4;
plot(x,cx,'k',x,sx,'k--')
grid
xlabel ('x')
ylabel ('Fresnel integrals: C(x); S(x)')
legend('C(x)','S(x)')
```

Listing 3.2. MATLAB Program "fig3_8.m"

```
% Use this program to reproduce Fig. 3.8 of text
close all
clear all
eps = 0.000001;
%Enter pulsewidth and bandwidth
B = 200.0e6; %200 MHZ bandwidth
T = 10.e-6; %10 micro second pulse;
% Compute alpha
mu = 2. * pi * B / T;
% Determine sampling times
delt = linspace(-T/2., T/2., 10001); % 1 nano second sampling interval
% Compute the complex LFM representation
Ichannal = cos(mu .* delt.^2 / 2.); % Real part
Qchannal = sin(mu .* delt.^2 / 2.); % Imaginary Part
LFM = Ichannal + sqrt(-1) .* Qchannal; % complex signal
%Compute the FFT of the LFM waveform
LFMFFT = fftshift(fft(LFM));
% Plot the real and Imaginary parts and the spectrum
```

```
freqlimit = 0.5 / 1.e-9;% the sampling interval 1 nano-second
freq = linspace(-freqlimit/1.e6,freqlimit/1.e6,10001);
figure(1)
plot(delt*1e6,Ichannal,'k');
axis([-1 1 -1 1])
grid
xlabel('Time - microsecs')
ylabel('Real part')
title('T = 10 Microsecond, B = 200 MHz')
figure(2)
plot(delt*1e6,Qchannal,'k');
axis([-1 1 -1 1])
grid
xlabel('Time - microsecs')
ylabel('Imaginary part')
title('T = 10 Microsecond, B = 200 MHz')
figure(3)
plot(freq, abs(LFMFFT),'k');
%axis tight
grid
xlabel('Frequency - MHz')
ylabel('Amplitude spectrum')
title('Spectrum for an LFM waveform and T = 10 Microsecond, ...
B = 200 MHZ')
```

Listing 3.3. MATLAB Function "hrr_profile.m"

```
function [hl] = hrr_profile (nscat, scat_range, scat_rcs, n, deltaf, prf, v,
                rnote,winid)
% Range or Time domain Profile
% Range_Profile returns the Range or Time domain plot of a simulated
% HRR SFWF returning from a predetermined number of targets with a prede-
                termined
% RCS for each target.
c=3.0e8;  % speed of light (m/s)
num_pulses  = n;
SNR_dB = 40;
nfft = 256;
%carrier_freq = 9.5e9; %Hz (10GHz)
freq_step   = deltaf; %Hz (10MHz)
V = v; % radial velocity (m/s)  -- (+)=towards radar (-)=away
PRI = 1. / prf; % (s)
if (nfft > 2*num_pulses)
   num_pulses = nfft/2;
```

```
end
Inphase = zeros((2*num_pulses),1);
Quadrature = zeros((2*num_pulses),1);
Inphase_tgt   = zeros(num_pulses,1);
Quadrature_tgt = zeros(num_pulses,1);
IQ_freq_domain = zeros((2*num_pulses),1);
Weighted_I_freq_domain = zeros((num_pulses),1);
Weighted_Q_freq_domain = zeros((num_pulses),1);
Weighted_IQ_time_domain = zeros((2*num_pulses),1);
Weighted_IQ_freq_domain = zeros((2*num_pulses),1);
abs_Weighted_IQ_time_domain = zeros((2*num_pulses),1);
dB_abs_Weighted_IQ_time_domain = zeros((2*num_pulses),1);
taur = 2. * rnote / c;
for jscat = 1:nscat
   ii = 0;
  for i = 1:num_pulses
    ii = ii+1;
    rec_freq = ((i-1)*freq_step);
    Inphase_tgt(ii) = Inphase_tgt(ii) + sqrt(scat_rcs(jscat)) * cos(-
                2*pi*rec_freq*...
      (2.*scat_range(jscat)/c - 2*(V/c)*((i-1)*PRI + taur/2 +
                2*scat_range(jscat)/c)));
    Quadrature_tgt(ii) = Quadrature_tgt(ii) + sqrt(scat_rcs(jscat))*sin(-
                2*pi*rec_freq*...
      (2*scat_range(jscat)/c - 2*(V/c)*((i-1)*PRI + taur/2 +
                2*scat_range(jscat)/c)));
  end
end
if(winid >= 0)
   window(1:num_pulses) = hamming(num_pulses);
else
   window(1:num_pulses) = 1;
end
Inphase = Inphase_tgt;
Quadrature = Quadrature_tgt;
Weighted_I_freq_domain(1:num_pulses) = Inphase(1:num_pulses).* window';
Weighted_Q_freq_domain(1:num_pulses) = Quadrature(1:num_pulses).* win-
                dow';
Weighted_IQ_freq_domain(1:num_pulses)= Weighted_I_freq_domain + ...
   Weighted_Q_freq_domain*j;
Weighted_IQ_freq_domain(num_pulses:2*num_pulses)=0.+0.i;
Weighted_IQ_time_domain = (ifft(Weighted_IQ_freq_domain));
abs_Weighted_IQ_time_domain = (abs(Weighted_IQ_time_domain));
```

```
dB_abs_Weighted_IQ_time_domain = 
                20.0*log10(abs_Weighted_IQ_time_domain)+SNR_dB;
% calculate the unambiguous range window size
Ru = c /2/deltaf;
hl = dB_abs_Weighted_IQ_time_domain;
numb = 2*num_pulses;
delx_meter = Ru / numb;
xmeter = 0:delx_meter:Ru-delx_meter;
plot(xmeter, dB_abs_Weighted_IQ_time_domain,'k')
xlabel ('relative distance - meters')
ylabel ('Range profile - dB')
grid
```

Listing 3.4. MATLAB Program "fig3_17.m"

```
% use this program to reproduce Fig. 3.17 of text
clear all
close all
nscat = 1;
scat_range = 912;
scat_rcs = 10;
n =64;
deltaf = 10e6;
prf = 10e3;
v = 15;
rnote = 900,
winid = 1;
count = 0;
for time = 0:.05:3
   count = count +1;
   hl = hrr_profile (nscat, scat_range, scat_rcs, n, deltaf, prf, v, rnote,winid);
   array(count,:) = transpose(hl);
   hl(1:end) = 0;
   scat_range =  scat_range - 2 * n * v / prf;
end
figure (1)
 numb = 2*256;% this number matches that used in hrr_profile.
 delx_meter = 15 / numb;
 xmeter = 0:delx_meter:15-delx_meter;
 imagesc(xmeter, 0:0.05:4,array)
 colormap(gray)
 ylabel ('Time in seconds')
 xlabel('Relative distance in meters')
```

Chapter 4

The Radar Ambiguity Function

4.1. Introduction

The radar ambiguity function represents the output of the matched filter, and it describes the interference caused by the range and/or Doppler shift of a target when compared to a reference target of equal RCS. The ambiguity function evaluated at $(\tau, f_d) = (0, 0)$ is equal to the matched filter output that is matched perfectly to the signal reflected from the target of interest. In other words, returns from the nominal target are located at the origin of the ambiguity function. Thus, the ambiguity function at nonzero τ and f_d represents returns from some range and Doppler different from those for the nominal target.

The radar ambiguity function is normally used by radar designers as a means of studying different waveforms. It can provide insight about how different radar waveforms may be suitable for the various radar applications. It is also used to determine the range and Doppler resolutions for a specific radar waveform. The three-dimensional (3-D) plot of the ambiguity function versus frequency and time delay is called the radar ambiguity diagram. The radar ambiguity function for the signal $s(t)$ is defined as the modulus squared of its 2-D correlation function, i.e., $|\chi(\tau;f_d)|^2$. More precisely,

$$|\chi(\tau;f_d)|^2 = \left| \int_{-\infty}^{\infty} s(t) s^*(t-\tau) e^{j2\pi f_d t} dt \right|^2 \tag{4.1}$$

In this notation, the target of interest is located at $(\tau, f_d) = (0, 0)$, and the ambiguity diagram is centered at the same point. Note that some authors define the ambiguity function as $|\chi(\tau;f_d)|$. In this book, $|\chi(\tau;f_d)|$ is called the uncertainty function.

Denote E as the energy of the signal $s(t)$,

$$E = \int_{-\infty}^{\infty} |s(t)|^2 dt \qquad (4.2)$$

The following list includes the properties for the radar ambiguity function:

1) The maximum value for the ambiguity function occurs at $(\tau, f_d) = (0, 0)$ and is equal to $4E^2$,

$$max\{|\chi(\tau;f_d)|^2\} = |\chi(0;0)|^2 = (2E)^2 \qquad (4.3)$$

$$|\chi(\tau;f_d)|^2 \leq |\chi(0;0)|^2 \qquad (4.4)$$

2) The ambiguity function is symmetric,

$$|\chi(\tau;f_d)|^2 = |\chi(-\tau;-f_d)|^2 \qquad (4.5)$$

3) The total volume under the ambiguity function is constant,

$$\iint |\chi(\tau;f_d)|^2 \, d\tau \, df_d = (2E)^2 \qquad (4.6)$$

4) If the function $S(f)$ is the Fourier transform of the signal $s(t)$, then by using Parseval's theorem we get

$$|\chi(\tau;f_d)|^2 = \left| \int S^*(f) S(f - f_d) e^{-j2\pi f \tau} df \right|^2 \qquad (4.7)$$

4.2. Examples of the Ambiguity Function

The ideal radar ambiguity function is represented by a spike of infinitesimally small width that peaks at the origin and is zero everywhere else, as illustrated in Fig. 4.1. An ideal ambiguity function provides perfect resolution between neighboring targets regardless of how close they may be to each other. Unfortunately, an ideal ambiguity function cannot physically exist. This is because the ambiguity function must have finite peak value equal to $(2E)^2$ and a finite volume also equal to $(2E)^2$. Clearly, the ideal ambiguity function cannot meet those two requirements.

4.2.1. Single Pulse Ambiguity Function

Consider the normalized rectangular pulse $s(t)$ defined by

Examples of the Ambiguity Function

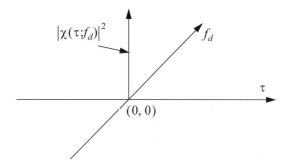

Figure 4.1. Ideal ambiguity function.

$$s(t) = \frac{1}{\sqrt{\tau'}} Rect\left(\frac{t}{\tau'}\right) \tag{4.8}$$

From Eq. (4.1) we have

$$\chi(\tau;f_d) = \int_{-\infty}^{\infty} s(t)s^*(t-\tau)e^{j2\pi f_d t} dt \tag{4.9}$$

Substituting Eq. (4.8) into Eq. (4.9) and performing the integration yield

$$|\chi(\tau;f_d)|^2 = \left|\left(1 - \frac{|\tau|}{\tau'}\right)\frac{\sin(\pi f_d(\tau' - |\tau|))}{\pi f_d(\tau' - |\tau|)}\right|^2 \quad |\tau| \leq \tau' \tag{4.10}$$

MATLAB Function "single_pulse_ambg.m"

The function *"single_pulse_ambg.m"* implements Eq. (4.10). It is given in Listing 4.1 in Section 4.6. The syntax is as follows:

single_pulse_ambg [taup]

taup is the pulsewidth. Fig 4.2 (a-d) show 3-D and contour plots of single pulse uncertainty and ambiguity functions. These plots can be reproduced using MATLAB program *"fig4_2.m"* given in Listing 4.2 in Section 4.6.

The ambiguity function cut along the time delay axis τ is obtained by setting $f_d = 0$. More precisely,

$$|\chi(\tau;0)|^2 = \left(1 - \frac{|\tau|}{\tau'}\right)^2 \quad |\tau| \leq \tau' \tag{4.11}$$

190 *MATLAB Simulations for Radar Systems Design*

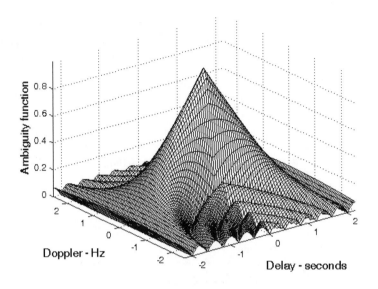

Figure 4.2a. Single pulse 3-D uncertainty plot. Pulsewidth is 2 seconds.

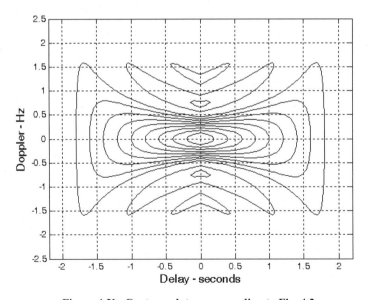

Figure 4.2b. Contour plot corresponding to Fig. 4.2a.

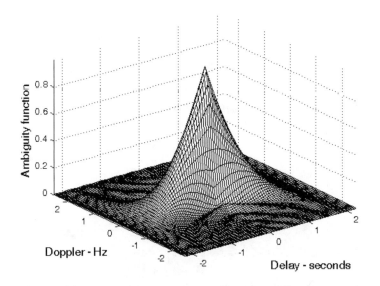

Figure 4.2c. Single pulse 3-D ambiguity plot. Pulsewidth is 2 seconds.

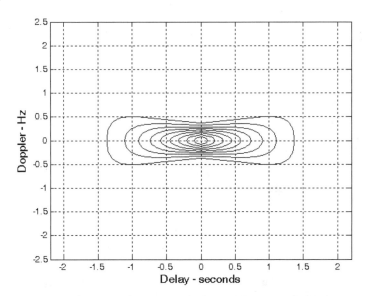

Figure 4.2d. Contour plot corresponding to Fig. 4.2c.

Note that the time autocorrelation function of the signal $s(t)$ is equal to $\chi(\tau;0)$. Similarly, the cut along the Doppler axis is

$$|\chi(0;f_d)|^2 = \left|\frac{\sin\pi\tau' f_d}{\pi\tau' f_d}\right|^2 \qquad (4.12)$$

Figs. 4.3 and 4.4, respectively, show the plots of the uncertainty function cuts defined by Eqs. (4.11) and (4.12). Since the zero Doppler cut along the time delay axis extends between $-\tau'$ and τ', then, close targets would be unambiguous if they are at least τ' seconds apart.

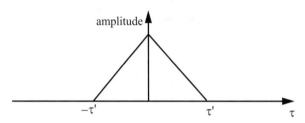

Figure 4.3. Zero Doppler uncertainty function cut along the time delay axis.

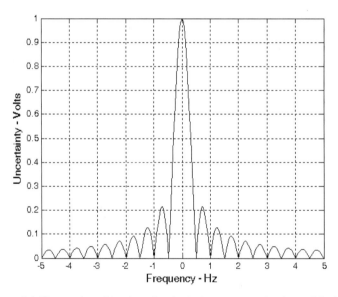

Figure 4.4. Uncertainty function of a single frequency pulse (zero delay). This plot can be reproduced using MATLAB program *"fig4_4.m"* given in Listing 4.3 in Section 4.6.

The zero time cut along the Doppler frequency axis has a $(\sin x / x)^2$ shape. It extends from $-\infty$ to ∞. The first null occurs at $f_d = \pm 1/\tau'$. Hence, it is possible to detect two targets that are shifted by $1/\tau'$, without any ambiguity.

We conclude that a single pulse range and Doppler resolutions are limited by the pulsewidth τ'. Fine range resolution requires that a very short pulse be used. Unfortunately, using very short pulses requires very large operating bandwidths, and may limit the radar average transmitted power to impractical values.

4.2.2. LFM Ambiguity Function

Consider the LFM complex envelope signal defined by

$$s(t) = \frac{1}{\sqrt{\tau'}} Rect\left(\frac{t}{\tau'}\right) e^{j\pi\mu t^2} \tag{4.13}$$

In order to compute the ambiguity function for the LFM complex envelope, we will first consider the case when $0 \leq \tau \leq \tau'$. In this case the integration limits are from $-\tau'/2$ to $(\tau'/2) - \tau$. Substituting Eq. (4.13) into Eq. (4.9) yields

$$\chi(\tau;f_d) = \frac{1}{\tau'} \int_{-\infty}^{\infty} Rect\left(\frac{t}{\tau'}\right) Rect\left(\frac{t-\tau}{\tau'}\right) e^{j\pi\mu t^2} e^{-j\pi\mu(t-\tau)^2} e^{j2\pi f_d t} dt \tag{4.14}$$

It follows that

$$\chi(\tau;f_d) = \frac{e^{-j\pi\mu\tau^2}}{\tau'} \int_{\frac{-\tau'}{2}}^{\frac{\tau'}{2}-\tau} e^{j2\pi(\mu\tau+f_d)t} dt \tag{4.15}$$

Finishing the integration process in Eq. (4.15) yields

$$\chi(\tau;f_d) = e^{j\pi\tau f_d}\left(1 - \frac{\tau}{\tau'}\right) \frac{\sin\left(\pi\tau'(\mu\tau+f_d)\left(1-\frac{\tau}{\tau'}\right)\right)}{\pi\tau'(\mu\tau+f_d)\left(1-\frac{\tau}{\tau'}\right)} \quad 0 \leq \tau \leq \tau' \tag{4.16}$$

Similar analysis for the case when $-\tau' \leq \tau \leq 0$ can be carried out, where in this case the integration limits are from $(-\tau'/2) - \tau$ to $\tau'/2$. The same result can be obtained by using the symmetry property of the ambiguity function ($|\chi(-\tau, -f_d)| = |\chi(\tau, f_d)|$). It follows that an expression for $\chi(\tau;f_d)$ that is valid for any τ is given by

$$\chi(\tau; f_d) = e^{j\pi\tau f_d}\left(1 - \frac{|\tau|}{\tau'}\right) \frac{\sin\left(\pi\tau'(\mu\tau + f_d)\left(1 - \frac{|\tau|}{\tau'}\right)\right)}{\pi\tau'(\mu\tau + f_d)\left(1 - \frac{|\tau|}{\tau'}\right)} \qquad |\tau| \leq \tau' \qquad (4.17)$$

and the LFM ambiguity function is

$$|\chi(\tau; f_d)|^2 = \left|\left(1 - \frac{|\tau|}{\tau'}\right) \frac{\sin\left(\pi\tau'(\mu\tau + f_d)\left(1 - \frac{|\tau|}{\tau'}\right)\right)}{\pi\tau'(\mu\tau + f_d)\left(1 - \frac{|\tau|}{\tau'}\right)}\right|^2 \qquad |\tau| \leq \tau' \qquad (4.18)$$

Again the time autocorrelation function is equal to $\chi(\tau, 0)$. The reader can verify that the ambiguity function for a down-chirp LFM waveform is given by

$$|\chi(\tau; f_d)|^2 = \left|\left(1 - \frac{|\tau|}{\tau'}\right) \frac{\sin\left(\pi\tau'(\mu\tau - f_d)\left(1 - \frac{|\tau|}{\tau'}\right)\right)}{\pi\tau'(\mu\tau - f_d)\left(1 - \frac{|\tau|}{\tau'}\right)}\right|^2 \qquad |\tau| \leq \tau' \qquad (4.19)$$

MATLAB Function "lfm_ambg.m"

The function *"lfm_ambg.m"* implements Eqs. (4.18) and (4.19). It is given in Listing 4.4 in Section 4.6. The syntax is as follows:

lfm_ambg [taup, b, up_down]

where

Symbol	Description	Units	Status
taup	pulsewidth	seconds	input
b	bandwidth	Hz	input
up_down	up_down = 1 for up chirp up_down = -1 for down chirp	none	input

Fig. 4.5 (a-d) shows 3-D and contour plots for the LFM uncertainty and ambiguity functions for

taup	b	up_down
1	10	1

These plots can be reproduced using MATLAB program *"fig4_5.m"* given in Listing 4.5 in Section 4.6. This function generates 3-D and contour plots of an LFM ambiguity function.

Examples of the Ambiguity Function 195

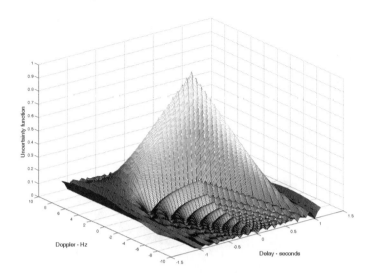

Figure 4.5a. Up-chirp LFM 3-D uncertainty plot. Pulsewidth is 1 second; and bandwidth is 10 Hz.

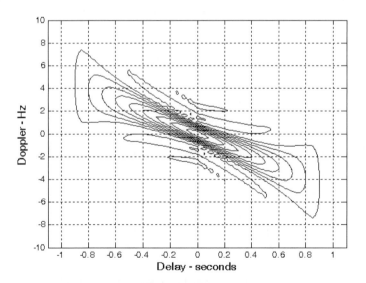

Figure 4.5b. Contour plot corresponding to Fig. 4.5a.

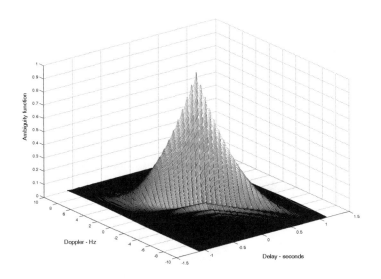

Figure 4.5c. Up-chirp LFM 3-D ambiguity plot. Pulsewidth is 1 second; and bandwidth is 10 Hz.

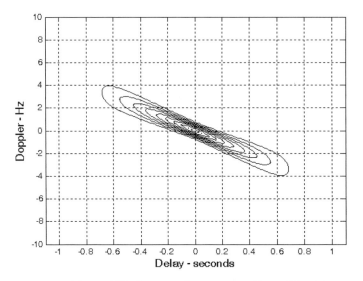

Figure 4.5d. Contour plot corresponding to Fig. 4.5c.

Examples of the Ambiguity Function

The up-chirp ambiguity function cut along the time delay axis τ is

$$|\chi(\tau;0)|^2 = \left|\left(1 - \frac{|\tau|}{\tau'}\right)\frac{\sin\left(\pi\mu\tau\tau'\left(1 - \frac{|\tau|}{\tau'}\right)\right)}{\pi\mu\tau\tau'\left(1 - \frac{|\tau|}{\tau'}\right)}\right|^2 \qquad |\tau| \leq \tau' \qquad (4.20)$$

Fig. 4.6 shows a plot for a cut in the uncertainty function corresponding to Eq. (4.20). Note that the LFM ambiguity function cut along the Doppler frequency axis is similar to that of the single pulse. This should not be surprising since the pulse shape has not changed (we only added frequency modulation). However, the cut along the time delay axis changes significantly. It is now much narrower compared to the unmodulated pulse cut. In this case, the first null occurs at

$$\tau_{n1} \approx 1/B \qquad (4.21)$$

which indicates that the effective pulsewidth (compressed pulsewidth) of the matched filter output is completely determined by the radar bandwidth. It follows that the LFM ambiguity function cut along the time delay axis is narrower than that of the unmodulated pulse by a factor

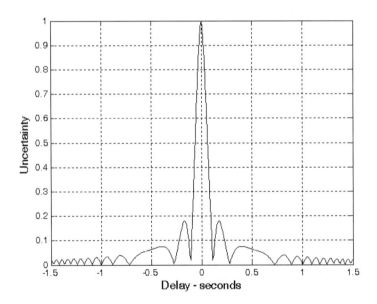

Figure 4.6. Zero Doppler uncertainty of an LFM pulse ($\tau' = 1$, $b = 20$). This plot can be reproduced using MATLAB program *"fig4_6.m"* given in Listing 4.6 in Section 4.6.

$$\xi = \frac{\tau'}{(1/B)} = \tau'B \qquad (4.22)$$

ξ is referred to as the compression ratio (also called time-bandwidth product and compression gain). All three names can be used interchangeably to mean the same thing. As indicated by Eq. (4.22) the compression ratio also increases as the radar bandwidth is increased.

Example:

Compute the range resolution before and after pulse compression corresponding to an LFM waveform with the following specifications: Bandwidth $B = 1 GHz$; and pulsewidth $\tau' = 10 ms$.

Solution:

The range resolution before pulse compression is

$$\Delta R_{uncomp} = \frac{c\tau'}{2} = \frac{3 \times 10^8 \times 10 \times 10^{-3}}{2} = 1.5 \times 10^6 \ meters$$

Using Eq. (4.21) yields

$$\tau_{n1} = \frac{1}{1 \times 10^9} = 1 \ ns$$

$$\Delta R_{comp} = \frac{c\tau_{n1}}{2} = \frac{3 \times 10^8 \times 1 \times 10^{-9}}{2} = 15 \ cm.$$

4.2.3. Coherent Pulse Train Ambiguity Function

Fig. 4.7 shows a plot of a coherent pulse train. The pulsewidth is denoted as τ' and the PRI is T. The number of pulses in the train is N; hence, the train's length is $(N-1)T$ seconds. A normalized individual pulse $s(t)$ is defined by

$$s_1(t) = \frac{1}{\sqrt{\tau'}} Rect\left(\frac{t}{\tau'}\right) \qquad (4.23)$$

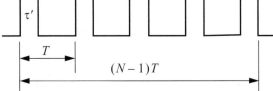

Figure 4.7. Coherent pulse train. N=5.

Examples of the Ambiguity Function

When coherency is maintained between the consecutive pulses, then an expression for the normalized train is

$$s(t) = \frac{1}{\sqrt{N}} \sum_{i=0}^{N-1} s_1(t - iT) \qquad (4.24)$$

The output of the matched filter is

$$\chi(\tau;f_d) = \int_{-\infty}^{\infty} s(t) s^*(t+\tau) e^{j2\pi f_d t} dt \qquad (4.25)$$

Substituting Eq. (4.24) into Eq. (4.25) and interchanging the summations and integration yield

$$\chi(\tau;f_d) = \frac{1}{N} \sum_{i=0}^{N-1} \sum_{j=0}^{N-1} \int_{-\infty}^{\infty} s_1(t-iT) \, s_1^*(t-jT-\tau) e^{j2\pi f_d t} dt \qquad (4.26)$$

Making the change of variable $t_1 = t - iT$ yields

$$\chi(\tau;f_d) = \frac{1}{N} \sum_{i=0}^{N-1} e^{j2\pi f_d iT} \sum_{j=0}^{N-1} \int_{-\infty}^{\infty} s_1(t_1) \, s_1^*(t_1 - [\tau - (i-j)T]) e^{j2\pi f_d t_1} dt_1 \qquad (4.27)$$

The integral inside Eq. (4.27) represents the output of the matched filter for a single pulse, and is denoted by χ_1. It follows that

$$\chi(\tau;f_d) = \frac{1}{N} \sum_{i=0}^{N-1} e^{j2\pi f_d iT} \sum_{j=0}^{N-1} \chi_1[\tau - (i-j)T; f_d] \qquad (4.28)$$

When the relation $q = i - j$ is used, then the following relation is true:[1]

$$\sum_{i=0}^{N} \sum_{m=0}^{N} = \sum_{q=-(N-1)}^{0} \sum_{i=0}^{N-1-|q|} \bigg|_{for \; j = i-q} + \sum_{q=1}^{N-1} \sum_{j=0}^{N-1-|q|} \bigg|_{for \; i = j+q} \qquad (4.29)$$

Using Eq. (4.29) into Eq. (4.28) gives

---

1. Rihaczek, A. W., *Principles of High Resolution Radar*, Artech House, 1994.

$$\chi(\tau;f_d) = \frac{1}{N} \sum_{q=-(N-1)}^{0} \left\{ \chi_1(\tau-qT;f_d) \sum_{i=0}^{N-1-|q|} e^{j2\pi f_d iT} \right\} \quad (4.30)$$

$$+ \frac{1}{N} \sum_{q=1}^{N-1} \left\{ e^{j2\pi f_d qT} \chi_1(\tau-qT;f_d) \sum_{j=0}^{N-1-|q|} e^{j2\pi f_d jT} \right\}$$

Setting $z = \exp(j2\pi f_d T)$, and using the relation

$$\sum_{j=0}^{N-1-|q|} z^j = \frac{1-z^{N-|q|}}{1-z} \quad (4.31)$$

yield

$$\sum_{i=0}^{N-1-|q|} e^{j2\pi f_d iT} = e^{[j\pi f_d(N-1-|q|T)]} \frac{\sin[\pi f_d(N-1-|q|T)]}{\sin(\pi f_d T)} \quad (4.32)$$

Using Eq. (4.32) in Eq. (4.30) yields two complementary sums for positive and negative q. Both sums can be combined as

$$\chi(\tau;f_d) = \frac{1}{N} \sum_{q=-(N-1)}^{N-1} \chi_1(\tau-qT;f_d) e^{[j\pi f_d(N-1+q)T]} \frac{\sin[\pi f_d(N-|q|T)]}{\sin(\pi f_d T)} \quad (4.33)$$

Finally, the ambiguity function associated with the coherent pulse train is computed as the modulus square of Eq. (4.33). For $\tau' < T/2$, the ambiguity function reduces to

$$\chi(\tau;f_d) = \frac{1}{N} \sum_{q=-(N-1)}^{N-1} |\chi_1(\tau-qT;f_d)| \left| \frac{\sin[\pi f_d(N-|q|T)]}{\sin(\pi f_d T)} \right| \quad (4.34)$$

Thus, the ambiguity function for a coherent pulse train is the superposition of the individual pulse's ambiguity functions. The ambiguity function cuts along the time delay and Doppler axes are, respectively, given by

$$|\chi(\tau;0)|^2 = \left| \sum_{q=-(N-1)}^{N-1} \left(1-\frac{|q|}{N}\right)\left(1-\frac{|\tau-qT|}{\tau'}\right) \right|^2 \quad ; \; |\tau-qT| < \tau' \quad (4.35)$$

Examples of the Ambiguity Function

$$|\chi(0;f_d)|^2 = \left| \frac{1}{N} \frac{\sin(\pi f_d \tau')}{\pi f_d \tau'} \frac{\sin(\pi f_d NT)}{\sin(\pi f_d T)} \right|^2 \qquad (4.36)$$

MATLAB Function "train_ambg.m"

The function *"train_ambg.m"* implements Eq. (4.34). It is given in Listing 4.7 in Section 4.6. The syntax is as follows:

train_ambg [taup, n, pri]

Symbol	Description	Units	Status
taup	pulsewidth	seconds	input
n	number of pulses in train	none	input
pri	pulse repetition interval	seconds	input

Fig. 4.8 (a-d) shows typical outputs of this function, for

taup	n	pri
0.2	5	1

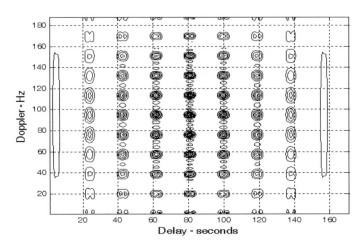

Figure 4.8a. Three-dimensional ambiguity plot for a five pulse equal amplitude coherent train. Pulsewidth is 0.2 seconds; and PRI is 1 second, N=5. This plot can be reproduced using MATLAB program *"fig4_8.m"* given in Listing 4.8 in Section 4.6.

202 *MATLAB Simulations for Radar Systems Design*

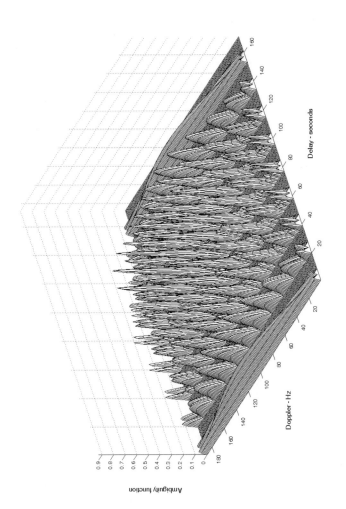

Figure 4.8b. 3-D plot corresponding to Fig. 4.8a.

Examples of the Ambiguity Function 203

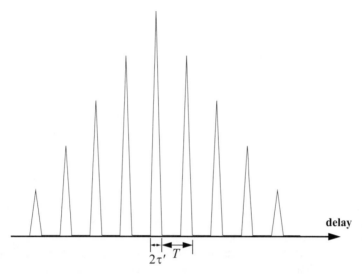

Figure 4.8c. Zero Doppler cut corresponding to Fig. 4.8a.

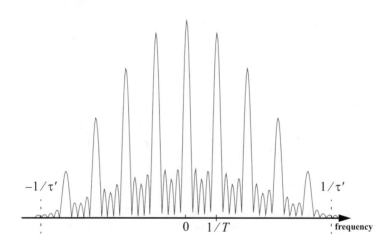

Figure 4.8d. Zero delay cut corresponding to Fig. 4.8a.

4.3. Ambiguity Diagram Contours

Plots of the ambiguity function are called ambiguity diagrams. For a given waveform, the corresponding ambiguity diagram is normally used to determine the waveform properties such as the target resolution capability, measurement (time and frequency) accuracy, and its response to clutter. Three-dimensional ambiguity diagrams are difficult to plot and interpret. This is the reason why contour plots of the 3-D ambiguity diagram are often used to study the characteristics of a waveform. An ambiguity contour is a 2-D plot (frequency/time) of a plane intersecting the 3-D ambiguity diagram that corresponds to some threshold value. The resultant plots are ellipses. It is customary to display the ambiguity contour plots that correspond to one half of the peak autocorrelation value.

Fig. 4.9 shows a sketch of typical ambiguity contour plots associated with a gated CW pulse. It indicates that narrow pulses provide better range accuracy than long pulses. Alternatively, the Doppler accuracy is better for a wider pulse than it is for a short one. This trade-off between range and Doppler measurements comes from the uncertainty associated with the time-bandwidth product of a single sinusoidal pulse, where the product of uncertainty in time (range) and uncertainty in frequency (Doppler) cannot be much smaller than unity. Note that an exact plot for Fig. 4.9 can be obtained using the function "single_pulse_ambg.m" and the MATLAB command *contour*.

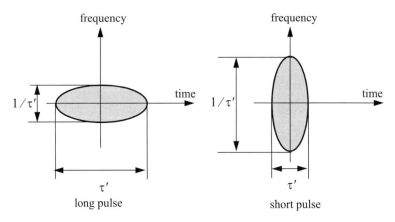

Figure 4.9. Ambiguity contour plot associated with a sinusoid modulated gated CW pulse. See Fig. 4.2.

Multiple ellipses in an ambiguity contour plot indicate the presence of multiple targets. Thus, it seems that one may improve the radar resolution by increasing the ambiguity diagram threshold value. This is illustrated in Fig.

4.10. However, in practice this is not possible for two reasons. First, in the presence of noise we lack knowledge of the peak correlation value; and second, targets in general will have different amplitudes.

Now consider the case of a coherent pulse train described in Fig. 4.7. For a pulse train, range accuracy is still determined by the pulsewidth, in the same manner as in the case of a single pulse, while Doppler accuracy is determined by the train length. Thus, time and frequency measurements can be made independently of each other. However, additional peaks appear in the ambiguity diagram which may cause range and Doppler uncertainties (see Fig. 4.11).

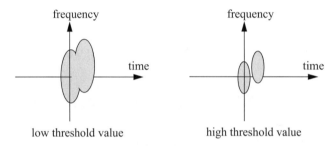

Figure 4.10. Effect of threshold value on resolution.

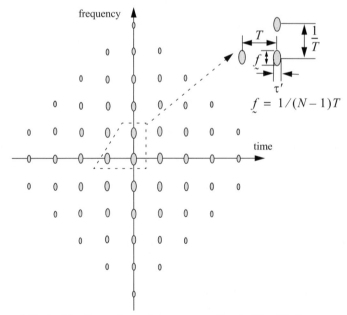

Figure 4.11. Ambiguity contour plot corresponding to Fig. 4.7. For an exact plot see Fig. 4.8a.

As one would expect, high PRF pulse trains (i.e., small T) lead to extreme uncertainty in range, while low PRF pulse trains have extreme ambiguity in Doppler. Medium PRF pulse trains have moderate ambiguity in both range and Doppler, which can be overcome by using multiple PRFs. It is possible to avoid ambiguities caused by pulse trains and still have reasonable independent control on both range and Doppler accuracies by using a single modulated pulse with a time-bandwidth product that is much larger than unity. Fig. 4.12 shows the ambiguity contour plot associated with an LFM waveform. In this case, τ' is the pulsewidth and B is the pulse bandwidth. The exact plots can be obtained using the function *"lfm_ambg.m"*.

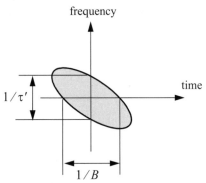

Figure 4.12. Ambiguity contour plot associated with an up-chirp LFM waveform. For an exact plot see Fig. 4.5b.

4.4. Digital Coded Waveforms

In this section we will briefly discuss the digital coded waveform. We will determine the waveform range and Doppler characteristics on the basis of its autocorrelation function, since in the absence of noise, the output of the matched filter is proportional to the code autocorrelation.

4.4.1. Frequency Coding (Costas Codes)

Construction of Costas codes can be understood from the construction process of Stepped Frequency Waveforms (SFW) described in Chapter 3. In SFW, a relatively long pulse of length τ' is divided into N subpulses, each of width τ_1 ($\tau' = N\tau_1$). Each group of N subpulses is called a burst. Within each burst the frequency is increased by Δf from one subpulse to the next. The overall burst bandwidth is $N\Delta f$. More precisely,

Digital Coded Waveforms

$$\tau_1 = \tau'/N \tag{4.37}$$

and the frequency for the ith subpulse is

$$f_i = f_0 + i\Delta f \quad ; \; i = 1, N \tag{4.38}$$

where f_0 is a constant frequency and $f_0 \gg \Delta f$. It follows that the time-bandwidth product of this waveform is

$$\Delta f \tau' = N^2 \tag{4.39}$$

Costas signals (or codes) are similar to SFW, except that the frequencies for the subpulses are selected in a random fashion, according to some predetermined rule or logic. For this purpose, consider the $N \times N$ matrix shown in Fig. 4.13b. In this case, the rows are indexed from $i = 1, 2, ..., N$ and the columns are indexed from $j = 0, 1, 2, ..., (N-1)$. The rows are used to denote the subpulses and the columns are used to denote the frequency. A *"dot"* indicates the frequency value assigned to the associated subpulse. In this fashion, Fig. 4.13a shows the frequency assignment associated with a SFW. Alternatively, the frequency assignments in Fig. 4.13b are chosen randomly. For a matrix of size $N \times N$, there are a total of $N!$ possible ways of assigning the *"dots"* (i.e., $N!$ possible codes).

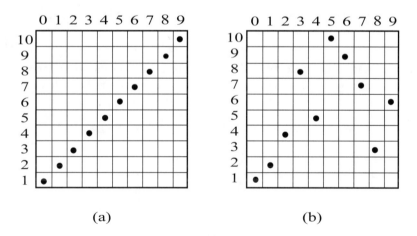

Figure 4.13. Frequency assignment for a burst of N subpulses. (a) SFW (stepped LFM); (b) Costas code of length Nc = 10.

The sequences of *"dot"* assignments for which the corresponding ambiguity function approaches an ideal or a *"thumbtack"* response are called Costas codes. A near thumbtack response was obtained by Costas[1] using the following logic: there is only one frequency per time slot (row) and per frequency slot (column). Therefore, for an $N \times N$ matrix the number of possible Costas codes is drastically less than $N!$. For example, there are $N_c = 4$ possible Costas codes for $N = 3$, and $N_c = 40$ possible codes for $N = 5$. It can be shown that the code density, defined as the ratio $N_c/N!$, gets significantly smaller as N becomes larger.

There are numerous analytical ways to generate Costas codes. In this section we will describe two of these methods. First, let q be an odd prime number, and choose the number of subpulses as

$$N = q - 1 \tag{4.40}$$

Define γ as the primitive root of q. A primitive root of q (an odd prime number) is defined as γ such that the powers $\gamma, \gamma^2, \gamma^3, \ldots, \gamma^{q-1}$ modulo q generate every integer from 1 to $q - 1$.

In the first method, for an $N \times N$ matrix, label the rows and columns, respectively, as

$$\begin{aligned} i &= 0, 1, 2, \ldots, (q-2) \\ j &= 1, 2, 3, \ldots, (q-1) \end{aligned} \tag{4.41}$$

Place a dot in the location (i, j) corresponding to f_i if and only if

$$i = (\gamma)^j \; (modulo \; q) \tag{4.42}$$

In the next method, Costas code is first obtained from the logic described above; then by deleting the first row and first column from the matrix a new code is generated. This method produces a Costas code of length $N = q - 2$.

Define the normalized complex envelope of the Costas signal as

$$s(t) = \frac{1}{\sqrt{N\tau_1}} \sum_{l=0}^{N-1} s_l(t - l\tau_1) \tag{4.43}$$

$$s_l(t) = \begin{pmatrix} \exp(j2\pi f_l t) & 0 \leq t \leq \tau_1 \\ 0 & elsewhere \end{pmatrix} \tag{4.44}$$

---

1. Costas, J. P., A Study of a Class of Detection Waveforms Having Nearly Ideal Range-Doppler Ambiguity Properties, *Proc. IEEE 72*, 1984, pp. 996-1009.

Costas showed that the output of the matched filter is

$$\chi(\tau, f_D) = \frac{1}{N} \sum_{l=0}^{N-1} \exp(j2\pi l f_D \tau) \left\{ \Phi_{ll}(\tau, f_D) + \sum_{\substack{q=0 \\ q \neq l}}^{N-1} \Phi_{lq}(\tau - (l-q)\tau_1, f_D) \right\} \quad (4.45)$$

$$\Phi_{lq}(\tau, f_D) = \left(\tau_1 - \frac{|\tau|}{\tau_1}\right) \frac{\sin \alpha}{\alpha} \exp(-j\beta - j2\pi f_q \tau) \quad , \ |\tau| \leq \tau_1 \quad (4.46)$$

$$\alpha = \pi(f_l - f_q - f_D)(\tau_1 - |\tau|) \quad (4.47)$$

$$\beta = \pi(f_l - f_q - f_D)(\tau_1 + |\tau|) \quad (4.48)$$

Three-dimensional plots of the ambiguity function of Costas signals show the near thumbtack response of the ambiguity function. All sidelobes, except for few around the origin, have amplitude $1/N$. Few sidelobes close to the origin have amplitude $2/N$, which is typical of Costas codes. The compression ratio of a Costas code is approximately N.

4.4.2. Binary Phase Codes

Consider the case of binary phase codes in which a relatively long pulse of width τ' is divided into N smaller pulses; each is of width $\Delta \tau = \tau'/N$. Then, the phase of each sub-pulse is randomly chosen as either 0 or π radians relative to some CW reference signal. It is customary to characterize a sub-pulse that has 0 phase (amplitude of +1 Volt) as either "1" or "+." Alternatively, a sub-pulse with phase equal to π (amplitude of -1 Volt) is characterized by either "0" or "-." The compression ratio associated with binary phase codes is equal to $\xi = \tau'/\Delta \tau$, and the peak value is N times larger than that of the long pulse. The goodness of a compressed binary phase code waveform depends heavily on the random sequence of the phases of the individual sub-pulses.

One family of binary phase codes that produces compressed waveforms with constant sidelobe levels equal to unity is the Barker code. Fig. 4.14 illustrates a Barker code of length seven. A Barker code of length n is denoted as B_n. There are only seven known Barker codes that share this unique property; they are listed in Table 4.1. Note that B_2 and B_4 have complementary forms that have the same characteristics. Since there are only seven Barker codes, they are not used when radar security is an issue.

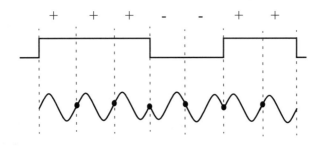

Figure 4.14. Binary phase code of length 7.

TABLE 4.1. Barker codes.

Code symbol	Code length	Code elements	Side lode reduction (dB)
B_2	2	1 -1	6.0
		1 1	
B_3	3	1 1 -1	9.5
B_4	4	1 1 -1 1	12.0
		1 1 1 -1	
B_5	5	1 1 1 -1 1	14.0
B_7	7	1 1 1 -1 -1 1 -1	16.9
B_{11}	11	1 1 1 -1 -1 -1 1 -1 -1 1 -1	20.8
B_{13}	13	1 1 1 1 1 -1 -1 1 1 -1 1 -1 1	22.3

In general, the autocorrelation function (which is an approximation of the matched filter output) for a B_N Barker code will be $2N\Delta\tau$ wide. The main lobe is $2\Delta\tau$ wide; the peak value is equal to N. There are $(N-1)/2$ side-lobes on either side of the main lobe. This is illustrated in Fig. 4.15 for a B_{13}. Notice that the main lobe is equal to 13, while all sidelobes are unity.

The most sidelobe reduction offered by a Barker code is $-22.3dB$, which may not be sufficient for the desired radar application. However, Barker codes can be combined to generate much longer codes. In this case, a B_m code can be used within a B_n code (m within n) to generate a code of length mn. The compression ratio for the combined B_{mn} code is equal to mn. As an example, a combined B_{54} is given by

$$B_{54} = \{11101, 11101, 00010, 11101\} \tag{4.49}$$

and is illustrated in Fig. 4.16. Unfortunately, the sidelobes of a combined Barker code autocorrelation function are no longer equal to unity.

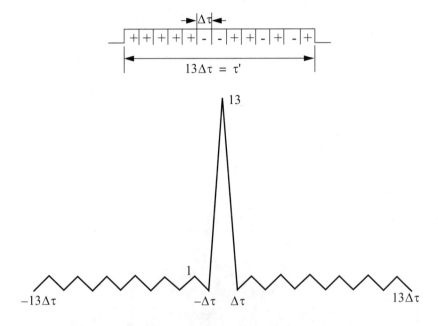

Figure 4.15. Barker code of length 13, and its corresponding autocorrelation function.

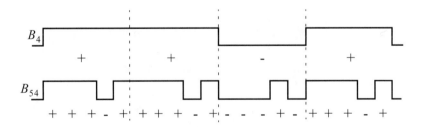

Figure 4.16. A combined B_{54} Barker code.

MATLAB Function "barker_ambig.m"

The MATLAB function *"barker_ambig.m"* calculates and plots the ambiguity function for Barker code. It is given in Listing 4.9 in Section 4.6. The syntax as follows:

$$[ambiguity] = barker\_ambig(u)$$

where u is a vector that defines the input code in terms of *"1's"* and *"-1's."* For example, using $u = [1\ \ 1\ \ 1\ \ -1\ \ -1\ \ 1\ \ -1]$ as an input, the function will calculate and plot the ambiguity function corresponding to B_7. Fig. 4.17 shows the output of this function when B_{13} is used as an input. Fig. 4.18 is similar to Fig. 4.17, except in this case B_7 is used as an input.

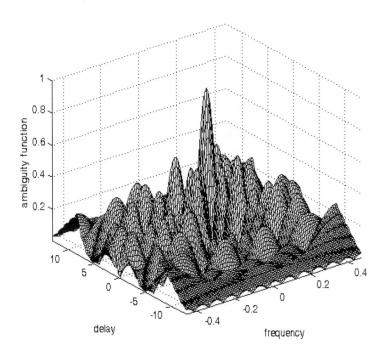

Figure 4.17a. Ambiguity function for B_{13} Barker code.

Digital Coded Waveforms 213

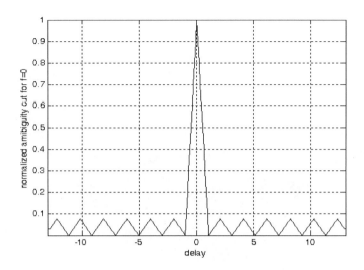

Figure 4.17b. Zero Doppler cut for the B_{13} ambiguity function.

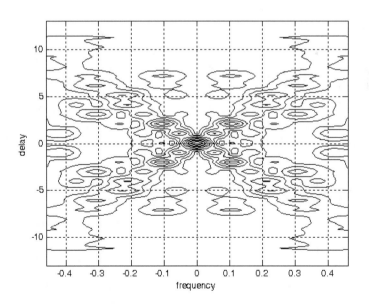

Figure 4.17c. Contour plot corresponding to Fig. 4.17a.

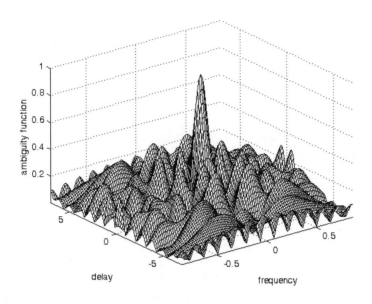

Figure 4.18a. Ambiguity function for B_7 Barker code.

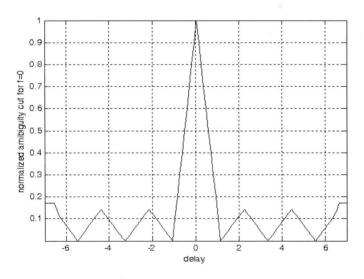

Figure 4.18b. Zero Doppler cut for the B_7 ambiguity function.

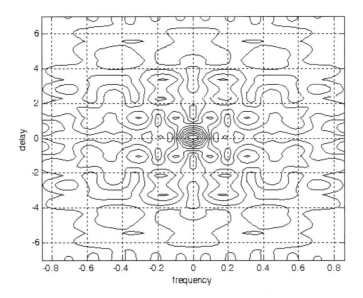

Figure 4.18c. Contour plot corresponding to Fig. 4.18a.

4.4.3. Pseudo-Random Number (PRN) Codes

Pseudo-Random Number (PRN) codes are also known as Maximal Length Sequences (MLS) codes. These codes are called pseudo-random because the statistics associated with their occurrence are similar to that associated with the coin-toss sequences. Maximum length sequences are periodic. The MLS codes have the following distinctive properties:

1. The number of ones per period is one more than the number of minus-ones.
2. Half the runs (consecutive states of the same kind) are of length one and one fourth are of length two.
3. Every maximal length sequence has the "shift and add" property. This means that, if a maximal length sequence is added (modulo 2) to a shifted version of itself, then the resulting sequence is a shifted version of the original sequence.
4. Every n-tuple of the code appears once and only once in one period of the sequence.
5. The correlation function is periodic and is given by

$$\phi(n) = \begin{cases} L & n = 0, \pm L, \pm 2L, \ldots \\ -1 & elsewhere \end{cases} \qquad (4.50)$$

Fig. 4.19 shows a typical sketch for an MLS autocorrelation function. Clearly these codes have the advantage that the compression ratio becomes very large as the period is increased. Additionally, adjacent peaks (grating lobes) become farther apart.

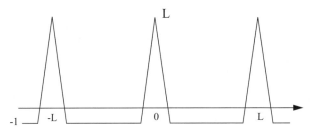

Figure 4.19. Typical autocorrelation of an MLS code of length L.

Linear Shift Register Generators

There are numerous ways to generate MLS codes. The most common is to use linear shift registers. When the binary sequence generated using a shift register implementation is periodic and has maximal length it is referred to as an MLS binary sequence with period L, where

$$L = 2^n - 1 \qquad (4.51)$$

n is the number of stages in the shift register generator.

A linear shift register generator basically consists of a shift register with modulo-two adders added to it. The adders can be connected to various stages of the register, as illustrated in Fig. 4.20 for $n = 4$ (i.e., $L = 15$). Note that the shift register initial state cannot be "zero."

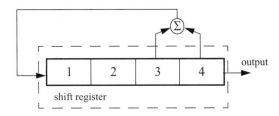

Figure 4.20. Circuit for generating an MLS sequence of length $L = 15$.

The feedback connections associated with a shift register generator determine whether the output sequence will be maximal or not. For a given size shift register, only few feedback connections lead to maximal sequence outputs. In order to illustrate this concept, consider the two 5-stage shift register generators shown in Fig. 4.21. The shift register generator shown in Fig. 4.21a generates a maximal length sequence, as clearly depicted by its state diagram. However, the shift register generator shown in Fig. 4.21b produces three non-maximal length sequences (depending on the initial state).

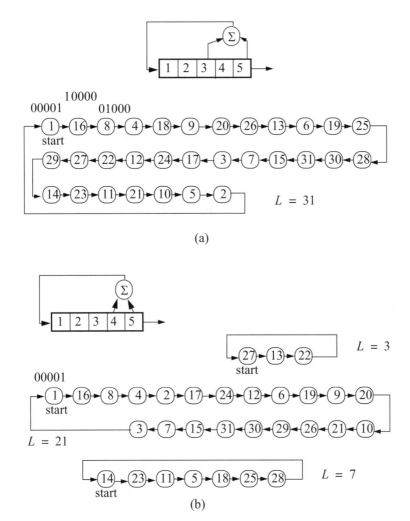

Figure 4.21. (a) A 5-stage shift register generator. (b) Non-maximal length 5 stage shift register generator.

Given an n-stage shift register generator, one would be interested in knowing how many feedback connections will yield maximal length sequences. Zierler[1] showed that the number of maximal length sequences possible for a given n-stage linear shift register generator is given by

$$N_L = \frac{\varphi(2^n - 1)}{n} \qquad (4.52)$$

where φ is the Euler's totient (or Euler's phi) function. Euler's phi function is defined by

$$\varphi(k) = k \prod_i \frac{(p_i - 1)}{p_i} \qquad (4.53)$$

where p_i are the prime factors of k. Note that when p_i has multiples, then only one of them is used (see example in Eq. (4.56)). Also note that when k is a prime number, then the Euler's phi function is

$$\varphi(k) = k - 1 \qquad (4.54)$$

For example, a 3-stage shift register generator will produce

$$N_L = \frac{\varphi(2^3 - 1)}{3} = \frac{\varphi(7)}{3} = \frac{7 - 1}{3} = 2 \qquad (4.55)$$

and a 6-stage shift register,

$$N_L = \frac{\varphi(2^6 - 1)}{6} = \frac{\varphi(63)}{6} = \frac{63}{6} \times \frac{(3-1)}{3} \times \frac{(7-1)}{7} = 6 \qquad (4.56)$$

Maximal Length Sequence Characteristic Polynomial

Consider an n-stage maximal length linear shift register whose feedback connections correspond to n, k, m, etc. This maximal length shift register can be described using its characteristic polynomial defined by

$$x^n + x^k + x^m + \ldots + 1 \qquad (4.57)$$

where the additions are modulo 2. Therefore, if the characteristic polynomial for an n-stage shift register is known, one can easily determine the register feedback connections and consequently deduce the corresponding maximal length sequence. For example, consider a 6-stage shift register whose characteristic polynomial is

---

1. Zierler, N., *Several Binary-Sequence Generators*, MIT Technical Report No. 95, Sept. 1955.

$$x^6 + x^5 + 1 \tag{4.58}$$

It follows that the shift register which generates a maximal length sequence is shown in Fig. 4.22.

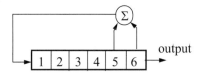

Figure 4.22. Linear shift register whose characteristic polynomial is $x^6 + x^5 + 1$.

One of the most important issues associated with generating a maximal length sequence using a linear shift register is determining the characteristic polynomial. This has been and continues to be a subject of research for many radar engineers and designers. It has been shown that polynomials which are both irreducible (not factorable) and primitive will produce maximal length shift register generators.

A polynomial of degree n is irreducible if it is not divisible by any polynomial of degree less than n. It follows that all irreducible polynomials must have an odd number of terms. Consequently, only linear shift register generators with an even number of feedback connections can produce maximal length sequences. An irreducible polynomial is primitive if and only if it divides $x^n - 1$ for no value of n less than $2^n - 1$.

MATLAB Function "prn_ambig.m"

The MATLAB function *"prn_ambig.m"* calculates and plots the ambiguity function associated with a given PRN code. It is given in Listing 4.10 in Section 4.6. The syntax is as follows:

$$[ambiguity] = prn\_ambig(u)$$

where u is a vector that defines the input maximal length code (sequence) in terms of *"1's"* and *"-1's."* Fig. 4.23 shows the output of this function for

$u31 = [1 -1 -1 -1 -1 1 -1 1 -1 1 1 1 -1 1 1 -1 -1 -1 1 1 1 1 1 -1 -1 1 1 -1 1 -1 -1]$

Fig. 4.24 is similar to Fig. 4.23, except in this case the input maximal length sequence is

$u15 = [1 -1 -1 -1 1 1 1 1 -1 1 -1 1 1 -1 -1]$

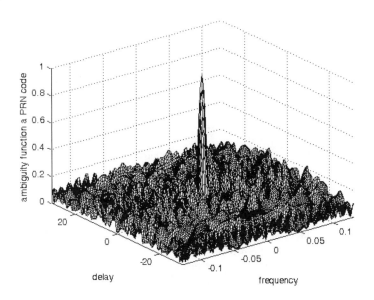

Figure 4.23a. Ambiguity function corresponding to a 31-bit PRN code.

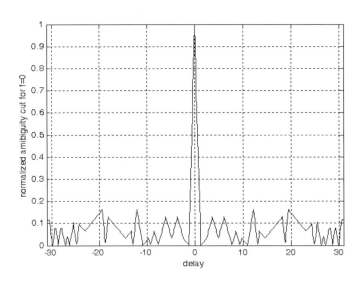

Figure 4.23b. Zero Doppler cut corresponding to Fig. 4.23a.

Digital Coded Waveforms 221

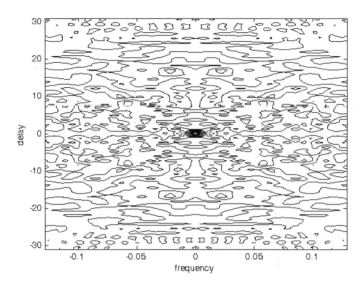

Figure 4.23c. Contour plot corresponding to Fig. 4.23a.

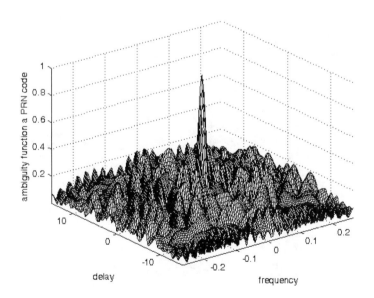

Figure 4.24a. Ambiguity function corresponding to a 15-bit PRN code.

222   *MATLAB Simulations for Radar Systems Design*

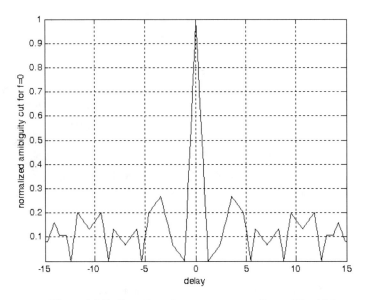

Figure 4.24b. Zero Doppler cut corresponding to Fig. 4.24a.

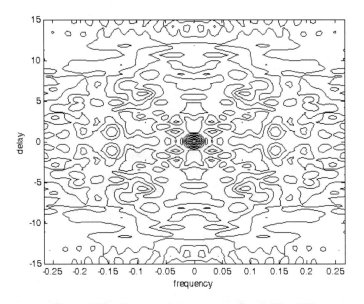

Figure 4.24c. Contour plot corresponding to Fig. 4.24a.

4.5. "MyRadar" Design Case Study - Visit 4

4.5.1. Problem Statement

Generate the ambiguity plots for the waveforms selected in Chapter 3 for this design case study.

4.5.2. A Design

In this section we will show the 3-D ambiguity diagram and the corresponding contour plot for only the search waveform. The user is advised to do the same for the track waveforms. For this purpose, use the MATLAB program *"myradar_visit4.m"*. It is given in Listing 4.11 in Section 4.6.

Figs. 4.25 and 4.26 show the output figures produced by the program *"myradar_visit4.m"* that correspond to the search waveform.

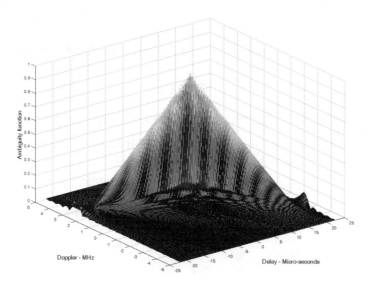

Figure 4.25. Ambiguity plot for *"MyRadar"* search waveform.

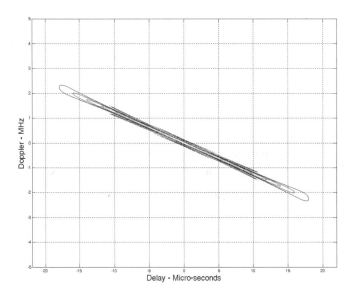

Figure 4.26. Contour of the ambiguity plot for *"MyRadar"* search waveform.

4.6. MATLAB Program and Function Listings

This section presents listings for all MATLAB programs/functions used in this chapter. The user is strongly advised to rerun the MATLAB programs in order to enhance his understanding of this chapter's material.

Listing 4.1. MATLAB Function *"single_pulse_ambg.m"*

```
function x = single_pulse_ambg (taup)
colormap (gray(1))
eps = 0.000001;
i = 0;
taumax = 1.1 * taup;
taumin = -taumax;
for tau = taumin:.05:taumax
   i = i + 1;
   j = 0;
   for fd = -5/taup:.05:5/taup %-2.5:.05:2.5
      j = j + 1;
```

```
    val1 = 1. - abs(tau) / taup;
    val2 = pi * taup * (1.0 - abs(tau) / taup) * fd;
    x(j,i) = abs( val1 * sin(val2+eps)/(val2+eps));
  end
end
```

Listing 4.2. MATLAB Program "fig4_2.m"

```
% Use this program to reproduce Fig. 4.2 of text
close all
clear all
eps = 0.000001;
taup = 2.;
taumin = -1.1 * taup;
taumax = -taumin;
x = single_pulse_ambg(taup);
taux = taumin:.05:taumax;
fdy = -5/taup:.05:5/taup;
figure(1)
mesh(taux,fdy,x);
xlabel ('Delay - seconds')
ylabel ('Doppler - Hz')
zlabel ('Ambiguity function')
colormap([.5 .5 .5])
colormap (gray)
figure(2)
contour(taux,fdy,x);
xlabel ('Delay - seconds')
ylabel ('Doppler - Hz')
colormap([.5 .5 .5])
colormap (gray)
grid
y = x.^2;
figure(3)
mesh(taux,fdy,y);
xlabel ('Delay - seconds')
ylabel ('Doppler - Hz')
zlabel ('Ambiguity function')
colormap([.5 .5 .5])
colormap (gray)
figure(4)
contour(taux,fdy,y);
xlabel ('Delay - seconds')
ylabel ('Doppler - Hz')
```

```
colormap([.5 .5 .5])
colormap (gray)
grid
```

Listing 4.3. MATLAB Program "fig4_4.m"

```
% Use this program to reproduce Fig 4.4 of text
close all
clear all
eps = 0.0001;
taup = 2.;
fd = -10./taup:.05:10./taup;
uncer = abs( sinc(taup .* fd));
ambg = uncer.^2;
plot(fd, ambg,'k')
xlabel ('Frequency - Hz')
ylabel ('Ambiguity - Volts')
grid
figure(2)
plot (fd, uncer,'k');
xlabel ('Frequency - Hz')
ylabel ('Uncertainty - Volts')
grid
```

Listing 4.4. MATLAB Function "lfm_ambg.m"

```
ffunction x = lfm_ambg(taup, b, up_down)
eps = 0.000001;
i = 0;
mu = up_down * b / 2. / taup;
delt = 2.2*taup/250;
delf = 2*b /250;
for tau = -1.1*taup:.05:1.1*taup
   i = i + 1;
   j = 0;
  for fd = -b:.05:b
    j = j + 1;
    val1 = 1. - abs(tau) / taup;
    val2 = pi * taup * (1.0 - abs(tau) / taup);
    val3 = (fd + mu * tau);
    val = val2 * val3;
    x(j,i) = abs( val1 * (sin(val+eps)/(val+eps))).^2;
  end
end
```

Listing 4.5. MATLAB Program "fig4_5.m"

```
% Use this program to reproduce Fig. 4.5 of text
close all
clear all
eps = 0.0001;
taup = 1.;
b =10.;
up_down = 1.;
x = lfm_ambg(taup, b, up_down);
taux = -1.1*taup:.05:1.1*taup;
fdy = -b:.05:b;
figure(1)
mesh(taux,fdy,x)
xlabel ('Delay - seconds')
ylabel ('Doppler - Hz')
zlabel ('Ambiguity function')
figure(2)
contour(taux,fdy,x)
xlabel ('Delay - seconds')
ylabel ('Doppler - Hz')
y = sqrt(x);
figure(3)
mesh(taux,fdy,y)
xlabel ('Delay - seconds')
ylabel ('Doppler - Hz')
zlabel ('Uncertainty function')
figure(4)
contour(taux,fdy,y)
xlabel ('Delay - seconds')
ylabel ('Doppler - Hz')
```

Listing 4.6. MATLAB Program "fig4_6.m"

```
% Use this program to reproduce Fig. 4.6 of text
close all
clear all
taup = 1;
b =20.;
up_down = 1.;
taux = -1.5*taup:.01:1.5*taup;
fd = 0.;
mu = up_down * b / 2. / taup;
ii = 0.;
```

```
for tau = -1.5*taup:.01:1.5*taup
   ii = ii + 1;
   val1 = 1. - abs(tau) / taup;
   val2 = pi * taup * (1.0 - abs(tau) / taup);
   val3 = (fd + mu * tau);
   val = val2 * val3;
   x(ii) = abs( val1 * (sin(val+eps)/(val+eps)));
end
figure(1)
plot(taux,x)
grid
xlabel ('Delay - seconds')
ylabel ('Uncertainty')
figure(2)
plot(taux,x.^2)
grid
xlabel ('Delay - seconds')
ylabel ('Ambiguity')
```

Listing 4.7. MATLAB Function "train_ambg.m"

```
function x = train_ambg (taup, n, pri)
if( taup > pri / 2.)
   'ERROR. Pulsewidth must be less than the PRI/2.'
   return
end
gap = pri - 2.*taup;
eps = 0.000001;
b = 1. / taup;
ii = 0.;
for q = -(n-1):1:n-1
   tauo = q - taup ;
   index = -1.;
   for tau1 = tauo:0.0533:tauo+gap+2.*taup
     index = index + 1;
     tau = -taup + index*.0533;
     ii = ii + 1;
     j = 0.;
     for fd = -b:.0533:b
       j = j + 1;
       if (abs(tau) <= taup)
         val1 = 1. -abs(tau) / taup;
         val2 = pi * taup * fd * (1.0 - abs(tau) / taup);
         val3 = abs(val1 * sin(val2+eps) /(val2+eps));
```

```
        val4 = abs((sin(pi*fd*(n-abs(q))*pri+eps))/(sin(pi*fd*pri+eps)));
        x(j,ii)=  val3 * val4 / n;
      else
        x(j,ii) = 0.;
      end
    end
  end
end
```

Listing 4.8. MATLAB Program "fig4_8.m"

```
% Use this program to reproduce Fig. 4.8 of text
close all
clear all
taup =0.2;
pri=1;
n=5;
x = train_ambg (taup, n, pri);
figure(1)
mesh(x)
xlabel ('Delay - seconds')
ylabel ('Doppler - Hz')
zlabel ('Ambiguity function')
figure(2)
contour(x);
xlabel ('Delay - seconds')
ylabel ('Doppler - Hz')
```

Listing 4.9. MATLAB Function "barker_ambig.m"

```
function [ambig] = barker_ambig(uinput)
% Compute and plot the ambiguity function for a Barker code
%Compute the ambiguity function
% by utilizing the FFT through combining multiple range cuts
N = size(uinput,2);
tau = N;
Barker_code = uinput;
samp_num = size(Barker_code,2) *10;
n = ceil(log(samp_num) / log(2));
nfft = 2^n;
u(1:nfft) = 0;
j = 0;
for index = 1:10:samp_num
   index;
```

```
    j = j+1;
    u(index:index+10-1) = Barker_code(j);
end
v = u;
delay = linspace(-tau, tau, nfft);
freq_del = 12 / tau /100;
j = 0;
vfft = fft(v,nfft);
for freq = -6/tau:freq_del:6/tau;
    j = j+1;
    exf = exp(sqrt(-1) * 2. * pi * freq .* delay);
    u_times_exf = u .* exf;
    ufft = fft(u_times_exf,nfft);
    prod = ufft .* conj(vfft);
    ambig(:,j) = fftshift(abs(ifft(prod))');
end
freq = -6/tau:freq_del:6/tau;
delay = linspace(-N,N,nfft);
figure (1)
mesh(freq,delay,ambig ./ max(max(ambig)))
colormap([.5 .5 .5])
colormap(gray)
axis tight
xlabel('frequency')
ylabel('delay')
zlabel('ambiguity function')
figure (2)
value = 10 * N ;
plot(delay,ambig(:,51)/value,'k')
xlabel('delay')
ylabel('normalized amibiguity cut for f=0')
grid
axis tight
figure (3)
contour(freq,delay,ambig ./ max(max(ambig)))
colormap([.5 .5 .5])
colormap (gray)
xlabel('frequency')
ylabel('delay')
grid on
```

Listing 4.10. MATLAB Function "prn_ambig.m"

```
function [ambig] = prn_ambig(uinput)
```

```
% Compute and plot the ambiguity function for a PRN code
% Compute the ambiguity function by utilizing the FFT
% through combining multiple range cuts

N = size(uinput,2);
tau = N;
PRN = uinput;
samp_num = size(PRN,2) * 10;
n = ceil(log(samp_num) / log(2));
nfft = 2^n;
u(1:nfft) = 0;
j = 0;
for index = 1:10:samp_num
   index;
   j = j+1;
   u(index:index+10-1) = PRN(j);
end
% set-up the array v
v = u;
delay = linspace(0,5*tau,nfft);
freq_del = 8 / tau /100;
j = 0;
vfft = fft(v,nfft);
for freq = -4/tau:freq_del:4/tau;
   j = j+1;
   exf = exp(sqrt(-1) * 2. * pi * freq .* delay);
   u_times_exf = u .* exf;
   ufft = fft(u_times_exf,nfft);
   prod = ufft .* conj(vfft);
   ambig(:,j) = fftshift(abs(ifft(prod))');
end
freq = -4/tau:freq_del:4/tau;
delay = linspace(-N,N,nfft);
figure(1)
mesh(freq,delay,ambig ./ max(max(ambig)))
colormap([.5 .5 .5])
colormap(gray)
axis tight
xlabel('frequency')
ylabel('delay')
zlabel('ambiguity function a PRN code')
figure(2)
plot(delay,ambig(:,51)/(max(max(ambig))),'k')
xlabel('delay')
```

```
ylabel('normalized amibiguity cut for f=0')
grid
axis tight
figure(3)
contour(freq,delay,ambig ./ max(max(ambig)))
axis tight
colormap([.5 .5 .5])
colormap(gray)
xlabel('frequency')
ylabel('delay')
```

Listing 4.11. MATLAB Program "myradar_visit4.m"

```
% Use this program to reproduce Figs. 4.25 to 4.27 of the text
close all
clear all
eps = 0.0001;
taup = 20.e-6;
b =1.e6;
up_down = 1.;
i = 0;
mu = up_down * b / 2. / taup;
delt = 2.2*taup /250;
delf = 2*b /300;
for tau = -1.1*taup:delt:1.1*taup
  i = i + 1;
  j = 0;
  for fd = -b:delf:b
    j = j + 1;
    val1 = 1. - abs(tau) / taup;
    val2 = pi * taup * (1.0 - abs(tau) / taup);
    val3 = (fd + mu * tau);
    val = val2 * val3;
    x(j,i) = abs( val1 * (sin(val+eps)/(val+eps))).^2;
  end
end
taux = linspace(-1.1*taup,1.1*taup,251).*1e6;
fdy = linspace(-b,b,301) .* 1e-6;
figure(1)
mesh(taux,fdy,sqrt(x))
xlabel ('Delay - Micro-seconds')
ylabel ('Doppler - MHz')
zlabel ('Ambiguity function')
figure(2)
```

```
contour(taux,fdy,sqrt(x))
xlabel ('Delay - Micro-seconds')
ylabel ('Doppler - MHz')
grid
```

Chapter 5 *Pulse Compression*

Range resolution for a given radar can be significantly improved by using very short pulses. Unfortunately, utilizing short pulses decreases the average transmitted power, which can hinder the radar's normal modes of operation, particularly for multi-function and surveillance radars. Since the average transmitted power is directly linked to the receiver SNR, it is often desirable to increase the pulsewidth (i.e., increase the average transmitted power) while simultaneously maintaining adequate range resolution. This can be made possible by using pulse compression techniques. Pulse compression allows us to achieve the average transmitted power of a relatively long pulse, while obtaining the range resolution corresponding to a short pulse. In this chapter, we will analyze analog and digital pulse compression techniques.

Two LFM pulse compression techniques are discussed in this chapter. The first technique is known as "correlation processing" which is predominantly used for narrow band and some medium band radar operations. The second technique is called "stretch processing" and is normally used for extremely wide band radar operations.

5.1. Time-Bandwidth Product

Consider a radar system that employs a matched filter receiver. Let the matched filter receiver bandwidth be denoted as B. Then the noise power available within the matched filter bandwidth is given by

$$N_i = 2 \frac{N_0}{2} B \qquad (5.1)$$

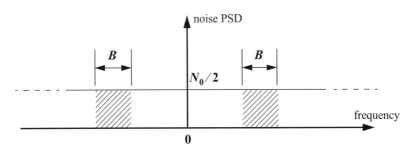

Figure 5.1. Input noise power.

where the factor of two is used to account for both negative and positive frequency bands, as illustrated in Fig. 5.1. The average input signal power over a pulse duration τ' is

$$S_i = \frac{E}{\tau'} \qquad (5.2)$$

E is the signal energy. Consequently, the matched filter input SNR is given by

$$(SNR)_i = \frac{S_i}{N_i} = \frac{E}{N_0 B \tau'} \qquad (5.3)$$

The output peak instantaneous SNR to the input SNR ratio is

$$\frac{SNR(t_0)}{(SNR)_i} = 2B\tau' \qquad (5.4)$$

The quantity $B\tau'$ is referred to as the "time-bandwidth product" for a given waveform or its corresponding matched filter. The factor $B\tau'$ by which the output SNR is increased over that at the input is called the matched filter gain, or simply the compression gain.

In general, the time-bandwidth product of an unmodulated pulse approaches unity. The time-bandwidth product of a pulse can be made much greater than unity by using frequency or phase modulation. If the radar receiver transfer function is perfectly matched to that of the input waveform, then the compression gain is equal to $B\tau'$. Clearly, the compression gain becomes smaller than $B\tau'$ as the spectrum of the matched filter deviates from that of the input signal.

5.2. Radar Equation with Pulse Compression

The radar equation for a pulsed radar can be written as

$$SNR = \frac{P_t \tau' G^2 \lambda^2 \sigma}{(4\pi)^3 R^4 k T_e F L} \qquad (5.5)$$

where P_t is peak power, τ' is pulsewidth, G is antenna gain, σ is target RCS, R is range, k is Boltzman's constant, T_e is effective noise temperature, F is noise figure, and L is total radar losses.

Pulse compression radars transmit relatively long pulses (with modulation) and process the radar echo into very short pulses (compressed). One can view the transmitted pulse as being composed of a series of very short subpulses (duty is 100%), where the width of each subpulse is equal to the desired compressed pulsewidth. Denote the compressed pulsewidth as τ_c. Thus, for an individual subpulse, Eq. (5.5) can be written as

$$(SNR)_{\tau_c} = \frac{P_t \tau_c G^2 \lambda^2 \sigma}{(4\pi)^3 R^4 k T_e F L} \qquad (5.6)$$

The SNR for the uncompressed pulse is then derived from Eq. (5.6) as

$$SNR = \frac{P_t (\tau' = n\tau_c) G^2 \lambda^2 \sigma}{(4\pi)^3 R^4 k T_e F L} \qquad (5.7)$$

where n is the number of subpulses. Equation (5.7) is denoted as the radar equation with pulse compression.

Observation of Eqs. (5.5) and (5.7) indicates the following (note that both equations have the same form): For a given set of radar parameters, and as long as the transmitted pulse remains unchanged, the SNR is also unchanged regardless of the signal bandwidth. More precisely, when pulse compression is used, the detection range is maintained while the range resolution is drastically improved by keeping the pulsewidth unchanged and by increasing the bandwidth. Remember that range resolution is proportional to the inverse of the signal bandwidth,

$$\Delta R = c/2B \qquad (5.8)$$

5.3. LFM Pulse Compression

Linear FM pulse compression is accomplished by adding frequency modulation to a long pulse at transmission, and by using a matched filter receiver in order to compress the received signal. As a result, the matched filter output is compressed by a factor $\xi = B\tau'$, where τ' is the pulsewidth and B is the bandwidth. Thus, by using long pulses and wideband LFM modulation large compression ratios can be achieved.

Figure 5.2 shows an ideal LFM pulse compression process. Part (a) shows the envelope for a wide pulse, part (b) shows the frequency modulation (in this case it is an upchirp LFM) with bandwidth $B = f_2 - f_1$. Part (c) shows the matched filter time-delay characteristic, while part (d) shows the compressed pulse envelope. Finally part (e) shows the Matched filter input / output waveforms.

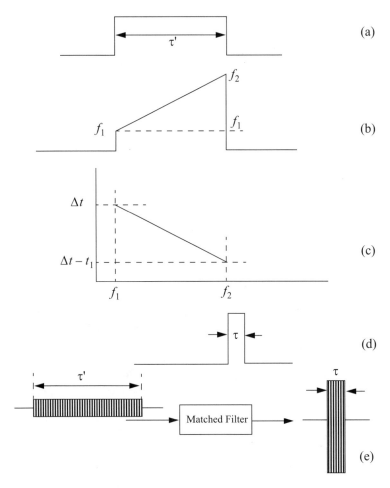

Figure 5.2 Ideal LFM pulse compression.

Fig. 5.3 illustrates the advantage of pulse compression using more realistic LFM waveform. In this example, two targets with RCS $\sigma_1 = 1m^2$ and $\sigma_2 = 0.5m^2$ are detected. The two targets are not separated enough in time to be resolved. Fig. 5.3a shows the composite echo signal from those targets.

Clearly, the target returns overlap and, thus, they are not resolved. However, after pulse compression the two pulses are completely separated and are resolved as two distinct targets. In fact, when using LFM, returns from neighboring targets are resolved as long as they are separated in time by τ_{n1}, the compressed pulsewidth. This figure can be reproduced using MATLAB program *"fig5_3.m"* given in Listing 5.1 in Section 5.5.

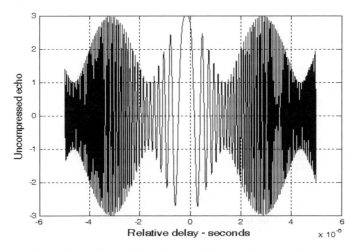

Figure 5.3a. Composite echo signal for two unresolved targets.

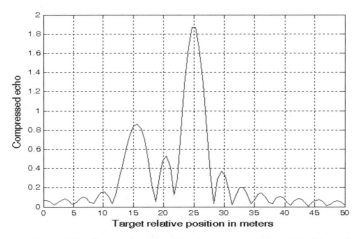

Figure 5.3b. Composite echo signal corresponding to Fig. 5.3a, after pulse compression.

5.3.1. Correlation Processor

Radar operations (search, track, etc.) are usually carried out over a specified range window, referred to as the receive window and defined by the difference between the radar maximum and minimum range. Returns from all targets within the receive window are collected and passed through matched filter circuitry to perform pulse compression. One implementation of such analog processors is the Surface Acoustic Wave (SAW) devices. Because of the recent advances in digital computer development, the correlation processor is often performed digitally using the FFT. This digital implementation is called Fast Convolution Processing (FCP) and can be implemented at base-band. The fast convolution process is illustrated in Fig. 5.4

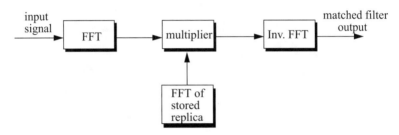

Figure 5.4. Computing the matched filter output using an FFT.

Since the matched filter is a linear time invariant system, its output can be described mathematically by the convolution between its input and its impulse response,

$$y(t) = s(t) \bullet h(t) \tag{5.9}$$

where $s(t)$ is the input signal, $h(t)$ is the matched filter impulse response (replica), and the $\bullet$ operator symbolically represents convolution. From the Fourier transform properties,

$$FFT\{s(t) \bullet h(t)\} = S(f) \cdot H(f) \tag{5.10}$$

and when both signals are sampled properly, the compressed signal $y(t)$ can be computed from

$$y = FFT^{-1}\{S \cdot H\} \tag{5.11}$$

where FFT^{-1} is the inverse FFT. When using pulse compression, it is desirable to use modulation schemes that can accomplish a maximum pulse compression ratio, and can significantly reduce the sidelobe levels of the compressed waveform. For the LFM case the first sidelobe is approximately $13.4 dB$ below the main peak, and for most radar applications this may not be

LFM Pulse Compression

sufficient. In practice, high sidelobe levels are not preferable because noise and/or jammers located at the sidelobes may interfere with target returns in the main lobe.

Weighting functions (windows) can be used on the compressed pulse spectrum in order to reduce the sidelobe levels. The cost associated with such an approach is a loss in the main lobe resolution, and a reduction in the peak value (i.e., loss in the SNR). Weighting the time domain transmitted or received signal instead of the compressed pulse spectrum will theoretically achieve the same goal. However, this approach is rarely used, since amplitude modulating the transmitted waveform introduces extra burdens on the transmitter.

Consider a radar system that utilizes a correlation processor receiver (i.e., matched filter). The receive window in meters is defined by

$$R_{rec} = R_{max} - R_{min} \tag{5.12}$$

where R_{max} and R_{min}, respectively, define the maximum and minimum range over which the radar performs detection. Typically R_{rec} is limited to the extent of the target complex. The normalized complex transmitted signal has the form

$$s(t) = \exp\left(j2\pi\left(f_0 t + \frac{\mu}{2}t^2\right)\right) \qquad 0 \leq t \leq \tau' \tag{5.13}$$

τ' is the pulsewidth, $\mu = B/\tau'$, and B is the bandwidth.

The radar echo signal is similar to the transmitted one with the exception of a time delay and an amplitude change that correspond to the target RCS. Consider a target at range R_1. The echo received by the radar from this target is

$$s_r(t) = a_1 \exp\left(j2\pi\left(f_0(t-\tau_1) + \frac{\mu}{2}(t-\tau_1)^2\right)\right) \tag{5.14}$$

where a_1 is proportional to target RCS, antenna gain, and range attenuation. The time delay τ_1 is given by

$$\tau_1 = 2R_1/c \tag{5.15}$$

The first step of the processing consists of removing the frequency f_0. This is accomplished by mixing $s_r(t)$ with a reference signal whose phase is $2\pi f_0 t$. The phase of the resultant signal, after low pass filtering, is then given by

$$\psi(t) = 2\pi\left(-f_0\tau_1 + \frac{\mu}{2}(t-\tau_1)^2\right) \tag{5.16}$$

and the instantaneous frequency is

$$f_i(t) = \frac{1}{2\pi}\frac{d}{dt}\psi(t) = \mu(t-\tau_1) = \frac{B}{\tau'}\left(t - \frac{2R_1}{c}\right) \tag{5.17}$$

The quadrature components are

$$\begin{pmatrix} x_I(t) \\ x_Q(t) \end{pmatrix} = \begin{pmatrix} \cos\psi(t) \\ \sin\psi(t) \end{pmatrix} \qquad (5.18)$$

Sampling the quadrature components is performed next. The number of samples, N, must be chosen so that foldover (ambiguity) in the spectrum is avoided. For this purpose, the sampling frequency, f_s (based on the Nyquist sampling rate), must be

$$f_s \geq 2B \qquad (5.19)$$

and the sampling interval is

$$\Delta t \leq 1/2B \qquad (5.20)$$

Using Eq. (5.17) it can be shown that (the proof is left as an exercise) the frequency resolution of the FFT is

$$\Delta f = 1/\tau' \qquad (5.21)$$

The minimum required number of samples is

$$N = \frac{1}{\Delta f \Delta t} = \frac{\tau'}{\Delta t} \qquad (5.22)$$

Equating Eqs. (5.20) and (5.22) yields

$$N \geq 2B\tau' \qquad (5.23)$$

Consequently, a total of $2B\tau'$ real samples, or $B\tau'$ complex samples, is sufficient to completely describe an LFM waveform of duration τ' and bandwidth B. For example, an LFM signal of duration $\tau' = 20$ μs and bandwidth $B = 5$ MHz requires 200 real samples to determine the input signal (100 samples for the I-channel and 100 samples for the Q-channel).

For better implementation of the FFT N is extended to the next power of two, by zero padding. Thus, the total number of samples, for some positive integer m, is

$$N_{FFT} = 2^m \geq N \qquad (5.24)$$

The final steps of the FCP processing include: (1) taking the FFT of the sampled sequence; (2) multiplying the frequency domain sequence of the signal with the FFT of the matched filter impulse response; and (3) performing the inverse FFT of the composite frequency domain sequence in order to generate the time domain compressed pulse (HRR profile). Of course, weighting, antenna gain, and range attenuation compensation must also be performed.

LFM Pulse Compression

Assume that I targets at ranges R_1, R_2, and so forth are within the receive window. From superposition, the phase of the down-converted signal is

$$\psi(t) = \sum_{i=1}^{I} 2\pi\left(-f_0\tau_i + \frac{\mu}{2}(t-\tau_i)^2\right) \tag{5.25}$$

The times $\{\tau_i = (2R_i/c); \; i = 1, 2, ..., I\}$ represent the two-way time delays, where τ_1 coincides with the start of the receive window.

MATLAB Function "matched_filter.m"

The function *"matched_filter.m"* performs fast convolution processing. It is given in Listing 5.2 in Section 5.5. The syntax is as follows:

[y] = matched_filter(nscat, taup, b, rrec, scat_range, scat_rcs, win)

where

Symbol	Description	Units	Status
nscat	number of point scatterers within the received window	none	input
rrec	receive window size	m	input
taup	uncompressed pulsewidth	seconds	input
b	chirp bandwidth	Hz	input
scat_range	vector of scatterers' relative range (within the receive window)	m	input
scat_rcs	vector of scatterers' RCS	m^2	input
win	0 = no window 1 = Hamming 2 = Kaiser with parameter pi 3 = Chebychev - sidelobes at -60dB	none	input
y	normalized compressed output	volts	output

The user can access this function either by a MATLAB function call, or by executing the MATLAB program *"matched_filter_gui.m"* which utilizes a MATLAB based GUI. The work space associated with this program is shown in Fig. 5.5. The outputs for this function include plots of the compressed and uncompressed signals as well as the replica used in the pulse compression process. This function utilizes the function *"power_integer_2.m"* which implements Eq. (5.24). It is given in Listing 5.3 in Section 5.5.

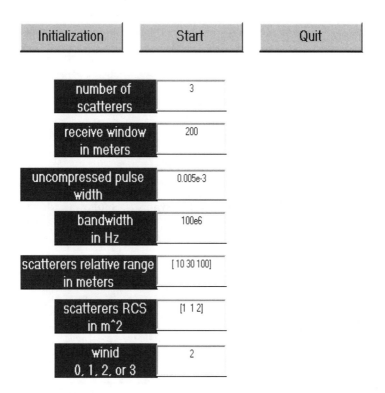

Figure 5.5. GUI workspace associated with the function *"matched_filter_gui.m"*.

As an example, consider the case where

nscat	3
rrec	200 m
taup	0.005 ms
b	100e6 Hz
scat_range	[10 75 120] m
scat_rcs	[1 2 1]m^2
win	2

Note that the compressed pulsed range resolution, without using a window, is $\Delta R = 1.5m$. Figs. 5.6 shows the real part and the amplitude spectrum for the replica used in the pulse compression. Fig. 5.7 shows the uncompressed echo, while Fig. 5.8 shows the compressed MF output. Note that the scatterer amplitude attenuation is a function of the inverse of the scatterer's range within the receive window.

LFM Pulse Compression 245

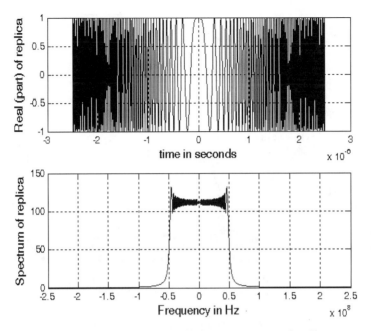

Figure 5.6. Real part and amplitude spectrum of replica.

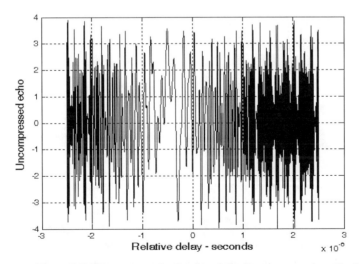

Figure 5.7. Uncompressed echo signal. Scatterers are not resolved.

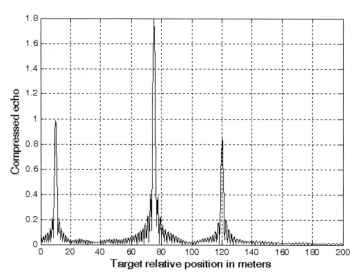

Figure 5.8. Compressed echo signal corresponding to Fig. 5.7. Scatterers are completely resolved.

Fig. 5.9 is similar to Fig. 5.8, except in this case the first and second scatterers are less than 1.5 meter apart (they are at 70 and 71 meters within the receive window).

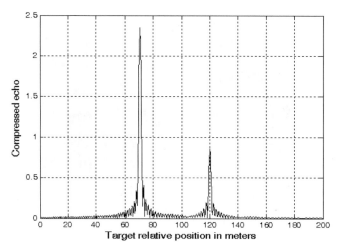

Figure 5.9. Compressed echo signal of three scatterers, two of which are not resolved.

5.3.2. Stretch Processor

Stretch processing, also known as *"active correlation,"* is normally used to process extremely high bandwidth LFM waveforms. This processing technique consists of the following steps: First, the radar returns are mixed with a replica (reference signal) of the transmitted waveform. This is followed by Low Pass Filtering (LPF) and coherent detection. Next, Analog to Digital (A/D) conversion is performed; and finally, a bank of Narrow Band Filters (NBFs) is used in order to extract the tones that are proportional to target range, since stretch processing effectively converts time delay into frequency. All returns from the same range bin produce the same constant frequency. Fig. 5.10a shows a block diagram for a stretch processing receiver. The reference signal is an LFM waveform that has the same LFM slope as the transmitted LFM signal. It exists over the duration of the radar "receive-window," which is computed from the difference between the radar maximum and minimum range. Denote the start frequency of the reference chirp as f_r.

Consider the case when the radar receives returns from a few close (in time or range) targets, as illustrated in Fig. 5.10a. Mixing with the reference signal and performing low pass filtering are effectively equivalent to subtracting the return frequency chirp from the reference signal. Thus, the LPF output consists of constant tones corresponding to the targets' positions. The normalized transmitted signal can be expressed by

$$s_1(t) = \cos\left(2\pi\left(f_0 t + \frac{\mu}{2}t^2\right)\right) \qquad 0 \leq t \leq \tau' \qquad (5.26)$$

where $\mu = B/\tau'$ is the LFM coefficient and f_0 is the chirp start frequency. Assume a point scatterer at range R. The signal received by the radar is

$$s_r(t) = a\cos\left[2\pi\left(f_0(t - \Delta\tau) + \frac{\mu}{2}(t - \Delta\tau)^2\right)\right] \qquad (5.27)$$

where a is proportional to target RCS, antenna gain, and range attenuation. The time delay $\Delta\tau$ is

$$\Delta\tau = 2R/c \qquad (5.28)$$

The reference signal is

$$s_{ref}(t) = 2\cos\left(2\pi\left(f_r t + \frac{\mu}{2}t^2\right)\right) \qquad 0 \leq t \leq T_{rec} \qquad (5.29)$$

The receive window in seconds is

$$T_{rec} = \frac{2(R_{max} - R_{min})}{c} = \frac{2R_{rec}}{c} \qquad (5.30)$$

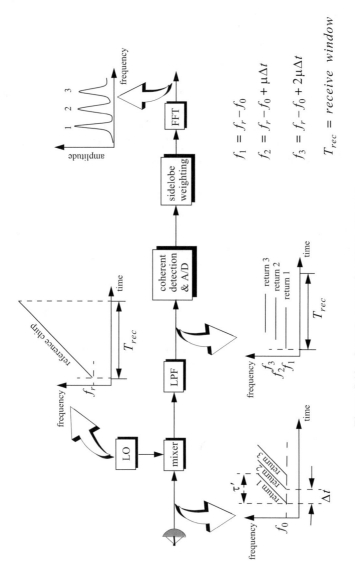

Figure 5.10a. Stretch processing block diagram.

It is customary to let $f_r = f_0$. The output of the mixer is the product of the received and reference signals. After low pass filtering the signal is

$$s_o(t) = a\cos(2\pi f_0 \Delta\tau + 2\pi\mu\Delta\tau t - \pi\mu(\Delta\tau)^2) \tag{5.31}$$

Substituting Eq. (5.28) into (5.31) and collecting terms yield

$$s_o(t) = a\cos\left[\left(\frac{4\pi BR}{c\tau'}\right)t + \frac{2R}{c}\left(2\pi f_0 - \frac{2\pi BR}{c\tau'}\right)\right] \tag{5.32}$$

and since $\tau' \gg 2R/c$, Eq. (5.32) is approximated by

$$s_o(t) \approx a\cos\left[\left(\frac{4\pi BR}{c\tau'}\right)t + \frac{4\pi R}{c}f_0\right] \tag{5.33}$$

The instantaneous frequency is

$$f_{inst} = \frac{1}{2\pi}\frac{d}{dt}\left(\frac{4\pi BR}{c\tau'}t + \frac{4\pi R}{c}f_0\right) = \frac{2BR}{c\tau'} \tag{5.34}$$

which clearly indicates that target range is proportional to the instantaneous frequency. Therefore, proper sampling of the LPF output and taking the FFT of the sampled sequence lead to the following conclusion: a peak at some frequency f_1 indicates presence of a target at range

$$R_1 = f_1 c\tau'/2B \tag{5.35}$$

Assume I close targets at ranges R_1, R_2, and so forth ($R_1 < R_2 < ... < R_I$). From superposition, the total signal is

$$s_r(t) = \sum_{i=1}^{I} a_i(t)\cos\left[2\pi\left(f_0(t-\tau_i) + \frac{\mu}{2}(t-\tau_i)^2\right)\right] \tag{5.36}$$

where $\{a_i(t); i = 1, 2, ..., I\}$ are proportional to the targets' cross sections, antenna gain, and range. The times $\{\tau_i = (2R_i/c); i = 1, 2, ..., I\}$ represent the two-way time delays, where τ_1 coincides with the start of the receive window. Using Eq. (5.32) the overall signal at the output of the LPF can then be described by

$$s_o(t) = \sum_{i=1}^{I} a_i \cos\left[\left(\frac{4\pi BR_i}{c\tau'}\right)t + \frac{2R_i}{c}\left(2\pi f_0 - \frac{2\pi BR_i}{c\tau'}\right)\right] \tag{5.37}$$

And hence, target returns appear as constant frequency tones that can be resolved using the FFT. Consequently, determining the proper sampling rate and FFT size is very critical. The rest of this section presents a methodology for computing the proper FFT parameters required for stretch processing.

Assume a radar system using a stretch processor receiver. The pulsewidth is τ' and the chirp bandwidth is B. Since stretch processing is normally used in extreme bandwidth cases (i.e., very large B), the receive window over which radar returns will be processed is typically limited to from a few meters to possibly less than 100 meters. The compressed pulse range resolution is computed from Eq. (5.8). Declare the FFT size to be N and its frequency resolution to be Δf. The frequency resolution can be computed using the following procedure: consider two adjacent point scatterers at range R_1 and R_2. The minimum frequency separation, Δf, between those scatterers so that they are resolved can be computed from Eq. (5.34). More precisely,

$$\Delta f = f_2 - f_1 = \frac{2B}{c\tau'}(R_2 - R_1) = \frac{2B}{c\tau'}\Delta R \tag{5.38}$$

Substituting Eq. (5.8) into Eq. (5.38) yields

$$\Delta f = \frac{2B}{c\tau'} \frac{c}{2B} = \frac{1}{\tau'} \tag{5.39}$$

The maximum frequency resolvable by the FFT is limited to the region $\pm N\Delta f/2$. Thus, the maximum resolvable frequency is

$$\frac{N\Delta f}{2} > \frac{2B(R_{max} - R_{min})}{c\tau'} = \frac{2BR_{rec}}{c\tau'} \tag{5.40}$$

Using Eqs. (5.30) and (5.39) into Eq. (5.40) and collecting terms yield

$$N > 2BT_{rec} \tag{5.41}$$

For better implementation of the FFT, choose an FFT of size

$$N_{FFT} \geq N = 2^m \tag{5.42}$$

m is a nonzero positive integer. The sampling interval is then given by

$$\Delta f = \frac{1}{T_s N_{FFT}} \Rightarrow T_s = \frac{1}{\Delta f N_{FFT}} \tag{5.43}$$

MATLAB Function "stretch.m"

The function *"stretch.m"* presents a digital implementation of stretch processing. It is given in Listing 5.4 in Section 5.5. The syntax is as follows:

[y] = stretch (nscat, taup, f0, b, scat_range, rrec, scat_rcs, win)

where

Symbol	Description	Units	Status
nscat	number of point scatterers within the received window	none	input
taup	uncompressed pulsewidth	seconds	input
f0	chirp start frequency	Hz	input
b	chirp bandwidth	Hz	input
scat_range	vector of scatterers' range	m	input
rrec	range receive window	m	input
scat_rcs	vector of scatterers' RCS	m^2	input
win	0 = no window 1 = Hamming 2 = Kaiser with parameter pi 3 = Chebychev - sidelobes at -60dB	none	input
y	compressed output	volts	output

The user can access this function either by a MATLAB function call or by executing the MATLAB program *"stretch_gui.m"* which utilizes MATLAB based GUI and is shown in Fig. 5.10b. The outputs of this function are the complex array y and plots of the uncompressed and compressed echo signal versus time. As an example, consider the case where

nscat	3
taup	10 ms
f0	5.6 GHz
b	1 GHz
rrec	30 m
scat_range	[2 5 10] m
scat_rcs	[1, 1, 2] m^2
win	2 (Kaiser)

Note that the compressed pulse range resolution, without using a window, is $\Delta R = 0.15 m$. Figs. 5.11 and 5.12, respectively, show the uncompressed and compressed echo signals corresponding to this example. Fig. 5.13 is similar to Figs. 5.11 and 5.12 except in this case two of the scatterers are less than 15 cm apart (i.e., unresolved targets at $R_{relative} = [3, 3.1] m$).

252 *MATLAB Simulations for Radar Systems Design*

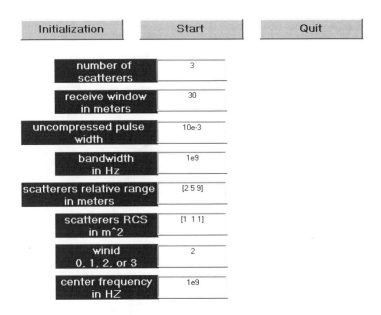

Figure 5.10b. GUI workspace associated with the function *"stretch_gui.m"*.

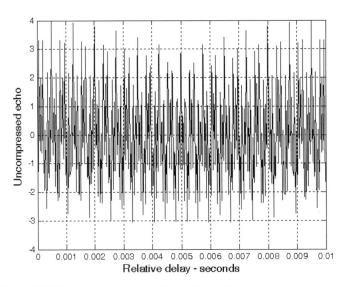

Figure 5.11. Uncompressed echo signal. Three targets are unresolved.

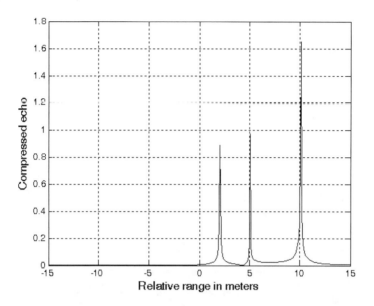

Figure 5.12. Compressed echo signal. Three targets are resolved.

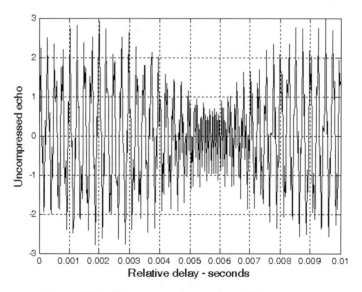

Figure 5.13a. Uncompressed echo signal. Three targets.

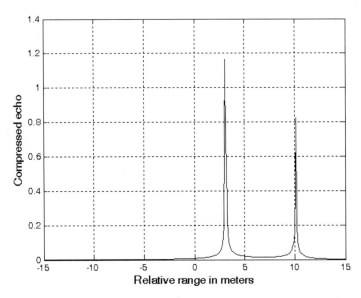

Figure 5.13b. Compressed echo signal. Three targets, two are not resolved.

5.3.3. Distortion Due to Target Velocity

Up to this point, we have analyzed pulse compression with no regard to target velocity. In fact, all analyses provided assumed stationary targets. Uncompensated target radial velocity, or equivalently Doppler shift, degrades the quality of the HRR profile generated by pulse compression. In Chapter 3, the effects of radial velocity on SFW were analyzed. Similar distortion in the HRR profile is also present with LFM waveforms when target radial velocity is not compensated for.

The two effects of target radial velocity (Doppler frequency) on the radar received pulse were developed in Chapter 1. When the target radial velocity is not zero, the received pulsewidth is expanded (or compressed) by the time dilation factor. Additionally, the received pulse center frequency is shifted by the amount of Doppler frequency. When these effects are not compensated for, the pulse compression processor output is distorted. This is illustrated in Fig. 5.14. Fig. 5.14a shows a typical output of the pulse compression processor with no distortion. Alternatively, Figs. 5.14b, 5.14c, and 5.14d show the output of the pulse compression processor when 5% shift of the chirp center frequency, 10% time dilation, and both are present.

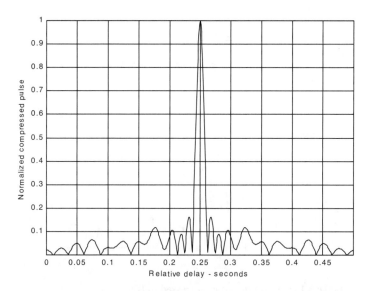

Figure 5.14a. Compressed pulse output of a pulse compression processor. No distortion is present. This figure can be reproduced using MATLAB program *"fig5_14"* given in Listing 5.5 in Section 5.5.

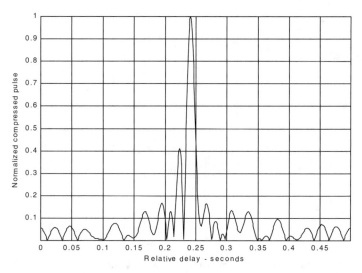

Figure 5.14b. Mismatched compressed pulse; 5% Doppler shift.

256   MATLAB Simulations for Radar Systems Design

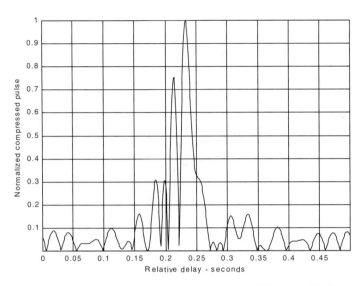

Figure 5.14c. Mismatched compressed pulse; 10% time dilation.

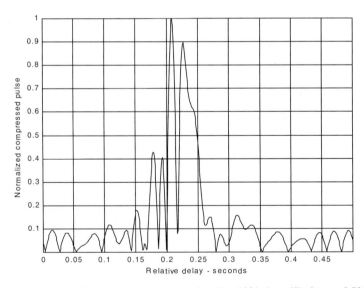

Figure 5.14d. Mismatched compressed pulse; 10% time dilation and 5% Doppler shift.

Correction for the distortion caused by the target radial velocity can be overcome by using the following approach. Over a period of a few pulses, the radar data processor estimates the radial velocity of the target under track. Then, the chirp slope and pulsewidth of the next transmitted pulse are changed to account for the estimated Doppler frequency and time dilation.

5.4. "MyRadar" Design Case Study - Visit 5

5.4.1. Problem Statement

Assume that the threat may consist of multiple aircraft and missiles. Show how the matched filter receiver can resolve multiple targets with a minimum range separation of 50 meters. Also verify that the waveforms selected in Chapter 3 are adequate to maintain proper detection and tracking (i.e., provide sufficient SNR).

5.4.2. A Design

It was determined in Chapter 3 that the pulsed compressed range resolutions during search and track are respectively given by

$$\Delta R_{search} = 30m; \quad B_{search} = 5MHz \tag{5.44}$$

$$\Delta R_{track} = 7.5m; \quad B_{track} = 20MHz \tag{5.45}$$

It was also determined that a single search waveform and 4 track waveforms would be used.

Assume that track is initiated once detection is declared. Aircraft target type are detected at $R_{max}^a = 90Km$ while the missile is detected at $R_{max}^m = 55Km$. It was shown in Section 2.10.2.2 that the minimum SNR at these ranges for both target types is $SNR \geq 4dB$ when 4-pulse non-coherent integration is utilized along with cumulative detection. It was also determined that a single pulse option was not desirable since it required prohibitive values for the peak power. At this point one should however take advantage of the increased SNR due to pulse compression. From Chapter 3, the pulse compression gain, for the selected waveforms, is equal to 100 (10 dB). One should investigate this SNR enhancement in the context of eliminating the need for pulse integration.

The pulsed compressed SNR can be computed using Eq. (5.7), which is repeated here as Eq. (5.46)

$$SNR = \frac{P_t \tau' G^2 \lambda^2 \sigma}{(4\pi)^3 R^4 k T_e FL} \tag{5.46}$$

where $G = 34.5dB$, $\lambda = 0.1m$, $T_e = 290 Kelvin$, $F = 6dB$, $L = 8dB$, $\sigma_m = 0.5m^2$, $\sigma_a = 4m^2$, and $P_t = 20KW$ (from Chapter 3). The search pulsewidth is $\tau' = 20\mu s$ and the track waveforms are $12.5\mu s \leq \tau'_i \leq 20\mu s$. First consider the missile case. The single pulse SNR at the maximum detection range $R^m_{max} = 55Km$ is given by

$$SNR_m = \frac{20 \times 10^3 \times 20 \times 10^{-6} \times (10^{3.45})^2 \times (0.1)^2 \times 0.5}{(4\pi)^3 \times (55 \times 10^3)^4 \times 1.38 \times 10^{-23} \times 290 \times 10^{0.8} \times 10^{0.6}} = \qquad (5.47)$$

$$8.7028 \Rightarrow SNR_m = 9.39dB$$

Alternatively, the single pulse SNR, with pulse compression, for the aircraft is

$$SNR_a = \frac{20 \times 10^3 \times 20 \times 10^{-6} \times (10^{3.45})^2 \times (0.1)^2 \times 4}{(4\pi)^3 \times (90 \times 10^3)^4 \times 1.38 \times 10^{-23} \times 290 \times 10^{0.8} \times 10^{0.6}} = \qquad (5.48)$$

$$9.7104 \Rightarrow SNR_m = 9.87dB$$

Using these calculated SNR values into the MATLAB program "myradar_visit2_2.m" (see Chapter 2) yields

$$P_{DC_{Aircarft}} = 0.999$$
$$P_{DC_{Missile}} = 0.9984 \qquad (5.49)$$

which clearly satisfies the design requirement of $P_D \geq 0.995$.

Next, consider the matched filter and its replicas and pulsed compressed outputs (due to different waveforms). For this purpose use the program "matched_filter_gui.m". Assume a receive window of 200 meters during search and 50 meters during track.

Fig. 5.15 shows the replica and the associated uncompressed and compressed signals. The targets consist of two aircraft separated by 50 meters. Fig. 5.16 is similar to Fig. 5.15, except it is for track waveform number 4 and the target separation is 20 m.

"MyRadar" Design Case Study - Visit 5   259

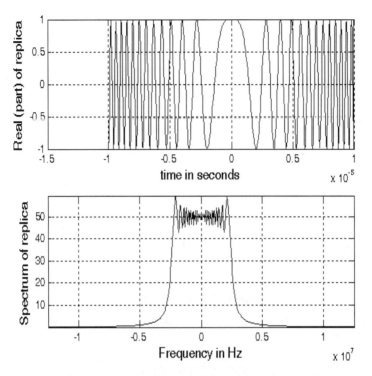

Figure 5.15a. Replica associated with search waveform.

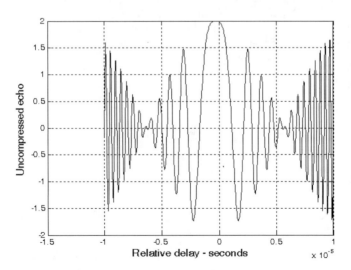

Figure 5.15b. Uncompressed signal of two aircraft separated by 50 m.

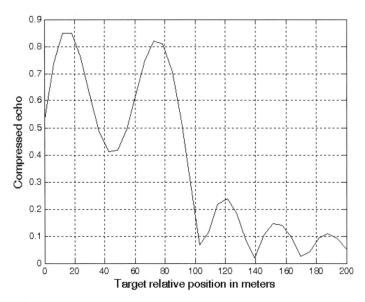

Figure 5.15c. Compressed signal corresponding to Fig. 5.15b. No window.

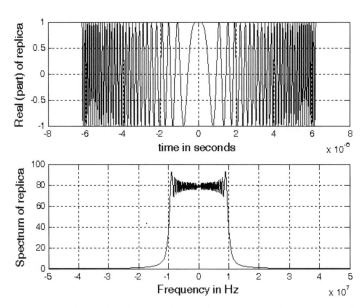

Figure 5.16a. Replica associated with track waveform number 4.

"MyRadar" Design Case Study - Visit 5   261

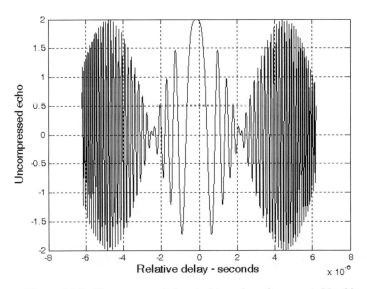

Figure 5.16b. Uncompressed signal of two aircraft separated by 20 m.

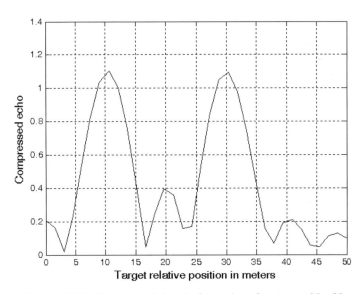

Figure 5.16b. Compressed signal of two aircraft separated by 20 m.

5.5. MATLAB Program and Function Listings

This section presents listings for all MATLAB programs/functions used in this chapter. The user is strongly advised to rerun the MATLAB programs in order to enhance his understanding of this chapter's material.

Listing 5.1. MATLAB Program "fig5_3.m"

```
% use this program to reproduce Fig. 5.3 of text
clear all
close all
nscat = 2; %two point scatterers
taup = 10e-6; % 10 microsecond uncompressed pulse
b = 50.0e6; % 50 MHz bandwidth
rrec = 50 ; % 50 meter processing window
scat_range = [15 25] ; % scatterers are 15 and 25 meters into window
scat_rcs = [1 2]; % RCS 1 m^2 and 2m^2
winid = 0; %no window used
[y] = matched_filter(nscat,taup,b,rrec,scat_range,scat_rcs,winid);
```

Listing 5.2. MATLAB Function "matched_filter.m"

```
function [y] = matched_filter(nscat,taup,b,rrec,scat_range,scat_rcs,winid)
eps = 1.0e-16;
% time bandwidth product
time_B_product = b * taup;
if(time_B_product < 5 )
   fprintf('********** Time Bandwidth product is TOO SMALL ********')
   fprintf('\n Change b and or taup')
   return
end
% speed of light
c = 3.e8;
% number of samples
n = fix(5 * taup * b)
% initialize input, output and replica vectors
x(nscat,1:n) = 0.;
y(1:n) = 0.;
replica(1:n) = 0.;
% determine proper window
if( winid == 0.)
  win(1:n) = 1.;
  win =win';
else
  if(winid == 1.)
```

```
      win = hamming(n);
   else
      if( winid == 2.)
         win = kaiser(n,pi);
      else
         if(winid == 3.)
            win = chebwin(n,60);
         end
      end
   end
end
% check to ensure that scatterers are within receive window
index = find(scat_range > rrec);
if (index ~= 0)
   'Error. Receive window is too large; or scatterers fall outside window'
   return
end
% calculate sampling interval
t = linspace(-taup/2,taup/2,n);
replica = exp(i * pi * (b/taup) .* t.^2);
figure(1)
subplot(2,1,1)
plot(t,real(replica))
ylabel('Real (part) of replica')
xlabel('time in seconds')
grid
subplot(2,1,2)
sampling_interval = taup / n;
freqlimit = 0.5/ sampling_interval;
freq = linspace(-freqlimit,freqlimit,n);
plot(freq,fftshift(abs(fft(replica))));
ylabel('Spectrum of replica')
xlabel('Frequency in Hz')
grid
 for j = 1:1:nscat
    range = scat_range(j) ;;
    x(j,:) = scat_rcs(j) .* exp(i * pi * (b/taup) .* (t +(2*range/c)).^2) ;
    y = x(j,:) + y;
end
figure(2)
plot(t,real(y),'k')
xlabel ('Relative delay - seconds')
ylabel ('Uncompressed echo')
grid
```

```
out =xcorr(replica, y);
out = out ./ n;
s = taup * c /2;
Npoints = ceil(rrec * n /s);
dist =linspace(0, rrec, Npoints);
delr = c/2/b
figure(3)
plot(dist,abs(out(n:n+Npoints-1)),'k')
xlabel ('Target relative position in meters')
ylabel ('Compressed echo')
grid
```

Listing 5.3. MATLAB Function "power_integer_2.m"

```
function n = power_integer_2 (x)
m = 0.;
for j = 1:30
  m = m + 1.;
  delta = x - 2.^m;
  if(delta < 0.)
    n = m;
    return
  else
  end
end
```

Listing 5.4. MATLAB Function "stretch.m"

```
function [y] = stretch(nscat,taup,f0,b,rrec,scat_range,scat_rcs,winid)
eps = 1.0e-16;
htau = taup / 2.;
c = 3.e8;
trec = 2. * rrec / c;
n = fix(2. * trec * b);
m = power_integer_2(n);
nfft = 2.^m;
x(nscat,1:n) = 0.;
y(1:n) = 0.;
if(winid == 0.)
  win(1:n) = 1.;
  win =win';
else
  if(winid == 1.)
    win = hamming(n);
  else
```

```
    if( winid == 2.)
       win = kaiser(n,pi);
    else
       if(winid == 3.)
          win = chebwin(n,60);
       end
    end
  end
end
deltar = c / 2. / b;
max_rrec = deltar * nfft / 2.;
maxr = max(scat_range);
if(rrec > max_rrec | maxr >= rrec )
   'Error. Receive window is too large; or scatterers fall outside window'
   return
end
t = linspace(0,taup,n);
for j = 1:1:nscat
   range = scat_range(j);% + rmin;
   psi1 = 4. * pi * range * f0 / c - ...
      4. * pi * b * range * range / c / c/ taup;
   psi2 = (2*4. * pi * b * range / c / taup) .* t;
   x(j,:) = scat_rcs(j) .* exp(i * psi1 + i .* psi2);
   y = y + x(j,:);
end
figure(1)
plot(t,real(y),'k')
xlabel ('Relative delay - seconds')
ylabel ('Uncompressed echo')
ywin = y .* win';
yfft = fft(y,n) ./ n;
out= fftshift(abs(yfft));
figure(2)
delinc = rrec/ n;
%dist = linspace(-delinc-rrec/2,rrec/2,n);
dist = linspace((-rrec/2), rrec/2,n);
plot(dist,out,'k')
xlabel ('Relative range in meters')
ylabel ('Compressed echo')
axis auto
grid
```

Listing 5.5. MATLAB Program "fig5_14.m"

```
% use this program to reproduce Fig. 5.14 of text
clear all
eps = 1.5e-5;
t = 0:0.001:.5;
y = chirp(t,0,.25,20);
figure(1)
plot(t,y);
yfft = fft(y,512) ;
ycomp = fftshift(abs(ifft(yfft .* conj(yfft))));
maxval = max (ycomp);
ycomp = eps + ycomp ./ maxval;
figure(1)
del = .5 /512.;
tt = 0:del:.5-eps;
plot (tt,ycomp,'k')
axis tight
xlabel ('Relative delay - seconds');
ylabel('Normalized compressed pulse')
grid
y1 = chirp (t,0,.25,21); % change center frequency
y1fft = fft(y1,512);
y1comp = fftshift(abs(ifft(y1fft .* conj(yfft))));
maxval = max (y1comp);
y1comp = eps + y1comp ./ maxval;
figure(2)
plot (tt,y1comp,'k')
axis tight
xlabel ('Relative delay - seconds');
ylabel('Normalized compressed pulse')
grid
t = 0:0.001:.45; % change pulsewidth
y2 = chirp (t,0,.225,20);
y2fft = fft(y2,512);
y2comp = fftshift(abs(ifft(y2fft .* conj(yfft))));
maxval = max (y2comp);
y2comp = eps + y2comp ./ maxval;
figure(3)
plot (tt,y2comp,'k')
axis tight
xlabel ('Relative delay - seconds');
ylabel('Normalized compressed pulse')
grid
```

Chapter 6 Surface and Volume Clutter

6.1. Clutter Definition

Clutter is a term used to describe any object that may generate unwanted radar returns that may interfere with normal radar operations. Parasitic returns that enter the radar through the antenna's main lobe are called main lobe clutter; otherwise they are called sidelobe clutter. Clutter can be classified into two main categories: surface clutter and airborne or volume clutter. Surface clutter includes trees, vegetation, ground terrain, man-made structures, and sea surface (sea clutter). Volume clutter normally has a large extent (size) and includes chaff, rain, birds, and insects. Surface clutter changes from one area to another, while volume clutter may be more predictable.

Clutter echoes are random and have thermal noise-like characteristics because the individual clutter components (scatterers) have random phases and amplitudes. In many cases, the clutter signal level is much higher than the receiver noise level. Thus, the radar's ability to detect targets embedded in high clutter background depends on the Signal-to-Clutter Ratio (SCR) rather than the SNR.

White noise normally introduces the same amount of noise power across all radar range bins, while clutter power may vary within a single range bin. Since clutter returns are target-like echoes, the only way a radar can distinguish target returns from clutter echoes is based on the target RCS σ_t, and the anticipated clutter RCS σ_c (via clutter map). Clutter RCS can be defined as the equivalent radar cross section attributed to reflections from a clutter area, A_c. The average clutter RCS is given by

$$\sigma_c = \sigma^0 A_c \tag{6.1}$$

where $\sigma^0(m^2/m^2)$ is the clutter scattering coefficient, a dimensionless quantity that is often expressed in dB. Some radar engineers express σ^0 in terms of squared centimeters per squared meter. In these cases, σ^0 is $40dB$ higher than normal.

6.2. Surface Clutter

Surface clutter includes both land and sea clutter, and is often called area clutter. Area clutter manifests itself in airborne radars in the look-down mode. It is also a major concern for ground-based radars when searching for targets at low grazing angles. The grazing angle ψ_g is the angle from the surface of the earth to the main axis of the illuminating beam, as illustrated in Fig. 6.1.

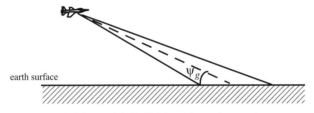

Figure 6.1. Definition of grazing angle.

Three factors affect the amount of clutter in the radar beam. They are the grazing angle, surface roughness, and the radar wavelength. Typically, the clutter scattering coefficient σ^0 is larger for smaller wavelengths. Fig. 6.2 shows a sketch describing the dependency of σ^0 on the grazing angle. Three regions are identified; they are the low grazing angle region, flat or plateau region, and the high grazing angle region.

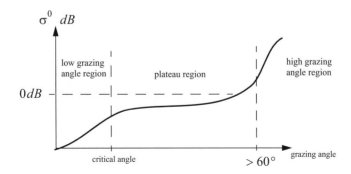

Figure 6.2. Clutter regions.

The low grazing angle region extends from zero to about the critical angle. The critical angle is defined by Rayleigh as the angle below which a surface is considered to be smooth, and above which a surface is considered to be rough; Denote the root mean square (rms) of a surface height irregularity as h_{rms}, then according to the Rayleigh criteria the surface is considered to be smooth if

$$\frac{4\pi h_{rms}}{\lambda}\sin\psi_g < \frac{\pi}{2} \quad (6.2)$$

Consider a wave incident on a rough surface, as shown in Fig. 6.3. Due to surface height irregularity (surface roughness), the "rough path" is longer than the "smooth path" by a distance $2h_{rms}\sin\psi_g$. This path difference translates into a phase differential $\Delta\psi$:

$$\Delta\psi = \frac{2\pi}{\lambda} 2h_{rms}\sin\psi_g \quad (6.3)$$

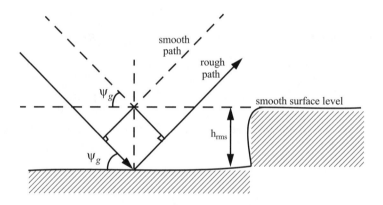

Figure 6.3. Rough surface definition.

The critical angle ψ_{gc} is then computed when $\Delta\psi = \pi$ (first null), thus

$$\frac{4\pi h_{rms}}{\lambda}\sin\psi_{gc} = \pi \quad (6.4)$$

or equivalently,

$$\psi_{gc} = \operatorname{asin}\frac{\lambda}{4h_{rms}} \quad (6.5)$$

In the case of sea clutter, for example, the rms surface height irregularity is

$$h_{rms} \approx 0.025 + 0.046 \ S_{state}^{1.72} \qquad (6.6)$$

where S_{state} is the sea state, which is tabulated in several cited references. The sea state is characterized by the wave height, period, length, particle velocity, and wind velocity. For example, $S_{state} = 3$ refers to a moderate sea state, where in this case the wave height is approximately between $0.9144 \ to \ 1.2192 \ m$, the wave period 6.5 to 4.5 seconds, wave length $1.9812 \ to \ 33.528 \ m$, wave velocity 20.372 to 25.928 Km/hr, and wind velocity 22.224 to 29.632 Km/hr.

Clutter at low grazing angles is often referred to as diffuse clutter, where there are a large number of clutter returns in the radar beam (non-coherent reflections). In the flat region the dependency of σ^0 on the grazing angle is minimal. Clutter in the high grazing angle region is more specular (coherent reflections) and the diffuse clutter components disappear. In this region the smooth surfaces have larger σ^0 than rough surfaces, opposite of the low grazing angle region.

6.2.1. Radar Equation for Area Clutter - Airborne Radar

Consider an airborne radar in the look-down mode shown in Fig. 6.4. The intersection of the antenna beam with the ground defines an elliptically shaped footprint. The size of the footprint is a function of the grazing angle and the antenna 3dB beamwidth θ_{3dB}, as illustrated in Fig. 6.5. The footprint is divided into many ground range bins each of size $(c\tau/2)\sec\psi_g$, where τ is the pulsewidth.

From Fig. 6.5, the clutter area A_c is

$$A_c \approx R\theta_{3dB} \frac{c\tau}{2} \sec\psi_g \qquad (6.7)$$

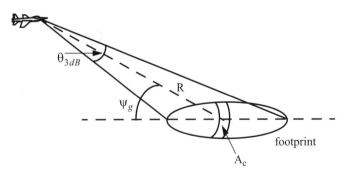

Figure 6.4. Airborne radar in the look-down mode.

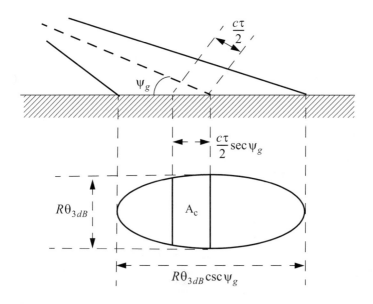

Figure 6.5. Footprint definition.

The power received by the radar from a scatterer within A_c is given by the radar equation as

$$S_t = \frac{P_t G^2 \lambda^2 \sigma_t}{(4\pi)^3 R^4} \tag{6.8}$$

where, as usual, P_t is the peak transmitted power, G is the antenna gain, λ is the wavelength, and σ_t is the target RCS. Similarly, the received power from clutter is

$$S_C = \frac{P_t G^2 \lambda^2 \sigma_c}{(4\pi)^3 R^4} \tag{6.9}$$

where the subscript C is used for area clutter. Substituting Eq. (6.1) for σ_c into Eq. (6.9), we can then obtain the SCR for area clutter by dividing Eq. (6.8) by Eq. (6.9). More precisely,

$$(SCR)_C = \frac{2\sigma_t \cos\psi_g}{\sigma^0 \theta_{3dB} R \tau} \tag{6.10}$$

Example:

Consider an airborne radar shown in Fig. 6.4. Let the antenna 3dB beamwidth be $\theta_{3dB} = 0.02\,rad$, the pulsewidth $\tau = 2\mu s$, range $R = 20\,Km$, and grazing angle $\psi_g = 20°$. The target RCS is $\sigma_t = 1m^2$. Assume that the clutter reflection coefficient is $\sigma^0 = 0.0136$. Compute the SCR.

Solution:

The SCR is given by Eq. (6.10) as

$$(SCR)_C = \frac{2\sigma_t \cos\psi_g}{\sigma^0 \theta_{3dB} Rc\tau} \Rightarrow$$

$$(SCR)_C = \frac{(2)(1)(\cos 20°)}{(0.0136)(0.02)(20000)(3\times 10^8)(2\times 10^{-6})} = 5.76 \times 10^{-4}$$

It follows that

$$(SCR)_C = -32.4 dB$$

Thus, for reliable detection the radar must somehow increase its SCR by at least $(32 + X)dB$, where X is on the order of 13 to 15 dB or better.

6.2.2. Radar Equation for Area Clutter - Ground Based Radar

Again the received power from clutter is also calculated using Eq. (6.9). However, in this case the clutter RCS σ_c is computed differently. It is

$$\sigma_c = \sigma_{MBc} + \sigma_{SLc} \qquad (6.11)$$

where σ_{MBc} is the main beam clutter RCS and σ_{SLc} is the sidelobe clutter RCS, as illustrated in Fig. 6.6.

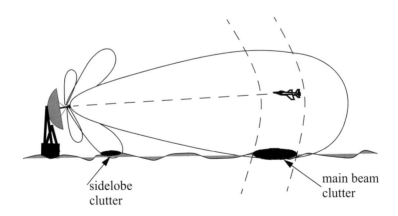

Figure 6.6. Geometry for ground based radar clutter

Surface Clutter

In order to calculate the total clutter RCS given in Eq. (6.11), one must first compute the corresponding clutter areas for both the main beam and the sidelobes. For this purpose, consider the geometry shown in Fig. 6.7. The angles θ_A and θ_E represent the antenna 3-dB azimuth and elevation beamwidths, respectively. The radar height (from the ground to the phase center of the antenna) is denoted by h_r, while the target height is denoted by h_t. The radar slant range is R, and its ground projection is R_g. The range resolution is ΔR and its ground projection is ΔR_g. The main beam clutter area is denoted by A_{MBc} and the sidelobe clutter area is denoted by A_{SLc}.

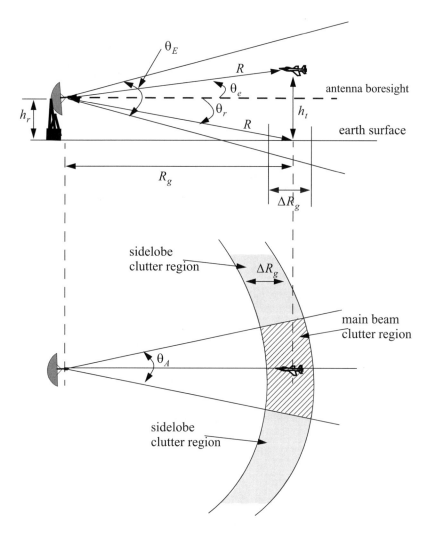

Figure 6.7. Clutter geometry for ground based radar. Side view and top view.

From Fig. 6.7 the following relations can be derived

$$\theta_r = \mathrm{asin}(h_r/R) \tag{6.12}$$

$$\theta_e = \mathrm{asin}((h_t - h_r)/R) \tag{6.13}$$

$$\Delta R_g = \Delta R \cos\theta_r \tag{6.14}$$

where ΔR is the radar range resolution. The slant range ground projection is

$$R_g = R\cos\theta_r \tag{6.15}$$

It follows that the main beam and the sidelobe clutter areas are

$$A_{MBc} = \Delta R_g \; R_g \; \theta_A \tag{6.16}$$

$$A_{SLc} = \Delta R_g \; \pi R_g \tag{6.17}$$

Assume a radar antenna beam $G(\theta)$ of the form

$$G(\theta) = \exp\left(-\frac{2.776\theta^2}{\theta_E^2}\right) \Rightarrow Gaussian \tag{6.18}$$

$$G(\theta) = \begin{cases} \left(\dfrac{\sin\left(2.78\dfrac{\theta}{\theta_E}\right)}{\left(2.78\dfrac{\theta}{\theta_E}\right)}\right)^2 & ; |\theta| \le \dfrac{\pi\theta_E}{2.78} \\ 0 & ; elsewhere \end{cases} \Rightarrow \left(\frac{\sin(x)}{x}\right)^2 \tag{6.19}$$

Then the main beam clutter RCS is

$$\sigma_{MBc} = \sigma^0 A_{MBc} G^2(\theta_e + \theta_r) = \sigma^0 \Delta R_g \; R_g \; \theta_A G^2(\theta_e + \theta_r) \tag{6.20}$$

and the sidelobe clutter RCS is

$$\sigma_{SLc} = \sigma^0 A_{SLc} (SL_{rms})^2 = \sigma^0 \Delta R_g \; \pi R_g (SL_{rms})^2 \tag{6.21}$$

where the quantity SL_{rms} is the root-mean-square (rms) for the antenna sidelobe level.

Finally, in order to account for the variation of the clutter RCS versus range, one can calculate the total clutter RCS as a function of range. It is given by

$$\sigma_c(R) = \frac{\sigma_{MBc} + \sigma_{SLc}}{(1 + (R/R_h)^4)} \tag{6.22}$$

where R_h is the radar range to the horizon calculated as

Surface Clutter

$$R_h = \sqrt{\frac{8h_r r_e}{3}} \qquad (6.23)$$

where r_e is the Earth's radius equal to $6371 Km$. The denominator in Eq. (6.22) is put in that format in order to account for refraction and for round (spherical) Earth effects.

The radar SNR due to a target at range R is

$$SNR = \frac{P_t G^2 \lambda^2 \sigma_t}{(4\pi)^3 R^4 k T_0 BFL} \qquad (6.24)$$

where, as usual, P_t is the peak transmitted power, G is the antenna gain, λ is the wavelength, σ_t is the target RCS, k is Boltzman's constant, T_0 is the effective noise temperature, B is the radar operating bandwidth, F is the receiver noise figure, and L is the total radar losses. Similarly, the Clutter-to-Noise (CNR) at the radar is

$$CNR = \frac{P_t G^2 \lambda^2 \sigma_c}{(4\pi)^3 R^4 k T_0 BFL} \qquad (6.25)$$

where the σ_c is calculated using Eq. (6.21).

When the clutter statistic is Gaussian, the clutter signal return and the noise return can be combined, and a new value for determining the radar measurement accuracy is derived from the Signal-to-Clutter+Noise-Ratio, denoted by SIR. It is given by

$$SIR = \frac{1}{\frac{1}{SNR} + \frac{1}{SCR}} \qquad (6.26)$$

Note that the SCR is computed by dividing Eq.(6.24) by Eq. (6.25).

MATLAB Function "clutter_rcs.m"

The function *"clutter_rcs.m"* implements Eq. (6.22); it is given in Listing 6.1 in Section 6.6. It also generates plots of the clutter RCS and the CNR versus the radar slant range. Its outputs include the clutter RCS in dBsm and the CNR in dB. The syntax is as follows:

[sigmaC,CNR] = clutter_rcs(sigma0, thetaE, thetaA, SL, range, hr, ht, pt, f0, b, t0, f, l, ant_id)

where

Symbol	Description	Units	Status
sigma0	clutter back scatterer coefficient	dB	input
thetaE	antenna 3dB elevation beamwidth	degrees	input
thetaA	antenna 3dB azimuth beamwidth	degrees	input
SL	antenna sidelobe level	dB	input
range	range; can be a vector or a single value	Km	input
hr	radar height	meters	input
ht	target height	meters	input
pt	radar peak power	KW	input
f0	radar operating frequency	Hz	input
b	bandwidth	Hz	input
t0	effective noise temperature	Kelvins	input
f	noise figure	dB	input
l	radar losses	dB	input
ant_id	1 for (sin(x)/x)^2 pattern 2 for Gaussian pattern	none	input
sigmac	clutter RCS; can be either vector or single value depending on "range"	dB	output
CNR	clutter to noise ratio; can be either vector or single value depending on "range"	dB	output

A GUI called *"clutter_rcs_gui"* was developed for this function. Executing this GUI generates plots of the σ_c and CNR versus range. Figure 6.8 shows typical plots produced by this GUI using the antenna pattern defined in Eq. (6.18). Figure 6.9 is similar to Fig. 6.8 except in this case Eq. (6.19) is used for the antenna pattern. Note that the dip in the clutter RCS (at very close range) occurs at the grazing angle corresponding to the null between the main beam and the first sidelobe. Fig. 6.9c shows the GUI workspace associated with this function.

In order to reproduce those two figures use the following MATLAB calls:

*[sigmaC,CNR] = clutter_rcs(-20, 2, 1, -20, linspace(2,50,100), 3, 100, 75,
5.6e9, 1e6, 290, 6, 10, 1)* **(6.27)**

*[sigmaC,CNR] = clutter_rcs(-20, 2, 1, -25, linspace(2,50,100), 3, 100, 100,
5.6e9, 1e6, 290, 6, 10, 2)* **(6.28)**

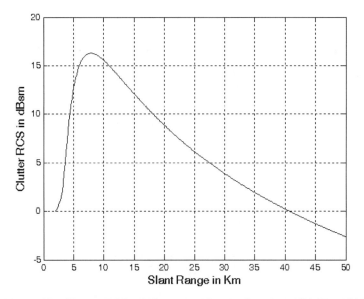

Figure 6.8a. Clutter RCS versus range using the function call in Eq. (6.27).

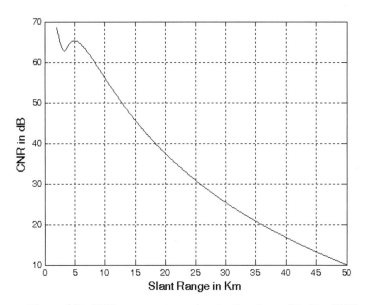

Figure 6.8b. CNR versus range using the function call in Eq. (6.27).

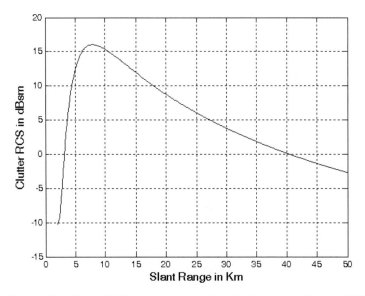

Figure 6.9a. Clutter RCS versus range using the function call in Eq. (6.28).

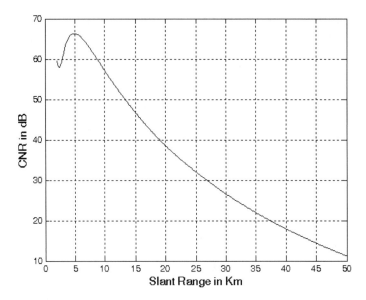

Figure 6.9b. CNR versus range using the function call in Eq. (6.28).

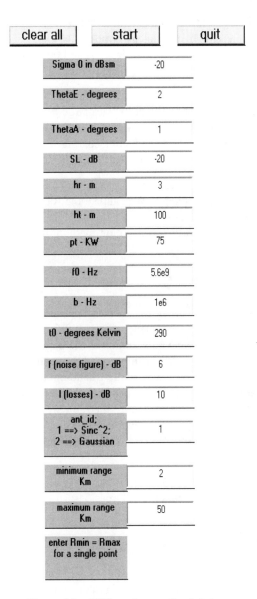

Figure 6.9c. GUI workspace for *"clutter_rcs_gui.m"*.

6.3. Volume Clutter

Volume clutter has large extents and includes rain (weather), chaff, birds, and insects. The volume clutter coefficient is normally expressed in square meters (RCS per resolution volume). Birds, insects, and other flying particles are often referred to as angle clutter or biological clutter.

As mentioned earlier, chaff is used as an ECM technique by hostile forces. It consists of a large number of dipole reflectors with large RCS values. Historically, chaff was made of aluminum foil; however, in recent years most chaff is made of the more rigid fiberglass with conductive coating. The maximum chaff RCS occurs when the dipole length L is one half the radar wavelength.

Weather or rain clutter is easier to suppress than chaff, since rain droplets can be viewed as perfect small spheres. We can use the Rayleigh approximation of a perfect sphere to estimate the rain droplets' RCS. The Rayleigh approximation, without regard to the propagation medium index of refraction is:

$$\sigma = 9\pi r^2 (kr)^4 \qquad r \ll \lambda \qquad (6.29)$$

where $k = 2\pi/\lambda$, and r is radius of a rain droplet.

Electromagnetic waves when reflected from a perfect sphere become strongly co-polarized (have the same polarization as the incident waves). Consequently, if the radar transmits, for example, a right-hand-circular (RHC) polarized wave, then the received waves are left-hand-circular (LHC) polarized, because they are propagating in the opposite direction. Therefore, the back-scattered energy from rain droplets retains the same wave rotation (polarization) as the incident wave, but has a reversed direction of propagation. It follows that radars can suppress rain clutter by co-polarizing the radar transmit and receive antennas.

Denote η as RCS per unit resolution volume V_W. It is computed as the sum of all individual scatterers RCS within the volume,

$$\eta = \sum_{i=1}^{N} \sigma_i \qquad (6.30)$$

where N is the total number of scatterers within the resolution volume. Thus, the total RCS of a single resolution volume is

$$\sigma_W = \sum_{i=1}^{N} \sigma_i V_W \qquad (6.31)$$

A resolution volume is shown in Fig. 6.10, and is approximated by

$$V_W \approx \frac{\pi}{8} \theta_a \theta_e R^2 c\tau \qquad (6.32)$$

where θ_a, θ_e are, respectively, the antenna azimuth and elevation beamwidths in radians, τ is the pulsewidth in seconds, c is speed of light, and R is range.

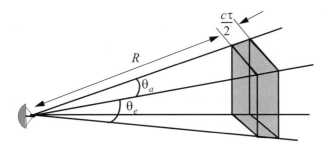

Figure 6.10. Definition of a resolution volume.

Consider a propagation medium with an index of refraction m. The ith rain droplet RCS approximation in this medium is

$$\sigma_i \approx \frac{\pi^5}{\lambda^4} K^2 D_i^6 \qquad (6.33)$$

where

$$K^2 = \left|\frac{m^2 - 1}{m^2 + 2}\right|^2 \qquad (6.34)$$

and D_i is the ith droplet diameter. For example, temperatures between $32°F$ and $68°F$ yield

$$\sigma_i \approx 0.93 \frac{\pi^5}{\lambda^4} D_i^6 \qquad (6.35)$$

and for ice Eq. (6.33) can be approximated by

$$\sigma_i \approx 0.2 \frac{\pi^5}{\lambda^4} D_i^6 \qquad (6.36)$$

Substituting Eq. (6.33) into Eq. (6.30) yields

$$\eta = \frac{\pi^5}{\lambda^4} K^2 Z \qquad (6.37)$$

where the weather clutter coefficient Z is defined as

$$Z = \sum_{i=1}^{N} D_i^6 \qquad (6.38)$$

In general, a rain droplet diameter is given in millimeters and the radar resolution volume is expressed in cubic meters; thus the units of Z are often expressed in $millimeter^6/m^3$.

6.3.1. Radar Equation for Volume Clutter

The radar equation gives the total power received by the radar from a σ_t target at range R as

$$S_t = \frac{P_t G^2 \lambda^2 \sigma_t}{(4\pi)^3 R^4} \qquad (6.39)$$

where all parameters in Eq. (6.39) have been defined earlier. The weather clutter power received by the radar is

$$S_W = \frac{P_t G^2 \lambda^2 \sigma_W}{(4\pi)^3 R^4} \qquad (6.40)$$

Using Eq. (6.31) and Eq. (6.32) in Eq. (6.40) and collecting terms yield

$$S_W = \frac{P_t G^2 \lambda^2}{(4\pi)^3 R^4} \frac{\pi}{8} R^2 \theta_a \theta_e c\tau \sum_{i=1}^{N} \sigma_i \qquad (6.41)$$

The SCR for weather clutter is then computed by dividing Eq. (6.39) by Eq. (6.41). More precisely,

$$(SCR)_V = \frac{S_t}{S_W} = \frac{8\sigma_t}{\pi \theta_a \theta_e c\tau R^2 \sum_{i=1}^{N} \sigma_i} \qquad (6.42)$$

where the subscript V is used to denote volume clutter.

Example:

A certain radar has target RCS $\sigma_t = 0.1 m^2$, pulsewidth $\tau = 0.2 \mu s$, antenna beamwidth $\theta_a = \theta_e = 0.02 radians$. Assume the detection range to be $R = 50 Km$, and compute the SCR if $\sum \sigma_i = 1.6 \times 10^{-8} (m^2/m^3)$.

Solution:

From Eq. (6.42) we have

$$(SCR)_V = \frac{8\sigma_t}{\pi \theta_a \theta_e c \tau R^2 \sum_{i=1}^{N} \sigma_i}$$

Substituting the proper values we get

$$(SCR)_V = \frac{(8)(0.1)}{\pi (0.02)^2 (3 \times 10^8)(0.2 \times 10^{-6})(50 \times 10^3)^2 (1.6 \times 10^{-8})} = 0.265$$

$$(SCR)_V = -5.76 dB.$$

6.4. Clutter Statistical Models

Since clutter within a resolution cell or volume is composed of a large number of scatterers with random phases and amplitudes, it is statistically described by a probability distribution function. The type of distribution depends on the nature of clutter itself (sea, land, volume), the radar operating frequency, and the grazing angle.

If sea or land clutter is composed of many small scatterers when the probability of receiving an echo from one scatterer is statistically independent of the echo received from another scatterer, then the clutter may be modeled using a Rayleigh distribution,

$$f(x) = \frac{2x}{x_0} \exp\left(\frac{-x^2}{x_0}\right) \quad ; \quad x \geq 0 \tag{6.43}$$

where x_0 is the mean squared value of x.

The log-normal distribution best describes land clutter at low grazing angles. It also fits sea clutter in the plateau region. It is given by

$$f(x) = \frac{1}{\sigma \sqrt{2\pi} \, x} \exp\left(-\frac{(\ln x - \ln x_m)^2}{2\sigma^2}\right) \quad ; \quad x > 0 \tag{6.44}$$

where x_m is the median of the random variable x, and σ is the standard deviation of the random variable $\ln(x)$.

The Weibull distribution is used to model clutter at low grazing angles (less than five degrees) for frequencies between 1 and $10 GHz$. The Weibull probability density function is determined by the Weibull slope parameter a (often tabulated) and a median scatter coefficient $\bar{\sigma}_0$, and is given by

$$f(x) = \frac{bx^{b-1}}{\bar{\sigma}_0} \exp\left(-\frac{x^b}{\bar{\sigma}_0}\right) \; ; \; x \geq 0 \tag{6.45}$$

where $b = 1/a$ is known as the shape parameter. Note that when $b = 2$ the Weibull distribution becomes a Rayleigh distribution.

6.5. "MyRadar" Design Case Study - Visit 6

6.5.1. Problem Statement

Analyze the impact of ground clutter on "MyRadar" design case study. Assume a Gaussian antenna pattern. Assume that the radar height is 5 meters. Consider an antenna sidelobe level $SL = -20$ dB and a ground clutter coefficient $\sigma^0 = -15$ dBsm. What conclusions can you draw about the radar's ability to maintain proper detection and track of both targets? Assume a radar height $h_r \geq 5m$.

6.5.2. A Design

From the design processes established in Chapters 1 and 2, it was determined that the minimum single pulse SNR required to accomplish the design objectives was $SNR \geq 4dB$ when non-coherent integration (4 pulses) and cumulative detection were used. Factoring in the surface clutter will degrade the SIR. However, one must maintain $SIR \geq 4dB$ in order to achieve the desired probability of detection.

Figure 6.11 shows a plot of the clutter RCS versus range corresponding to "MyRadar" design requirements. This figure can be reproduced using the MATLAB GUI *"clutter_rcs_gui"* with the following inputs:

Symbol	Value	Units
sigma0	-15	dB
thetaE	11 (see page 45)	degrees

Symbol	Value	Units
thetaA	1.33 (see page 45)	degrees
SL	-20	dB
range	linspace(10,120,1000)	Km
hr	5	meter
ht	2000 for missile; 10000 for aircraft	meter
pt	20	KW
f0	3e9	Hz
b	5e6	Hz
t0	290	Kelvins
f	6	dB
l	8	dB
ant_id	2 for Gaussian pattern	none

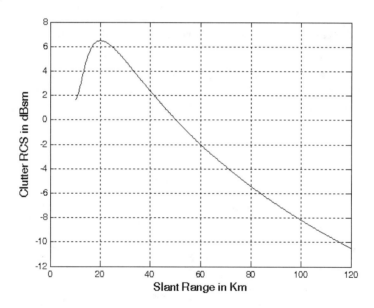

Figure 6.11a. Clutter RCS entering the radar for the missile case.

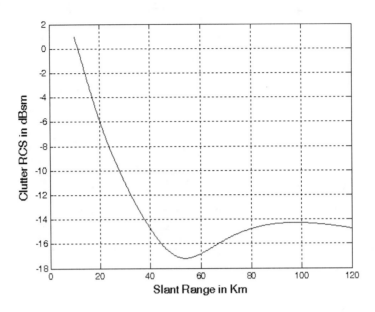

Figure 6.11b. Clutter RCS entering the radar for the aircraft case.

The MATLAB program *"myradar_visit6.m"* was developed to calculate and plot the CNR and SIR for *"MyRadar"* design case study. It is given in Listing 6.2 in Section 6.6. This program assumes the design parameters derived in Chapters 1 and 2. More precisely:

Symbol	Description	Value
σ^0	clutter backscatter coefficient	-15 dBsm
SL	antenna sidelobe level	-20 dB
σ_m	missile RCS	$0.5 m^2$
σ_a	aircraft RCS	$4 m^2$
θ_E	antenna elevation beamwidth	11 deg
θ_A	antenna azimuth beamwidth	1.33 deg
hr	radar height	5 m
hta	target height (aircraft)	10 Km
htm	target height (missile)	2 Km

Symbol	Description	Value
P_t	radar peak power	20 KW
f_0	radar operating frequency	3 GHz
T_0	effective noise temperature	290 degrees Kelvin
F	noise figure	6 dB
L	radar total losses	8 dB
τ'	Uncompressed pulsewidth	20 microseconds

Figure 6.12 shows a plot of the CNR and the SIR associated with the missile. Figure 6.13 is similar to Fig. 6.12 except it is for the aircraft case. It is clear from these figures that the required SIR has been degraded significantly for the missile case and not as much for the aircraft case. This should not be surprising, since the missile's altitude is much smaller than that of the aircraft. Without clutter mitigation, the missile would not be detected at all. Alternatively, the aircraft detection is compromised at $R \leq 80 Km$. Clutter mitigation is the subject of the next chapter.

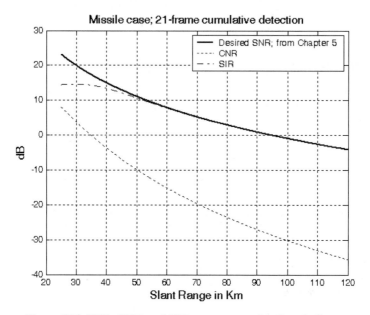

Figure 6.12. SNR, CNR, and SIR versus range for the missile case.

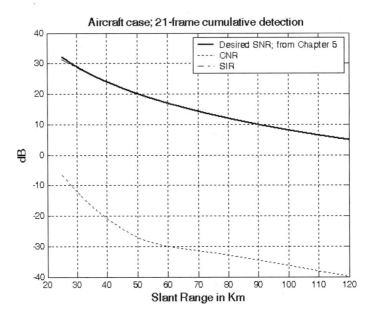

Figure 6.13. SNR, CNR and SIR versus range for the aircraft case.

6.6. MATLAB Program and Function Listings

This section presents listings for all MATLAB programs/functions used in this chapter. The user is advised to rerun these programs with different input parameters.

Listing 6.1. MATALB Function "clutter_rcs.m"

function [sigmaC,CNR] = clutter_rcs(sigma0, thetaE, thetaA, SL, range, hr, ht, pt, f0, b, t0, f, l,ant_id)
% This function calculates the clutter RCS and the CNR for a ground based radar.
clight = 3.e8; % speed of light in meters per second
lambda = clight /f0;
thetaA_deg = thetaA;
thetaE_deg = thetaE;
*thetaA = thetaA_deg * pi /180; % antenna azimuth beamwidth in radians*
*thetaE = thetaE_deg * pi /180.; % antenna elevation beamwidth in radians*
re = 6371000; % earth radius in meters
*rh = sqrt(8.0*hr*re/3.); % range to horizon in meters*

```
SLv = 10.0^(SL/10); % radar rms sidelobes in volts
sigma0v = 10.0^(sigma0/10); % clutter backscatter coefficient
tau = 1/b; % pulsewidth
deltar = clight * tau / 2.; % range resolution for unmodulated pulse
%%%%%%%%%%%%%%%%%%%%%%%%%%%%%%%%%%%%%
range_m = 1000 .* range;  % range in meters
%%%%%%%%%%%%%%%%%%%%%%%%%%%%%%%%%%%%%
thetar = asin(hr ./ range_m);
thetae = asin((ht-hr) ./ range_m);
propag_atten = 1. + ((range_m ./ rh).^4); % propagation attenuation due to
round earth
Rg = range_m .* cos(thetar);
deltaRg = deltar .* cos(thetar);
theta_sum = thetae + thetar;
% use sinc^2 antenna pattern when ant_id=1
% use Gaussian antenna pattern when ant_id=2
if(ant_id ==1) % use sinc^2 antenna pattern
   ant_arg = (2.78 * theta_sum ) ./ (pi*thetaE);
   gain = (sinc(ant_arg)).^2;
else
   gain = exp(-2.776 .*(theta_sum./thetaE).^2);
end
% compute sigmac
sigmac = (sigma0v .* Rg .* deltaRg) .* (pi * SLv * SLv + thetaA .* gain.^2) ./
propag_atten;
sigmaC = 10*log10(sigmac);
%%%%%%%%%%%%%%%%%%%%%%%%%%%%%%%%%%%%%
if (size(range,2)==1)
   fprintf('Sigma_Clutter='); sigmaC
else
   figure(1)
   plot(range, sigmaC)
   grid
   xlabel('Slant Range in Km')
   ylabel('Clutter RCS in dBsm')
end
%%%%%%%%%%%%%%%%%%%%%%%%%%%%%%%%%%%%%
% Calculate CNR
pt = pt * 1000;
g = 26000 / (thetaA_deg*thetaE_deg); % antenna gain
F = 10.^(f/10); % noise figure is 6 dB
Lt = 10.^(l/10); % total radar losses 13 dB
k = 1.38e-23; % Boltzman's constant
T0 = t0; % noise temperature 290K
```

```
argnumC = 10*log10(pt*g*g*lambda*lambda*tau .* sigmac);
argdem = 10*log10(((4*pi)^3)*k*T0*Lt*F .*(range_m).^4);
CNR = argnumC - argdem;
%%%%%%%%%%%%%%%%%%%%%%%%%%%%%%%%%%%%%
if (size(range,2) ==1)
   fprintf('Cluuter_to_Noise_ratio='); CNR
else
   figure(2)
   plot(range, CNR,'r')
   grid
   xlabel('Slant Range in Km')
   ylabel('CNR in dB')
end
```

Listing 6.2. MATLAB Program "myradar_visit6.m"

```
clear all
close all
thetaA= 1.33; % antenna azimuth beamwidth in degrees
thetaE = 11; % antenna elevation beamwidth in degrees
hr = 5.; % radar height to center of antenna (phase reference) in meters
htm = 2000.; % target (missile) high in meters
hta = 10000.; % target (aircraft) high in meters
SL = -20; % radar rms sidelobes in dB
sigma0 = -15; % clutter backscatter coefficient
b = 1.0e6; %1-MHz bandwidth
t0 = 290; % noise temperature 290 degrees Kelvin
f0 = 3e9; % 3 GHz center frequency
pt = 114.6; % radar peak power in KW
f = 6; % 6 dB noise figure
l = 8; % 8 dB radar losses
range = linspace(25,120,500); % radar slant range 25 to 120 Km, 500 points
% calculate the clutter RCS and the associated CNR for both targets
[sigmaCa,CNRa] = clutter_rcs(sigma0, thetaE, thetaA, SL, range, hr, hta, pt,
f0, b, t0, f, l, 2);
[sigmaCm,CNRm] = clutter_rcs(sigma0, thetaE, thetaA, L, range, hr, htm, pt,
f0, b, t0, f, l, 2);
close all
%%%%%%%%%%%%%%%%%%%%%%%%%%%%%%
np = 4;
pfa = 1e-7;
pdm = 0.99945;
pda = 0.99812;
% calculate the improvement factor
```

```
Im = improv_fac(np,pfa, pdm);
Ia = improv_fac(np, pfa, pda);
% calculate the integration loss
Lm = 10*log10(np) - Im;
La = 10*log10(np) - Ia;
pt = pt * 1000; % peak power in watts
range_m = 1000 .* range; % range in meters
g = 34.5139; % antenna gain in dB
sigmam = 0.5; % missile RCS m squared
sigmaa = 4; % aircraft RCS m squared
nf = f; %noise figure in dB
loss = l; % radar losses in dB
losstm = loss + Lm; % total loss for missile
lossta = loss + La; % total loss for aircraft
% modify pt by np*pt to account for pulse integration
SNRm = radar_eq(np*pt, f0, g, sigmam, t0, b, nf, losstm, range_m);
SNRa = radar_eq(np*pt, f0, g, sigmaa, t0, b, nf, lossta, range_m);
snrm = 10.^(SNRm./10);
snra = 10.^(SNRa./10);
cnrm = 10.^(CNRm./10);
cnra = 10.^(CNRa./10);
SIRm = 10*log10(snrm ./ (1+cnrm));
SIRa = 10*log10(snra ./ (1+cnra));
%%%%%%%%%%%%%%%%%%%%%%%%%%%%%%%%%%%%
figure(3)
plot(range, SNRm,'k', range, CNRm,'k :', range,SIRm,'k -.')
grid
legend('Desired SNR; from Chapter 5','CNR','SIR')
xlabel('Slant Range in Km')
ylabel('dB')
title('Missile case; 21-frame cumulative detection')
%%%%%%%%%%%%%%%%%%%%%%%%%%%%%%%%%%%%%%%%%%%
%%%%%%
figure(4)
plot(range, SNRa,'k', range, CNRa,'k :', range,SIRa,'k -.')
grid
legend('Desired SNR; from Chapter 5','CNR','SIR')
xlabel('Slant Range in Km')
ylabel('dB')
title('Aircraft case; 21-frame cumulative detection')
```

Chapter 7
Moving Target Indicator (MTI) and Clutter Mitigation

7.1. Clutter Spectrum

The power spectrum of stationary clutter (zero Doppler) can be represented by a delta function. However, clutter is not always stationary; it actually exhibits some Doppler frequency spread because of wind speed and motion of the radar scanning antenna. In general, the clutter spectrum is concentrated around $f = 0$ and integer multiples of the radar PRF f_r, and may exhibit a small amount of spreading.

The clutter power spectrum can be written as the sum of fixed (stationary) and random (due to frequency spreading) components. For most cases, the random component is Gaussian. If we denote the stationary-to-random power ratio by W^2, then we can write the clutter spectrum as

$$S_c(\omega) = \bar{\sigma}_0 \left(\frac{W^2}{1+W^2} \right) \delta(\omega_0) + \frac{\bar{\sigma}_0}{(1+W^2)\sqrt{2\pi\sigma_\omega^2}} \exp\left(-\frac{(\omega-\omega_0)^2}{2\sigma_\omega^2} \right) \quad (7.1)$$

where $\omega_0 = 2\pi f_0$ is the radar operating frequency in radians per second, σ_ω is the rms frequency spread component (determines the Doppler frequency spread), and $\bar{\sigma}_0$ is the Weibull parameter.

The first term of the right-hand side of Eq. (7.1) represents the PSD for stationary clutter, while the second term accounts for the frequency spreading. Nevertheless, since most of the clutter power is concentrated around zero Doppler with some spreading (typically less than 100 Hz), it is customary to model clutter using a Gaussian-shaped power spectrum (which is easier to analyze than Eq. (7.1)). More precisely,

$$S_c(\omega) = \frac{P_c}{\sqrt{2\pi\sigma_\omega^2}} \exp\left(-\frac{(\omega-\omega_0)^2}{2\sigma_\omega^2}\right) \tag{7.2}$$

where P_c is the total clutter power; σ_ω^2 and ω_0 were defined earlier. Fig. 7.1 shows a typical PSD sketch of radar returns when both target and clutter are present. Note that the clutter power is concentrated around DC and integer multiples of the PRF.

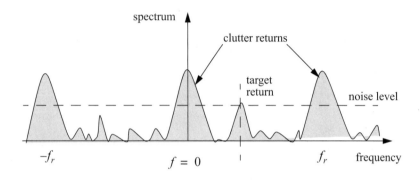

Figure 7.1. Typical radar return PSD when clutter and target are present.

7.2. Moving Target Indicator (MTI)

The clutter spectrum is normally concentrated around DC ($f = 0$) and multiple integers of the radar PRF f_r, as illustrated in Fig. 7.2a. In CW radars, clutter is avoided or suppressed by ignoring the receiver output around DC, since most of the clutter power is concentrated about the zero frequency band. Pulsed radar systems may utilize special filters that can distinguish between slowly moving or stationary targets and fast moving ones. This class of filter is known as the Moving Target Indicator (MTI). In simple words, the purpose of an MTI filter is to suppress target-like returns produced by clutter, and allow returns from moving targets to pass through with little or no degradation. In order to effectively suppress clutter returns, an MTI filter needs to have a deep stop-band at DC and at integer multiples of the PRF. Fig. 7.2b shows a typical sketch of an MTI filter response, while Fig. 7.2c shows its output when the PSD shown in Fig. 7.2a is the input.

MTI filters can be implemented using delay line cancelers. As we will show later in this chapter, the frequency response of this class of MTI filter is periodic, with nulls at integer multiples of the PRF. Thus, targets with Doppler frequencies equal to nf_r are severely attenuated. Since Doppler is proportional to target velocity ($f_d = 2v/\lambda$), target speeds that produce Doppler frequencies equal to integer multiples of f_r are known as blind speeds. More precisely,

$$v_{blind} = \frac{\lambda f_r}{2} \; ; \; n \geq 0 \tag{7.3}$$

Radar systems can minimize the occurrence of blind speeds by either employing multiple PRF schemes (PRF staggering) or by using high PRFs where in this case the radar may become range ambiguous. The main difference between PRF staggering and PRF agility is that the pulse repetition interval (within an integration interval) can be changed between consecutive pulses for the case of PRF staggering.

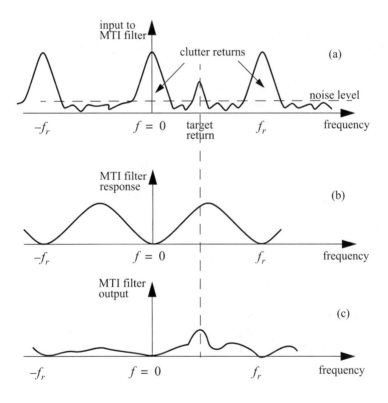

Figure 7.2. (a) Typical radar return PSD when clutter and target are present. (b) MTI filter frequency response. (c) Output from an MTI filter.

Fig. 7.3 shows a block diagram of a coherent MTI radar. Coherent transmission is controlled by the STAble Local Oscillator (STALO). The outputs of the STALO, f_{LO}, and the COHerent Oscillator (COHO), f_C, are mixed to produce the transmission frequency, $f_{LO} + f_C$. The Intermediate Frequency (IF), $f_C \pm f_d$, is produced by mixing the received signal with f_{LO}. After the IF amplifier, the signal is passed through a phase detector and is converted into a base band. Finally, the video signal is inputted into an MTI filter.

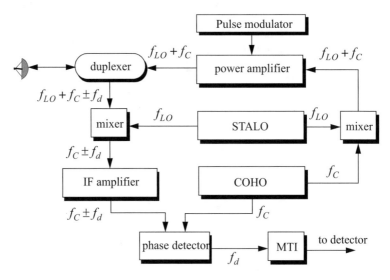

Figure 7.3. Coherent MTI radar block diagram.

7.3. Single Delay Line Canceler

A single delay line canceler can be implemented as shown in Fig. 7.4. The canceler's impulse response is denoted as $h(t)$. The output $y(t)$ is equal to the convolution between the impulse response $h(t)$ and the input $x(t)$. The single delay canceler is often called a "two-pulse canceler" since it requires two distinct input pulses before an output can be read.

The delay T is equal to the PRI of the radar ($1/f_r$). The output signal $y(t)$ is

$$y(t) = x(t) - x(t - T) \tag{7.4}$$

The impulse response of the canceler is given by

$$h(t) = \delta(t) - \delta(t - T) \tag{7.5}$$

Single Delay Line Canceler

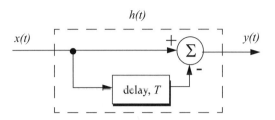

Figure 7.4. Single delay line canceler.

where $\delta(\cdot)$ is the delta function. It follows that the Fourier transform (FT) of $h(t)$ is

$$H(\omega) = 1 - e^{-j\omega T} \qquad (7.6)$$

where $\omega = 2\pi f$.

In the z-domain, the single delay line canceler response is

$$H(z) = 1 - z^{-1} \qquad (7.7)$$

The power gain for the single delay line canceler is given by

$$|H(\omega)|^2 = H(\omega)H^*(\omega) = (1 - e^{-j\omega T})(1 - e^{j\omega T}) \qquad (7.8)$$

It follows that

$$|H(\omega)|^2 = 1 + 1 - (e^{j\omega T} + e^{-j\omega T}) = 2(1 - \cos\omega T) \qquad (7.9)$$

and using the trigonometric identity $(2 - 2\cos 2\vartheta) = 4(\sin\vartheta)^2$ yields

$$|H(\omega)|^2 = 4(\sin(\omega T/2))^2 \qquad (7.10)$$

MATLAB Function "single_canceler.m"

The function *"single_canceler.m"* computes and plots (as a function of f/f_r) the amplitude response for a single delay line canceler. It is given in Listing 7.1 in Section 7.11. The syntax is as follows:

$$[resp] = single\_canceler\ (fofr)$$

where *fofr* is the number of periods desired. Typical output of the function *"single_canceler.m"* is shown in Fig. 7.5. Clearly, the frequency response of a single canceler is periodic with a period equal to f_r. The peaks occur at $f = (2n+1)/(2f_r)$, and the nulls are at $f = nf_r$, where $n \geq 0$.

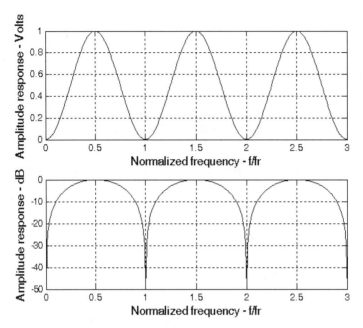

Figure 7.5. Single canceler frequency response.

In most radar applications the response of a single canceler is not acceptable since it does not have a wide notch in the stop-band. A double delay line canceler has better response in both the stop- and pass-bands, and thus it is more frequently used than a single canceler. In this book, we will use the names "single delay line canceler" and "single canceler" interchangeably.

7.4. Double Delay Line Canceler

Two basic configurations of a double delay line canceler are shown in Fig. 7.6. Double cancelers are often called "three-pulse cancelers" since they require three distinct input pulses before an output can be read. The double line canceler impulse response is given by

$$h(t) = \delta(t) - 2\delta(t-T) + \delta(t-2T) \tag{7.11}$$

Again, the names "double delay line" canceler and "double canceler" will be used interchangeably. The power gain for the double delay line canceler is

Double Delay Line Canceler

$$|H(\omega)|^2 = |H_1(\omega)|^2|H_1(\omega)|^2 \tag{7.12}$$

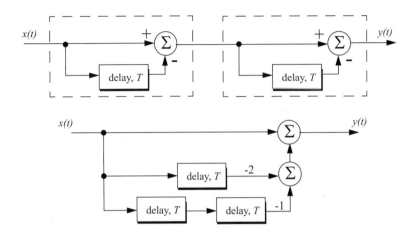

Figure 7.6. Two configurations for a double delay line canceler.

where $|H_1(\omega)|^2$ is the single line canceler power gain given in Eq. (7.10). It follows that

$$|H(\omega)|^2 = 16\left(\sin\left(\omega\frac{T}{2}\right)\right)^4 \tag{7.13}$$

And in the z-domain, we have

$$H(z) = (1 - z^{-1})^2 = 1 - 2z^{-1} + z^{-2} \tag{7.14}$$

MATLAB Function "double_canceler.m"

The function *"double_canceler.m"* computes and plots (as a function of f/f_r) the amplitude response for a double delay line canceler. It is given in Listing 7.2 in Section 7.11. The syntax is as follows:

$$[resp] = double\_canceler\ (fofr)$$

where *fofr* is the number of periods desired.

Fig. 7.7 shows typical output from this function. Note that the double canceler has a better response than the single canceler (deeper notch and flatter pass-band response).

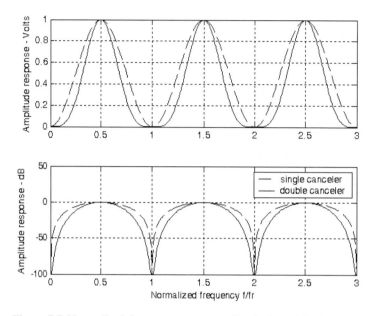

Figure 7.7. Normalized frequency responses for single and double cancelers.

7.5. *Delay Lines with Feedback (Recursive Filters)*

Delay line cancelers with feedback loops are known as recursive filters. The advantage of a recursive filter is that through a feedback loop we will be able to shape the frequency response of the filter. As an example, consider the single canceler shown in Fig. 7.8. From the figure we can write

$$y(t) = x(t) - (1-K)w(t) \qquad (7.15)$$

$$v(t) = y(t) + w(t) \qquad (7.16)$$

$$w(t) = v(t-T) \qquad (7.17)$$

Applying the z-transform to the above three equations yields

$$Y(z) = X(z) - (1-K)W(z) \qquad (7.18)$$

$$V(z) = Y(z) + W(z) \qquad (7.19)$$

$$W(z) = z^{-1}V(z) \qquad (7.20)$$

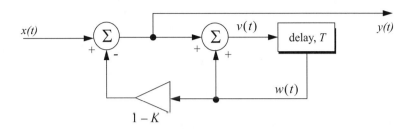

Figure 7.8. MTI recursive filter.

Solving for the transfer function $H(z) = Y(z)/X(z)$ yields

$$H(z) = \frac{1-z^{-1}}{1-Kz^{-1}} \quad (7.21)$$

The modulus square of $H(z)$ is then equal to

$$|H(z)|^2 = \frac{(1-z^{-1})(1-z)}{(1-Kz^{-1})(1-Kz)} = \frac{2-(z+z^{-1})}{(1+K^2)-K(z+z^{-1})} \quad (7.22)$$

Using the transformation $z = e^{j\omega T}$ yields

$$z + z^{-1} = 2\cos\omega T \quad (7.23)$$

Thus, Eq. (7.22) can now be rewritten as

$$|H(e^{j\omega T})|^2 = \frac{2(1-\cos\omega T)}{(1+K^2)-2K\cos(\omega T)} \quad (7.24)$$

Note that when $K = 0$, Eq. (7.24) collapses to Eq. (7.10) (single line canceler). Fig. 7.9 shows a plot of Eq. (7.24) for $K = 0.25, 0.7, 0.9$. Clearly, by changing the gain factor K one can control the filter response.

In order to avoid oscillation due to the positive feedback, the value of K should be less than unity. The value $(1-K)^{-1}$ is normally equal to the number of pulses received from the target. For example, $K = 0.9$ corresponds to ten pulses, while $K = 0.98$ corresponds to about fifty pulses.

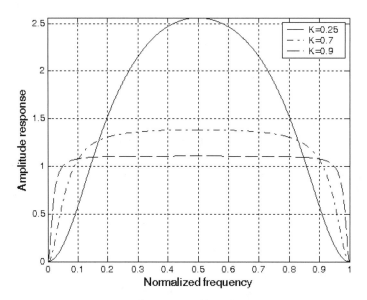

Figure 7.9. Frequency response corresponding to Eq. (7.24). This plot can be reproduced using MATLAB program "fig7_9.m" given in Listing 7.3 in Section 7.11.

7.6. PRF Staggering

Target velocities that correspond to multiple integers of the PRF are referred to as blind speeds. This terminology is used since an MTI filter response is equal to zero at these values (see Fig. 7.7). Blind speeds can pose serious limitations on the performance of MTI radars and their ability to perform adequate target detection. Using PRF agility by changing the pulse repetition interval between consecutive pulses can extend the first blind speed to tolerable values. In order to show how PRF staggering can alleviate the problem of blind speeds, let us first assume that two radars with distinct PRFs are utilized for detection. Since blind speeds are proportional to the PRF, the blind speeds of the two radars would be different. However, using two radars to alleviate the problem of blind speeds is a very costly option. A more practical solution is to use a single radar with two or more different PRFs.

For example, consider a radar system with two interpulse periods T_1 and T_2, such that

$$\frac{T_1}{T_2} = \frac{n_1}{n_2} \tag{7.25}$$

where n_1 and n_2 are integers. The first true blind speed occurs when

$$\frac{n_1}{T_1} = \frac{n_2}{T_2} \tag{7.26}$$

This is illustrated in Fig. 7.10 for $n_1 - 4$ and $n_2 - 5$. Note that if $n_2 = n_1 + 1$, then the process of PRF staggering is similar to that discussed in Chapter 3. The ratio

$$k_s = \frac{n_1}{n_2} \tag{7.27}$$

is known as the stagger ratio. Using staggering ratios closer to unity pushes the first true blind speed farther out. However, the dip in the vicinity of $1/T_1$ becomes deeper, as illustrated in Fig. 7.11 for stagger ratio $k_s = 63/64$. In general, if there are N PRFs related by

$$\frac{n_1}{T_1} = \frac{n_2}{T_2} = \ldots = \frac{n_N}{T_N} \tag{7.28}$$

and if the first blind speed to occur for any of the individual PRFs is v_{blind1}, then the first true blind speed for the staggered waveform is

$$v_{blind} = \frac{n_1 + n_2 + \ldots + n_N}{N} v_{blind1} \tag{7.29}$$

7.7. MTI Improvement Factor

In this section two quantities that are normally used to define the performance of MTI systems are introduced. They are "Clutter Attenuation (CA)" and the MTI "Improvement Factor." The MTI CA is defined as the ratio between the MTI filter input clutter power C_i to the output clutter power C_o,

$$CA = C_i/C_o \tag{7.30}$$

The MTI improvement factor is defined as the ratio of the Signal to Clutter (SCR) at the output to the SCR at the input,

$$I = \left(\frac{S_o}{C_o}\right) / \left(\frac{S_i}{C_i}\right) \tag{7.31}$$

which can be rewritten as

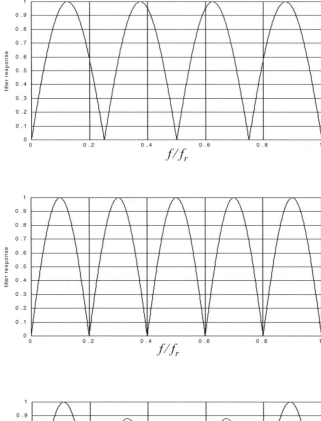

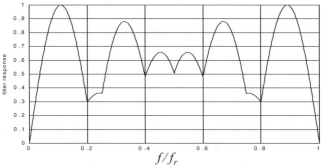

Figure 7.10. Frequency responses of a single canceler. Top plot corresponds to T_1, middle plot corresponds to T_2, bottom plot corresponds to stagger ratio $T_1/T_2 = 4/3$. This plot can be reproduced using MATLAB program "fig7_10.m" given in Listing 7.4 in Section 7.11.

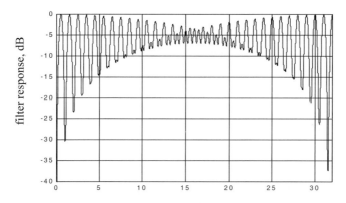

target velocity relative to first blind speed; 63/64

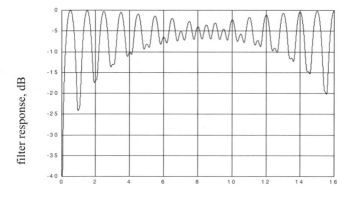

target velocity relative to first blind speed; 33/34

Figure 7.11. MTI responses, staggering ratio 63/64. This plot can be reproduced using MATLAB program *"fig7_11.m"* given in Listing 7.5 in Section 7.11.

$$I = \frac{S_o}{S_i} CA \qquad (7.32)$$

The ratio S_o/S_i is the average power gain of the MTI filter, and it is equal to $|H(\omega)|^2$. In this section, a closed form expression for the improvement factor using a Gaussian-shaped power spectrum is developed. A Gaussian-shaped clutter power spectrum is given by

$$W(f) = \frac{P_c}{\sqrt{2\pi}\,\sigma_t} \exp(-f^2/2\sigma_t^2) \tag{7.33}$$

where P_c is the clutter power (constant), and σ_t is the clutter rms frequency (which describes the clutter spectrum spread in the frequency domain). It is given by

$$\sigma_t = \sqrt{\sigma_v^2 + \sigma_s^2 + \sigma_w^2} \tag{7.34}$$

σ_v is the standard deviation for the clutter spectrum spread due to wind velocity; σ_s is the standard deviation for the clutter spectrum spread due to antenna scanning; and σ_v is the standard deviation for the clutter spectrum spread due to the radar platform motion (if applicable). It can be shown that[1]

$$\sigma_v = \frac{2\sigma_w}{\lambda} \tag{7.35}$$

$$\sigma_s = 0.265 \left(\frac{2\pi}{\Theta_a T_{scan}}\right) \tag{7.36}$$

$$\sigma_s \approx \frac{v}{\lambda} \sin\theta \tag{7.37}$$

where λ is the wavelength and σ_w is the wind rms velocity; Θ_a is the antenna 3-db azimuth beamwidth (in radians); T_{scan} is the antenna scan time; v is the platform velocity; and θ is the azimuth angle (in radians) relative to the direction of motion.

The clutter power at the input of an MTI filter is

$$C_i = \int_{-\infty}^{\infty} \frac{P_c}{\sqrt{2\pi}\,\sigma_t} \exp\left(-\frac{f^2}{2\sigma_t^2}\right) df \tag{7.38}$$

Factoring out the constant P_c yields

$$C_i = P_c \int_{-\infty}^{\infty} \frac{1}{\sqrt{2\pi}\sigma_t} \exp\left(-\frac{f^2}{2\sigma_t^2}\right) df \tag{7.39}$$

It follows that

---

1. Berkowtiz, R. S., *Modern Radar, Analysis, Evaluation, and System Deign*, John Wiley & Sons, New York, 1965.

MTI Improvement Factor

$$C_i = P_c \tag{7.40}$$

The clutter power at the output of an MTI is

$$C_o = \int_{-\infty}^{\infty} W(f)|H(f)|^2 \, df \tag{7.41}$$

7.7.1. Two-Pulse MTI Case

In this section we will continue the analysis using a ***single delay line canceler***. The frequency response for a single delay line canceler is given by Eq. (7.6). The single canceler power gain is given in Eq. (7.10), which will be repeated here, in terms of f rather than ω, as Eq. (7.42),

$$|H(f)|^2 = 4\left(\sin\left(\frac{\pi f}{f_r}\right)\right)^2 \tag{7.42}$$

It follows that

$$C_o = \int_{-\infty}^{\infty} \frac{P_c}{\sqrt{2\pi}\,\sigma_t} \exp\left(-\frac{f^2}{2\sigma_t^2}\right) 4\left(\sin\left(\frac{\pi f}{f_r}\right)\right)^2 df \tag{7.43}$$

Now, since clutter power will only be significant for small f, then the ratio f/f_r is very small (i.e., $\sigma_t \ll f_r$). Consequently, by using the small angle approximation, Eq. (7.43) is approximated by

$$C_o \approx \int_{-\infty}^{\infty} \frac{P_c}{\sqrt{2\pi}\,\sigma_t} \exp\left(-\frac{f^2}{2\sigma_t^2}\right) 4\left(\frac{\pi f}{f_r}\right)^2 df \tag{7.44}$$

which can be rewritten as

$$C_o = \frac{4P_c \pi^2}{f_r^2} \int_{-\infty}^{\infty} \frac{1}{\sqrt{2\pi\sigma_t^2}} \exp\left(-\frac{f^2}{2\sigma_t^2}\right) f^2 \, df \tag{7.45}$$

The integral part in Eq. (7.45) is the second moment of a zero mean Gaussian distribution with variance σ_t^2. Replacing the integral in Eq. (7.45) by σ_t^2 yields

$$C_o = \frac{4P_c \pi^2}{f_r^2} \sigma_t^2 \tag{7.46}$$

Substituting Eqs. (7.46) and (7.40) into Eq. (7.30) produces

$$CA = \frac{C_i}{C_o} = \left(\frac{f_r}{2\pi\sigma_t}\right)^2 \qquad (7.47)$$

It follows that the improvement factor for a single canceler is

$$I = \left(\frac{f_r}{2\pi\sigma_t}\right)^2 \frac{S_o}{S_i} \qquad (7.48)$$

The power gain ratio for a single canceler is (remember that $|H(f)|$ is periodic with period f_r)

$$\frac{S_o}{S_i} = |H(f)|^2 = \frac{1}{f_r}\int_{-f_r/2}^{f_r/2} 4\left(\sin\frac{\pi f}{f_r}\right)^2 df \qquad (7.49)$$

Using the trigonometric identity $(2 - 2\cos 2\vartheta) = 4(\sin\vartheta)^2$ yields

$$|H(f)|^2 = \frac{1}{f_r}\int_{-f_r/2}^{f_r/2}\left(2 - 2\cos\frac{2\pi f}{f_r}\right)df = 2 \qquad (7.50)$$

It follows that

$$I = 2\left(\frac{f_r}{2\pi\sigma_t}\right)^2 \qquad (7.51)$$

The expression given in Eq. (7.51) is an approximation valid only for $\sigma_t \ll f_r$. When the condition $\sigma_t \ll f_r$ is not true, then the autocorrelation function needs to be used in order to develop an exact expression for the improvement factor.

Example:

A certain radar has $f_r = 800 Hz$. If the clutter rms is $\sigma_v = 6.4 Hz$ (wooded hills with $\sigma_w = 1.16311 Km/hr$), find the improvement factor when a single delay line canceler is used.

Solution:

In this case $\sigma_t = \sigma_v$. It follows that the clutter attenuation CA is

$$CA = \left(\frac{f_r}{2\pi\sigma_t}\right)^2 = \left(\frac{800}{(2\pi)(6.4)}\right)^2 = 395.771 = 25.974 dB$$

and since $S_o/S_i = 2 = 3dB$ we get

$$I_{dB} = (CA + S_o/S_i)_{dB} = 3 + 25.97 = 28.974 dB.$$

7.7.2. The General Case

A general expression for the improvement factor for the n-pulse MTI (shown for a 2-pulse MTI in Eq. (7.51)) is given by

$$I = \frac{1}{Q^2(2(n-1)-1)!!}\left(\frac{f_r}{2\pi\sigma_t}\right)^{2(n-1)} \tag{7.52}$$

where the double factorial notation is defined by

$$(2n-1)!! = 1 \times 3 \times 5 \times \ldots \times (2n-1) \tag{7.53}$$

$$(2n)!! = 2 \times 4 \times \ldots \times 2n \tag{7.54}$$

Of course $0!! = 1$; Q is defined by

$$Q^2 = \frac{1}{\sum_{i=1}^{n} A_i^2} \tag{7.55}$$

where A_i are the Binomial coefficients for the MTI filter. It follows that Q^2 for a 2-pulse, 3-pulse, and 4-pulse MTI are respectively

$$\left\{\frac{1}{2}, \frac{1}{20}, \frac{1}{70}\right\} \tag{7.56}$$

Using this notation, then the improvement factor for a 3-pulse and 4-pulse MTI are respectively given by

$$I_{3-pulse} = 2\left(\frac{f_r}{2\pi\sigma_t}\right)^4 \tag{7.57}$$

$$I_{4-pulse} = \frac{4}{3}\left(\frac{f_r}{2\pi\sigma_t}\right)^6 \tag{7.58}$$

7.8. "MyRadar" Design Case Study - Visit 7

7.8.1. Problem Statement

The impact of surface clutter on the "MyRadar" design case study was analyzed. Assume that the wind rms velocity $\sigma_w = 0.45 m/s$. Propose a clutter mitigation process utilizing a 2-pulse and a 3-pulse MTI. All other parameters are as calculated in the previous chapters.

7.8.2. A Design

In earlier chapters we determined that the wavelength is $\lambda = 0.1m$, the PRF is $f_r = 1KHz$, the scan rate is $T_{scan} = 2s$, and the antenna azimuth 3-db beamwidth is $\Theta_a = 1.3°$. It follows that

$$\sigma_v = \frac{2\sigma_w}{\lambda} = \frac{2 \times 0.45}{0.1} = 9Hz \tag{7.59}$$

$$\sigma_s = 0.265\left(\frac{2\pi}{\Theta_a T_{scan}}\right) = 0.265 \times \frac{2 \times \pi}{1.32 \times \frac{\pi}{180} \times 2} = 36.136 Hz \tag{7.60}$$

Thus, the total clutter rms spectrum spread is

$$\sigma_t = \sqrt{\sigma_v^2 + \sigma_s^2} = \sqrt{81 + 1305.810} = \sqrt{1386.810} = 37.24 Hz \tag{7.61}$$

The expected clutter attenuation using a 2-pulse and a 3-pulse MTI are respectively given by

$$I_{2pulse} = 2\left(\frac{f_r}{2\pi\sigma_t}\right)^2 = 2 \times \left(\frac{1000}{2 \times \pi \times 37.24}\right)^2 = 36.531\frac{W}{W} \Rightarrow 15.63 dB \tag{7.62}$$

$$I_{3pulse} = 2\left(\frac{f_r}{2\pi\sigma_t}\right)^4 = 2 \times \left(\frac{1000}{2 \times \pi \times 37.24}\right)^4 = 667.247\frac{W}{W} \Rightarrow 28.24 dB \tag{7.63}$$

To demonstrate the effect of a 2-pulse and 3-pulse MTI on *"MyRadar"* design case study, the MATLAB program *"myradar_visit7.m"* has been developed. It is given in Listing 7.6 in Section 7.5. This program utilizes the radar equation with pulse compression. In this case, the peak power was established in Chapter 5 as $P_t \leq 10KW$. Figs. 7.12 and 7.13 show the desired SNR and the calculated SIR using a 2-pulse and a 3-pulse MTI filter respectively, for the missile case. Figs. 7.14 and 7.15 show similar output for the aircraft case.

One may argue, depending on the tracking scheme adopted by the radar, that for a tracking radar

$$\sigma_t = \sigma_v = 9Hz \tag{7.64}$$

since $\sigma_s = 0$ for a radar that employs a monopulse tracking option. In this design, we will assume a Kalman filter tracker. For more details the reader is advised to visit Chapter 9.

"MyRadar" Design Case Study - Visit 7   311

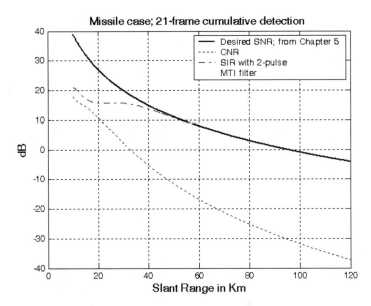

Figure 7.12. SIR for the missile case using a 2-pulse MTI filter.

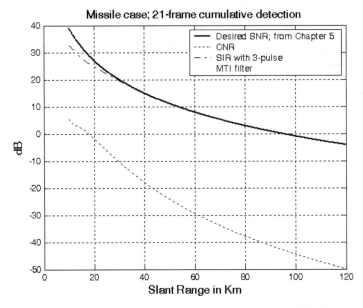

Figure 7.13. SIR for the missile case using a 3-pulse MTI filter.

312   *MATLAB Simulations for Radar Systems Design*

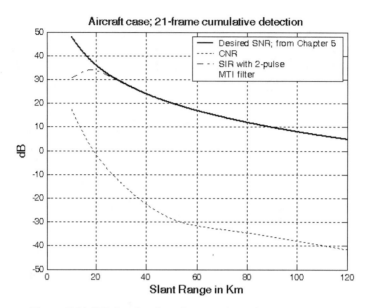

Figure 7.14. SIR for the aircraft case using a 2-pulse MTI filter.

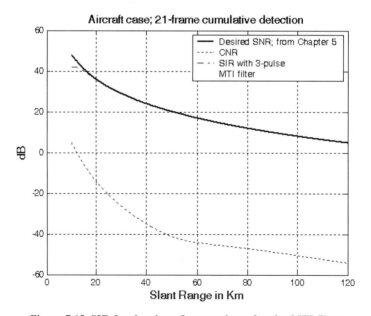

Figure 7.15. SIR for the aircraft case using a 3-pulse MTI filter.

As clearly indicated by the previous four figures, a 3-pulse MTI filter would provide adequate clutter rejection for both target types. However, if we assume that targets are detected at maximum range (90 Km for aircraft and 55 Km for missile) and then are tracked for the rest of the flight, then 2-pulse MTI may be adequate. This is true since the SNR would be expected to be larger during track than it is during detection, especially when pulse compression is used. Nonetheless, in this design a 3-pulse MTI filter is adopted.

7.9. MATLAB Program and Function Listings

This section contains listings of all MATLAB programs and functions used in this chapter. Users are encouraged to rerun this code with different inputs in order to enhance their understanding of the theory.

Listing 7.1. MATLAB Function "single_canceler.m"

```
function [resp] = single_canceler (fofr1)
eps = 0.00001;
fofr = 0:0.01:fofr1;
arg1 = pi .* fofr;
resp = 4.0 .*((sin(arg1)).^2);
max1 = max(resp);
resp = resp ./ max1;
subplot(2,1,1)
plot(fofr,resp,'k')
xlabel ('Normalized frequency - f/fr')
ylabel( 'Amplitude response - Volts')
grid
subplot(2,1,2)
resp=10.*log10(resp+eps);
plot(fofr,resp,'k');
axis tight
grid
xlabel ('Normalized frequency - f/fr')
ylabel( 'Amplitude response - dB')
```

Listing 7.2. MATLAB Function "double_canceler.m"

```
function [resp] = double_canceler(fofr1)
eps = 0.00001;
fofr = 0:0.01:fofr1;
arg1 = pi .* fofr;
```

```
resp = 4.0 .* ((sin(arg1)).^2);
max1 = max(resp);
resp = resp ./ max1;
resp2 = resp .* resp;
subplot(2,1,1);
plot(fofr,resp,'k--',fofr, resp2,'k');
ylabel ('Amplitude response - Volts')
resp2 = 20. .* log10(resp2+eps);
resp1 = 20. .* log10(resp+eps);
subplot(2,1,2)
plot(fofr,resp1,'k--',fofr,resp2,'k');
legend ('single canceler','double canceler')
xlabel ('Normalized frequency f/fr')
ylabel ('Amplitude response - dB')
```

Listing 7.3. MATLAB Program "fig7_9.m"

```
clear all
fofr = 0:0.001:1;
arg = 2.*pi.*fofr;
nume = 2.*(1.-cos(arg));
den11 = (1. + 0.25 * 0.25);
den12 = (2. * 0.25) .* cos(arg);
den1 = den11 - den12;
den21 = 1.0 + 0.7 * 0.7;
den22 = (2. * 0.7) .* cos(arg);
den2 = den21 - den22;
den31 = (1.0 + 0.9 * 0.9);
den32 = ((2. * 0.9) .* cos(arg));
den3 = den31 - den32;
resp1 = nume ./ den1;
resp2 = nume ./ den2;
resp3 = nume ./ den3;
plot(fofr,resp1,'k',fofr,resp2,'k-.',fofr,resp3,'k--');
xlabel('Normalized frequency')
ylabel('Amplitude response')
legend('K=0.25','K=0.7','K=0.9')
grid
axis tight
```

Listing 7.4. MATLAB Program "fig7_10.m"

```
clear all
fofr = 0:0.001:1;
```

```
f1 = 4.0 .* fofr;
f2 = 5.0 .* fofr;
arg1 = pi .* f1;
arg2 = pi .* f2;
resp1 = abs(sin(arg1));
resp2 = abs(sin(arg2));
resp = resp1+resp2;
max1 = max(resp);
resp = resp./max1;
plot(fofr,resp1,fofr,resp2,fofr,resp);
xlabel('Normalized frequency f/fr')
ylabel('Filter response')
```

Listing 7.5. MATLAB Program "fig7_11.m"

```
clear all
fofr = 0.01:0.001:32;
a = 63.0 / 64.0;
term1 = (1. - 2.0 .* cos(a*2*pi*fofr) + cos(4*pi*fofr)).^2;
term2 = (-2. .* sin(a*2*pi*fofr) + sin(4*pi*fofr)).^2;
resp = 0.25 .* sqrt(term1 + term2);
resp = 10. .* log(resp);
plot(fofr,resp);
axis([0 32 -40 0]);
grid
```

Listing 7.6. MATLAB Program "myradar_visit7.m"

```
clear all
close all
clutter_attenuation = 28.24;
thetaA = 1.33; % antenna azimuth beamwidth in degrees
thetaE = 11; % antenna elevation beamwidth in degrees
hr = 5.; % radar height to center of antenna (phase reference) in meters
htm = 2000.; % target (missile) height in meters
hta = 10000.; % target (aircraft) height in meters
SL = -20; % radar rms sidelobes in dB
sigma0 = -15; % clutter backscatter coefficient in dB
b = 1.0e6; %1-MHz bandwidth
t0 = 290; % noise temperature 290 degrees Kelvin
f0 = 3e9; % 3 GHz center frequency
pt = 114.6; % radar peak power in KW
f = 6; % 6 dB noise figure
l = 8; % 8 dB radar losses
```

```
range = linspace(25,120,500); % radar slant range 25 to 120 Km, 500 points
% calculate the clutter RCS and the associated CNR for both targets
[sigmaCa,CNRa] = clutter_rcs(sigma0, thetaE, thetaA, SL, range, hr, hta, pt,
f0, b, t0, f, l,2);
[sigmaCm,CNRm] = clutter_rcs(sigma0, thetaE, thetaA, SL, range, hr, htm, pt,
f0, b, t0, f, l,2);
close all
%%%%%%%%%%%%%%%%%%%%%%%%%%%%%
np = 4;
pfa = 1e-7;
pdm = 0.99945;
pda = 0.99812;
% calculate the improvement factor
Im = improv_fac(np,pfa, pdm);
Ia = improv_fac(np, pfa, pda);
% caculate the integration loss
Lm = 10*log10(np) - Im;
La = 10*log10(np) - Ia;
pt = pt * 1000; % peak power in watts
range_m = 1000 .* range; % range in meters
g = 34.5139; % antenna gain in dB
sigmam = 0.5; % missile RCS m squared
sigmaa = 4; % aircraft RCS m squared
nf = f; %noise figure in dB
loss = l; % radar losses in dB
losstm = loss + Lm; % total loss for missile
lossta = loss + La; % total loss for aircraft
% modify pt by np*pt to account for pulse integration
SNRm = radar_eq(np*pt, f0, g, sigmam, t0, b, nf, losstm, range_m);
SNRa = radar_eq(np*pt, f0, g, sigmaa, t0, b, nf, lossta, range_m);
snrm = 10.^(SNRm./10);
snra = 10.^(SNRa./10);
CNRm = CNRm - clutter_attenuation;
CNRa = CNRa - clutter_attenuation;
cnrm = 10.^(CNRm./10);
cnra = 10.^(CNRa./10);
SIRm = 10*log10(snrm ./ (1+cnrm));
SIRa = 10*log10(snra ./ (1+cnra));
%%%%%%%%%%%%%%%%%%%%%%%%%%%%%%%%%%%
figure(3)
plot(range, SNRm,'k', range, CNRm,'k :', range,SIRm,'k -.')
grid
legend('Desired SNR; from Chapter 5','CNR','SIR with 3-pulse','MTI filter')
xlabel('Slant Range in Km')
```

```
ylabel('dB')
title('Missile case;  21-frame cumulative detection')
%%%%%%%%%%%%%%%%%%%%%%%%%%%%%%%%%%%%%%%%%%
figure(4)
plot(range, SNRa,'k', range, CNRa,'k :', range,SIRa,'k -.')
grid
legend('Desired SNR; from Chapter 5','CNR','SIR with 3-pulse','MTI filter')
xlabel('Slant Range in Km')
ylabel('dB')
title('Aircraft case; 21-frame cumulative detection')
```

Chapter 8 *Phased Arrays*

8.1. Directivity, Power Gain, and Effective Aperture

Radar antennas can be characterized by the directive gain G_D, power gain G, and effective aperture A_e. Antenna gain is a term used to describe the ability of an antenna to concentrate the transmitted energy in a certain direction. Directive gain, or simply directivity, is more representative of the antenna radiation pattern, while power gain is normally used in the radar equation. Plots of the power gain and directivity, when normalized to unity, are called *antenna radiation pattern*. The directivity of a transmitting antenna can be defined by

$$G_D = \frac{maximum\ radiation\ intensity}{average\ radiation\ intensity} \qquad (8.1)$$

The radiation intensity is the power per unit solid angle in the direction (θ, ϕ) and denoted by $P(\theta, \phi)$. The average radiation intensity over 4π radians (solid angle) is the total power divided by 4π. Hence, Eq. (8.1) can be written as

$$G_D = \frac{4\pi(maximum\ radiated\ power/unit\ solid\ angle)}{total\ radiated\ power} \qquad (8.2)$$

It follows that

$$G_D = 4\pi \frac{P(\theta, \phi)_{max}}{\int_0^{2\pi}\int_0^{\pi} P(\theta, \phi)\,d\theta\,d\phi} \qquad (8.3)$$

As an approximation, it is customary to rewrite Eq. (8.3) as

$$G_D \approx \frac{4\pi}{\theta_3 \phi_3} \qquad (8.4)$$

where θ_3 and ϕ_3 are the antenna half-power (3-dB) beamwidths in either direction.

The antenna power gain and its directivity are related by

$$G = \rho_r G_D \qquad (8.5)$$

where ρ_r is the radiation efficiency factor. In this book, the antenna power gain will be denoted as *gain*. The radiation efficiency factor accounts for the ohmic losses associated with the antenna. Therefore, the definition for the antenna gain is also given in Eq. (8.1). The antenna effective aperture A_e is related to gain by

$$A_e = \frac{G\lambda^2}{4\pi} \qquad (8.6)$$

where λ is the wavelength. The relationship between the antenna's effective aperture A_e and the physical aperture A is

$$A_e = \rho A \qquad (8.7)$$
$$0 \le \rho \le 1$$

ρ is referred to as the aperture efficiency, and good antennas require $\rho \to 1$ (in this book $\rho = 1$ is always assumed, i.e., $A_e = A$).

Using simple algebraic manipulations of Eqs. (8.4) through (8.6) (assuming that $\rho_r = 1$) yields

$$G = \frac{4\pi A_e}{\lambda^2} \approx \frac{4\pi}{\theta_3 \phi_3} \qquad (8.8)$$

Consequently, the angular cross section of the beam is

$$\theta_3 \phi_3 \approx \frac{\lambda^2}{A_e} \qquad (8.9)$$

Eq. (8.9) indicates that the antenna beamwidth decreases as $\sqrt{A_e}$ increases. It follows that, in surveillance operations, the number of beam positions an antenna will take on to cover a volume V is

$$N_{Beams} > \frac{V}{\theta_3 \phi_3} \qquad (8.10)$$

and when V represents the entire hemisphere, Eq. (8.10) is modified to

$$N_{Beams} > \frac{2\pi}{\theta_3 \phi_3} \approx \frac{2\pi A_e}{\lambda^2} \approx \frac{G}{2} \tag{8.11}$$

8.2. Near and Far Fields

The electric field intensity generated from the energy emitted by an antenna is a function of the antenna physical aperture shape and the electric current amplitude and phase distribution across the aperture. Plots of the modulus of the electric field intensity of the emitted radiation, $|E(\theta, \phi)|$, are referred to as the *intensity pattern* of the antenna. Alternatively, plots of $|E(\theta, \phi)|^2$ are called the *power radiation pattern* (the same as $P(\theta, \phi)$).

Based on the distance from the face of the antenna, where the radiated electric field is measured, three distinct regions are identified. They are the near field, Fresnel, and the Fraunhofer regions. In the near field and the Fresnel regions, rays emitted from the antenna have spherical wavefronts (equi-phase fronts). In the Fraunhofer regions the wavefronts can be locally represented by plane waves. The near field and the Fresnel regions are normally of little interest to most radar applications. Most radar systems operate in the Fraunhofer region, which is also known as the far field region. In the far field region, the electric field intensity can be computed from the aperture Fourier transform.

Construction of the far criterion can be developed with the help of Fig. 8.1. Consider a radiating source at point O that emits spherical waves. A receiving antenna of length d is at distance r away from the source. The phase difference between a spherical wave and a local plane wave at the receiving antenna can be expressed in terms of the distance δr. The distance δr is given by

$$\delta r = \overline{AO} - \overline{OB} = \sqrt{r^2 + \left(\frac{d}{2}\right)^2} - r \tag{8.12}$$

and since in the far field $r \gg d$, Eq. (8.12) is approximated via binomial expansion by

$$\delta r = r\left(\sqrt{1 + \left(\frac{d}{2r}\right)^2} - 1\right) \approx \frac{d^2}{8r} \tag{8.13}$$

It is customary to assume far field when the distance δr corresponds to less than $1/16$ of a wavelength (i.e., $22.5°$). More precisely, if

$$\delta r = d^2/8r \leq \lambda/16 \tag{8.14}$$

then a useful expression for far field is

$$r \geq 2d^2/\lambda \tag{8.15}$$

Note that far field is a function of both the antenna size and the operating wavelength.

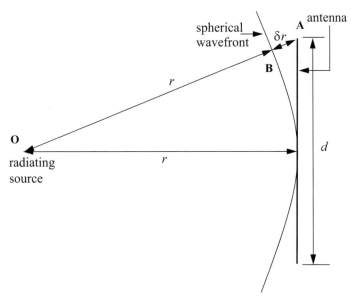

Figure 8.1. Construction of far field criterion.

8.3. General Arrays

An array is a composite antenna formed from two or more basic radiators. Each radiator is denoted as an element. The elements forming an array could be dipoles, dish reflectors, slots in a wave guide, or any other type of radiator. Array antennas synthesize narrow directive beams that may be steered, mechanically or electronically, in many directions. Electronic steering is achieved by controlling the phase of the current feeding the array elements. Arrays with electronic beam steering capability are called phased arrays. Phased array antennas, when compared to other simple antennas such as dish reflectors, are costly and complicated to design. However, the inherent flexibility of phased array antennas to steer the beam electronically and also the need for specialized multi-function radar systems have made phased array antennas attractive for radar applications.

General Arrays

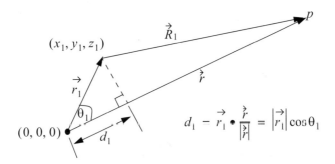

Figure 8.2 Geometry for an array antenna. Single element

Fig. 8.2 shows the geometrical fundamentals associated with this problem. In general, consider the radiation source located at (x_1, y_1, z_1) with respect to a phase reference at $(0, 0, 0)$. The electric field measured at far field point P is

$$E(\theta, \phi) = I_0 \frac{e^{-jkR_1}}{R_1} f(\theta, \phi) \tag{8.16}$$

where I_0 is the complex amplitude, $k = 2\pi/\lambda$ is the wave number, and $f(\theta, \phi)$ is the radiation pattern.

Now, consider the case where the radiation source is an array made of many elements, as shown in Fig. 8.3. The coordinates of each radiator with respect to the phase reference is (x_i, y_i, z_i), and the vector from the origin to the *ith* element is given by

$$\vec{r}_i = \hat{a}_x x_i + \hat{a}_y y_i + \hat{a}_z z_i \tag{8.17}$$

The far field components that constitute the total electric field are

$$E_i(\theta, \phi) = I_i \frac{e^{-jkR_i}}{R_i} f(\theta_i, \phi_i) \tag{8.18}$$

where

$$R_i = |\vec{R}_i| = |\vec{r} - \vec{r}_i| = \sqrt{(x - x_i)^2 + (y - y_i)^2 + (z - z_i)^2}$$
$$= r\sqrt{1 + (x_i^2 + y_i^2 + z_i^2)/r^2 - 2(xx_i + yy_i + zz_i)/r^2} \tag{8.19}$$

Using spherical coordinates, where $x = r\sin\theta\cos\varphi$, $y = r\sin\theta\sin\varphi$, and $z = r\cos\theta$ yields

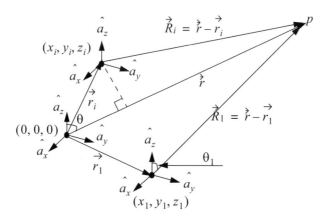

Figure 8.3 Geometry for an array antenna.

$$\frac{(x_i^2 + y_i^2 + z_i^2)}{r^2} = \frac{|\vec{r_i}|}{r^2} \ll 1 \tag{8.20}$$

Thus, a good approximation (using binomial expansion) for Eq. (8.19) is

$$R_i = r - r(x_i \sin\theta \cos\phi + y_i \sin\theta \sin\phi + z_i \cos\theta) \tag{8.21}$$

It follows that the phase contribution at the far field point from the *ith* radiator with respect to the phase reference is

$$e^{-jkR_i} = e^{-jkr} e^{jk(x_i \sin\theta \cos\phi + y_i \sin\theta \sin\phi + z_i \cos\theta)} \tag{8.22}$$

Remember, however, that the unit vector $\hat{r}_0$ along the vector $\vec{r}$ is

$$\hat{r}_0 = \frac{\vec{r}}{|\vec{r}|} = \hat{a}_x \sin\theta \cos\phi + \hat{a}_y \sin\theta \sin\phi + \hat{a}_z \cos\theta \tag{8.23}$$

Hence, we can rewrite Eq. (8.22) as

$$e^{-jkR_i} = e^{-jkr} e^{jk(\vec{r}_i \cdot \hat{r}_0)} = e^{-jkr} e^{j\Psi_i(\theta, \phi)} \tag{8.24}$$

Finally, by virtue of superposition, the total electric field is

$$E(\theta, \phi) = \sum_{i=1}^{N} I_i e^{j\Psi_i(\theta, \phi)} \tag{8.25}$$

which is known as the array factor for an array antenna where the complex current for the *ith* element is I_i.

Linear Arrays

In general, an array can be fully characterized by its array factor. This is true since knowing the array factor provides the designer with knowledge of the array's (1) 3-dB beamwidth; (2) null-to-null beamwidth; (3) distance from the main peak to the first sidelobe; (4) height of the first sidelobe as compared to the main beam; (5) location of the nulls; (6) rate of decrease of the sidelobes; and (7) grating lobes' locations.

8.4. Linear Arrays

Fig. 8.4 shows a linear array antenna consisting of N identical elements. The element spacing is d (normally measured in wavelength units). Let element #1 serve as a phase reference for the array. From the geometry, it is clear that an outgoing wave at the nth element leads the phase at the $(n+1)th$ element by $kd\sin\psi$, where $k = 2\pi/\lambda$. The combined phase at the far field observation point P is independent of ϕ and is computed from Eq. (8.24) as

$$\Psi(\psi, \phi) = k(\vec{r}_i \bullet \vec{r}_0) = (n-1)kd\sin\psi \tag{8.26}$$

Thus, from Eq. (8.25), the electric field at a far field observation point with direction-sine equal to $\sin\psi$ (assuming isotropic elements) is

$$E(\sin\psi) = \sum_{n=1}^{N} e^{j(n-1)(kd\sin\psi)} \tag{8.27}$$

Expanding the summation in Eq. (8.27) yields

$$E(\sin\psi) = 1 + e^{jkd\sin\psi} + \ldots + e^{j(N-1)(kd\sin\psi)} \tag{8.28}$$

The right-hand side of Eq. (8.29) is a geometric series, which can be expressed in the form

$$1 + a + a^2 + a^3 + \ldots + a^{(N-1)} = \frac{1-a^N}{1-a} \tag{8.29}$$

Replacing a by $e^{jkd\sin\psi}$ yields

$$E(\sin\psi) = \frac{1 - e^{jNkd\sin\psi}}{1 - e^{jkd\sin\psi}} = \frac{1-(\cos Nkd\sin\psi)-j(\sin Nkd\sin\psi)}{1-(\cos kd\sin\psi)-j(\sin kd\sin\psi)} \tag{8.30}$$

The far field array intensity pattern is then given by

$$|E(\sin\psi)| = \sqrt{E(\sin\psi)E^*(\sin\psi)} \tag{8.31}$$

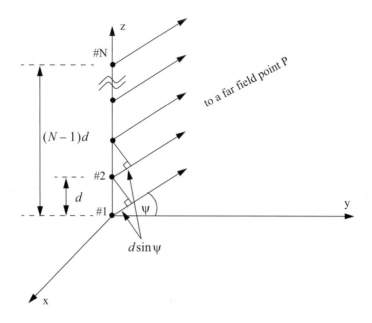

Figure 8.4. Linear array of equally spaced elements.

Substituting Eq. (8.30) into Eq. (8.31) and collecting terms yield

$$|E(\sin\psi)| = \sqrt{\frac{(1-\cos Nkd\sin\psi)^2 + (\sin Nkd\sin\psi)^2}{(1-\cos kd\sin\psi)^2 + (\sin kd\sin\psi)^2}} \qquad (8.32)$$

$$= \sqrt{\frac{1-\cos Nkd\sin\psi}{1-\cos kd\sin\psi}}$$

and using the trigonometric identity $1 - \cos\theta = 2(\sin\theta/2)^2$ yields

$$|E(\sin\psi)| = \left|\frac{\sin(Nkd\sin\psi/2)}{\sin(kd\sin\psi/2)}\right| \qquad (8.33)$$

which is a periodic function of $kd\sin\psi$, with a period equal to 2π.

The maximum value of $|E(\sin\psi)|$, which occurs at $\psi = 0$, is equal to N. It follows that the normalized intensity pattern is equal to

$$|E_n(\sin\psi)| = \frac{1}{N}\left|\frac{\sin((Nkd\sin\psi)/2)}{\sin((kd\sin\psi)/2)}\right| \qquad (8.34)$$

The normalized two-way array pattern (radiation pattern) is given by

Linear Arrays

$$G(\sin\psi) = |E_n(\sin\psi)|^2 = \frac{1}{N^2}\left(\frac{\sin((Nkd\sin\psi)/2)}{\sin((kd\sin\psi)/2)}\right)^2 \qquad (8.35)$$

Fig. 8.5 shows a plot of Eq. (8.35) versus $\sin\theta$ for $N = 8$. The radiation pattern $G(\sin\psi)$ has cylindrical symmetry about its axis ($\sin\psi = 0$), and is independent of the azimuth angle. Thus, it is completely determined by its values within the interval $(0 < \psi < \pi)$. This plot can be reproduced using MATLAB program "fig8_5.m" given in Listing 8.1 in Section 8.8.

The main beam of an array can be steered electronically by varying the phase of the current applied to each array element. Steering the main beam into the direction-sine $\sin\psi_0$ is accomplished by making the phase difference between any two adjacent elements equal to $kd\sin\psi_0$. In this case, the normalized radiation pattern can be written as

$$G(\sin\psi) = \frac{1}{N^2}\left(\frac{\sin[(Nkd/2)(\sin\psi - \sin\psi_0)]}{\sin[(kd/2)(\sin\psi - \sin\psi_0)]}\right)^2 \qquad (8.36)$$

If $\psi_0 = 0$ then the main beam is perpendicular to the array axis, and the array is said to be a broadside array. Alternatively, the array is called an endfire array when the main beam points along the array axis.

The radiation pattern maxima are computed using L'Hopital's rule when both the denominator and numerator of Eq. (8.35) are zeros. More precisely,

$$\frac{kd\sin\psi}{2} = \pm m\pi \quad ; \ m = 0, 1, 2, \ldots \qquad (8.37)$$

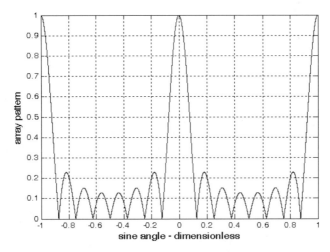

Figure 8.5a. Normalized radiation pattern for a linear array; $N = 8$; $d = \lambda$.

328   *MATLAB Simulations for Radar Systems Design*

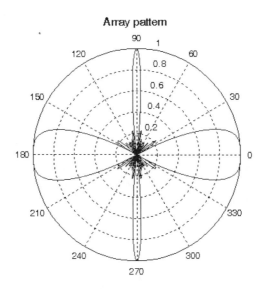

Figure 8.5b. Polar plot for the array pattern in Fig. 8.5a.

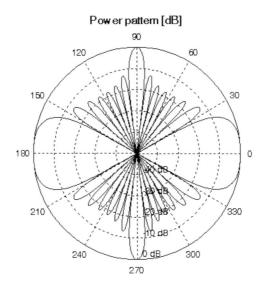

Figure 8.5c. Polar plot for the power pattern in Fig. 8.5a.

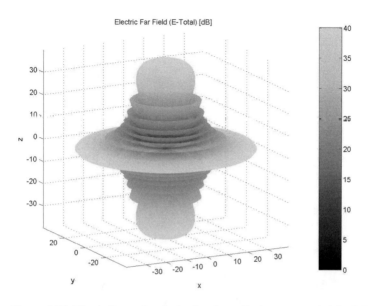

Figure 8.5d. Three-dimensional plot for the radiation pattern in Fig. 8.5a.

Solving for ψ yields

$$\psi_m = \operatorname{asin}\left(\pm\frac{\lambda m}{d}\right) \quad ; \; m = 0, 1, 2, \ldots \quad (8.38)$$

where the subscript m is used as a maxima indicator. The first maximum occurs at $\psi_0 = 0$, and is denoted as the main beam (lobe). Other maxima occurring at $|m| \geq 1$ are called grating lobes. Grating lobes are undesirable and must be suppressed. The grating lobes occur at non-real angles when the absolute value of the arc-sine argument in Eq. (8.38) is greater than unity; it follows that $d < \lambda$. Under this condition, the main lobe is assumed to be at $\psi = 0$ (broadside array). Alternatively, when electronic beam steering is considered, the grating lobes occur at

$$|\sin\psi - \sin\psi_0| = \pm\frac{\lambda n}{d} \quad ; \; n = 1, 2, \ldots \quad (8.39)$$

Thus, in order to prevent the grating lobes from occurring between $\pm 90°$, the element spacing should be $d < \lambda/2$.

The radiation pattern attains secondary maxima (sidelobes) when the numerator of Eq. (8.35) is maximum, or equivalently

$$\frac{Nkd\sin\psi}{2} = \pm(2l+1)\frac{\pi}{2} \quad ; l = 1, 2, \ldots \quad (8.40)$$

Solving for ψ yields

$$\psi_l = \operatorname{asin}\left(\pm\frac{\lambda}{2d}\frac{2l+1}{N}\right) \quad ; l = 1, 2, \ldots \quad (8.41)$$

where the subscript l is used as an indication of sidelobe maxima. The nulls of the radiation pattern occur when only the numerator of Eq. (8.36) is zero. More precisely,

$$\frac{N}{2}kd\sin\psi = \pm n\pi \quad ; \begin{array}{l} n = 1, 2, \ldots \\ n \ne N, 2N, \ldots \end{array} \quad (8.42)$$

Again solving for ψ yields

$$\psi_n = \operatorname{asin}\left(\pm\frac{\lambda}{d}\frac{n}{N}\right) \quad ; \begin{array}{l} n = 1, 2, \ldots \\ n \ne N, 2N, \ldots \end{array} \quad (8.43)$$

where the subscript n is used as a null indicator. Define the angle which corresponds to the half power point as ψ_h. It follows that the half power (3 dB) beamwidth is $2|\psi_m - \psi_h|$. This occurs when

$$\frac{N}{2}kd\sin\psi_h = 1.391 \; radians \Rightarrow \psi_h = \operatorname{asin}\left(\frac{\lambda}{2\pi d}\frac{2.782}{N}\right) \quad (8.44)$$

8.4.1. Array Tapering

Fig. 8.6a shows a normalized two-way radiation pattern of a uniformly excited linear array of size $N = 8$, element spacing $d = \lambda/2$. The first sidelobe is about 13.46 dB below the main lobe, and for most radar applications this may not be sufficient. Fig. 8.6b shows the 3-D plot for the radiation pattern shown in Fig. 8.6.a.

In order to reduce the sidelobe levels, the array must be designed to radiate more power towards the center, and much less at the edges. This can be achieved through tapering (windowing) the current distribution over the face of the array. There are many possible tapering sequences that can be used for this purpose. However, as known from spectral analysis, windowing reduces sidelobe levels at the expense of widening the main beam. Thus, for a given radar application, the choice of the tapering sequence must be based on the trade-off between sidelobe reduction and main beam widening. The MATLAB signal processing toolbox provides users with a wide variety of built-in windows. This list includes: *"Bartlett, Barthannwin, Blackmanharris, Bohmanwin, Chebwin, Gausswin, Hamming, Hann, Kaiser, Nuttallwin, Rectwin, Triang, and Tukeywin."*

Linear Arrays 331

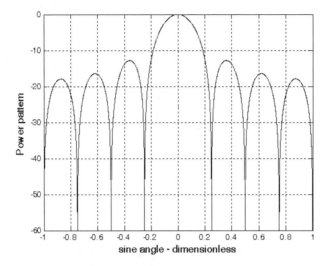

Figure 8.6a. Normalized pattern for a linear array. $N = 8$, $d = \lambda/2$.

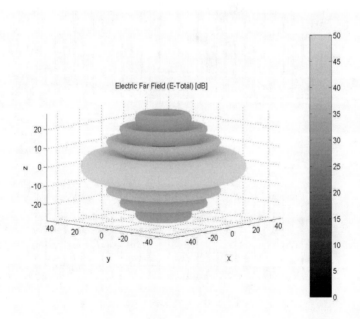

Figure 8.6b. Three-dimensional plot for the radiation pattern in Fig. 8.6a.

Table 8.1 summarizes the impact of most common windows on the array pattern in terms of main beam widening and peak reduction. Note that the rectangular window is used as the baseline. This is also illustrated in Fig. 8.7.

TABLE 8.1. Common windows.

Window	Null-to-null Beamwidth	Peak Reduction
Rectangular	1	1
Hamming	2	0.73
Hanning	2	0.664
Blackman	6	0.577
Kaiser ($\beta = 6$)	2.76	0.683
Kaiser ($\beta = 3$)	1.75	0.882

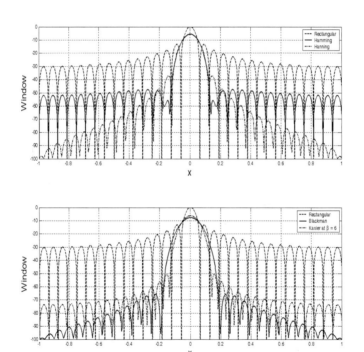

Figure 8.7. Most common windows. This figure can be reproduced using MATLAB program *"fig8_7.m"* given in Listing 8.2 in Section 8.8.

8.4.2. Computation of the Radiation Pattern via the DFT

Fig. 8.8 shows a linear array of size N, element spacing d, and wavelength λ. The radiators are circular dishes of diameter d. Let $w(n)$ and $\Phi(n)$, respectively, denote the tapering and phase shifting sequences. The normalized electric field at a far field point in the direction-sine $\sin\psi$ is

$$E(\sin\psi) = \sum_{n=0}^{N-1} w(n) e^{j\Delta\phi\left(n - \left(\frac{N-1}{2}\right)\right)} \tag{8.45}$$

where in this case the phase reference is taken as the physical center of the array, and

$$\Delta\phi = \frac{2\pi d}{\lambda} \sin\psi \tag{8.46}$$

Expanding Eq. (8.45) and factoring the common phase term $\exp[j(N-1)\Delta\phi/2]$ yield

$$E(\sin\psi) = e^{j(N-1)\Delta\phi/2} \{ w(0) e^{-j(N-1)\Delta\phi} + w(1) e^{-j(N-2)\Delta\phi} + \ldots + w(N-1) \} \tag{8.47}$$

By using the symmetry property of a window sequence (remember that a window must be symmetrical about its central point), we can rewrite Eq. (8.47) as

$$E(\sin\psi) = e^{j\phi_0} \{ w(N-1) e^{-j(N-1)\Delta\phi} + w(N-2) e^{-j(N-2)\Delta\phi} + \ldots + w(0) \} \tag{8.48}$$

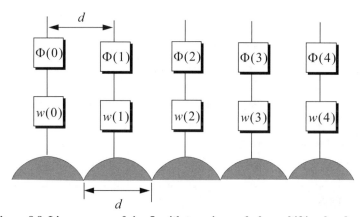

Figure 8.8. Linear array of size 5, with tapering and phase shifting hardware.

where $\phi_0 = (N-1)\Delta\phi/2$.

Define $\{V_1^n = \exp(-j\Delta\phi n); n = 0, 1, \ldots, N-1\}$. It follows that

$$E(\sin\psi) = e^{j\phi_0}[w(0) + w(1)V_1^1 + \ldots + w(N-1)V_1^{N-1}] \qquad (8.49)$$

$$= e^{j\phi_0} \sum_{n=0}^{N-1} w(n)V_1^n$$

The discrete Fourier transform of the sequence $w(n)$ is defined as

$$W(q) = \sum_{n=0}^{N-1} w(n) e^{-\frac{(j2\pi nq)}{N}} \quad ; q = 0, 1, \ldots, N-1 \qquad (8.50)$$

The set $\{\sin\psi_q\}$ which makes V_1 equal to the DFT kernel is

$$\sin\psi_q = \frac{\lambda q}{Nd} \quad ; q = 0, 1, \ldots, N-1 \qquad (8.51)$$

Then by using Eq. (8.51) in Eq. (8.50) yields

$$E(\sin\psi) = e^{j\phi_0} W(q) \qquad (8.52)$$

The one-way array pattern is computed as the modulus of Eq. (8.52). It follows that the one-way radiation pattern of a tapered linear array of circular dishes is

$$G(\sin\psi) = G_e |W(q)| \qquad (8.53)$$

where G_e is the element pattern.

In practice, phase shifters are normally implemented as part of the Transmit/Receive (TR) modules, using a finite number of bits. Consequently, due to the quantization error (difference between desired phase and actual quantized phase) the sidelobe levels are affected.

MATLAB Function "linear_array.m"

The function *"linear_array.m"* computes and plots the linear array gain pattern as a function of real sine-space (sine the steering angle). It is given in Listing 8.3 in Section 8.8. The syntax is as follows:

[theta, patternr, patterng] = linear_array(Nr, dolr, theta0, winid, win, nbits)

where

Symbol	Description	Units	Status
Nr	number of elements in array	none	input
dolr	element spacing in lambda units	wavelengths	input
theta0	steering angle	degrees	input
winid	-1: No weighting is used 1: Use weighting defined in win	none	input
win	window for sidelobe control	none	input
nbits	negative #: perfect quantization positive #: use 2^{nbits} quantization levels	none	input
theta	real angle available for steering	degrees	output
patternr	array pattern	dB	output
patterng	gain pattern	dB	output

A MATLAB based GUI workspace called *"linear_array_gui.m"*[1] was developed for this function. It shown in Fig. 8.9.

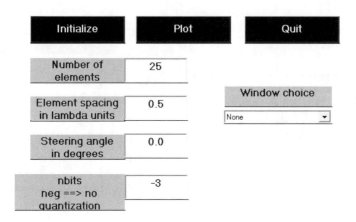

Figure 8.9. MATLAB GUI workspace associated with the function *"linear_array.m"*.

---

1. The MATLAB "Signal Processing" Toolbox is required to execute this program.

Figs. 8.10 through 8. 18 respectively show plots of the array gain pattern versus steering angle for the following cases:

[theta, patternr, patterng] = linear_array(25, 0.5, 0, -1, -1, -3);

[theta, patternr, patterng] = linear_array(25, 0.5, 0, 1, 'Hamming', -3);

[theta, patternr, patterng] = linear_array(25, 0.5, 5, -1, -1, 3);

[theta, patternr, patterng] = linear_array(25, 0.5, 5, 1, 'Hamming', 3);

[theta, patternr, patterng] = linear_array(25, 0.5, 25, 1, 'Hamming', 3);

[theta, patternr, patterng] = linear_array(25, 1.5, 40, -1, -1, -3);

[theta, patternr, patterng] = linear_array(25, 1.5, 40, 1, 'Hamming', -3);

[theta, patternr, patterng] = linear_array(25, 1.5, -40, -1, -1, 3);

[theta, patternr, patterng] = linear_array(25, 1.5, -40, 1, 'Hamming', 3);

Users are advised to utilize the GUI developed for this function and test a few cases of their own.

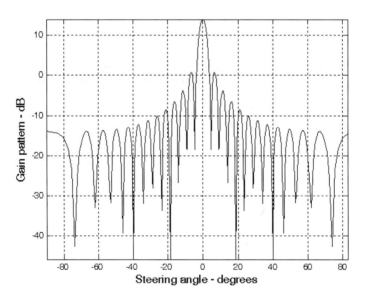

Figure 8.10. Array gain pattern: $Nr = 25$; $dolr = 0.5$; $\theta_0 = 0°$; $win = none$; $nbits = -3$.

Linear Arrays

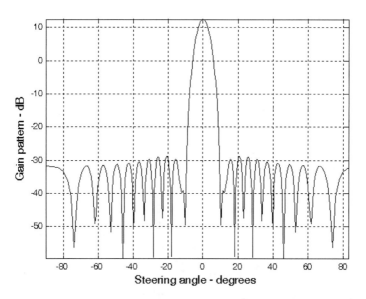

Figure 8.11. Array gain pattern: $Nr = 25$; $dolr = 0.5$; $\theta_0 = 0°$; $win = Hamming$; $nbits = -3$

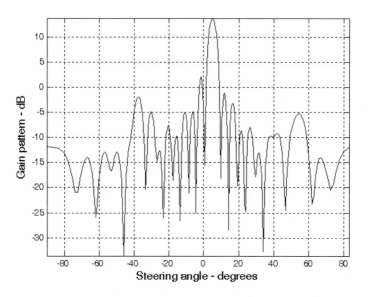

Figure 8.12. Array gain pattern: $Nr = 25$; $dolr = 0.5$; $\theta_0 = 5°$; $win = none$; $nbits = 3$

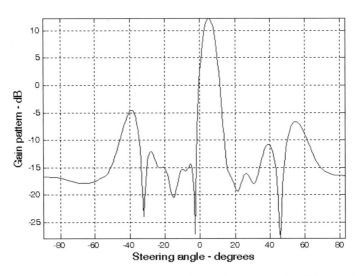

Figure 8.13. Array gain pattern: $Nr = 25$; $dolr = 0.5$; $\theta_0 = 5°$; $win = Hamming$; $nbits = 3$

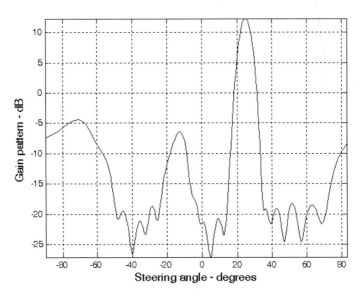

Figure 8.14. Array gain pattern: $Nr = 25$; $dolr = 0.5$; $\theta_0 = 25°$; $win = Hamming$; $nbits = 3$

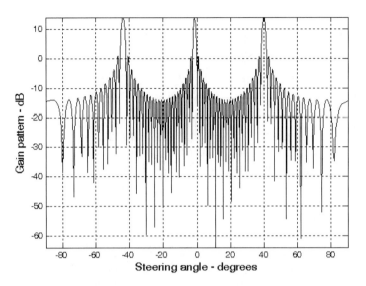

Figure 8.15. Array gain pattern: $Nr = 25$; $dolr = 1.5$; $\theta_0 = 40°$; $win = none$; $nbits = -3$

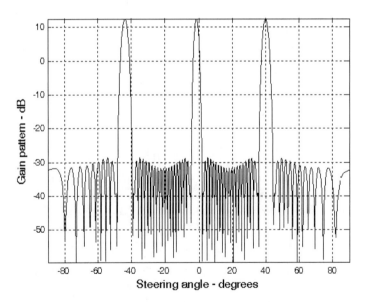

Figure 8.16. Array gain pattern: $Nr = 25$; $dolr = 1.5$; $\theta_0 = 40°$; $win = Hamming$; $nbits = -3$

340   MATLAB Simulations for Radar Systems Design

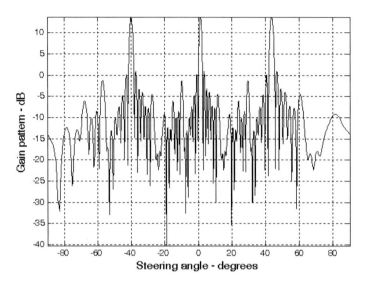

Figure 8.17. Array gain pattern: $Nr = 25$; $dolr = 1.5$; $\theta_0 = -40°$; $win = none$; $nbits = 3$

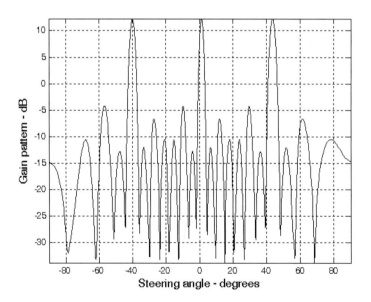

Figure 8.18. Array gain pattern: $Nr = 25$; $dolr = 1.5$; $\theta_0 = -40°$; $win = Hamming$; $nbits = 3$

8.5. Planar Arrays

Planar arrays are a natural extension of linear arrays. Planar arrays can take on many configurations, depending on the element spacing and distribution defined by a "grid." Examples include rectangular, rectangular with circular boundary, hexagonal with circular boundary, circular, and concentric circular grids, as illustrated in Fig. 8.19.

Planar arrays can be steered in elevation and azimuth $((0, \phi))$, as illustrated in Fig. 8.20 for a rectangular grid array. The element spacing along the x- and y-directions are respectively denoted by d_x and d_y. The total electric field at a far field observation point for any planar array can be computed using Eqs. (8.24) and (8.25).

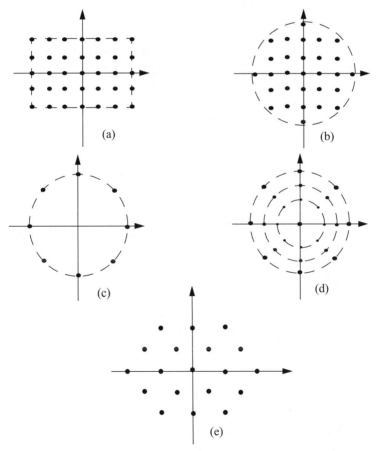

Figure 8.19. Planar array grids. (a) Rectangular; (b) Rectangular with circular boundary; (c) Circular; (d) Concentric circular; and (e) Hexagonal.

Rectangular Grid Arrays

Consider the $N \times M$ rectangular grid as shown in Fig. 8.20. The dot product $\vec{r}_i \bullet \vec{r}_0$, where the vector $\vec{r}_i$ is the vector to the *ith* element in the array and $\vec{r}_0$ is the unit vector to the far field observation point, can be broken linearly into its x- and y-components. It follows that the electric field components due to the elements distributed along the x- and y-directions are respectively, given by

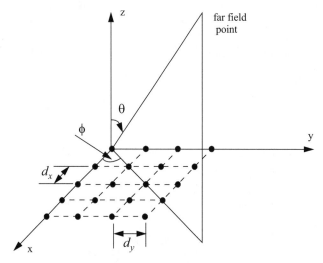

Figure 8.20. Rectangular array geometry.

$$E_x(\theta, \phi) = \sum_{n=1}^{N} I_{x_n} e^{j(n-1)kd_x \sin\theta \cos\phi} \quad (8.54)$$

$$E_y(\theta, \phi) = \sum_{m=1}^{N} I_{y_m} e^{j(m-1)kd_y \sin\theta \sin\phi} \quad (8.55)$$

The total electric field at the far field observation point is then given by

$$E(\theta, \phi) = E_x(\theta, \phi) E_y(\theta, \phi) = \quad (8.56)$$

$$\left(\sum_{m=1}^{N} I_{y_m} e^{j(m-1)kd_y \sin\theta \sin\phi} \right) \left(\sum_{n=1}^{N} I_{x_n} e^{j(n-1)kd_x \sin\theta \cos\phi} \right)$$

Planar Arrays

Eq. (8.56) can be expressed in terms of the directional cosines

$$u = \sin\theta\cos\phi$$
$$v = \sin\theta\sin\phi \qquad \text{(8.57a)}$$

$$\phi = \operatorname{atan}\left(\frac{u}{v}\right)$$
$$\theta = \operatorname{asin}\sqrt{u^2 + v^2} \qquad \text{(8.57b)}$$

The visible region is then defined by

$$\sqrt{u^2 + v^2} \le 1 \qquad \text{(8.58)}$$

It is very common to express a planar array's ability to steer the beam in space in terms of the U, V space instead of the angles θ, ϕ. Fig. 8.21 shows how a beam steered in a certain θ, ϕ direction is translated into U, V space.

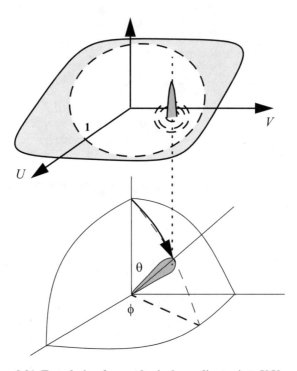

Figure 8.21. Translation from spherical coordinates into U,V space.

The rectangular array one-way intensity pattern is then equal to the product of the individual patterns. More precisely for a uniform excitation ($I_{y_m} = I_{x_n} = const$),

$$E(\theta, \phi) = \left|\frac{\sin((Nkd_x\sin\theta\cos\phi)/2)}{\sin((kd_x\sin\theta\cos\phi)/2)}\right| \left|\frac{\sin((Nkd_y\sin\theta\sin\phi)/2)}{\sin((kd_y\sin\theta\sin\phi)/2)}\right| \quad (8.59)$$

The radiation pattern maxima, nulls, sidelobes, and grating lobes in both the x- and y-axes are computed in a similar fashion to the linear array case. Additionally, the same conditions for grating lobe control are applicable. Note the symmetry is about the angle ϕ.

Circular Grid Arrays

The geometry of interest is shown in Fig. 8.19c. In this case, N elements are distributed equally on the outer circle whose radius is a. For this purpose consider the geometry shown in Fig. 8.22. From the geometry

$$\Phi_n = \frac{2\pi}{N}n \quad ; n = 1, 2, ..., N \quad (8.60)$$

The coordinates of the *nth* element are

$$x_n = a\cos\Phi_n$$
$$y_n = a\sin\Phi_n \quad (8.61)$$
$$z_n = 0$$

It follows that

$$k(\vec{r}_n \bullet \vec{r}_0) = \Psi_n = k(a\sin\theta\cos\phi\cos\Phi_n + a\sin\theta\sin\phi\sin\Phi_n + 0) \quad (8.62)$$

which can be rearranged as

$$\Psi_n = ak\sin\theta(\cos\phi\cos\Phi_n + \sin\phi\sin\Phi_n) \quad (8.63)$$

Then by using the identity $\cos(A-B) = \cos A\cos B + \sin A\sin B$, Eq.(8.63) collapses to

$$\Psi_n = ak\sin\theta\cos(\Phi_n - \phi) \quad (8.64)$$

Finally by using Eq. (8.25), the far field electric field is then given by

$$E(\theta, \phi; a) = \sum_{n=1}^{N} I_n \exp\left\{j\frac{2\pi a}{\lambda}\sin\theta\cos(\Phi_n - \phi)\right\} \quad (8.65)$$

where I_n represents the complex current distribution for the *nth* element. When the array main beam is directed in the (θ_0, ϕ_0), Eq. (8.65) takes on the following form

$$E(\theta, \phi; a) = \sum_{n=1}^{N} I_n \exp\left\{j\frac{2\pi a}{\lambda}[\sin\theta\cos(\Phi_n - \phi) - \sin\theta_0\cos(\Phi_n - \phi_0)]\right\} \quad (8.66)$$

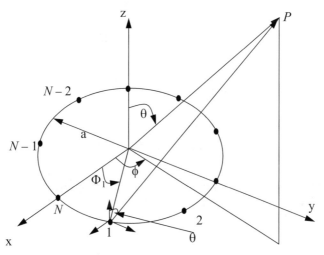

Figure 8.22. Geometry for a circular array.

MATLAB program "circular_array.m"

The MATLAB program *"circular_array.m"* calculates and plots the rectangular and polar array patterns for a circular array versus θ and ϕ constant planes. It is given in Listing 8.4 in Section 8.8. The input parameters to this program include:

Symbol	Description	Units
a	Circular array radius	λ
N	number of elements	none
theta0	main direction in θ	degrees
phi0	main direction in ϕ	degrees
Variations	'Theta'; or 'Phi'	none

Symbol	Description	Units
phid	constant ϕ plane	degrees
thetad	constant θ plane	degrees

Consider the case when the inputs are:

a	1.5
N	10 dipole antennas
theta0	$\theta_0 = 45°$
phi0	$\phi_0 = 60°$
Variations	'Theta'
phid	$\phi_d = 60°$
thetad	$\theta_d = 45°$

Fig.s 8.23 and 8.24 respectively show the array pattern in relative amplitude and the power pattern versus the angle θ. Figs. 8.25 and 8.26 are similar to Figs. 8.23 and 8.24 except in this case the patterns are plotted in polar coordinates.

Fig. 8.27 shows a plot of the normalized single element pattern (upper left corner), the normalized array factor (upper right corner), and the total array pattern (lower left corner). Fig. 8.28 shows the 3-D pattern for this example in the θ, ϕ space.

Figs. 8.29 through 8.33 are similar to those in Figs. 8.23 through 8.27, except in this case the input parameters are given by:

a	1.5
N	10 dipole antennas
theta0	$\theta_0 = 45°$
phi0	$\phi_0 = 60°$
Variations	'Phi'
phid	$\phi_d = 60°$
thetad	$\theta_d = 45°$

Planar Arrays 347

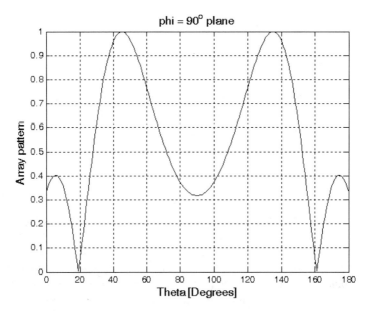

Figure 8.23. Array factor pattern for a circular array, using the parameters defined in the table on top of page 346 (rectangular coordinates).

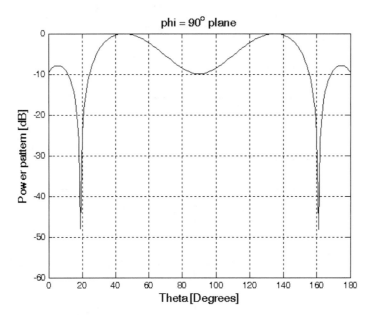

Figure 8.24. Same as Fig. 8.23 using dB scale.

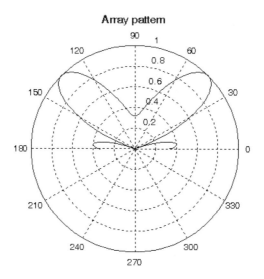

Figure 8.25. Array factor pattern for a circular array, using the parameters defined in the table on top of page 346 (polar coordinates).

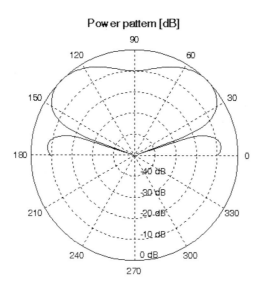

Figure 8.26. Same as Fig. 8.25 using dB scale.

Planar Arrays

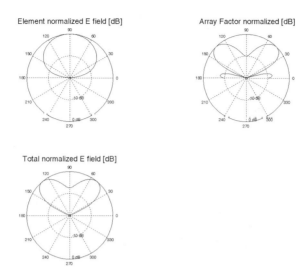

Figure 8.27. Element, array factor, and total pattern for the circular array defined in the table on top of page 346.

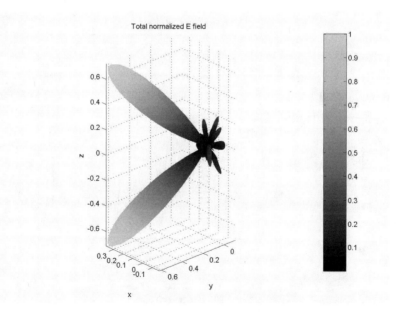

Figure 8.28. 3-D total array pattern (in θ, ϕ space) for the circular array defined in the table on top of page 346.

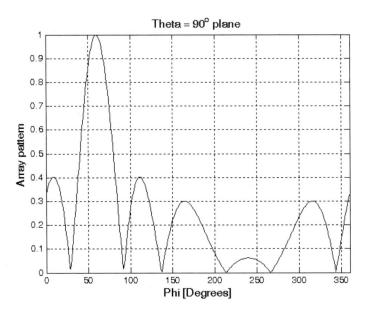

Figure 8.29. Array factor pattern for a circular array, using the parameters defined in the table on bottom of page 346 (rectangular coordinates).

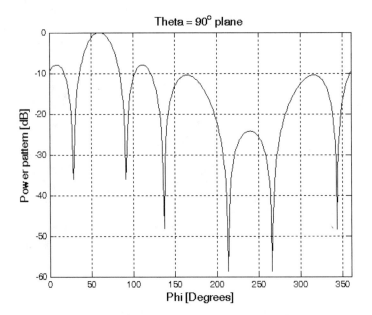

Figure 8.30. Same as Fig. 8.29 using dB scale.

Planar Arrays 351

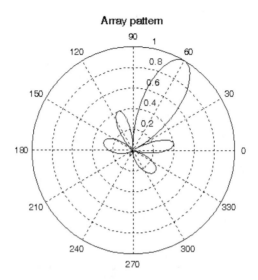

Figure 8.31. Array factor pattern for a circular array, using the parameters defined in the table on bottom of page 346 (polar coordinates).

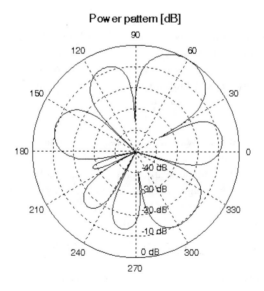

Figure 8.32. Same as Fig. 8.31 using dB scale.

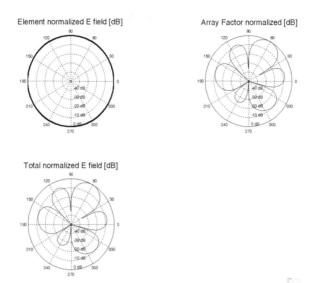

Figure 8.33. Element, array factor, and total pattern for the circular array defined in the table on bottom of page 346.

Concentric Grid Circular Arrays

The geometry of interest is shown in Fig. 8.19d and Fig. 8.34. In this case, N_2 elements are distributed equally on the outer circle whose radius is a_2, while other N_1 elements are linearly distributed on the inner circle whose radius is a_1. The element located on the center of both circles is used as the phase reference. In this configuration, there are $N_1 + N_2 + 1$ total elements in the array.

The array pattern is derived in two steps. First, the array pattern corresponding to the linearly distributed concentric circular arrays with N_1 and N_2 elements and the center element are computed separately. Second, the overall array pattern corresponding to the two concentric arrays and the center element are added. The element pattern of the identical antenna elements are considered in the first step. Thus, the total pattern becomes,

$$E(\theta, \phi) = E_0(\theta, \phi) + E_1(\theta, \phi; a_1) + E_2(\theta, \phi; a_2) \tag{8.67}$$

Fig. 8.35 shows a 3-D plot for concentric circular array in the θ, ϕ space for the following parameters:

a_1	N_1	a_2	N_2
1λ	$8\ (\lambda/2\ dipoles)$	2λ	$8\ (\lambda/2\ dipoles)$

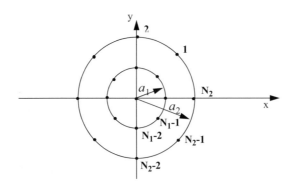

Figure 8.34. Concentric circular array geometry.

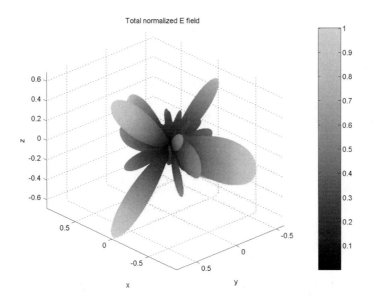

Figure 8.35. 3-D array pattern for a concentric circular array; $\theta = 45°$ and $\phi = 90°$

Rectangular Grid with Circular Boundary Arrays

The far field electric field associated with this configuration can be easily obtained from that corresponding to a rectangular grid. In order to accomplish this task follow these steps: First, select the desired maximum number of elements along the diameter of the circle and denote it by N_d. Also select the associated element spacings d_x, d_y. Define a rectangular array of size $N_d \times N_d$. Draw a circle centered at $(x, y) = (0, 0)$ with radius r_d where

$$r_d = \frac{N_d - 1}{2} + \Delta x \qquad (8.68)$$

and $\Delta x \leq d_x/4$. Finally, modify the weighting function across the rectangular array by multiplying it with the two-dimensional sequence $a(m, n)$, where

$$a(m, n) = \begin{cases} 1 & , \text{ if dis to } (m, n)\text{th element} < r_d \\ 0 & ; \text{ elsewhere} \end{cases} \qquad (8.69)$$

where distance, dis, is measured from the center of the circle. This is illustrated in Fig. 8.36.

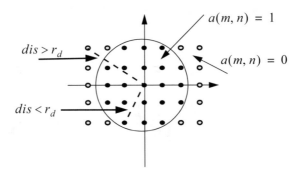

Figure 8.36. Elements with solid dots have $a(m, n) = 1$; other elements have $a(m, n) = 0$.

Hexagonal Grid Arrays

The analysis provided in this section is limited to hexagonal arrays with circular boundaries. The horizontal element spacing is denoted as d_x and the vertical element spacing is

$$d_y = \frac{\sqrt{3}}{2} d_x \qquad (8.70)$$

Planar Arrays

The array is assumed to have the maximum number of identical elements along the x-axis ($y = 0$). This number is denoted by N_x, where N_x is an odd number in order to obtain a symmetric array, where an element is present at $(x, y) = (0, 0)$. The number of rows in the array is denoted by M. The horizontal rows are indexed by m which varies from $-(N_x - 1)/2$ to $(N_x - 1)/2$. The number of elements in the mth row is denoted by N_r and is defined by

$$N_r = N_x - |m| \tag{8.71}$$

The electric field at a far field observation point is computed using Eq. (8.24) and (8.25). The phase associated with $(m, n)th$ location is

$$\psi_{m,n} = \frac{2\pi d_x}{\lambda} \sin\theta \left[\left(m + \frac{n}{2}\right) \cos\phi + n\frac{\sqrt{3}}{2} \sin\phi \right] \tag{8.72}$$

MATLAB Function "rect_array.m"

The function "rect_array.m" computes and plots the rectangular antenna gain pattern in the visible U, V space. This function is given in Listing 8.5 in Section 8.8. The syntax is as follows:

[pattern] = rect_array(Nxr, Nyr, dolxr, dolyr, theta0, phi0, winid, win, nbits)

where

Symbol	Description	Units	Status
Nxr	number of elements along x	none	input
Nyr	number of elements along y	none	input
dolxr	element spacing in lambda units along x	wavelengths	input
dolyr	element spacing in lambda units along y	wavelengths	input
theta0	elevation steering angle	degrees	input
phi0	azimuth steering angle	degrees	input
winid	-1: No weighting is used 1: Use weighting defined in win	none	input
win	window for sidelobe control	none	input
nbits	negative #: perfect quantization positive #: use 2^{nbits} quantization levels	none	input
pattern	gain pattern	dB	output

A MATLAB based GUI workspace called *"array.m"* was developed for this function. It shown in Fig. 8.37. The user is advised to use this MATLAB GUI[1] workspace to generate array gain patterns that match this requirement.

Fig.s 8.38 through 8.43 respectively show plots of the array gain pattern in the *U-V* space, for the following cases:

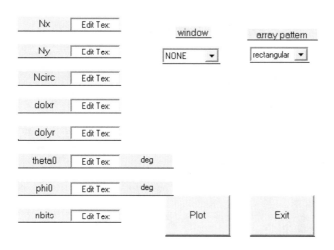

Figure 8.37. MATLAB GUI workspace *"array.m."*

[pattern] = rect_array(15, 15, 0.5, 0.5, 0, 0, -1, -1, -3)	**(8.73)**
[pattern] = rect_array(15, 15, 0.5, 0.5, 20, 30, -1, -1, -3)	**(8.74)**
[pattern] = rect_array(15, 15, 0.5, 0.5, 30, 30, 1, 'Hamming', -3)	**(8.75)**
[pattern] = rect_array(15, 15, 0.5, 0.5, 30, 30, -1, -1, 3)	**(8.76)**
[pattern] = rect_array(15, 15, 1, 0.5, 10, 30, -1, -1, -3)	**(8.77)**
[pattern] = rect_array(15, 15, 1, 1, 0, 0, -1, -1, -3)	**(8.78)**

---

1. This GUI was developed by Mr. David J. Hall, Consultant to Decibel Research, Inc., Huntsville, Alabama.

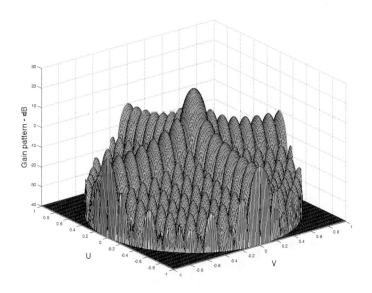

Figure 8.38a. 3-D gain pattern corresponding to Eq. (8.73).

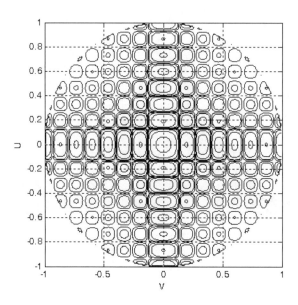

Figure 8.38b. Contour plot corresponding to Eq. (8.73).

358 *MATLAB Simulations for Radar Systems Design*

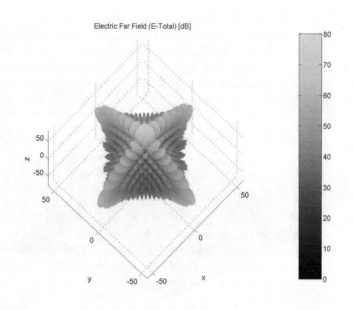

Figure 8.38c. Three-dimensional plot (θ, ϕ space) corresponding to Eq. (8.73).

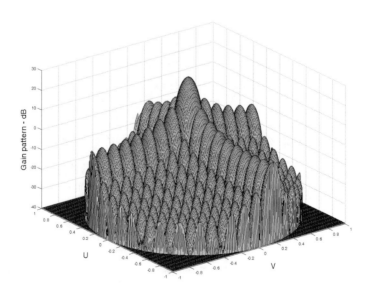

Figure 8.39a. 3-D gain pattern corresponding to Eq. (8.74).

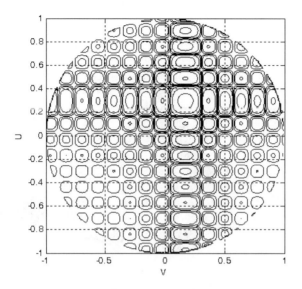

Figure 8.39b. Contour plot corresponding to Eq. (8.74).

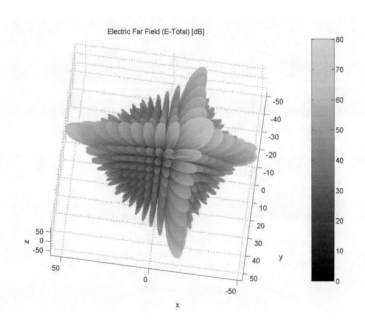

Figure 8.39c. 3-D plot (θ, ϕ space) corresponding to Eq. (8.74).

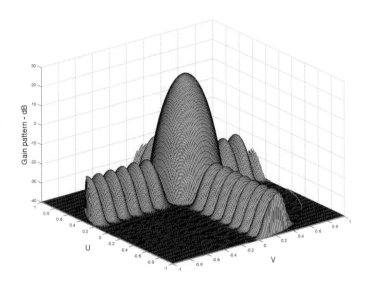

Figure 8.40a. 3-D gain pattern corresponding to Eq. (8.75).

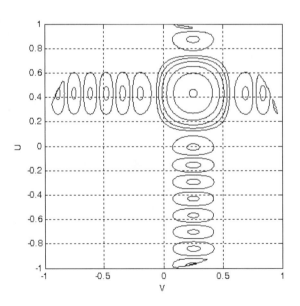

Figure 8.40b. Contour plot corresponding to Eq. (8.75).

Planar Arrays

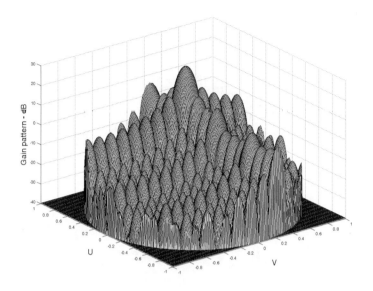

Figure 8.41a. 3-D gain pattern corresponding to Eq. (8.76).

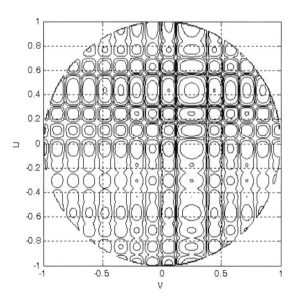

Figure 8.41b. Contour plot corresponding to Eq. (8.76).

362 *MATLAB Simulations for Radar Systems Design*

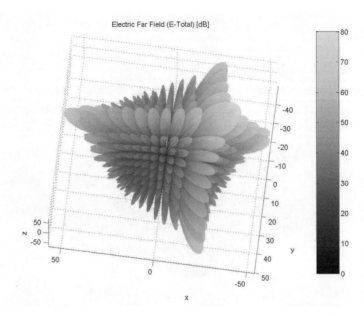

Figure 8.41c. 3-D plot (θ, ϕ space) corresponding to Eq. (8.76).

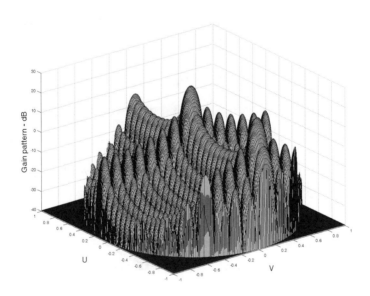

Figure 8.42a. 3-D gain pattern corresponding to Eq. (8.77).

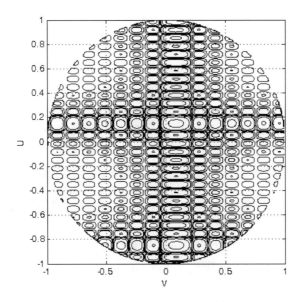

Figure 8.42b. Contour plot corresponding to Eq. (8.77).

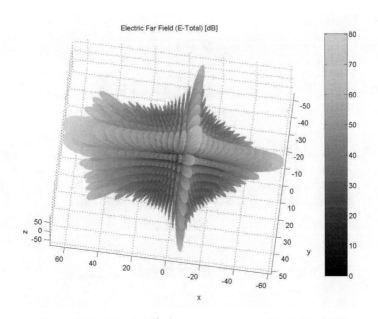

Figure 8.42c. 3-D plot (θ, ϕ space) corresponding to Eq. (8.77).

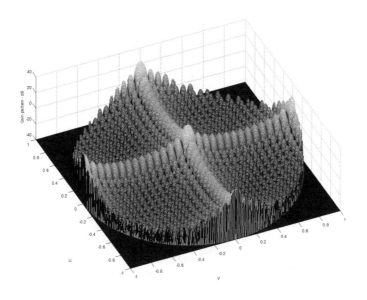

Figure 8.43a. 3-D gain pattern corresponding to Eq. (8.78).

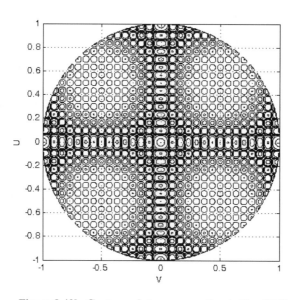

Figure 8.43b. Contour plot corresponding to Eq. (8.78).

Planar Arrays 365

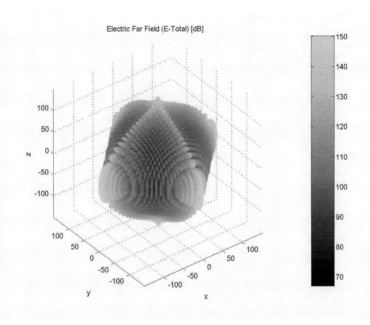

Figure 8.43c. 3-D plot (θ, ϕ space) corresponding to Eq. (8.78).

MATLAB Function "circ_array.m"

The function *"circ_array.m"* computes and plots the rectangular grid with a circular array boundary antenna gain pattern in the visible U,V space. This function is given in Listing 8.6 in Section 8.8. The syntax is as follows:

[pattern, amn] = circ_array(N, dolxr, dolyr, theta0, phi0, winid, win, nbits);

where

Symbol	Description	Units	Status
N	number of elements along diameter	none	input
dolxr	element spacing in lambda units along x	wavelengths	input
dolyr	element spacing in lambda units along y	wavelengths	input
theta0	elevation steering angle	degrees	input
phi0	azimuth steering angle	degrees	input
winid	-1: No weighting is used 1: Use weighting defined in win	none	input

Symbol	Description	Units	Status
win	window for sidelobe control	none	input
nbits	negative #: perfect quantization positive #: use 2^{nbits} quantization levels	none	input
patterng	gain pattern	dB	output
amn	a(m,n) sequence defined in Eq. (8.68)	none	output

Figs. 8.44 through 8.49 respectively show plots of the array gain pattern versus steering for the following cases:

$$[pattern, amn] = circ\_array(15, 0.5, 0.5, 0, 0, -1, -1, -3) \quad (8.79)$$

$$[pattern, amn] = circ\_array(15, 0.5, 0.5, 20, 30, -1, -1, -3) \quad (8.80)$$

$$[pattern, amn] = circ\_array(15, 0.5, 0.5, 30, 30, 1, \text{'Hamming'}, -3) \quad (8.81)$$

$$[pattern, amn] = circ\_array(15, 0.5, 0.5, 30, 30, -1, -1, 3) \quad (8.82)$$

$$[pattern, amn] = circ\_array(15, 1, 0.5, 10, 30, -1, -1, -3) \quad (8.83)$$

$$[pattern, amn] = circ\_array(15, 1, 1, 0, 0, -1, -1, -3) \quad (8.84)$$

Note the function *"circ_array.m"* uses the function *"rec_to_circ.m"*, which computes the array $a(m, n)$. It is given in Listing 8.7 in Section 8.8.

The MATLAB GUI workspace defined in *"array.m"* can be used to execute this function.

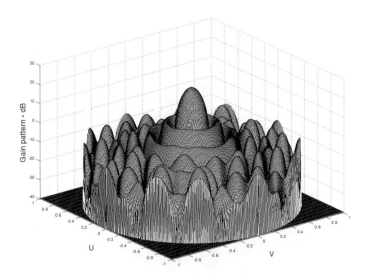

Figure 8.44a. 3-D gain pattern corresponding to Eq. (8.79).

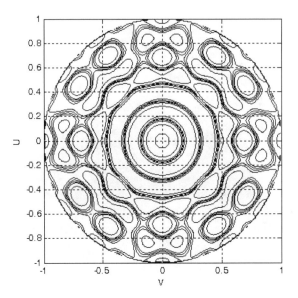

Figure 8.44b. Contour plot corresponding to Eq. (8.79).

368 *MATLAB Simulations for Radar Systems Design*

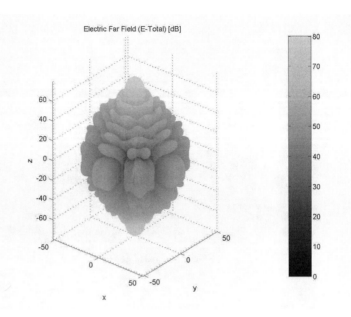

Figure 8.44c. 3-D plot (θ, ϕ space) corresponding to Eq. (8.79).

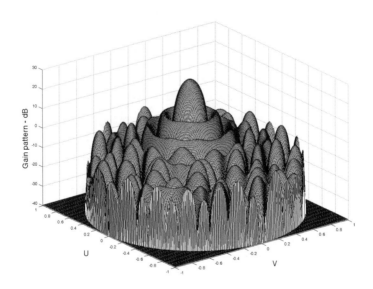

Figure 8.45a. 3-D gain pattern corresponding to Eq. (8.80).

Planar Arrays 369

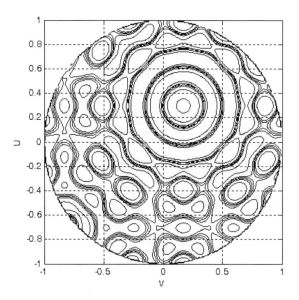

Figure 8.45b. Contour plot corresponding to Eq. (8.80).

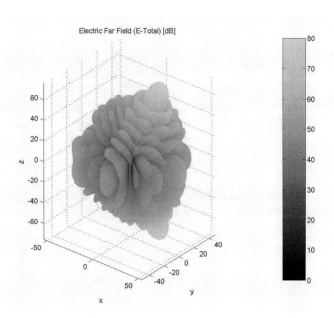

Figure 8.45c. 3-D plot (θ, ϕ space) corresponding to Eq. (8.80).

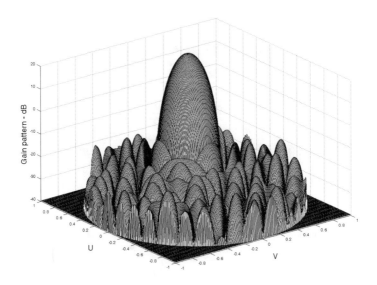

Figure 8.46a. 3-D gain pattern corresponding to Eq. (8.81).

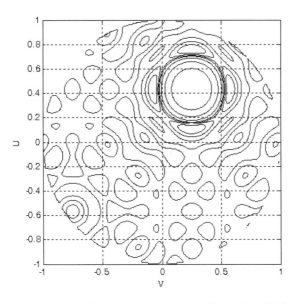

Figure 8.46b. Contour plot corresponding to Eq. (8.81).

Planar Arrays 371

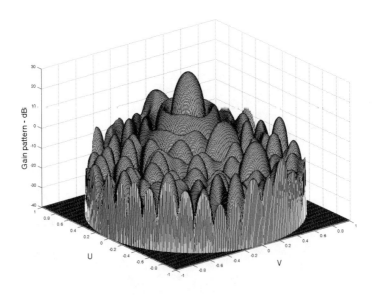

Figure 8.47a. 3-D gain pattern corresponding to Eq. (8.82).

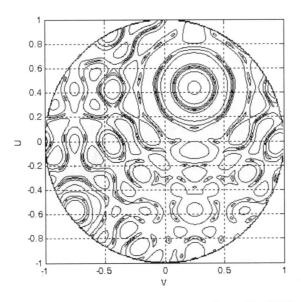

Figure 8.47b. Contour plot corresponding to Eq. (8.82).

372 *MATLAB Simulations for Radar Systems Design*

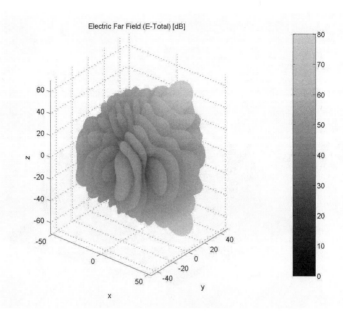

Figure 8.47c. 3-D plot (θ, ϕ space) corresponding to Eq. (8.82).

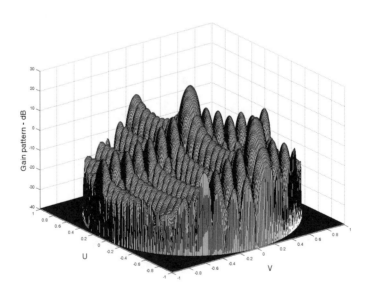

Figure 8.48a. 3-D gain pattern corresponding to Eq. (8.83).

Planar Arrays 373

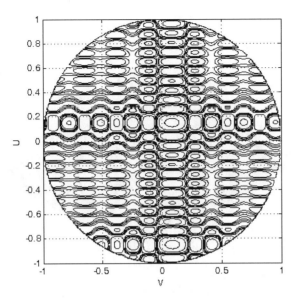

Figure 8.48b. Contour plot corresponding to Eq. (8.83).

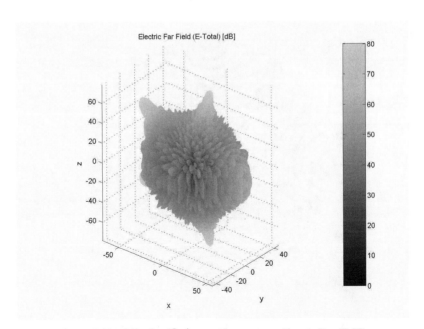

Figure 8.48c. 3-D plot (θ, ϕ space) corresponding to Eq. (8.83).

374 *MATLAB Simulations for Radar Systems Design*

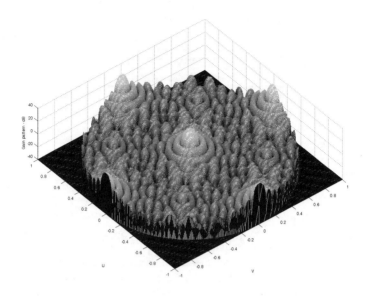

Figure 8.49a. 3-D gain pattern corresponding to Eq. (8.84).

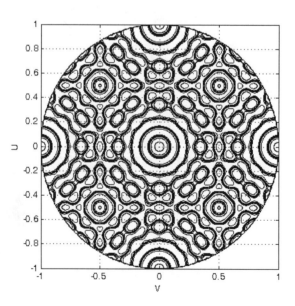

Figure 8.49b. Contour plot corresponding to Eq. (8.84).

Array Scan Loss

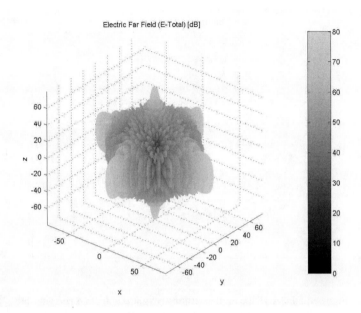

Figure 8.49c. 3-D plot (θ, ϕ space) corresponding to Eq. (8.84).

The program *"array.m"* also plots the array's element spacing pattern. Figs. 8.50a and 8.50b show two examples. The *"x's"* indicate the location of actual active array elements, while the *"o's"* indicate the location of dummy or virtual elements created merely for computational purposes. More precisely, Fig. 8.50a shows a rectangular grid with circular boundary as defined in Eqs. (8.67) and (8.68) with $d_x = d_y = 0.5\lambda$ and $a = 0.35\lambda$. Fig. 8.50b shows a similar configuration except that an element spacing $d_x = 1.5\lambda$ and $d_y = 0.5\lambda$.

8.6. Array Scan Loss

Phased arrays experience gain loss when the beam is steered away from the array boresight, or zenith (normal to the face of the array). This loss is due to the fact that the array effective aperture becomes smaller and consequently the array beamwidth is broadened, as illustrated in Fig. 8.51. This loss in antenna gain is called scan loss, L_{scan}, where

$$L_{scan} = \left(\frac{A}{A_\theta}\right)^2 = \left(\frac{G}{G_\theta}\right)^2 \tag{8.85}$$

A_θ is effective aperture area at scan angle θ, and G_θ is effective array gain at the same angle.

376 *MATLAB Simulations for Radar Systems Design*

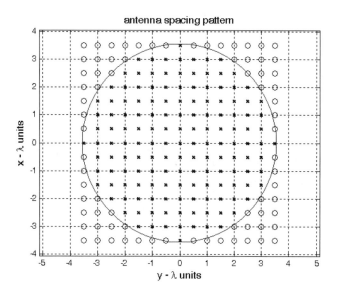

Figure 8.50a. A 15 element circular array made from a rectangular array with circular boundary. Element spacing $d_x = 0.5\lambda = d_y$.

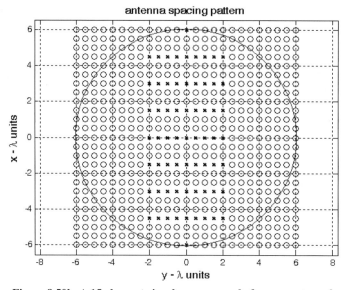

Figure 8.50b. A 15 element circular array made from a rectangular array with circular boundary. Element spacing $d_y = 0.5\lambda$ and $d_x = 1.5\lambda$.

Array Scan Loss

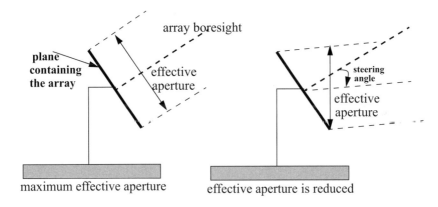

Figure 8.51. Reduction in array effective aperture due to electronic scanning.

The beamwidth at scan angle θ is

$$\Theta_\theta = \frac{\Theta_{broadside}}{\cos\theta} \tag{8.86}$$

due to the increased scan loss at large scanning angles. In order to limit the scan loss to under some acceptable practical values, most arrays do not scan electronically beyond about $\theta = 60°$. Such arrays are called Full Field Of View (FFOV). FFOV arrays employ element spacing of 0.6λ or less to avoid grating lobes. FFOV array scan loss is approximated by

$$L_{scan} \approx (\cos\theta)^{2.5} \tag{8.87}$$

Arrays that limit electronic scanning to under $\theta = 60°$ are referred to as Limited Field of View (LFOV) arrays. In this case the scan loss is

$$L_{scan} = \left[\frac{\sin\left(\frac{\pi d}{\lambda}\sin\theta\right)}{\frac{\pi d}{\lambda}\sin\theta}\right]^{-4} \tag{8.88}$$

Fig. 8.52 shows a plot for scan loss versus scan angle. This figure can be reproduced using MATLAB program *"fig8_52.m"* given in Listing 8.8.

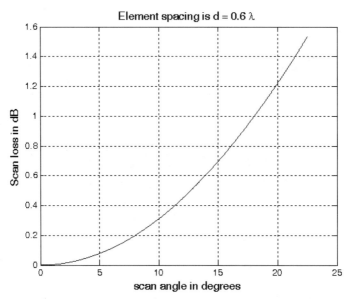

Figure 8.52. Scan loss versus scan angle, based on Eq. (8.87).

8.7. "MyRadar" Design Case Study - Visit 8

8.7.1. Problem Statement

Modify the "MyRadar" design case study such that we employ a phased array antenna. For this purpose, modify the design requirements such that the search volume is now defined by $\Theta_e = 10°$ and $\Theta_a \leq 45°$. Assume X-band, if possible. Design an electronically steered radar (ESR). Non-coherent integration of a few pulses may be used, if necessary. Size the radar so that it can fulfill this mission. Calculate the antenna gain, aperture size, missile and aircraft detection range, number of elements in the array, etc. All other design requirements are as defined in the previous chapters.

8.7.2. A Design

The search volume is

$$\Omega = \frac{10° \times 45°}{(57.296)^2} = 0.1371 \ \ steradian \tag{8.89}$$

"MyRadar" Design Case Study - Visit 8

For an X-band radar, choose $f_o = 9GHz$, then

$$\lambda = \frac{3 \times 10^8}{9 \times 10^9} = 0.0333m \tag{8.90}$$

Assume an aperture size $A_e = 2.25m^2$; thus

$$G = \frac{4\pi A_e}{\lambda^2} = \frac{4 \times \pi \times 2.25}{(0.0333)^2} = 25451.991 \Rightarrow G = 44dB \tag{8.91}$$

Assume square aperture. It follows that the aperture 3-dB beamwidth is calculated from

$$\left(G = \frac{4\pi}{\theta_{3db}^2}\right) \Rightarrow \theta_{3dB} = \sqrt{\frac{4 \times \pi \times 180^2}{25451.991 \times \pi^2}} = 1.3° \tag{8.92}$$

The number of beams required to fill the search volume is

$$n_b = k_p \frac{\Omega}{(1.3/57.296)^2}\bigg|_{k_p = 1.5} \Rightarrow n_b = 399.5 \Rightarrow choose\ n_b = 400 \tag{8.93}$$

Note that the packing factor k_p is used to allow for beam overlap in order to avoid gaps in the beam coverage. The search scan rate is 2 seconds. Thus, the minimum PRF should correspond to 200 beams per second (i.e., $f_r = 200Hz$). This PRF will allow the radar to visit each beam position only once during a complete scan.

It was determined in Chapter 2 that 4-pulse non-coherent integration along with a cumulative detection scheme are required to achieve the desired probability of detection. It was also determined that the single pulse energy for the missile and aircraft cases are respectively given by (see page 118)

$$E_m = 0.1147 Joules \tag{8.94}$$

$$E_a = 0.1029 Joules \tag{8.95}$$

However, these values were derived using $\lambda = 0.1m$ and $G = 2827.4$. The new wavelength is $\lambda = 0.0333m$ and the new gain is $G = 25451.99$. Thus, the missile and aircraft single pulse energy, assuming the same single pulse SNR as derived in Chapter 2 (i.e., $SNR = 4dB$) are

$$E_m = 0.1147 \times \frac{0.1^2 \times 2827.4^2}{0.0333^2 \times 25452^2} = 0.012765 Joules \tag{8.96}$$

$$E_a = 0.1029 \times \frac{0.1^2 \times 2827.4^2}{0.0333^2 \times 25452^2} = 0.01145 Joules \qquad (8.97)$$

The single pulse peak power that will satisfy detection for both target types is

$$P_t = \frac{0.012765}{20 \times 10^{-6}} = 638.25 W \qquad (8.98)$$

where $\tau = 20\mu s$ is used.

Note that since a 4-pulse non-coherent integration is adopted, the minimum PRF is increased to

$$f_r = 200 \times 4 = 800 Hz \qquad (8.99)$$

and the total number of beams is $n_b = 1600$. Consequently the unambiguous range is

$$R_u \leq \frac{3 \times 10^8}{2 \times 800} = 187.5 Km \qquad (8.100)$$

$$(1.101)$$

Since the effective aperture is $A_e = 2.25 m^2$, then by assuming an array efficiency $\rho = 0.8$ the actual array size is

$$A = \frac{2.25}{0.8} = 2.8125 m^2 \qquad (8.102)$$

It follows that the physical array sides are $1.68m \times 1.68m$. Thus, by selecting the array element spacing $d = 0.6\lambda$ an array of size 84×84 elements satisfies the design requirements.

Since the field of view is less than $\pm 22.5°$, one can use element spacing as large as $d = 1.5\lambda$ without introducing any grating lobes into the array FOV. Using this option yields an array of size $34 \times 34 = 1156$ elements. Hence, the required power per element is less than $0.6 W$.

8.8. MATLAB Program and Function Listings

This section contains listings of all MATLAB programs and functions used in this chapter. Users are encouraged to rerun this code with different inputs in order to enhance their understanding of the theory.

Listing 8.1. MATLAB Program "fig8_5.m"

```
% Use this code to produce figure 8.5a and 8.5b based on equation 8.34
clear all
close all
eps = 0.00001;
k = 2*pi;
theta = -pi : pi / 10791 : pi;
var = sin(theta);
nelements = 8;
d = 1;      %  d = 1;
num = sin((nelements * k * d * 0.5) .* var);

if(abs(num) <= eps)
   num = eps;
end
den = sin((k* d * 0.5) .* var);
if(abs(den) <= eps)
   den = eps;
end

pattern = abs(num ./ den);
maxval = max(pattern);
pattern = pattern ./ maxval;

figure(1)
plot(var,pattern)
xlabel('sine angle - dimensionless')
ylabel('Array pattern')
grid

figure(2)
plot(var,20*log10(pattern))
axis ([-1 1 -60 0])
xlabel('sine angle - dimensionless')
ylabel('Power pattern [dB]')
grid;

figure(3)
theta = theta +pi/2;
polar(theta,pattern)
title ('Array pattern')

figure(4)
```

polardb(theta,pattern)
title ('Power pattern [dB]')

Listing 8.2. MATLAB Program "fig8_7.m"

% Use this program to reproduce Fig. 8.7 of text
clear all
close all
eps =0.00001;
N = 32;
rect(1:32) = 1;
ham = hamming(32);
han = hanning(32);
blk = blackman(32);
k3 = kaiser(32,3);
k6 = kaiser(32,6);
RECT = 20*log10(abs(fftshift(fft(rect, 512)))./32 +eps);
HAM = 20*log10(abs(fftshift(fft(ham, 512)))./32 +eps);
HAN = 20*log10(abs(fftshift(fft(han, 512)))./32+eps);
BLK = 20*log10(abs(fftshift(fft(blk, 512)))./32+eps);
K6 = 20*log10(abs(fftshift(fft(k6, 512)))./32+eps);
x = linspace(-1,1,512);
figure
subplot(2,1,1)
plot(x,RECT,'k--',x,HAM,'k',x,HAN,'k-.');
xlabel('x')
ylabel('Window')
grid
axis tight
legend('Rectangular','Hamming','Hanning')
subplot(2,1,2)
plot(x,RECT,'k--',x,BLK,'k',x,K6,'K-.')
xlabel('x')
ylabel('Window')
legend('Rectangular','Blackman','Kasier at \beta = 6')
grid
axis tight

Listing 8.3. MATLAB Function "linear_array.m"

function [theta,patternr,patterng] =
linear_array(Nr,dolr,theta0,winid,win,nbits);
% This function computes and returns the gain radiation pattern for a linear array

% It uses the FFT to compute the pattern
%%%%%%%%%% ********** INPUTS *********** %%%%%%%%%%%%
% Nr ==> number of elements; dolr ==> element spacing (d) in lambda units divided by lambda
% theta0 ==> steering angle in degrees; winid ==> use winid negative for no window, winid positive to enter your window of size(Nr)
% win is input window, NOTE that win must be an NrX1 row vector; nbits ==> number of bits used in the phase shifters
% negative nbits mean no quantization is used
%%%%%% ********** OUTPUTS ********** %%%%%%%%%%%%%%
% theta ==> real-space angle; patternr ==> array radiation pattern in dBs
% patterng ==> array directive gain pattern in dBs
%%%%%%%%%%%%%%% *************** %%%%%%%%%%%%%%
eps = 0.00001;
n = 0:Nr-1;
i = sqrt(-1);
%if dolr is > 0.5 then; choose dol = 0.25 and compute new N
if(dolr <=0.5)
  dol = dolr;
  N = Nr;
else
  ratio = ceil(dolr/.25);
  N = Nr * ratio;
  dol = 0.25;
end
% choose proper size fft, for minimum value choose 256
Nrx = 10 * N;
nfft = 2^(ceil(log(Nrx)/log(2)));
if nfft < 256
  nfft = 256;
end
% convert steering angle into radians; and compute the sine of angle
theta0 = theta0 *pi /180.;
sintheta0 = sin(theta0);
% determine and compute quantized steering angle
if nbits < 0
  phase0 = exp(i*2.0*pi .* n * dolr * sintheta0);
else
  % compute and add the phase shift terms (WITH nbits quantization)
  % Use formula theta1 = (2*pi*n*dol) * sin(theta0) divided into 2^nbits
  % and rounded to the nearest qunatization level
  levels = 2^nbits;
  qlevels = 2.0 * pi / levels; % compute quantization levels

% compute the phase level and round it to the closest quantization level at each array element
    angleq = round(dolr .* n * sintheta0 * levels) .* qlevels; % vector of possible angles
    phase0 = exp(i*angleq);
end
% generate array of elements with or without window
if winid < 0
    wr(1:Nr) = 1;
else
    wr = win';
end
% add the phase shift terms
 wr = wr .* phase0;
 % determine if interpolation is needed (i.e., N > Nr)
 if N > Nr
    w(1:N) = 0;
    w(1:ratio:N) = wr(1:Nr);
else
    w = wr;
end
% compute the sine(theta) in real space that corresponds to the FFT index
arg = [-nfft/2:(nfft/2)-1] ./ (nfft*dol);
idx = find(abs(arg) <= 1);
sinetheta = arg(idx);
theta = asin(sinetheta);
% convert angle into degrees
theta = theta .* (180.0 / pi);
% Compute fft of w (radiation pattern)
patternv = (abs(fftshift(fft(w,nfft)))).^2;
% convert radiationa pattern to dBs
patternr = 10*log10(patternv(idx) ./Nr + eps);
% Compute directive gain pattern
rbarr = 0.5 *sum(patternv(idx)) ./ (nfft * dol);
patterng = 10*log10(patternv(idx) + eps) - 10*log10(rbarr + eps);
return

Listing 8.4. MATLAB Program "circular_array.m"

% Circular Array in the x-y plane
% Element is a short dipole antenna parallel to the z axis
% 2D Radiation Patterns for fixed phi or fixed theta
% dB polar plots uses the polardb.m file
% Last modified: July 13, 2003

```
%
%%%% Element expression needs to be modified if different
%%%% than a short dipole antenna along the z axis
clear all
clf
% close all
% ====  Input Parameters  ====
a = 1.;        % radius of the circle
N = 20;        % number of Elements of the circular array
theta0 = 45;   % main beam Theta direction
phi0 = 60;     % main beam Phi direction
% Theta or Phi variations for the calculations of the far field pattern
Variations = 'Phi';  % Correct selections are 'Theta' or 'Phi'
phid = 60;     % constant phi plane for theta variations
thetad = 45;   % constant theta plane for phi variations
% ====  End of Input parameters section  ====
dtr = pi/180;       % conversion factors
rtd = 180/pi;
phi0r = phi0*dtr;
theta0r = theta0*dtr;
lambda = 1;
k = 2*pi/lambda;
ka = k*a;          % Wavenumber times the radius
jka = j*ka;
I(1:N) = 1;        % Elements excitation Amplitude and Phase
alpha(1:N) =0;
for n = 1:N        % Element positions Uniformly distributed along the circle
   phin(n) = 2*pi*n/N;
end
switch Variations
case 'Theta'
   phir = phid*dtr;   % Pattern in a constant Phi plane
   i = 0;
   for theta = 0.001:1:181
      i = i+1;
      thetar(i) = theta*dtr;
      angled(i) = theta;  angler(i) = thetar(i);
      Arrayfactor(i) = 0;
      for n = 1:N
         Arrayfactor(i) = Arrayfactor(i) + I(n)*exp(j*alpha(n)) ...
                 * exp( jka*(sin(thetar(i))*cos(phir -phin(n))) ...
                     -jka*(sin(theta0r )*cos(phi0r-phin(n))) );
      end
      Arrayfactor(i) = abs(Arrayfactor(i));
```

```
        Element(i) = abs(sin(thetar(i)+0*dtr));  % use the abs function to avoid
    end
case 'Phi'
    thetar = thetad*dtr;  % Pattern in a constant Theta plane
    i = 0;
    for phi = 0.001:1:361
        i = i+1;
        phir(i)   = phi*dtr;
        angled(i) = phi;  angler(i) = phir(i);
        Arrayfactor(i) = 0;
        for n = 1:N
            Arrayfactor(i) = Arrayfactor(i) + I(n)*exp(j*alpha(n)) ...
                    * exp( jka*(sin(thetar )*cos(phir(i)-phin(n))) ...
                        -jka*(sin(theta0r)*cos(phi0r -phin(n))) );
        end
        Arrayfactor(i) = abs(Arrayfactor(i));
        Element(i) = abs(sin(thetar+0*dtr));  % use the abs function to avoid
    end
end
angler = angled*dtr;
Element = Element/max(Element);
Array = Arrayfactor/max(Arrayfactor);
ArraydB = 20*log10(Array);
EtotalR =(Element.*Arrayfactor)/max(Element.*Arrayfactor);
figure(1)
plot(angled,Array)
ylabel('Array pattern')
grid
switch Variations
case 'Theta'
  axis ([0 180 0 1 ])
% theta = theta +pi/2;
  xlabel('Theta [Degrees]')
  title ( 'phi = 90^o plane')
case 'Phi'
axis ([0 360 0 1 ])
  xlabel('Phi [Degrees]')
  title ( 'Theta = 90^o plane')
end
figure(2)
plot(angled,ArraydB)
%axis ([-1 1 -60 0])
ylabel('Power pattern [dB]')
grid;
```

```
switch Variations
case 'Theta'
  axis ([0 180 -60 0 ])
   xlabel('Theta [Degrees]')
     title ( 'phi = 90^o plane')
case 'Phi'
axis ([0 360 -60 0 ])
   xlabel('Phi [Degrees]')
      title ( 'Theta = 90^o plane')
end
figure(3)
polar(angler,Array)
title ('Array pattern')
figure(4)
polardb(angler,Array)
title ('Power pattern [dB]')
% the plots provided above are for the array factor based on the circular
% array plots for other patterns such as those for the antenna element
% (Element)or the total pattern (Etotal based on Element*Arrayfactor) can
% also be displayed by the user as all these patterns are already computed
% above.
figure(10)
subplot(2,2,1)
polardb (angler,Element,'b-'); % rectangular plot of element pattern
title('Element normalized E field [dB]')
subplot(2,2,2)
polardb(angler,Array,'b-')
title(' Array Factor normalized [dB]')
subplot(2,2,3)
polardb(angler,EtotalR,'b-');  % polar plot
title('Total normalized E field [dB]')
%%%%%%%%%%%%%%%%%%%%%%%%%%%%%%%
%%%%%%%%%%%%%%%%%%%%%%%%%%%%%%%
function polardb(theta,rho,line_style)
%  POLARDB  Polar coordinate plot.
%  POLARDB(THETA, RHO) makes a plot using polar coordinates of
%  the angle THETA, in radians, versus the radius RHO in dB.
%  The maximum value of RHO should not exceed 1. It should not be
%  normalized, however (i.e., its max. value may be less than 1).
%  POLAR(THETA,RHO,S) uses the linestyle specified in string S.
%  See PLOT for a description of legal linestyles.
if nargin < 1
   error('Requires 2 or 3 input arguments.')
elseif nargin == 2
```

```
   if isstr(rho)
      line_style = rho;
      rho = theta;
      [mr,nr] = size(rho);
      if mr == 1
         theta = 1:nr;
      else
         th = (1:mr)';
         theta = th(:,ones(1,nr));
      end
   else
      line_style = 'auto';
   end
elseif nargin == 1
   line_style = 'auto';
   rho = theta;
   [mr,nr] = size(rho);
   if mr == 1
      theta = 1:nr;
   else
      th = (1:mr)';
      theta = th(:,ones(1,nr));
   end
end
if isstr(theta) | isstr(rho)
   error('Input arguments must be numeric.');
end
if ~isequal(size(theta),size(rho))
   error('THETA and RHO must be the same size.');
end
% get hold state
cax = newplot;
next = lower(get(cax,'NextPlot'));
hold_state = ishold;

% get x-axis text color so grid is in same color
tc = get(cax,'xcolor');
ls = get(cax,'gridlinestyle');
% Hold on to current Text defaults, reset them to the
% Axes' font attributes so tick marks use them.
fAngle  = get(cax, 'DefaultTextFontAngle');
fName   = get(cax, 'DefaultTextFontName');
fSize   = get(cax, 'DefaultTextFontSize');
fWeight = get(cax, 'DefaultTextFontWeight');
```

```
fUnits  = get(cax, 'DefaultTextUnits');
set(cax, 'DefaultTextFontAngle',  get(cax, 'FontAngle'), ...
    'DefaultTextFontName',   get(cax, 'FontName'), ...
    'DefaultTextFontSize',   get(cax, 'FontSize'), ...
    'DefaultTextFontWeight', get(cax, 'FontWeight'), ...
    'DefaultTextUnits','data')
% make a radial grid
    hold on;
    maxrho =1;
    hhh=plot([-maxrho -maxrho maxrho maxrho],[-maxrho maxrho maxrho -maxrho]);
    set(gca,'dataaspectratio',[1 1 1],'plotboxaspectratiomode','auto')
    v = [get(cax,'xlim') get(cax,'ylim')];
    ticks = sum(get(cax,'ytick')>=0);
    delete(hhh);
% check radial limits and ticks
    rmin = 0; rmax = v(4); rticks = max(ticks-1,2);
    if rticks > 5   % see if we can reduce the number
        if rem(rticks,2) == 0
            rticks = rticks/2;
        elseif rem(rticks,3) == 0
            rticks = rticks/3;
        end
    end
% only do grids if hold is off
if ~hold_state
% define a circle
    th = 0:pi/50:2*pi;
    xunit = cos(th);
    yunit = sin(th);
% now really force points on x/y axes to lie on them exactly
    inds = 1:(length(th)-1)/4:length(th);
    xunit(inds(2:2:4)) = zeros(2,1);
    yunit(inds(1:2:5)) = zeros(3,1);
% plot background if necessary
    if ~isstr(get(cax,'color')),
        patch('xdata',xunit*rmax,'ydata',yunit*rmax, ...
              'edgecolor',tc,'facecolor',get(gca,'color'),...
              'handlevisibility','off');
    end
% draw radial circles with dB ticks
    c82 = cos(82*pi/180);
    s82 = sin(82*pi/180);
    rinc = (rmax-rmin)/rticks;
```

```matlab
    tickdB=-10*(rticks-1);    % the innermost tick dB value
    for i=(rmin+rinc):rinc:rmax
        hhh = plot(xunit*i,yunit*i,ls,'color',tc,'linewidth',1,...
            'handlevisibility','off');
        text((i+rinc/20)*c82*0,-(i+rinc/20)*s82, ...
            [' ' num2str(tickdB) ' dB'],'verticalalignment','bottom',...
            'handlevisibility','off')
        tickdB=tickdB+10;
    end
    set(hhh,'linestyle','-') % Make outer circle solid
% plot spokes
    th = (1:6)*2*pi/12;
    cst = cos(th); snt = sin(th);
    cs = [-cst; cst];
    sn = [-snt; snt];
    plot(rmax*cs,rmax*sn,ls,'color',tc,'linewidth',1,...
        'handlevisibility','off')
% annotate spokes in degrees
    rt = 1.1*rmax;
    for i = 1:length(th)
        text(rt*cst(i),rt*snt(i),int2str(i*30),...
            'horizontalalignment','center',...
            'handlevisibility','off');
        if i == length(th)
            loc = int2str(0);
        else
            loc = int2str(180+i*30);
        end
        text(-rt*cst(i),-rt*snt(i),loc,'horizontalalignment','center',...
            'handlevisibility','off')
    end
% set view to 2-D
    view(2);
% set axis limits
    axis(rmax*[-1 1 -1.15 1.15]);
end
% Reset defaults.
set(cax, 'DefaultTextFontAngle', fAngle , ...
    'DefaultTextFontName',   fName , ...
    'DefaultTextFontSize',   fSize, ...
    'DefaultTextFontWeight', fWeight, ...
    'DefaultTextUnits', fUnits );
% Transfrom data to dB scale
rmin = 0; rmax=1;
```

```
rinc = (rmax-rmin)/rticks;
rhodb=zeros(1,length(rho));
for i=1:length(rho)
   if rho(i)==0
      rhodb(i)=0;
   else
      rhodb(i)=rmax+2*log10(rho(i))*rinc;
   end
   if rhodb(i)<-0
      rhodb(i)=0;
   end
end
% transform data to Cartesian coordinates.
xx = rhodb.*cos(theta);
yy = rhodb.*sin(theta);
% plot data on top of grid
if strcmp(line_style,'auto')
   q = plot(xx,yy);
else
   q = plot(xx,yy,line_style);
end
if nargout > 0
   hpol = q;
end
if ~hold_state
   set(gca,'dataaspectratio',[1 1 1]), axis off; set(cax,'NextPlot',next);
end
set(get(gca,'xlabel'),'visible','on')
set(get(gca,'ylabel'),'visible','on')
```

Listing 8.5. MATLAB Function "rect_array.m"

```
function [pattern] =
rect_array(Nxr,Nyr,dolxr,dolyr,theta0,phi0,winid,win,nbits);
%%%%%%%%%% ********************* %%%%%%%%%%
% This function computes the 3-D directive gain patterns for a planar array
% This function uses the fft2 to compute its output
%%%%%%%% *********** INPUTS ************ %%%%%%%%%%
% Nxr ==> number of along x-axis; Nyr ==> number of elements along y-
axis
% dolxr ==> element spacing in x-direction; dolyr ==> element spacing in y-
direction Both are in lambda units
% theta0 ==> elevation steering angle in degrees, phi0 ==> azimuth steering
angle in degrees
```

```
% winid ==> window identifier; winid negative ==> no window ; winid posi-
tive ==> use window given by win
% win ==> input window function (2-D window) MUST be of size (Nxr X Nyr)
% nbits is the number of nbits used in phase quantization; nbits negative ==>
NO quantization
%%%%% ********** OUTPUTS ************* %%%%%%%
% pattern ==> directive gain pattern
%%%%%%% *********************** %%%%%%%%%%%%%
eps = 0.0001;
nx = 0:Nxr-1;
ny = 0:Nyr-1;
i = sqrt(-1);
% check that window size is the same as the array size
[nw,mw] = size(win);
if winid >0
   if nw ~= Nxr
   fprintf('STOP == Window size must be the same as the array')
   return
end
if mw ~= Nyr
   fprintf('STOP == Window size must be the same as the array')
   return
end
end

%if dol is > 0.5 then; choose dol = 0.5 and compute new N
if(dolxr <=0.5)
   ratiox = 1 ;
   dolx = dolxr ;
   Nx = Nxr ;
else
   ratiox = ceil(dolxr/.5) ;
   Nx = (Nxr -1 ) * ratiox + 1 ;
   dolx = 0.5 ;
end
if(dolyr <=0.5)
   ratioy = 1 ;
   doly = dolyr ;
   Ny = Nyr ;
else
   ratioy = ceil(dolyr/.5) ;
   Ny = (Nyr -1) * ratioy + 1 ;
   doly = 0.5 ;
end
```

```
% choose proper size fft, for minimum value choose 256X256
Nrx = 10 * Nx;
Nry = 10 * Ny;
nfftx = 2^(ceil(log(Nrx)/log(2)));
nffty = 2^(ceil(log(Nry)/log(2)));
if nfftx < 256
  nfftx = 256;
end
if nffty < 256
  nffty = 256;
end
% generate array of elements with or without window
if winid < 0
  array = ones(Nxr,Nyr);
else
  array = win;
end
% convert steering angles (theta0, phi0) to radians
theta0 = theta0 * pi / 180;
phi0 = phi0 * pi / 180;
% convert steering angles (theta0, phi0) to U-V sine-space
u0 = sin(theta0) * cos(phi0);
v0 = sin(theta0) * sin(phi0);
% Use formula thetal = (2*pi*n*dol) * sin(theta0) divided into 2^m levels
% and rounded to the nearest qunatization level
if nbits < 0
  phasem = exp(i*2*pi*dolx*u0 .* nx *ratiox);
  phasen = exp(i*2*pi*doly*v0 .* ny *ratioy);
else
  levels = 2^nbits;
  qlevels = 2.0*pi / levels; % compute quantization levels
  sinthetaq = round(dolx .* nx * u0 * levels * ratiox) .* qlevels; % vector of possible angles
  sinphiq = round(doly .* ny * v0 * levels *ratioy) .* qlevels; % vector of possible angles
  phasem = exp(i*sinthetaq);
  phasen = exp(i*sinphiq);
end
% add the phase shift terms
array = array .* (transpose(phasem) * phasen);
% determine if interpolation is needed (i.e., N > Nr)
if (Nx > Nxr )| (Ny > Nyr)
  for xloop = 1 : Nxr
    temprow = array(xloop, :) ;
```

```
   w( (xloop-1)*ratiox+1, 1:ratioy:Ny) = temprow ;
  end
  array = w;
else
  w = array ;
%   w(1:Nx, :) = array(1:N,:);
end
% Compute array pattern
arrayfft = abs(fftshift(fft2(w,nfftx,nffty))).^2 ;
%compute [su,sv] matrix
U = [-nfftx/2:(nfftx/2)-1] ./(dolx*nfftx);
indexx = find(abs(U) <= 1);
U = U(indexx);
V = [-nffty/2:(nffty/2)-1] ./(doly*nffty);
indexy = find(abs(V) <= 1);
V = V(indexy);
%Normalize to generate gain patern
rbar=sum(sum(arrayfft(indexx,indexy))) / dolx/doly/4./nfftx/nffty;
arrayfft = arrayfft(indexx,indexy) ./rbar;
[SU,SV] = meshgrid(V,U);
indx = find((SU.^2 + SV.^2) >1);
arrayfft(indx) = eps/10;
pattern = 10*log10(arrayfft +eps);
figure(1)
mesh(V,U,pattern);
xlabel('V')
ylabel('U');
zlabel('Gain pattern - dB')
figure(2)
contour(V,U,pattern)
grid
axis image
xlabel('V')
ylabel('U');
axis([-1 1 -1 1])
figure(3)
x0 = (Nx+1)/2 ;
y0 = (Ny+1)/2 ;
radiusx = dolx*((Nx-1)/2) ;
radiusy = doly*((Ny-1)/2) ;
[xxx, yyy]=find(abs(array)>eps);
xxx = xxx-x0 ;
yyy = yyy-y0 ;
plot(yyy*doly, xxx*dolx,'rx')
```

```
hold on
axis([-radiusy-0.5 radiusy+0.5 -radiusx-0.5 radiusx+0.5]);
grid
title('antenna spacing pattern');
xlabel('y - \lambda units')
ylabel('x - \lambda units')
[xxx0, yyy0]=find(abs(array)<=eps);
xxx0 = xxx0-x0 ;
yyy0 = yyy0-y0 ;
plot(yyy0*doly, xxx0*dolx,'co')
axis([-radiusy-0.5 radiusy+0.5 -radiusx-0.5 radiusx+0.5]);
hold off
return
```

Listing 8.6. MATLAB Function "circ_array.m"

```
function [pattern,amn] =
circ_array(N,dolxr,dolyr,theta0,phi0,winid,win,nbits);
%%%%%%%%% ************************ %%%%%%%%%%%%
% This function computes the 3-D directive gain patterns for a circular planar array
% This function uses the fft2 to compute its output. It assumes that there are the same number of elements along the major x- and y-axes
%%%%%%%%% ************ INPUTS ************ %%%%%%%%%
% N ==> number of elements along x-aixs or y-axis
% dolxr ==> element spacing in x-direction; dolyr ==> element spacing in y-direction. Both are in lambda units
% theta0 ==> elevation steering angle in degrees, phi0 ==> azimuth steering angle in degrees
% This function uses the function (rec_to_circ) which computes the circular array from a square
% array (of size NXN) using the notation developed by ALLEN,J.L.,"The Theory of Array Antennas
% (with Emphasis on Radar Application)" MIT-LL Technical Report No. 323, July, 25 1965.
% winid ==> window identifier; winid negative ==> no window ; winid positive ==> use window given by win
% win ==> input window function (2-D window) MUST be of size (Nxr X Nyr)
% nbits is the number of nbits used in phase quantization; nbits negative ==> NO quantization
%%%%%%% *********** OUTPUTS ************ %%%%%%%%%
% amn ==> array of ones and zeros; ones indicate true element location on the grid
% zeros mean no elements at that location; pattern ==> directive gain pattern
```

```
%%%%%%%%% ************************ %%%%%%%%%%%%%
eps = 0.0001;
nx = 0:N-1;
ny = 0:N-1;
i = sqrt(-1);
% check that window size is the same as the array size
[nw,mw] = size(win);
if winid >0
  if mw ~= N
    fprintf('STOP == Window size must be the same as the array')
    return
  end
  if nw ~= N
    fprintf('STOP == Window size must be the same as the array')
    return
  end
end
%if dol is > 0.5 then; choose dol = 0.5 and compute new N
if(dolxr <=0.5)
  ratiox = 1 ;
  dolx = dolxr ;
  Nx = N ;
else
  ratiox = ceil(dolxr/.5) ;
  Nx = (N-1) * ratiox + 1 ;
  dolx = 0.5 ;
end
if(dolyr <=0.5)
  ratioy = 1 ;
  doly = dolyr ;
  Ny = N ;
else
  ratioy = ceil(dolyr/.5);
  Ny = (N-1)*ratioy + 1 ;
  doly = 0.5 ;
end
% choose proper size fft, for minimum value choose 256X256
Nrx = 10 * Nx;
Nry = 10 * Ny;
nfftx = 2^(ceil(log(Nrx)/log(2)));
nffty = 2^(ceil(log(Nry)/log(2)));
if nfftx < 256
  nfftx = 256;
end
```

```
if nffty < 256
    nffty = 256;
end
% generate array of elements with or without window
if winid < 0
    array = ones(N,N);
else
    array = win;
end
% convert steering angles (theta0, phi0) to radians
theta0 = theta0 * pi / 180;
phi0 = phi0 * pi / 180;
% convert steering angles (theta0, phi0) to U-V sine-space
u0 = sin(theta0) * cos(phi0);
v0 = sin(theta0) * sin(phi0);
% Use formula theta1 = (2*pi*n*dol) * sin(theta0) divided into 2^m levels
% and rounded to the nearest qunatization level
if nbits < 0
    phasem = exp(i*2*pi*dolx*u0 .* nx * ratiox);
    phasen = exp(i*2*pi*doly*v0 .* ny * ratioy);
else
    levels = 2^nbits;
    qlevels = 2.0*pi / levels; % compute quantization levels
    sinthetaq = round(dolx .* nx * u0 * levels * ratiox) .* qlevels; % vector of possible angles
    sinphiq = round(doly .* ny * v0 * levels *ratioy) .* qlevels; % vector of possible angles
    phasem = exp(i*sinthetaq);
    phasen = exp(i*sinphiq);
end
% add the phase shift terms
array = array .* (transpose(phasem) * phasen) ;

% determine if interpolation is needed (i.e., N > Nr)
if (Nx > N )| (Ny > N)
    for xloop = 1 : N
        temprow = array(xloop, :) ;
        w( (xloop-1)*ratiox+1, 1:ratioy:Ny) = temprow ;
    end
    array = w;
else
    w(1:Nx, :) = array(1:N,:);
end
% Convert rectangular array into circular using function rec_to_circ
```

```
[m,n] = size(w) ;
NC = max(m,n);  % Use Allens algorithm
if Nx == Ny
   temp_array = w;
else
   midpoint = (NC-1)/2 +1 ;
   midwm = (m-1)/2 ;
   midwn = (n-1)/2 ;
   temp_array = zeros(NC,NC);
   temp_array(midpoint-midwm:midpoint+midwm, midpoint-midwn:midpoint+midwn) = w ;
end
amn = rec_to_circ(NC);  % must be rectangular array (Nx=Ny)
amn = temp_array .* amn ;

% Compute array pattern
arrayfft = abs(fftshift(fft2(amn,nfftx,nffty))).^2 ;
%compute [su,sv] matrix
U = [-nfftx/2:(nfftx/2)-1] ./(dolx*nfftx);
indexx = find(abs(U) <= 1);
U = U(indexx);
V = [-nffty/2:(nffty/2)-1] ./(doly*nffty);
indexy = find(abs(V) <= 1);
V = V(indexy);
[SU,SV] = meshgrid(V,U);
indx = find((SU.^2 + SV.^2) >1);
arrayfft(indx) = eps/10;
%Normalize to generate gain pattern
rbar=sum(sum(arrayfft(indexx,indexy))) / dolx/doly/4./nfftx/nffty;
arrayfft = arrayfft(indexx,indexy) ./rbar;
[SU,SV] = meshgrid(V,U);
indx = find((SU.^2 + SV.^2) >1);
arrayfft(indx) = eps/10;
pattern = 10*log10(arrayfft +eps);
figure(1)
mesh(V,U,pattern);
xlabel('V')
ylabel('U');
zlabel('Gain pattern - dB')
figure(2)
contour(V,U,pattern)
axis image
grid
xlabel('V')
```

```
ylabel('U');
axis([-1 1 -1 1])
figure(3)
x0 = (NC+1)/2 ;
y0 = (NC+1)/2 ;
radiusx = dolx*((NC-1)/2 + 0.05/dolx) ;
radiusy = doly*((NC-1)/2 + 0.05/dolx) ;
theta = 5 ;
[xxx, yyy]=find(abs(amn)>0);
xxx = xxx-x0 ;
yyy = yyy-y0 ;
plot(yyy*doly, xxx*dolx,'rx')
axis equal
hold on
axis([-radiusy-0.5 radiusy+0.5 -radiusx-0.5  radiusx+0.5]);
grid
title('antenna spacing pattern');
xlabel('y - \lambda units')
ylabel('x - \lambda units')
[x, y]= makeellip( 0, 0, radiusx, radiusy, theta) ;
plot(y, x) ;
axis([-radiusy-0.5 radiusy+0.5 -radiusx-0.5  radiusx+0.5]);
[xxx0, yyy0]=find(abs(amn)<=0);
xxx0 = xxx0-x0 ;
yyy0 = yyy0-y0 ;
plot(yyy0*doly, xxx0*dolx,'co')
axis([-radiusy-0.5 radiusy+0.5 -radiusx-0.5  radiusx+0.5]);
axis equal
hold off ;
return
```

Listing 8.7. MATLAB Function "rec_to_circ.m"

```
function amn = rec_to_circ(N)
midpoint = (N-1)/2 + 1;
amn = zeros(N);
array1(midpoint,midpoint) = N;
x0 = midpoint;
y0 = x0;
for i = 1:N
   for j = 1:N
      distance(i,j) = sqrt((x0-i)^2 + (y0-j)^2);
   end
end
```

```
idx = find(distance < (N-1)/2 + .4);
amn (idx) = 1;
return
```

Listing 8.8. MATLAB Program "fig8_52.m"

```
%Use this program to reproduce Fig. 8.40. Based on Eq. (8.87)
clear all
close all
d = 0.6; % element spacing in lambda units
betadeg = linspace(0,22.5,1000);
beta = betadeg .*pi ./180;
den = pi*d .* sin(beta);
numarg = den;
num = sin(numarg);
lscan = (num./den).^-4;
LSCAN = 10*log10(lscan+eps);
figure (1)
plot(betadeg,LSCAN)
xlabel('scan angle in degrees')
ylabel('Scan loss in dB')
grid
title('Element spacing is d = 0.6 \lambda ')
```

Chapter 9   *Target Tracking*

Single Target Tracking

Tracking radar systems are used to measure the target's relative position in range, azimuth angle, elevation angle, and velocity. Then, by using and keeping track of these measured parameters the radar can predict their future values. Target tracking is important to military radars as well as to most civilian radars. In military radars, tracking is responsible for fire control and missile guidance; in fact, missile guidance is almost impossible without proper target tracking. Commercial radar systems, such as civilian airport traffic control radars, may utilize tracking as a means of controlling incoming and departing airplanes.

Tracking techniques can be divided into range/velocity tracking and angle tracking. It is also customary to distinguish between continuous single-target tracking radars and multi-target track-while-scan (TWS) radars. Tracking radars utilize pencil beam (very narrow) antenna patterns. It is for this reason that a separate search radar is needed to facilitate target acquisition by the tracker. Still, the tracking radar has to search the volume where the target's presence is suspected. For this purpose, tracking radars use special search patterns, such as helical, T.V. raster, cluster, and spiral patterns, to name a few.

9.1. Angle Tracking

Angle tracking is concerned with generating continuous measurements of the target's angular position in the azimuth and elevation coordinates. The accuracy of early generation angle tracking radars depended heavily on the size of the pencil beam employed. Most modern radar systems achieve very fine angular measurements by utilizing monopulse tracking techniques.

Tracking radars use the angular deviation from the antenna main axis of the target within the beam to generate an error signal. This deviation is normally measured from the antenna's main axis. The resultant error signal describes how much the target has deviated from the beam main axis. Then, the beam position is continuously changed in an attempt to produce a zero error signal. If the radar beam is normal to the target (maximum gain), then the target angular position would be the same as that of the beam. In practice, this is rarely the case.

In order to be able to quickly change the beam position, the error signal needs to be a linear function of the deviation angle. It can be shown that this condition requires the beam's axis to be squinted by some angle (squint angle) off the antenna's main axis.

9.1.1. Sequential Lobing

Sequential lobing is one of the first tracking techniques that was utilized by the early generation of radar systems. Sequential lobing is often referred to as lobe switching or sequential switching. It has a tracking accuracy that is limited by the pencil beamwidth used and by the noise caused by either mechanical or electronic switching mechanisms. However, it is very simple to implement. The pencil beam used in sequential lobing must be symmetrical (equal azimuth and elevation beamwidths).

Tracking is achieved (in one coordinate) by continuously switching the pencil beam between two pre-determined symmetrical positions around the antenna's Line of Sight (LOS) axis. Hence, the name sequential lobing is adopted. The LOS is called the radar tracking axis, as illustrated in Fig. 9.1.

As the beam is switched between the two positions, the radar measures the returned signal levels. The difference between the two measured signal levels is used to compute the angular error signal. For example, when the target is tracked on the tracking axis, as the case in Fig. 9.1a, the voltage difference is zero. However, when the target is off the tracking axis, as in Fig. 9.1b, a non-zero error signal is produced. The sign of the voltage difference determines the direction in which the antenna must be moved. Keep in mind, the goal here is to make the voltage difference be equal to zero.

In order to obtain the angular error in the orthogonal coordinate, two more switching positions are required for that coordinate. Thus, tracking in two coordinates can be accomplished by using a cluster of four antennas (two for each coordinate) or by a cluster of five antennas. In the latter case, the middle antenna is used to transmit, while the other four are used to receive.

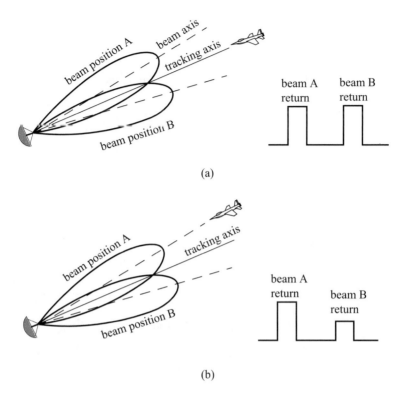

Figure 9.1. Sequential lobing. (a) Target is located on track axis. (b) Target is off track axis.

9.1.2. Conical Scan

Conical scan is a logical extension of sequential lobing where, in this case, the antenna is continuously rotated at an offset angle, or has a feed that is rotated about the antenna's main axis. Fig. 9.2 shows a typical conical scan beam. The beam scan frequency, in radians per second, is denoted as ω_s. The angle between the antenna's LOS and the rotation axis is the squint angle φ. The antenna's beam position is continuously changed so that the target will always be on the tracking axis.

Fig. 9.3 shows a simplified conical scan radar system. The envelope detector is used to extract the return signal amplitude and the Automatic Gain Control (AGC) tries to hold the receiver output to a constant value. Since the AGC operates on large time constants, it can hold the average signal level constant and still preserve the signal rapid scan variation. It follows that the tracking error signals (azimuth and elevation) are functions of the target's RCS; they are functions of its angular position off the main beam axis.

In order to illustrate how conical scan tracking is achieved, we will first consider the case shown in Fig. 9.4. In this case, as the antenna rotates around the tracking axis all target returns have the same amplitude (zero error signal). Thus, no further action is required.

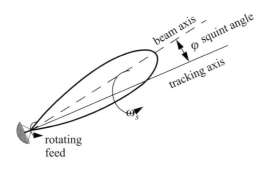

Figure 9.2. Conical scan beam.

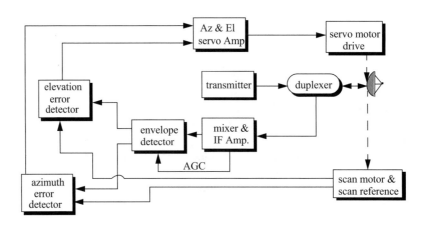

Figure 9.3. Simplified conical scan radar system.

Angle Tracking

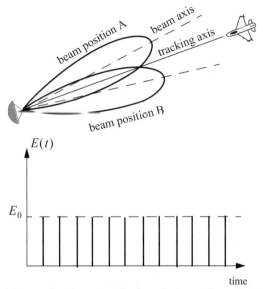

Figure 9.4. Error signal produced when the target is on the tracking axis for conical scan.

Next, consider the case depicted by Fig. 9.5. Here, when the beam is at position B, returns from the target will have maximum amplitude, and when the antenna is at position A, returns from the target have minimum amplitude. Between those two positions, the amplitude of the target returns will vary between the maximum value at position B, and the minimum value at position A. In other words, Amplitude Modulation (AM) exists on top of the returned signal. This AM envelope corresponds to the relative position of the target within the beam. Thus, the extracted AM envelope can be used to derive a servo-control system in order to position the target on the tracking axis.

Now, let us derive the error signal expression that is used to drive the servo-control system. Consider the top view of the beam axis location shown in Fig. 9.6. Assume that $t = 0$ is the starting beam position. The locations for maximum and minimum target returns are also identified. The quantity ε defines the distance between the target location and the antenna's tracking axis. It follows that the azimuth and elevation errors are, respectively, given by

$$\varepsilon_a = \varepsilon \sin \varphi \qquad (9.1)$$

$$\varepsilon_e = \varepsilon \cos \varphi \qquad (9.2)$$

These are the error signals that the radar uses to align the tracking axis on the target.

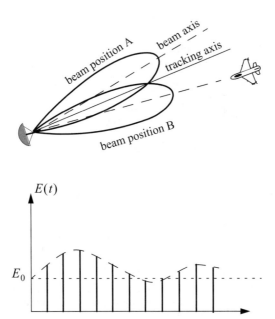

Figure 9.5. Error signal produced when the target is off the tracking axis for conical scan.

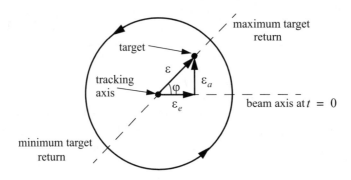

Figure 9.6. Top view of beam axis for a complete scan.

The AM signal $E(t)$ can then be written as

$$E(t) = E_0 \cos(\omega_s t - \varphi) = E_0 \varepsilon_e \cos\omega_s t + E_0 \varepsilon_a \sin\omega_s t \tag{9.3}$$

where E_0 is a constant called the error slope, ω_s is the scan frequency in radians per seconds, and φ is the angle already defined. The scan reference is the signal that the radar generates to keep track of the antenna's position around a complete path (scan). The elevation error signal is obtained by mixing the signal $E(t)$ with $\cos\omega_s t$ (the reference signal) followed by low pass filtering. More precisely,

$$E_e(t) = E_0 \cos(\omega_s t - \varphi)\cos\omega_s t = -\frac{1}{2}E_0 \cos\varphi + \frac{1}{2}\cos(2\omega_s t - \varphi) \tag{9.4}$$

and after low pass filtering we get

$$E_e(t) = -\frac{1}{2}E_0 \cos\varphi \tag{9.5}$$

Negative elevation error drives the antenna beam downward, while positive elevation error drives the antenna beam upward. Similarly, the azimuth error signal is obtained by multiplying $E(t)$ by $\sin\omega_s t$ followed by low pass filtering. It follows that

$$E_a(t) = \frac{1}{2}E_0 \sin\varphi \tag{9.6}$$

The antenna scan rate is limited by the scanning mechanism (mechanical or electronic), where electronic scanning is much faster and more accurate than mechanical scan. In either case, the radar needs at least four target returns to be able to determine the target azimuth and elevation coordinates (two returns per coordinate). Therefore, the maximum conical scan rate is equal to one fourth of the PRF. Rates as high as 30 scans per seconds are commonly used.

The conical scan squint angle needs to be large enough so that a good error signal can be measured. However, due to the squint angle, the antenna gain in the direction of the tracking axis is less than maximum. Thus, when the target is in track (located on the tracking axis), the SNR suffers a loss equal to the drop in the antenna gain. This loss is known as the squint or crossover loss. The squint angle is normally chosen such that the two-way (transmit and receive) crossover loss is less than a few decibels.

9.2. Amplitude Comparison Monopulse

Amplitude comparison monopulse tracking is similar to lobing in the sense that four squinted beams are required to measure the target's angular position. The difference is that the four beams are generated simultaneously rather than sequentially. For this purpose, a special antenna feed is utilized such that the four beams are produced using a single pulse, hence the name "monopulse." Additionally, monopulse tracking is more accurate and is not susceptible to lobing anomalies, such as AM jamming and gain inversion ECM. Finally, in sequential and conical lobing, variations in the radar echoes degrade the tracking accuracy; however, this is not a problem for monopulse techniques since a single pulse is used to produce the error signals. Monopulse tracking radars can employ both antenna reflectors as well as phased array antennas.

Fig. 9.7 show a typical monopulse antenna pattern. The four beams A, B, C, and D represent the four conical scan beam positions. Four feeds, mainly horns, are used to produce the monopulse antenna pattern. Amplitude monopulse processing requires that the four signals have the same phase and different amplitudes.

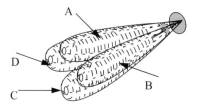

Figure 9.7. Monopulse antenna pattern.

A good way to explain the concept of amplitude monopulse technique is to represent the target echo signal by a circle centered at the antenna's tracking axis, as illustrated by Fig. 9.8a, where the four quadrants represent the four beams. In this case, the four horns receive an equal amount of energy, which indicates that the target is located on the antenna's tracking axis. However, when the target is off the tracking axis (Figs. 9.8b-d), an imbalance of energy occurs in the different beams. This imbalance of energy is used to generate an error signal that drives the servo-control system. Monopulse processing consists of computing a sum Σ and two difference Δ (azimuth and elevation) antenna patterns. Then by dividing a Δ channel voltage by the Σ channel voltage, the angle of the signal can be determined.

The radar continuously compares the amplitudes and phases of all beam returns to sense the amount of target displacement off the tracking axis. It is critical that the phases of the four signals be constant in both transmit and receive modes. For this purpose, either digital networks or microwave comparator circuitry are utilized. Fig. 9.9 shows a block diagram for a typical microwave comparator, where the three receiver channels are declared as the sum channel, elevation angle difference channel, and azimuth angle difference channel.

Amplitude Comparison Monopulse

(a)   (b)   (c)   (d)

Figure 9.8. Illustration of monopulse concept. (a) Target is on the tracking axis. (b) - (d) Target is off the tracking axis.

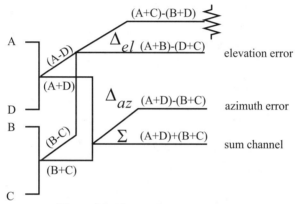

Figure 9.9. Monopulse comparator.

To generate the elevation difference beam, one can use the beam difference (A−D) or (B−C). However, by first forming the sum patterns (A+B) and (D+C) and then computing the difference (A+B)−(D+C), we achieve a stronger elevation difference signal, Δ_{el}. Similarly, by first forming the sum patterns (A+D) and (B+C) and then computing the difference (A+D)−(B+C), a stronger azimuth difference signal, Δ_{az}, is produced.

A simplified monopulse radar block diagram is shown in Fig. 9.10. The sum channel is used for both transmit and receive. In the receive mode the sum channel provides the phase reference for the other two difference channels. Range measurements can also be obtained from the sum channel. In order to illustrate how the sum and difference antenna patterns are formed, we will assume a $\sin\varphi/\varphi$ single element antenna pattern and squint angle φ_0. The sum signal in one coordinate (azimuth or elevation) is then given by

$$\Sigma(\varphi) = \frac{\sin(\varphi - \varphi_0)}{(\varphi - \varphi_0)} + \frac{\sin(\varphi + \varphi_0)}{(\varphi + \varphi_0)} \qquad (9.7)$$

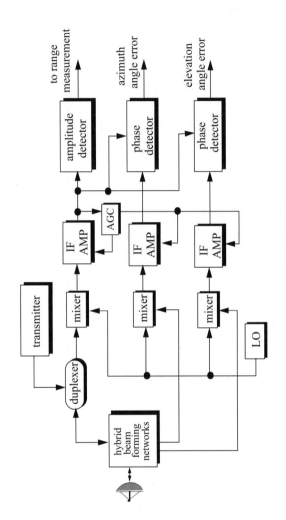

Figure 9.10. Simplified amplitude comparison monopulse radar block diagram.

and a difference signal in the same coordinate is

$$\Delta(\varphi) = \frac{\sin(\varphi - \varphi_0)}{(\varphi - \varphi_0)} - \frac{\sin(\varphi + \varphi_0)}{(\varphi + \varphi_0)} \tag{9.8}$$

MATLAB Function "mono_pulse.m"

The function "mono_pulse.m" implements Eqs. (9.7) and (9.8). Its output includes plots of the sum and difference antenna patterns as well as the difference to-sum ratio. It is given in Listing 9.1 in Section 9.11. The syntax is as follows:

mono_pulse (phi0)

where *phi0* is the squint angle in radians.

Fig. 9.11 (a-c) shows the corresponding plots for the sum and difference patterns for $\varphi_0 = 0.15$ radians. Fig. 9.12 (a-c) is similar to Fig. 9.11, except in this case $\varphi_0 = 0.75$ radians. Clearly, the sum and difference patterns depend heavily on the squint angle. Using a relatively small squint angle produces a better sum pattern than that resulting from a larger angle. Additionally, the difference pattern slope is steeper for the small squint angle.

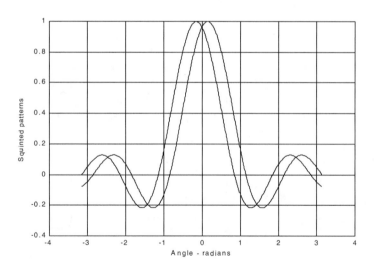

Figure 9.11a. Two squinted patterns. Squint angle is $\varphi_0 = 0.15$ radians.

412   *MATLAB Simulations for Radar Systems Design*

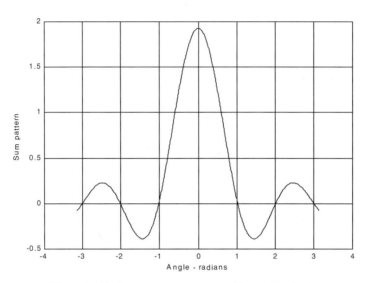

Figure 9.11b. Sum pattern corresponding to Fig. 9.11a.

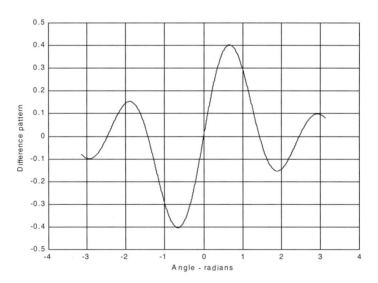

Figure 9.11c. Difference pattern corresponding to Fig. 9.11a.

Amplitude Comparison Monopulse 413

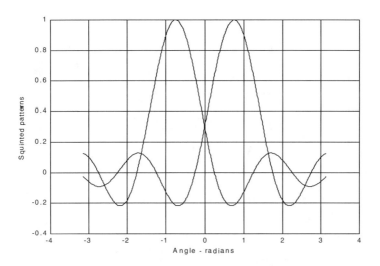

Figure 9.12a. Two squinted patterns. Squint angle is $\varphi_0 = 0.75$ radians.

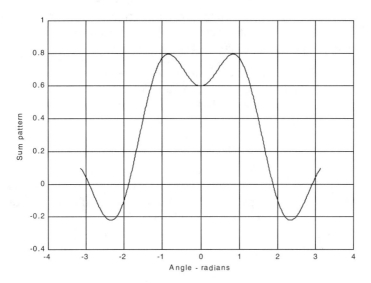

Figure 9.12b. Sum pattern corresponding to Fig. 9.12a.

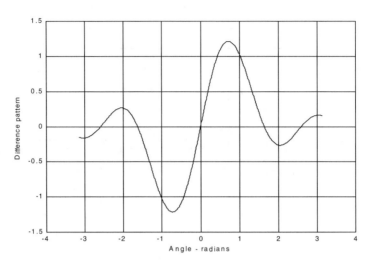

Figure 9.12c. Difference pattern corresponding to Fig. 9.12a.

The difference channels give us an indication of whether the target is on or off the tracking axis. However, this signal amplitude depends not only on the target angular position, but also on the target's range and RCS. For this reason the ratio Δ/Σ (delta over sum) can be used to accurately estimate the error angle that only depends on the target's angular position.

Let us now address how the error signals are computed. First, consider the azimuth error signal. Define the signals S_1 and S_2 as

$$S_1 = A + D \tag{9.9}$$

$$S_2 = B + C \tag{9.10}$$

The sum signal is $\Sigma = S_1 + S_2$, and the azimuth difference signal is $\Delta_{az} = S_1 - S_2$. If $S_1 \geq S_2$, then both channels have the same phase $0°$ (since the sum channel is used for phase reference). Alternatively, if $S_1 < S_2$, then the two channels are $180°$ out of phase. Similar analysis can be done for the elevation channel, where in this case $S_1 = A + B$ and $S_2 = D + C$. Thus, the error signal output is

$$\varepsilon_\varphi = \frac{|\Delta|}{|\Sigma|} \cos\xi \tag{9.11}$$

where ξ is the phase angle between the sum and difference channels and it is equal to $0°$ or $180°$. More precisely, if $\xi = 0$, then the target is on the tracking axis; otherwise it is off the tracking axis. Fig. 9.13 (a,b) shows a plot for the ratio Δ/Σ for the monopulse radar whose sum and difference patterns are in Figs. 9.11 and 9.12.

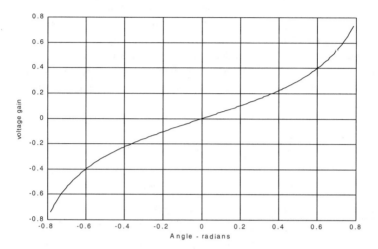

Figure 9.13a. Difference-to-sum ratio corresponding to Fig. 9.11a.

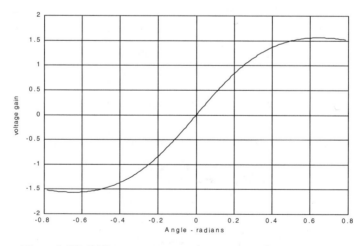

Figure 9.13b. Difference-to-sum ratio corresponding to Fig. 9.12a.

9.3. Phase Comparison Monopulse

Phase comparison monopulse is similar to amplitude comparison monopulse in the sense that the target angular coordinates are extracted from one sum and two difference channels. The main difference is that the four signals produced in amplitude comparison monopulse will have similar phases but different amplitudes; however, in phase comparison monopulse the signals have the same amplitude and different phases. Phase comparison monopulse tracking radars use a minimum of a two-element array antenna for each coordinate (azimuth and elevation), as illustrated in Fig. 9.14. A phase error signal (for each coordinate) is computed from the phase difference between the signals generated in the antenna elements.

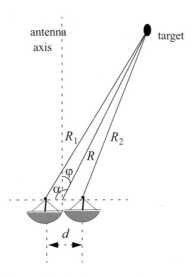

Figure 9.14. Single coordinate phase comparison monopulse antenna.

Consider Fig. 9.14; since the angle α is equal to $\varphi + \pi/2$, it follows that

$$R_1^2 = R^2 + \left(\frac{d}{2}\right)^2 - 2\frac{d}{2}R\cos\left(\varphi + \frac{\pi}{2}\right) \tag{9.12}$$

$$= R^2 + \frac{d^2}{4} - dR\sin\varphi$$

and since $d \ll R$ we can use the binomial series expansion to get

$$R_1 \approx R\left(1 + \frac{d}{2R}\sin\varphi\right) \tag{9.13}$$

Similarly,

$$R_2 \approx R\left(1 - \frac{d}{2R}\sin\varphi\right) \tag{9.14}$$

The phase difference between the two elements is then given by

$$\phi = \frac{2\pi}{\lambda}(R_1 - R_2) = \frac{2\pi}{\lambda}d\sin\varphi \tag{9.15}$$

where λ is the wavelength. The phase difference ϕ is used to determine the angular target location. Note that if $\phi = 0$, then the target would be on the antenna's main axis. The problem with this phase comparison monopulse technique is that it is quite difficult to maintain a stable measurement of the off boresight angle φ, which causes serious performance degradation. This problem can be overcome by implementing a phase comparison monopulse system as illustrated in Fig. 9.15.

The (single coordinate) sum and difference signals are, respectively, given by

$$\Sigma(\varphi) = S_1 + S_2 \tag{9.16}$$

$$\Delta(\varphi) = S_1 - S_2 \tag{9.17}$$

where the S_1 and S_2 are the signals in the two elements. Now, since S_1 and S_2 have similar amplitude and are different in phase by ϕ, we can write

$$S_1 = S_2 e^{-j\phi} \tag{9.18}$$

It follows that

$$\Delta(\varphi) = S_2(1 - e^{-j\phi}) \tag{9.19}$$

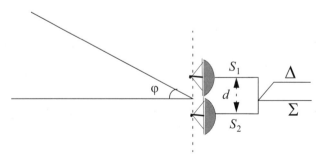

Figure 9.15. Single coordinate phase monopulse antenna, with sum and difference channels.

$$\Sigma(\varphi) = S_2(1 + e^{-j\phi}) \tag{9.20}$$

The phase error signal is computed from the ratio Δ/Σ. More precisely,

$$\frac{\Delta}{\Sigma} = \frac{1 - e^{-j\phi}}{1 + e^{-j\phi}} = j\tan\left(\frac{\phi}{2}\right) \tag{9.21}$$

which is purely imaginary. The modulus of the error signal is then given by

$$\frac{|\Delta|}{|\Sigma|} = \tan\left(\frac{\phi}{2}\right) \tag{9.22}$$

This kind of phase comparison monopulse tracker is often called the half-angle tracker.

9.4. Range Tracking

Target range is measured by estimating the round-trip delay of the transmitted pulses. The process of continuously estimating the range of a moving target is known as range tracking. Since the range to a moving target is changing with time, the range tracker must be constantly adjusted to keep the target locked in range. This can be accomplished using a split gate system, where two range gates (early and late) are utilized. The concept of split gate tracking is illustrated in Fig. 9.16, where a sketch of a typical pulsed radar echo is shown in the figure. The early gate opens at the anticipated starting time of the radar echo and lasts for half its duration. The late gate opens at the center and closes at the end of the echo signal. For this purpose, good estimates of the echo duration and the pulse center time must be reported to the range tracker so that the early and late gates can be placed properly at the start and center times of the expected echo. This reporting process is widely known as the "designation process."

The early gate produces positive voltage output while the late gate produces negative voltage output. The outputs of the early and late gates are subtracted, and the difference signal is fed into an integrator to generate an error signal. If both gates are placed properly in time, the integrator output will be equal to zero. Alternatively, when the gates are not timed properly, the integrator output is not zero, which gives an indication that the gates must be moved in time, left or right depending on the sign of the integrator output.

Range Tracking

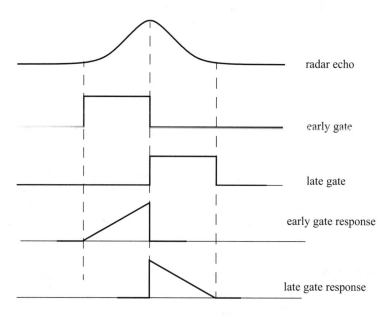

Figure 9.16. Illustration of split-range gate.

Multiple Target Tracking

Track-while-scan radar systems sample each target once per scan interval, and use sophisticated smoothing and prediction filters to estimate the target parameters between scans. To this end, the Kalman filter and the Alpha-Beta-Gamma ($\alpha\beta\gamma$) filter are commonly used. Once a particular target is detected, the radar may transmit up to a few pulses to verify the target parameters, before it establishes a track file for that target. Target position, velocity, and acceleration comprise the major components of the data maintained by a track file.

The principles of recursive tracking and prediction filters are presented in this part. First, an overview of state representation for Linear Time Invariant (LTI) systems is discussed. Then, second and third order one-dimensional fixed gain polynomial filter trackers are developed. These filters are, respectively, known as the $\alpha\beta$ and $\alpha\beta\gamma$ filters (also known as the g-h and g-h-k filters). Finally, the equations for an *n*-dimensional multi-state Kalman filter are introduced and analyzed. As a matter of notation, small case letters, with an underbar, are used.

9.5. Track-While-Scan (TWS)

Modern radar systems are designed to perform multi-function operations, such as detection, tracking, and discrimination. With the aid of sophisticated computer systems, multi-function radars are capable of simultaneously tracking many targets. In this case, each target is sampled once (mainly range and angular position) during a dwell interval (scan). Then, by using smoothing and prediction techniques future samples can be estimated. Radar systems that can perform multi-tasking and multi-target tracking are known as Track-While-Scan (TWS) radars.

Once a TWS radar detects a new target it initiates a separate track file for that detection; this ensures that sequential detections from that target are processed together to estimate the target's future parameters. Position, velocity, and acceleration comprise the main components of the track file. Typically, at least one other confirmation detection (verify detection) is required before the track file is established.

Unlike single target tracking systems, TWS radars must decide whether each detection (observation) belongs to a new target or belongs to a target that has been detected in earlier scans. And in order to accomplish this task, TWS radar systems utilize correlation and association algorithms. In the correlation process each new detection is correlated with all previous detections in order to avoid establishing redundant tracks. If a certain detection correlates with more than one track, then a pre-determined set of association rules is exercised so that the detection is assigned to the proper track. A simplified TWS data processing block diagram is shown in Fig. 9.17.

Choosing a suitable tracking coordinate system is the first problem a TWS radar has to confront. It is desirable that a fixed reference of an inertial coordinate system be adopted. The radar measurements consist of target range, velocity, azimuth angle, and elevation angle. The TWS system places a gate around the target position and attempts to track the signal within this gate. The gate dimensions are normally azimuth, elevation, and range. Because of the uncertainty associated with the exact target position during the initial detections, a gate has to be large enough so that targets do not move appreciably from scan to scan; more precisely, targets must stay within the gate boundary during successive scans. After the target has been observed for several scans the size of the gate is reduced considerably.

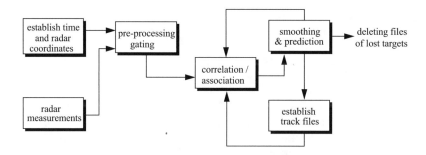

Figure. 9.17. Simplified block diagram of TWS data processing.

Gating is used to decide whether an observation is assigned to an existing track file, or to a new track file (new detection). Gating algorithms are normally based on computing a statistical error distance between a measured and an estimated radar observation. For each track file, an upper bound for this error distance is normally set. If the computed difference for a certain radar observation is less than the maximum error distance of a given track file, then the observation is assigned to that track.

All observations that have an error distance less than the maximum distance of a given track are said to correlate with that track. For each observation that does not correlate with any existing tracks, a new track file is established accordingly. Since new detections (measurements) are compared to all existing track files, a track file may then correlate with no observations or with one or more observations. The correlation between observations and all existing track files is identified using a correlation matrix. Rows of the correlation matrix represent radar observations, while columns represent track files. In cases where several observations correlate with more than one track file, a set of predetermined association rules can be utilized so that a single observation is assigned to a single track file.

9.6. State Variable Representation of an LTI System

A linear time invariant system (continuous or discrete) can be described mathematically using three variables. They are the input, output, and the state variables. In this representation, any LTI system has observable or measurable objects (abstracts). For example, in the case of a radar system, range may be an object measured or observed by the radar tracking filter. States can be derived in many different ways. For the scope of this book, states of an object or an abstract are the components of the vector that contains the object and its time derivatives. For example, a third-order one-dimensional (in this case range) state vector representing range can be given by

$$\underline{x} = \begin{bmatrix} R \\ \dot{R} \\ \ddot{R} \end{bmatrix} \qquad (9.23)$$

where R, $\dot{R}$, and $\ddot{R}$ are, respectively, the range measurement, range rate (velocity), and acceleration. The state vector defined in Eq. (9.23) can be representative of continuous or discrete states. In this book, the emphasis is on discrete time representation, since most radar signal processing is executed using digital computers. For this purpose, an n-dimensional state vector has the following form:

$$\underline{x} = \begin{bmatrix} x_1 & \dot{x}_1 & \ldots & x_2 & \dot{x}_2 & \ldots & x_n & \dot{x}_n & \ldots \end{bmatrix}^t \qquad (9.24)$$

where the superscript indicates the transpose operation.

The LTI system of interest can be represented using the following state equations:

$$\underline{\dot{x}}(t) = \underline{A}\ \underline{x}(t) + \underline{B}\underline{w}(t) \qquad (9.25)$$

$$\underline{y}(t) = \underline{C}\ \underline{x}(t) + \underline{D}\underline{w}(t) \qquad (9.26)$$

where: $\underline{\dot{x}}$ is the value of the $n \times 1$ state vector; $\underline{y}$ is the value of the $p \times 1$ output vector; $\underline{w}$ is the value of the $m \times 1$ input vector; $\underline{A}$ is an $n \times n$ matrix; $\underline{B}$ is an $n \times m$ matrix; $\underline{C}$ is $p \times n$ matrix; and $\underline{D}$ is an $p \times m$ matrix. The

State Variable Representation of an LTI System

homogeneous solution (i.e., $\underline{w} = 0$) to this linear system, assuming known initial condition $\underline{x}(0)$ at time t_0, has the form

$$\underline{x}(t) = \underline{\Phi}(t - t_0)\underline{x}(t - t_0) \tag{9.27}$$

The matrix $\underline{\Phi}$ is known as the state transition matrix, or fundamental matrix, and is equal to

$$\underline{\Phi}(t - t_0) = e^{\underline{A}(t - t_0)} \tag{9.20}$$

Eq. (9.28) can be expressed in series format as

$$\underline{\Phi}(t - t_0)\big|_{t_0 = 0} = e^{\underline{A}(t)} = \underline{I} + \underline{A}t + \underline{A}^2 \frac{t^2}{2!} + \ldots = \sum_{k=0}^{\infty} \underline{A}^k \frac{t^k}{k!} \tag{9.29}$$

Example:

Compute the state transition matrix for an LTI system when

$$\underline{A} = \begin{bmatrix} 0 & 1 \\ -0.5 & -1 \end{bmatrix}$$

Solution:

The state transition matrix can be computed using Eq. (9.29). For this purpose, compute $\underline{A}^2$ and $\underline{A}^3$ .... It follows

$$\underline{A}^2 = \begin{bmatrix} -\frac{1}{2} & -1 \\ \frac{1}{2} & \frac{1}{2} \end{bmatrix} \qquad \underline{A}^3 = \begin{bmatrix} \frac{1}{2} & \frac{1}{2} \\ -\frac{1}{4} & 0 \end{bmatrix} \qquad \ldots$$

Therefore,

$$\underline{\Phi} = \begin{bmatrix} 1 + 0t - \dfrac{\frac{1}{2}t^2}{2!} + \dfrac{\frac{1}{2}t^3}{3!} + \ldots & 0 + t - \dfrac{t^2}{2!} + \dfrac{\frac{1}{2}t^3}{3!} + \ldots \\ 0 - \dfrac{1}{2}t + \dfrac{\frac{1}{2}t^2}{2!} - \dfrac{\frac{1}{4}t^3}{3!} + \ldots & 1 - t + \dfrac{\frac{1}{2}t^2}{2!} + \dfrac{0t^3}{3!} + \ldots \end{bmatrix}$$

The state transition matrix has the following properties (the proof is left as an exercise):

1. *Derivative property*

$$\frac{\partial}{\partial t}\underline{\Phi}(t-t_0) = \underline{A}\underline{\Phi}(t-t_0) \tag{9.30}$$

2. Identity property

$$\underline{\Phi}(t_0-t_0) = \underline{\Phi}(0) = \underline{I} \tag{9.31}$$

3. Initial value property

$$\left.\frac{\partial}{\partial t}\underline{\Phi}(t-t_0)\right|_{t=t_0} = \underline{A} \tag{9.32}$$

4. Transition property

$$\underline{\Phi}(t_2-t_0) = \underline{\Phi}(t_2-t_1)\underline{\Phi}(t_1-t_0) \quad ; \ t_0 \le t_1 \le t_2 \tag{9.33}$$

5. Inverse property

$$\underline{\Phi}(t_0-t_1) = \underline{\Phi}^{-1}(t_1-t_0) \tag{9.34}$$

6. Separation property

$$\underline{\Phi}(t_1-t_0) = \underline{\Phi}(t_1)\underline{\Phi}^{-1}(t_0) \tag{9.35}$$

The general solution to the system defined in Eq. (9.25) can be written as

$$\underline{x}(t) = \underline{\Phi}(t-t_0)\underline{x}(t_0) + \int_{t_0}^{t} \underline{\Phi}(t-\tau)\underline{B}\underline{w}(\tau)d\tau \tag{9.36}$$

The first term of the right-hand side of Eq. (9.36) represents the contribution from the system response to the initial condition. The second term is the contribution due to the driving force $\underline{w}$. By combining Eqs. (9.26) and (9.36) an expression for the output is computed as

$$\underline{y}(t) = \underline{C}e^{\underline{A}(t-t_0)}\underline{x}(t_0) + \int_{t_0}^{t}[\underline{C}e^{\underline{A}(t-\tau)}\underline{B} - \underline{D}\delta(t-\tau)]\underline{w}(\tau)d\tau \tag{9.37}$$

Note that the system impulse response is equal to $\underline{C}e^{\underline{A}t}\underline{B} - \underline{D}\delta(t)$.

The difference equations describing a discrete time system, equivalent to Eqs. (9.25) and (9.26), are

$$\underline{x}(n+1) = \underline{A}\ \underline{x}(n) + \underline{B}\underline{w}(n) \tag{9.38}$$

$$\underline{y}(n) = \underline{C}\ \underline{x}(n) + \underline{D}\underline{w}(n) \tag{9.39}$$

State Variable Representation of an LTI System

where n defines the discrete time nT and T is the sampling interval. All other vectors and matrices were defined earlier. The homogeneous solution to the system defined in Eq. (9.38), with initial condition $\underline{x}(n_0)$, is

$$\underline{x}(n) = \underline{A}^{n-n_0}\underline{x}(n_0) \tag{9.40}$$

In this case the state transition matrix is an $n \times n$ matrix given by

$$\underline{\Phi}(n, n_0) = \underline{\Phi}(n - n_0) = \underline{A}^{n-n_0} \tag{9.41}$$

The following is the list of properties associated with the discrete transition matrix

$$\underline{\Phi}(n + 1 - n_0) = \underline{A}\,\underline{\Phi}(n - n_0) \tag{9.42}$$

$$\underline{\Phi}(n_0 - n_0) = \underline{\Phi}(0) = \underline{I} \tag{9.43}$$

$$\underline{\Phi}(n_0 + 1 - n_0) = \underline{\Phi}(1) = \underline{A} \tag{9.44}$$

$$\underline{\Phi}(n_2 - n_0) = \underline{\Phi}(n_2 - n_1)\underline{\Phi}(n_1 - n_0) \tag{9.45}$$

$$\underline{\Phi}(n_0 - n_1) = \underline{\Phi}^{-1}(n_1 - n_0) \tag{9.46}$$

$$\underline{\Phi}(n_1 - n_0) = \underline{\Phi}(n_1)\underline{\Phi}^{-1}(n_0) \tag{9.47}$$

The solution to the general case (i.e., non-homogeneous system) is given by

$$\underline{x}(n) = \underline{\Phi}(n - n_0)\underline{x}(n_0) + \sum_{m=n_0}^{n-1} \underline{\Phi}(n - m - 1)\underline{B}\underline{w}(m) \tag{9.48}$$

It follows that the output is given by

$$\underline{y}(n) = \underline{C}\,\underline{\Phi}(n - n_0)\underline{x}(n_0) + \sum_{m=n_0}^{n-1} \underline{C}\,\underline{\Phi}(n - m - 1)\underline{B}\underline{w}(m) + \underline{D}\underline{w}(n) \tag{9.49}$$

where the system impulse response is given by

$$\underline{h}(n) = \sum_{m=n_0}^{n-1} \underline{C}\,\underline{\Phi}(n - m - 1)\underline{B}\delta(m) + \underline{D}\delta(n) \tag{9.50}$$

Taking the Z-transform for Eqs. (9.38) and (9.39) yields

$$z\underline{x}(z) = \underline{A}\underline{x}(z) + \underline{B}\underline{w}(z) + z\underline{x}(0) \tag{9.51}$$

$$\underline{y}(z) = \underline{C}\underline{x}(z) + \underline{D}\underline{w}(z) \tag{9.52}$$

Manipulating Eqs. (9.51) and (9.52) yields

$$\underline{x}(z) = [z\underline{I} - \underline{A}]^{-1}\underline{B}\underline{w}(z) + [z\underline{I} - \underline{A}]^{-1}z\underline{x}(0) \tag{9.53}$$

$$\underline{y}(z) = \{\underline{C}[z\underline{I} - \underline{A}]^{-1}\underline{B} + \underline{D}\}\underline{w}(z) + \underline{C}[z\underline{I} - \underline{A}]^{-1}z\underline{x}(0) \tag{9.54}$$

It follows that the state transition matrix is

$$\underline{\Phi}(z) = z[z\underline{I} - \underline{A}]^{-1} = [\underline{I} - z^{-1}\underline{A}]^{-1} \tag{9.55}$$

and the system impulse response in the z-domain is

$$\underline{h}(z) = \underline{C}\underline{\Phi}(z)z^{-1}\underline{B} + \underline{D} \tag{9.56}$$

9.7. The LTI System of Interest

For the purpose of establishing the framework necessary for the Kalman filter development, consider the LTI system shown in Fig. 9.18. This system (which is a special case of the system described in the previous section) can be described by the following first order differential vector equations

$$\underline{\dot{x}}(t) = \underline{A}\ \underline{x}(t) + \underline{u}(t) \tag{9.57}$$

$$\underline{y}(t) = \underline{G}\ \underline{x}(t) + \underline{v}(t) \tag{9.58}$$

where $\underline{y}$ is the observable part of the system (i.e., output), $\underline{u}$ is a driving force, and $\underline{v}$ is the measurement noise. The matrices $\underline{A}$ and $\underline{G}$ vary depending on the system. The noise observation $\underline{v}$ is assumed to be uncorrelated. If the initial condition vector is $\underline{x}(t_0)$, then from Eq. (9.36) we get

$$\underline{x}(t) = \underline{\Phi}(t - t_0)\underline{x}(t_0) + \int_{t_0}^{t} \underline{\Phi}(t - \tau)\underline{u}(\tau)d\tau \tag{9.59}$$

The object (abstract) is observed only at discrete times determined by the system. These observation times are declared by discrete time nT where T is the sampling interval. Using the same notation adopted in the previous section, the discrete time representations of Eqs. (9.57) and (9.58) are

$$\underline{x}(n) = \underline{A}\ \underline{x}(n-1) + \underline{u}(n) \tag{9.60}$$

The LTI System of Interest

$$\underline{y}(n) = \underline{G}\,\underline{x}(n) + \underline{v}(n) \tag{9.61}$$

The homogeneous solution to this system is given in Eq. (9.27) for continuous time, and in Eq. (9.40) for discrete time.

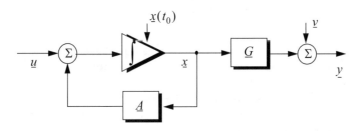

Figure 9.18. An LTI system.

The state transition matrix corresponding to this system can be obtained using Taylor series expansion of the vector $\underline{x}$. More precisely,

$$\begin{aligned} x &= x + T\dot{x} + \frac{T^2}{2!}\ddot{x} + \ldots \\ \dot{x} &= \dot{x} + T\ddot{x} + \ldots \\ \ddot{x} &= \ddot{x} + \ldots \end{aligned} \tag{9.62}$$

It follows that the elements of the state transition matrix are defined by

$$\underline{\Phi}[ij] = \begin{cases} T^{j-i} \div (j-i)! & 1 \le i, j \le n \\ 0 & j < i \end{cases} \tag{9.63}$$

Using matrix notation, the state transition matrix is then given by

$$\underline{\Phi} = \begin{bmatrix} 1 & T & \frac{T^2}{2!} & \ldots \\ 0 & 1 & T & \ldots \\ 0 & 0 & 1 & \ldots \\ \ldots & \ldots & \ldots & \ldots \end{bmatrix} \tag{9.64}$$

The matrix given in Eq. (9.64) is often called the Newtonian matrix.

9.8. Fixed-Gain Tracking Filters

This class of filters (or estimators) is also known as "Fixed-Coefficient" filters. The most common examples of this class of filters are the $\alpha\beta$ and $\alpha\beta\gamma$ filters and their variations. The $\alpha\beta$ and $\alpha\beta\gamma$ trackers are one-dimensional second and third order filters, respectively. They are equivalent to special cases of the one-dimensional Kalman filter. The general structure of this class of estimators is similar to that of the Kalman filter.

The standard $\alpha\beta\gamma$ filter provides smoothed and predicted data for target position, velocity (Doppler), and acceleration. It is a polynomial predictor/corrector linear recursive filter. This filter can reconstruct position, velocity, and constant acceleration based on position measurements. The $\alpha\beta\gamma$ filter can also provide a smoothed (corrected) estimate of the present position which can be used in guidance and fire control operations.

Notation:

For the purpose of the discussion presented in the remainder of this chapter, the following notation is adopted: $x(n|m)$ represents the estimate during the *nth* sampling interval, using all data up to and including the *mth* sampling interval; y_n is the *nth* measured value; and e_n is the *nth* residual (error).

The fixed-gain filter equation is given by

$$\underline{x}(n|n) = \underline{\Phi x}(n-1|n-1) + \underline{K}[y_n - \underline{G\Phi x}(n-1|n-1)] \quad (9.65)$$

Since the transition matrix assists in predicting the next state,

$$\underline{x}(n+1|n) = \underline{\Phi x}(n|n) \quad (9.66)$$

Substituting Eq. (9.66) into Eq. (9.65) yields

$$\underline{x}(n|n) = \underline{x}(n|n-1) + \underline{K}[y_n - \underline{Gx}(n|n-1)] \quad (9.67)$$

The term enclosed within the brackets on the right hand side of Eq. (9.67) is often called the residual (error) which is the difference between the measured input and predicted output. Eq. (9.67) means that the estimate of $\underline{x}(n)$ is the sum of the prediction and the weighted residual. The term $\underline{Gx}(n|n-1)$ represents the prediction state. In the case of the $\alpha\beta\gamma$ estimator, $\underline{G}$ is the row vector given by

$$\underline{G} = \begin{bmatrix} 1 & 0 & 0 & \ldots \end{bmatrix} \quad (9.68)$$

and the gain matrix $\underline{K}$ is given by

Fixed-Gain Tracking Filters

$$\underline{K} = \begin{bmatrix} \alpha \\ \beta/T \\ \gamma/T^2 \end{bmatrix} \qquad (9.69)$$

One of the main objectives of a tracking filter is to decrease the effect of the noise observation on the measurement. For this purpose the noise covariance matrix is calculated. More precisely, the noise covariance matrix is

$$\underline{C}(n|n) = E\{(\underline{x}(n|n)\,)\underline{x}^t(n|n)\} \qquad ; y_n = v_n \qquad (9.70)$$

where E indicates the expected value operator. Noise is assumed to be a zero mean random process with variance equal to σ_v^2. Additionally, noise measurements are also assumed to be uncorrelated,

$$E\{v_n v_m\} = \begin{cases} \delta \sigma_v^2 & n = m \\ 0 & n \neq m \end{cases} \qquad (9.71)$$

Eq. (9.65) can be written as

$$\underline{x}(n|n) = \underline{A}\underline{x}(n-1|n-1) + \underline{K}y_n \qquad (9.72)$$

where

$$\underline{A} = (\underline{I} - \underline{K}\underline{G})\Phi \qquad (9.73)$$

Substituting Eqs. (9.72) and (9.73) into Eq. (9.70) yields

$$\underline{C}(n|n) = E\{(\underline{A}\underline{x}(n-1|n-1) + \underline{K}y_n)(\underline{A}\underline{x}(n-1|n-1) + \underline{K}y_n)^t\} \qquad (9.74)$$

Expanding the right hand side of Eq. (9.74) and using Eq. (9.71) give

$$\underline{C}(n|n) = \underline{A}\,\underline{C}(n-1|n-1)\underline{A}^t + \underline{K}\sigma_v^2 \underline{K}^t \qquad (9.75)$$

Under the steady state condition, Eq. (9.75) collapses to

$$\underline{C}(n|n) = \underline{A}\,\underline{C}\,\underline{A}^t + \underline{K}\sigma_v^2 \underline{K}^t \qquad (9.76)$$

where $\underline{C}$ is the steady state noise covariance matrix. In the steady state,

$$\underline{C}(n|n) = \underline{C}(n-1|n-1) = \underline{C} \qquad \text{for any } n \qquad (9.77)$$

Several criteria can be used to establish the performance of the fixed-gain tracking filter. The most commonly used technique is to compute the Variance Reduction Ratio (VRR). The VRR is defined only when the input to the tracker is noise measurements. It follows that in the steady state case, the VRR is the steady state ratio of the output variance (auto-covariance) to the input measurement variance.

In order to determine the stability of the tracker under consideration, consider the Z-transform for Eq. (9.72),

$$\underline{x}(z) = \underline{A}z^{-1}\underline{x}(z) + \underline{K}y_n(z) \tag{9.78}$$

Rearranging Eq. (9.78) yields the following system transfer functions:

$$\underline{h}(z) = \frac{\underline{x}(z)}{y_n(z)} = (\underline{I} - \underline{A}z^{-1})^{-1}\underline{K} \tag{9.79}$$

where $(\underline{I} - \underline{A}z^{-1})$ is called the characteristic matrix. Note that the system transfer functions can exist only when the characteristic matrix is a non-singular matrix. Additionally, the system is stable if and only if the roots of the characteristic equation are within the unit circle in the z-plane,

$$\left|(\underline{I} - \underline{A}z^{-1})\right| = 0 \tag{9.80}$$

The filter's steady state errors can be determined with the help of Fig. 9.19. The error transfer function is

$$\underline{e}(z) = \frac{y(z)}{1 + \underline{h}(z)} \tag{9.81}$$

and by using Abel's theorem, the steady state error is

$$\underline{e}_\infty = \lim_{t \to \infty} \underline{e}(t) = \lim_{z \to 1}\left(\frac{z-1}{z}\right)\underline{e}(z) \tag{9.82}$$

Substituting Eq. (9.82) into (9.81) yields

$$\underline{e}_\infty = \lim_{z \to 1} \frac{z-1}{z} \frac{y(z)}{1 + \underline{h}(z)} \tag{9.83}$$

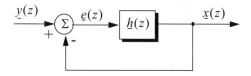

Figure 9.19. Steady state error computation.

9.8.1. The $\alpha\beta$ Filter

The $\alpha\beta$ tracker produces, on the nth observation, smoothed estimates for position and velocity, and a predicted position for the $(n+1)$th observation. Fig. 9.20 shows an implementation of this filter. Note that the subscripts "p" and "s" are used to indicate, respectively, the predicated and smoothed values. The $\alpha\beta$ tracker can follow an input ramp (constant velocity) with no steady state errors. However, a steady state error will accumulate when constant acceleration is present in the input. Smoothing is done to reduce errors in the predicted position through adding a weighted difference between the measured and predicted values to the predicted position, as follows:

$$x_s(n) = x(n|n) = x_p(n) + \alpha(x_0(n) - x_p(n)) \tag{9.84}$$

$$\dot{x}_s(n) = \dot{x}'(n|n) = \dot{x}_s(n-1) + \frac{\beta}{T}(x_0(n) - x_p(n)) \tag{9.85}$$

x_0 is the position input samples. The predicted position is given by

$$x_p(n) = x_s(n|n-1) = x_s(n-1) + T\dot{x}_s(n-1) \tag{9.86}$$

The initialization process is defined by

$$x_s(1) = x_p(2) = x_0(1)$$

$$\dot{x}_s(1) = 0$$

$$\dot{x}_s(2) = \frac{x_0(2) - x_0(1)}{T}$$

A general form for the covariance matrix was developed in the previous section, and is given in Eq. (9.75). In general, a second order one-dimensional covariance matrix (in the context of the $\alpha\beta$ filter) can be written as

$$\underline{C}(n|n) = \begin{bmatrix} C_{xx} & C_{x\dot{x}} \\ C_{\dot{x}x} & C_{\dot{x}\dot{x}} \end{bmatrix} \tag{9.87}$$

where, in general, C_{xy} is

$$C_{xy} = E\{xy^t\} \tag{9.88}$$

By inspection, the $\alpha\beta$ filter has

$$\underline{A} = \begin{bmatrix} 1-\alpha & (1-\alpha)T \\ -\beta/T & (1-\beta) \end{bmatrix} \tag{9.89}$$

$$\underline{K} = \begin{bmatrix} \alpha \\ \beta/T \end{bmatrix} \quad (9.90)$$

$$\underline{G} = \begin{bmatrix} 1 & 0 \end{bmatrix} \quad (9.91)$$

$$\underline{\Phi} = \begin{bmatrix} 1 & T \\ 0 & 1 \end{bmatrix} \quad (9.92)$$

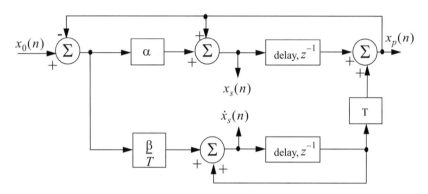

Figure 9.20. An implementation of an $\alpha\beta$ tracker.

Finally, using Eqs. (9.89) through (9.92) in Eq. (9.72) yields the steady state noise covariance matrix,

$$\underline{C} = \frac{\sigma_v^2}{\alpha(4 - 2\alpha - \beta)} \begin{bmatrix} 2\alpha^2 - 3\alpha\beta + 2\beta & \dfrac{\beta(2\alpha - \beta)}{T} \\ \dfrac{\beta(2\alpha - \beta)}{T} & \dfrac{2\beta^2}{T^2} \end{bmatrix} \quad (9.93)$$

It follows that the position and velocity VRR ratios are, respectively, given by

$$(VRR)_x = C_{xx}/\sigma_v^2 = \frac{2\alpha^2 - 3\alpha\beta + 2\beta}{\alpha(4 - 2\alpha - \beta)} \quad (9.94)$$

$$(VRR)_{\dot{x}} = C_{\dot{x}\dot{x}}/\sigma_v^2 = \frac{1}{T^2} \frac{2\beta^2}{\alpha(4 - 2\alpha - \beta)} \quad (9.95)$$

The stability of the $\alpha\beta$ filter is determined from its system transfer functions. For this purpose, compute the roots for Eq. (9.80) with $\underline{A}$ from Eq. (9.89),

Fixed-Gain Tracking Filters

$$|\underline{I} - \underline{A}z^{-1}| = 1 - (2 - \alpha - \beta)z^{-1} + (1 - \alpha)z^{-2} = 0 \tag{9.96}$$

Solving Eq. (9.96) for z yields

$$z_{1,2} = 1 - \frac{\alpha + \beta}{2} \pm \frac{1}{2}\sqrt{(\alpha - \beta)^2 - 4\beta} \tag{9.97}$$

and in order to guarantee stability

$$|z_{1,2}| < 1 \tag{9.98}$$

Two cases are analyzed. First, $z_{1,2}$ are real. In this case (the details are left as an exercise),

$$\beta > 0 \quad ; \quad \alpha > -\beta \tag{9.99}$$

The second case is when the roots are complex; in this case we find

$$\alpha > 0 \tag{9.100}$$

The system transfer functions can be derived by using Eqs. (9.79), (9.89), and (9.90),

$$\begin{bmatrix} h_x(z) \\ h_{\dot{x}}(z) \end{bmatrix} = \frac{1}{z^2 - z(2 - \alpha - \beta) + (1 - \alpha)} \begin{bmatrix} \alpha z\left(z - \frac{(\alpha - \beta)}{\alpha}\right) \\ \frac{\beta z(z - 1)}{T} \end{bmatrix} \tag{9.101}$$

Up to this point all relevant relations concerning the $\alpha\beta$ filter were made with no regard to how to choose the gain coefficients (α and β). Before considering the methodology of selecting these coefficients, consider the main objective behind using this filter. The twofold purpose of the $\alpha\beta$ tracker can be described as follows:

1. The tracker must reduce the measurement noise as much as possible.
2. The filter must be able to track maneuvering targets, with as little residual (tracking error) as possible.

The reduction of measurement noise is normally determined by the VRR ratios. However, the maneuverability performance of the filter depends heavily on the choice of the parameters α and β.

A special variation of the $\alpha\beta$ filter was developed by Benedict and Bordner[1], and is often referred to as the Benedict-Bordner filter. The main advan-

---

1. Benedict, T. R. and Bordner, G. W., Synthesis of an Optimal Set of Radar Track-While-Scan Smoothing Equations, *IRE Transaction on Automatic Control*, AC-7, July 1962, pp. 27-32.

tage of the Benedict-Bordner is reducing the transient errors associated with the $\alpha\beta$ tracker. This filter uses both the position and velocity VRR ratios as measures of performance. It computes the sum of the squared differences between the input (position) and the output when the input has a unit step velocity at time zero. Additionally, it computes the squared differences between the real velocity and the velocity output when the input is as described earlier. Both error differences are minimized when

$$\beta = \frac{\alpha^2}{2-\alpha} \tag{9.102}$$

In this case, the position and velocity VRR ratios are, respectively, given by

$$(VRR)_x = \frac{\alpha(6-5\alpha)}{\alpha^2 - 8\alpha + 8} \tag{9.103}$$

$$(VRR)_{\dot{x}} = \frac{2}{T^2} \frac{\alpha^3/(2-\alpha)}{\alpha^2 - 8\alpha + 8} \tag{9.104}$$

Another important sub-class of the $\alpha\beta$ tracker is the critically damped filter, often called the fading memory filter. In this case, the filter coefficients are chosen on the basis of a smoothing factor ξ, where $0 \le \xi \le 1$. The gain coefficients are given by

$$\alpha = 1 - \xi^2 \tag{9.105}$$

$$\beta = (1-\xi)^2 \tag{9.106}$$

Heavy smoothing means $\xi \to 1$ and little smoothing means $\xi \to 0$. The elements of the covariance matrix for a fading memory filter are

$$C_{xx} = \frac{1-\xi}{(1+\xi)^3} (1 + 4\xi + 5\xi^2) \sigma_v^2 \tag{9.107}$$

$$C_{x\dot{x}} = C_{\dot{x}x} = \frac{1}{T} \frac{1-\xi}{(1+\xi)^3} (1 + 2\xi + 3\xi^2) \sigma_v^2 \tag{9.108}$$

$$C_{\dot{x}\dot{x}} = \frac{2}{T^2} \frac{1-\xi}{(1+\xi)^3} (1-\xi)^2 \sigma_v^2 \tag{9.109}$$

9.8.2. The $\alpha\beta\gamma$ Filter

The $\alpha\beta\gamma$ tracker produces, for the *nth* observation, smoothed estimates of position, velocity, and acceleration. It also produces the predicted position and

Fixed-Gain Tracking Filters

velocity for the $(n+1)th$ observation. An implementation of the $\alpha\beta\gamma$ tracker is shown in Fig. 9.21.

The $\alpha\beta\gamma$ tracker will follow an input whose acceleration is constant with no steady state errors. Again, in order to reduce the error at the output of the tracker, a weighted difference between the measured and predicted values is used in estimating the smoothed position, velocity, and acceleration as follows:

$$x_s(n) = x_p(n) + \alpha(x_0(n) - x_p(n)) \tag{9.110}$$

$$\dot{x}_s(n) = \dot{x}_s(n-1) + T\ddot{x}_s(n-1) + \frac{\beta}{T}(x_0(n) - x_p(n)) \tag{9.111}$$

$$\ddot{x}_s(n) = \ddot{x}_s(n-1) + \frac{2\gamma}{T^2}(x_0(n) - x_p(n)) \tag{9.112}$$

$$x_p(n+1) = x_s(n) + T\dot{x}_s(n) + \frac{T^2}{2}\ddot{x}_s(n) \tag{9.113}$$

and the initialization process is

$$x_s(1) = x_p(2) = x_0(1)$$

$$\dot{x}_s(1) = \ddot{x}_s(1) = \ddot{x}_s(2) = 0$$

$$\dot{x}_s(2) = \frac{x_0(2) - x_0(1)}{T}$$

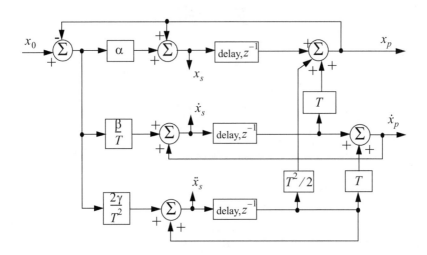

Figure 9.21. An implementation for an $\alpha\beta\gamma$ tracker.

$$\ddot{x}_s(3) = \frac{x_0(3) + x_0(1) - 2x_0(2)}{T^2}$$

Using Eq. (9.63) the state transition matrix for the $\alpha\beta\gamma$ filter is

$$\underline{\Phi} = \begin{bmatrix} 1 & T & \frac{T^2}{2} \\ 0 & 1 & T \\ 0 & 0 & 1 \end{bmatrix} \quad (9.114)$$

The covariance matrix (which is symmetric) can be computed from Eq. (9.76). For this purpose, note that

$$\underline{K} = \begin{bmatrix} \alpha \\ \beta/T \\ \gamma/T^2 \end{bmatrix} \quad (9.115)$$

$$\underline{G} = \begin{bmatrix} 1 & 0 & 0 \end{bmatrix} \quad (9.116)$$

and

$$\underline{A} = (\underline{I} - \underline{K}\underline{G})\underline{\Phi} = \begin{bmatrix} 1-\alpha & (1-\alpha)T & (1-\alpha)T^2/2 \\ -\beta/T & -\beta+1 & (1-\beta/2)T \\ -2\gamma/T^2 & -2\gamma/T & (1-\gamma) \end{bmatrix} \quad (9.117)$$

Substituting Eq. (9.117) into (9.76) and collecting terms the VRR ratios are computed as

$$(VRR)_x = \frac{2\beta(2\alpha^2 + 2\beta - 3\alpha\beta) - \alpha\gamma(4 - 2\alpha - \beta)}{(4 - 2\alpha - \beta)(2\alpha\beta + \alpha\gamma - 2\gamma)} \quad (9.118)$$

$$(VRR)_{\dot{x}} = \frac{4\beta^3 - 4\beta^2\gamma + 2\gamma^2(2-\alpha)}{T^2(4 - 2\alpha - \beta)(2\alpha\beta + \alpha\gamma - 2\gamma)} \quad (9.119)$$

$$(VRR)_{\ddot{x}} = \frac{4\beta\gamma^2}{T^4(4 - 2\alpha - \beta)(2\alpha\beta + \alpha\gamma - 2\gamma)} \quad (9.120)$$

As in the case of any discrete time system, this filter will be stable if and only if all of its poles fall within the unit circle in the z-plane.

The $\alpha\beta\gamma$ characteristic equation is computed by setting

Fixed-Gain Tracking Filters

$$|I - Az^{-1}| = 0 \tag{9.121}$$

Substituting Eq. (9.117) into (9.121) and collecting terms yield the following characteristic function:

$$f(z) = z^3 + (-3\alpha + \beta + \gamma)z^2 + (3 - \beta - 2\alpha + \gamma)z - (1 - \alpha) \tag{9.122}$$

The $\alpha\beta\gamma$ becomes a Benedict-Bordner filter when

$$2\beta - \alpha\left(\alpha + \beta + \frac{\gamma}{2}\right) = 0 \tag{9.123}$$

Note that for $\gamma = 0$ Eq. (9.123) reduces to Eq. (9.102). For a critically damped filter the gain coefficients are

$$\alpha = 1 - \xi^3 \tag{9.124}$$

$$\beta = 1.5(1 - \xi^2)(1 - \xi) = 1.5(1 - \xi)^2(1 + \xi) \tag{9.125}$$

$$\gamma = (1 - \xi)^3 \tag{9.126}$$

Note that heavy smoothing takes place when $\xi \to 1$, while $\xi = 0$ means that no smoothing is present.

MATLAB Function "ghk_tracker.m"

The function *"ghk_tracker.m"* implements the steady state $\alpha\beta\gamma$ filter. It is given in Listing 9.2 in Section 9.11. The syntax is as follows:

[residual, estimate] = ghk_tracker (X0, smoocof, inp, npts, T, nvar)

where

Symbol	Description	Status
X0	initial state vector	input
smoocof	desired smoothing coefficient	input
inp	array of position measurements	input
npts	number of points in input position	input
T	sampling interval	input
nvar	desired noise variance	input
residual	array of position error (residual)	output
estimate	array of predicted position	output

Note that *"ghk_tracker.m"* uses MATLAB's function *"normrnd.m"* to generate zero mean Gaussian noise, which is part of MATLAB's Statistics Toolbox. If this toolbox is not available to the user, then *"ghk_tracker.m"* function-call must be modified to

[residual, estimate] = ghk_tracker1 (X0, smoocof, inp, npts, T)

which is also part of Listing 9.2. In this case, noise measurements are either to be considered unavailable or are part of the position input array.

To illustrate how to use the functions *ghk_tracker.m* and *ghk_tracker1.m*, consider the inputs shown in Figs. 9.22 and 9.23. Fig. 9.22 assumes an input with lazy maneuvering, while Fig. 9.23 assumes an aggressive maneuvering case. For this purpose, the program called *"fig9_21.m"* was written. It is given in Listing 9.3 in Section 9.11.

Figs. 9.24 and 9.25 show the residual error and predicted position corresponding (generated using the program *"fig9_21.m"*) to Fig. 9.22 for two cases: heavy smoothing and little smoothing with and without noise. The noise is white Gaussian with zero mean and variance of $\sigma_v^2 = 0.05$. Figs. 9.26 and 9.27 show the residual error and predicted position corresponding (generated using the program *"fig9_20.m"*) to Fig. 9.23 with and without noise.

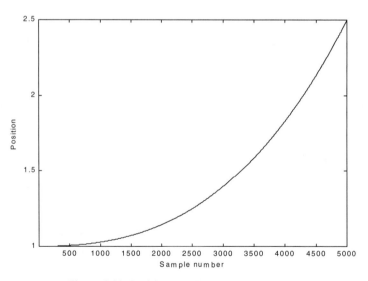

Figure 9.22. Position (truth-data); lazy maneuvering.

Fixed-Gain Tracking Filters 439

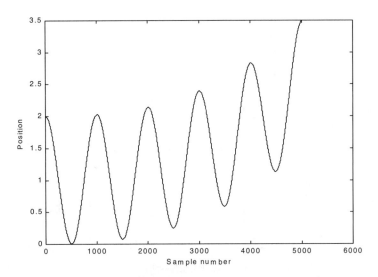

Figure 9.23. Position (truth-data); aggressive maneuvering.

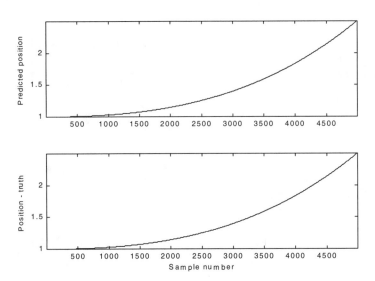

Figure 9.24a-1. Predicted and true position. $\xi = 0.1$ (i.e., large gain coefficients). No noise present.

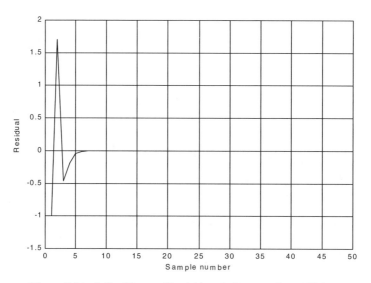

Figure 9.24a-2. Position residual (error). Large gain coefficients. No noise. The error settles to zero fairly quickly.

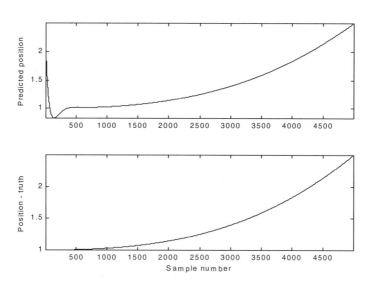

Figure 9.24b-1. Predicted and true position. $\xi = 0.9$ (i.e., small gain coefficients). No noise present.

Fixed-Gain Tracking Filters

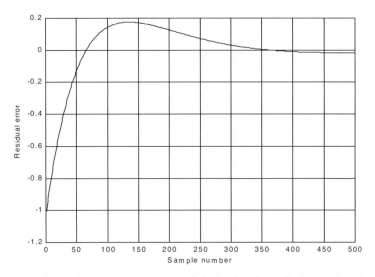

Figure 9.24b-2. Position residual (error). Small gain coefficients. No noise. It takes the filter longer time for the error to settle down.

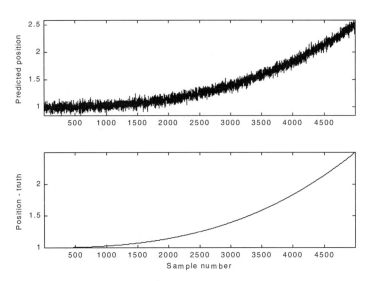

Figure 9.25a-1. Predicted and true position. $\xi = 0.1$ (i.e., large gain coefficients). Noise is present.

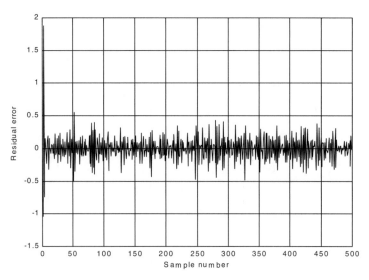

Figure 9.25a-2. Position residual (error). Large gain coefficients. Noise present. The error settles down quickly. The variation is due to noise.

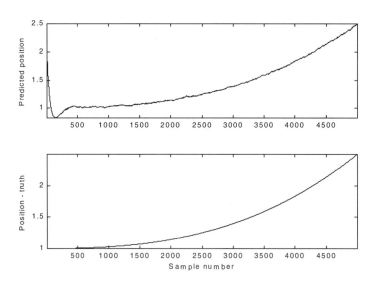

Figure 9.25b-1. Predicted and true position. $\xi = 0.9$ (i.e., small gain coefficients). Noise is present.

Fixed-Gain Tracking Filters 443

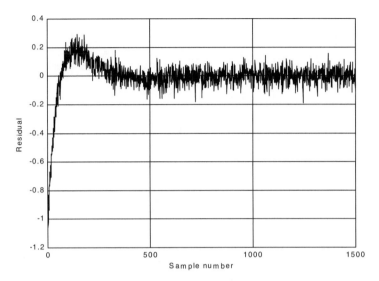

Figure 9.25b-2. Position residual (error). Small gain coefficients. Noise present. The error requires more time before settling down. The variation is due to noise.

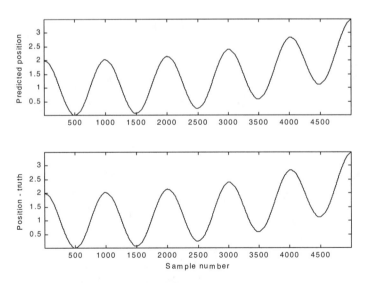

Figure 9.26a. Predicted and true position. $\xi = 0.1$ (i.e., large gain coefficients). Noise is present.

444   *MATLAB Simulations for Radar Systems Design*

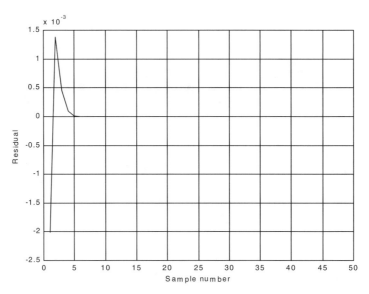

Figure 9.26b. Position residual (error). Large gain coefficients. No noise. The error settles down quickly.

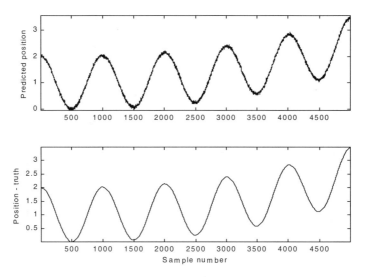

Figure 9.27a. Predicted and true position. $\xi = 0.8$ (i.e., small gain coefficients). Noise is present.

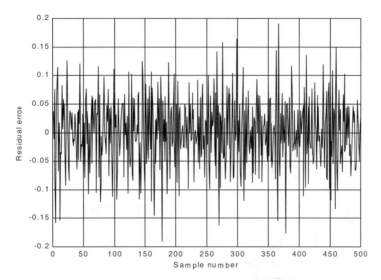

Figure 9.27b. Position residual (error). Small gain coefficients. Noise present. The error stays fairly large; however, its average is around zero. The variation is due to noise.

9.9. The Kalman Filter

The Kalman filter is a linear estimator that minimizes the mean squared error as long as the target dynamics are modeled accurately. All other recursive filters, such as the $\alpha\beta\gamma$ and the Benedict-Bordner filters, are special cases of the general solution provided by the Kalman filter for the mean squared estimation problem. Additionally, the Kalman filter has the following advantages:

1. *The gain coefficients are computed dynamically. This means that the same filter can be used for a variety of maneuvering target environments.*
2. *The Kalman filter gain computation adapts to varying detection histories, including missed detections.*
3. *The Kalman filter provides an accurate measure of the covariance matrix. This allows for better implementation of the gating and association processes.*
4. *The Kalman filter makes it possible to partially compensate for the effects of mis-correlation and mis-association.*

Many derivations of the Kalman filter exist in the literature; only results are provided in this chapter. Fig. 9.28 shows a block diagram for the Kalman filter.

The Kalman filter equations can be deduced from Fig. 9.28. The filtering equation is

$$x(n|n) = \underline{x}_s(n) = \underline{x}(n|n-1) + \underline{K}(n)[\underline{y}(n) - \underline{G}\underline{x}(n|n-1)] \quad (9.127)$$

The measurement vector is

$$\underline{y}(n) = \underline{G}\underline{x}(n) + \underline{v}(n) \quad (9.128)$$

where $\underline{v}(n)$ is zero mean, white Gaussian noise with covariance $\Re_c$,

$$\Re_c = E\{\underline{y}(n)\ \underline{y}^t(n)\} \quad (9.129)$$

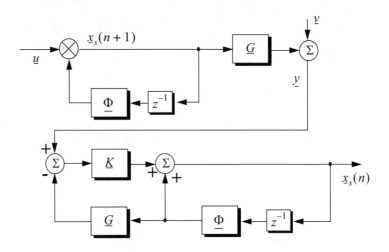

Figure 9.28. Structure of the Kalman filter.

The gain (weight) vector is dynamically computed as

$$\underline{K}(n) = \underline{P}(n|n-1)\underline{G}^t[\underline{G}\underline{P}(n|n-1)\underline{G}^t + \Re_c]^{-1} \quad (9.130)$$

where the measurement noise matrix $\underline{P}$ represents the predictor covariance matrix, and is equal to

$$\underline{P}(n+1|n) = E\{\underline{x}_s(n+1)\underline{x}^*_s(n)\} = \underline{\Phi}\underline{P}(n|n)\underline{\Phi}^t + \underline{Q} \quad (9.131)$$

where $\underline{Q}$ is the covariance matrix for the input $\underline{u}$,

$$\underline{Q} = E\{\underline{u}(n)\ \underline{u}^t(n)\} \quad (9.132)$$

The corrector equation (covariance of the smoothed estimate) is

$$\underline{P}(n|n) = [\underline{I} - \underline{K}(n)\underline{G}]\underline{P}(n|n-1) \tag{9.133}$$

Finally, the predictor equation is

$$\underline{x}(n+1|n) = \underline{\Phi}\underline{x}(n|n) \tag{9.134}$$

9.9.1. The Singer $\alpha\beta\gamma$-Kalman Filter

The Singer[1] filter is a special case of the Kalman where the filter is governed by a specified target dynamic model whose acceleration is a random process with autocorrelation function given by

$$E\{\ddot{x}(t)\ \ddot{x}(t+t_1)\} = \sigma_a^2\ e^{-\frac{|t_1|}{\tau_m}} \tag{9.135}$$

where τ_m is the correlation time of the acceleration due to target maneuvering or atmospheric turbulence. The correlation time τ_m may vary from as low as 10 seconds for aggressive maneuvering to as large as 60 seconds for lazy maneuvering cases.

Singer defined the random target acceleration model by a first order Markov process given by

$$\ddot{x}(n+1) = \rho_m\ \ddot{x}(n) + \sqrt{1-\rho_m^2}\ \sigma_m\ w(n) \tag{9.136}$$

where $w(n)$ is a zero mean, Gaussian random variable with unity variance, σ_m is the maneuver standard deviation, and the maneuvering correlation coefficient ρ_m is given by

$$\rho_m = e^{-\frac{T}{\tau_m}} \tag{9.137}$$

The continuous time domain system that corresponds to these conditions is the same as the Wiener-Kolmogorov whitening filter which is defined by the differential equation

$$\frac{d}{dt}v(t) = -\beta_m v(t) + w(t) \tag{9.138}$$

---

1. Singer, R. A., Estimating Optimal Tracking Filter Performance for Manned Maneuvering Targets, *IEEE Transaction on Aerospace and Electronics, AES-5*, July, 1970, pp. 473-483.

where β_m is equal to $1/\tau_m$. The maneuvering variance using Singer's model is given by

$$\sigma_m^2 = \frac{A_{max}^2}{3}[1 + 4P_{max} - P_0] \tag{9.139}$$

A_{max} is the maximum target acceleration with probability P_{max} and the term P_0 defines the probability that the target has no acceleration.

The transition matrix that corresponds to the Singer filter is given by

$$\underline{\Phi} = \begin{bmatrix} 1 & T & \frac{1}{\beta_m^2}(-1 + \beta_m T + \rho_m) \\ 0 & 1 & \frac{1}{\beta_m}(1 - \rho_m) \\ 0 & 0 & \rho_m \end{bmatrix} \tag{9.140}$$

Note that when $T\beta_m = T/\tau_m$ is small (the target has constant acceleration), then Eq. (9.140) reduces to Eq. (9.114). Typically, the sampling interval T is much less than the maneuvering time constant τ_m; hence, Eq. (9.140) can be accurately replaced by its second order approximation. More precisely,

$$\underline{\Phi} = \begin{bmatrix} 1 & T & T^2/2 \\ 0 & 1 & T(1 - T/2\tau_m) \\ 0 & 0 & \rho_m \end{bmatrix} \tag{9.141}$$

The covariance matrix was derived by Singer, and it is equal to

$$\underline{C} = \frac{2\sigma_m^2}{\tau_m} \begin{bmatrix} C_{11} & C_{12} & C_{13} \\ C_{21} & C_{22} & C_{23} \\ C_{31} & C_{32} & C_{33} \end{bmatrix} \tag{9.142}$$

where

$$C_{11} = \sigma_x^2 = \frac{1}{2\beta_m^5}\left[1 - e^{-2\beta_m T} + 2\beta_m T + \frac{2\beta_m^3 T^3}{3} - 2\beta_m^2 T^2 - 4\beta_m T e^{-\beta_m T}\right] \tag{9.143}$$

$$C_{12} = C_{21} = \frac{1}{2\beta_m^4}[e^{-2\beta_m T} + 1 - 2e^{-\beta_m T} + 2\beta_m T e^{-\beta_m T} - 2\beta_m T + \beta_m^2 T^2] \tag{9.144}$$

$$C_{13} = C_{31} = \frac{1}{2\beta_m^3}[1 - e^{-2\beta_m T} - 2\beta_m T e^{-\beta_m T}] \tag{9.145}$$

The Kalman Filter

$$C_{22} = \frac{1}{2\beta_m^3}[4e^{-\beta_m T} - 3 - e^{-2\beta_m T} + 2\beta_m T] \tag{9.146}$$

$$C_{23} = C_{32} = \frac{1}{2\beta_m^2}[e^{-2\beta_m T} + 1 - 2e^{-\beta_m T}] \tag{9.147}$$

$$C_{33} = \frac{1}{2\beta_m}[1 - e^{-2\beta_m T}] \tag{9.148}$$

Two limiting cases are of interest:

1. *The short sampling interval case* ($T \ll \tau_m$),

$$\lim_{\beta_m T \to 0} \underline{C} = \frac{2\sigma_m^2}{\tau_m} \begin{bmatrix} T^5/20 & T^4/8 & T^3/6 \\ T^4/8 & T^3/3 & T^2/2 \\ T^3/6 & T^2/2 & T \end{bmatrix} \tag{9.149}$$

and the state transition matrix is computed from Eq. (9.141) as

$$\lim_{\beta_m T \to 0} \underline{\Phi} = \begin{bmatrix} 1 & T & T^2/2 \\ 0 & 1 & T \\ 0 & 0 & 1 \end{bmatrix} \tag{9.150}$$

which is the same as the case for the $\alpha\beta\gamma$ filter (constant acceleration).

2. *The long sampling interval* ($T \gg \tau_m$). *This condition represents the case when acceleration is a white noise process. The corresponding covariance and transition matrices are, respectively, given by*

$$\lim_{\beta_m T \to \infty} \underline{C} = \sigma_m^2 \begin{bmatrix} \frac{2T^3 \tau_m}{3} & T^2 \tau_m & \tau_m^2 \\ T^2 \tau_m & 2T\tau_m & \tau_m \\ \tau_m^2 & \tau_m & 1 \end{bmatrix} \tag{9.151}$$

$$\lim_{\beta_m T \to \infty} \underline{\Phi} = \begin{bmatrix} 1 & T & T\tau_m \\ 0 & 1 & \tau_m \\ 0 & 0 & 0 \end{bmatrix} \tag{9.152}$$

Note that under the condition that $T \gg \tau_m$, the cross correlation terms C_{13} and C_{23} become very small. It follows that estimates of acceleration are no longer available, and thus a two state filter model can be used to replace the three state model. In this case,

$$\underline{C} = 2\sigma_m^2 \tau_m \begin{bmatrix} T^3/3 & T^2/2 \\ T^2/2 & T \end{bmatrix} \tag{9.153}$$

$$\underline{\Phi} = \begin{bmatrix} 1 & T \\ 0 & 1 \end{bmatrix} \tag{9.154}$$

9.9.2. Relationship between Kalman and $\alpha\beta\gamma$ Filters

The relationship between the Kalman filter and the $\alpha\beta\gamma$ filters can be easily obtained by using the appropriate state transition matrix $\underline{\Phi}$, and gain vector $\underline{K}$ corresponding to the $\alpha\beta\gamma$ in Eq. (9.127). Thus,

$$\begin{bmatrix} x(n|n) \\ \dot{x}(n|n) \\ \ddot{x}(n|n) \end{bmatrix} = \begin{bmatrix} x(n|n-1) \\ \dot{x}(n|n-1) \\ \ddot{x}(n|n-1) \end{bmatrix} + \begin{bmatrix} k_1(n) \\ k_2(n) \\ k_3(n) \end{bmatrix} [x_0(n) - x(n|n-1)] \tag{9.155}$$

with (see Fig. 9.21)

$$x(n|n-1) = x_s(n-1) + T\dot{x}_s(n-1) + \frac{T^2}{2}\ddot{x}_s(n-1) \tag{9.156}$$

$$\dot{x}(n|n-1) = \dot{x}_s(n-1) + T\ddot{x}_s(n-1) \tag{9.157}$$

$$\ddot{x}(n|n-1) = \ddot{x}_s(n-1) \tag{9.158}$$

Comparing the previous three equations with the $\alpha\beta\gamma$ filter equations yields

$$\begin{bmatrix} \alpha \\ \dfrac{\beta}{T} \\ \dfrac{\gamma}{T^2} \end{bmatrix} = \begin{bmatrix} k_1 \\ k_2 \\ k_3 \end{bmatrix} \tag{9.159}$$

Additionally, the covariance matrix elements are related to the gain coefficients by

$$\begin{bmatrix} k_1 \\ k_2 \\ k_3 \end{bmatrix} = \frac{1}{C_{11} + \sigma_v^2} \begin{bmatrix} C_{11} \\ C_{12} \\ C_{13} \end{bmatrix} \qquad (9.160)$$

Eq. (9.160) indicates that the first gain coefficient depends on the estimation error variance of the total residual variance, while the other two gain coefficients are calculated through the covariances between the second and third states and the first observed state.

MATLAB Function "kalman_filter.m"

The function *"kalman_filter.m"* implements a state Singer-$\alpha\beta\gamma$ Kalman filter. It is given in Listing 9.4 in Section 9.11. The syntax is as follows:

[residual, estimate] = kalman_filter(npts, T, X0, inp, R, nvar)

where

Symbol	Description	Status
npts	number of points in input position	input
T	sampling interval	input
X0	initial state vector	input
inp	input array	input
R	noise variance see Eq. (9-129)	input
nvar	desired state noise variance	input
residual	array of position error (residual)	output
estimate	array of predicted position	output

Note that *"kalman_filter.m"* uses MATLAB's function *"normrnd.m"* to generate zero mean Gaussian noise, which is part of MATLAB's Statistics Toolbox.

To illustrate how to use the functions *"kalman_filter.m"*, consider the inputs shown in Figs. 9.22 and 9.23. Figs. 9.29 and 9.30 show the residual error and predicted position corresponding to Figs. 9.22 and 9.23. These plots can be reproduced using the program *"fig9_28.m"* given in Listing 9.5 in Section 9.11.

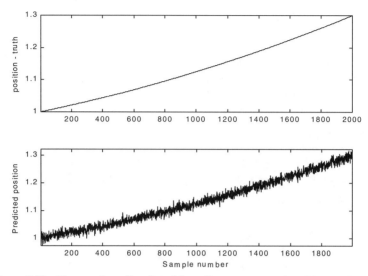

Figure 9.29a. True and predicted positions. Lazy maneuvering. Plot produced using the function *"kalman_filter.m"*.

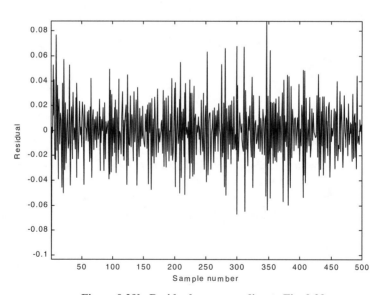

Figure 9.29b. Residual corresponding to Fig. 9.29a.

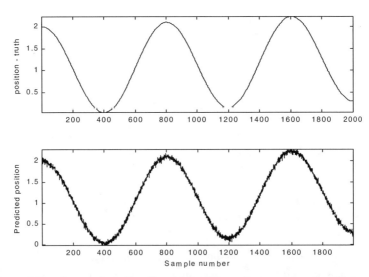

Figure 9.30a. True and predicted positions. Aggressive maneuvering. Plot produced using the function *"kalman_filter.m"*.

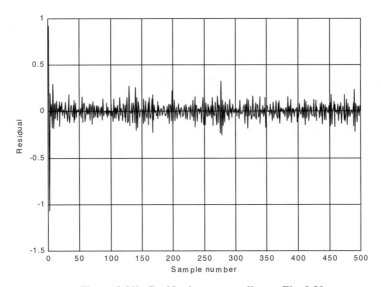

Figure 9.30b. Residual corresponding to Fig. 9.30a.

9.10. "MyRadar" Design Case Study - Visit 9

9.10.1. Problem Statement

Implement a Kalman filter tracker into the "MyRadar" design case study.

9.10.2. A Design[1]

For this purpose, the MATLAB GUI workspace entitled *"kalman_gui.m"* was developed. It is shown in Fig. 9.31. In this design, the inputs can be initialized to correspond to either target type (aircraft and missile). For example, when you click on the button *"ResetMissile,"* the initial *x*-, *y*-, and *z*-detection coordinates for the missile are loaded into the *"Starting Location"* field. The corresponding target velocity is also loaded in the *"velocity in x direction"* field. Finally, all other fields associated with the Kalman filter are also loaded using default values that are appropriate for this design case study. Note that the user can alter these entries as appropriate.

This program generates a fictitious trajectory for the selected target type. This is accomplished using the function *"maketraj.m"*. It is given in Listing 9.6 in Section 9.11. The user can either use this program, or import their own specific trajectory. The function *"maketraj.m"* assumes constant altitude, and generates a manuevering trajectory in the *x-y* plane, as shown in Fig. 9.32. This trajectory can be changed using the different fields in the *"trajectory Parameter"* fields.

Next the program corrupts the trajectory by adding white Guassian noise to it. This is accomplished by the function *"addnoise.m"* which is given in Listing 9.7 in Section 9.11. A six-state Kalman filter named *"kalfilt.m"* is then utilized to perform the tracking task. This function is given in Listing 9.8.

The azimuth, elevation, and range errors are input to the program using their corresponding fields on the GUI. In this example, these entries are assumed constant throughout the simulation. In practice, this is not true and these values will change. They are caluclated by the radar signal processor on a "per processing interval" basis and then are input into the tracker. For example, the standard deviation of the error in the range measurement is

$$\sigma_R = \frac{\Delta R}{\sqrt{2 \times SNR}} = \frac{c}{2B\sqrt{2 \times SNR}} \qquad (9.161)$$

---

[1]. The MATLAB code in this section was developed by Mr. David Hall, Consultant to Decibel Research, Inc., Huntsville, Alabama.

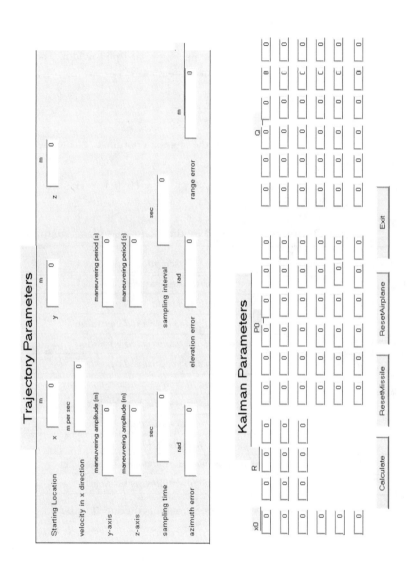

Figure 9.31. MATLAB GUI workspace associated with the *"MyRadar"* design case study- visit 9.

where ΔR is the range resolution, c is the speed of light, B is the bandwidth, and SNR is the measurement SNR.

The standard deviation of the error in the velocity measurement is

$$\sigma_v = \frac{\lambda}{2\tau\sqrt{2 \times SNR}} \tag{9.162}$$

where λ is the wavelength and τ is the uncompressed pulsewidth. The standard deviation of the error in the angle measurement is

$$\sigma_a = \frac{\Theta}{1.6\sqrt{2 \times SNR}} \tag{9.163}$$

where Θ is the antenna beamwidth of the angular coordinate of the measurement (azimuth and elevation).

In this example, the radar is located at $(x, y, z) = (0, 0, 0)$. This simulation calculates and plots the following outputs:

TABLE 9.1. Output list generated by the *"kalman_gui.m"* simulation

Figure #	Description
9.32	uncorrupted input trajectory
9.33	corrupted input trajectory
9.34	corrupted and uncorrupted x-position
9.35	corrupted and uncorrupted y-position
9.36	corrupted and uncorrupted z-position
9.37	corrupted and filtered x-, y- and z-positions
9.38	predicted x-, y-, and z- velocities
9.39	position residuals
9.40	velocity residuals
9.41	covariance matrix components versus time
9.42	Kalman filter gains versus time

Fig. 9.32 through Fig. 9.42 shows typical outputs produced using this simulation for the missile.

"MyRadar" Design Case Study - Visit 9

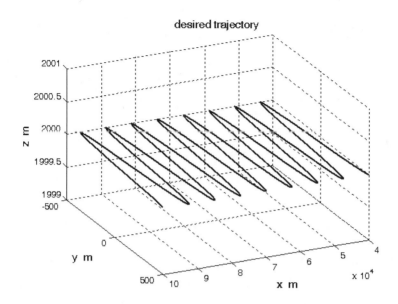

Figure 9.32. Missile uncorrupted trajectory.

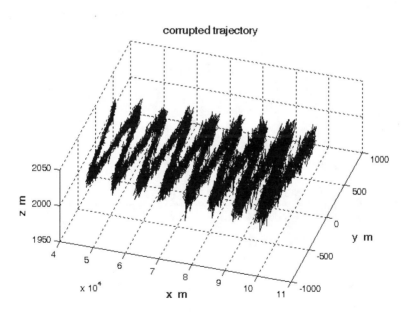

Figure 9.33. Missile corrupted trajectory.

458　*MATLAB Simulations for Radar Systems Design*

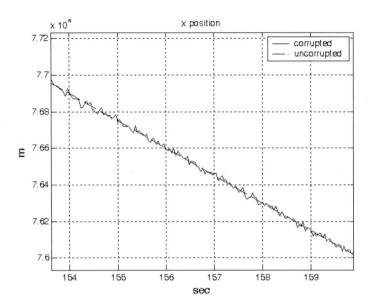

Figure 9.34. Missile x-position from 153 to 160 seconds.

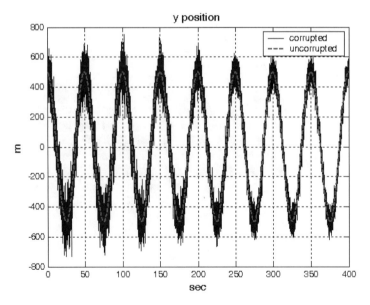

Figure 9.35. Missile y-position.

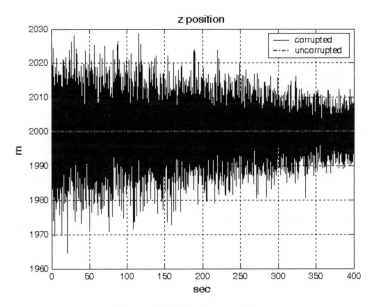

Figure 9.36. Missile z-position.

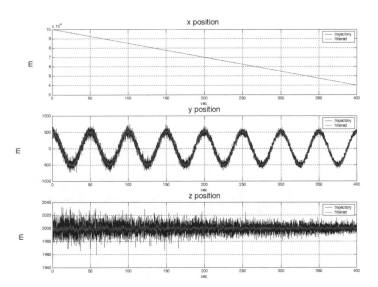

Figure 9.37. Missile trajectory and filtered trajectory.

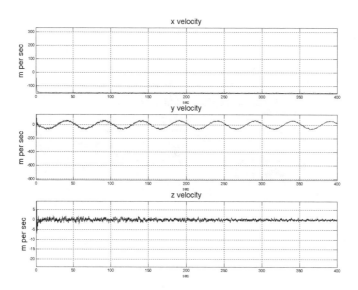

Figure 9.38. Missile velocity filtered.

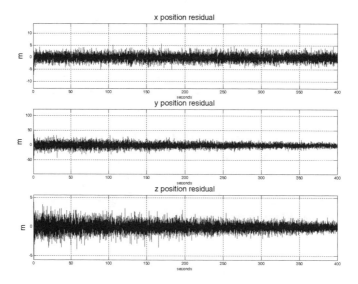

Figure 9.39. Missile position residuals.

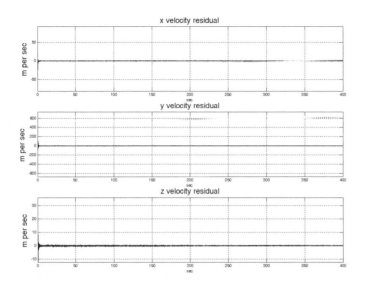

Figure 9.40. Missile velocity residuals.

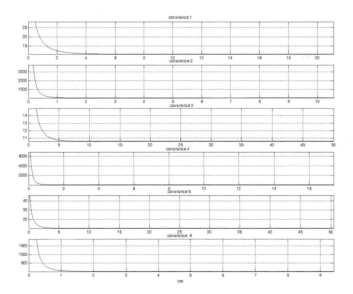

Figure 9.41. Missile covariance matrix components versus time.

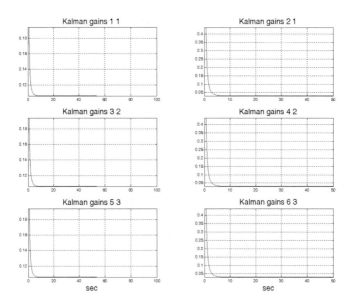

Figure 9.42. Kalman filter gains versus time.

9.11. MATLAB Program and Function Listings

This section contains listings of all MATLAB programs and functions used in this chapter. Users are encouraged to rerun this code with different inputs in order to enhance their understanding of the theory.

Listing 9.1. MATLAB Function "mono_pulse.m"

```
function mono_pulse(phi0)
eps = 0.0000001;
angle = -pi:0.01:pi;
y1 = sinc(angle + phi0);
y2 = sinc((angle - phi0));
ysum = y1 + y2;
ydif = -y1 + y2;
figure (1)
plot (angle,y1,'k',angle,y2,'k');
grid;
xlabel ('Angle - radians')
ylabel ('Squinted patterns')
```

```
figure (2)
plot(angle,ysum,'k');
grid;
xlabel ('Angle - radians')
ylabel ('Sum pattern')
figure (3)
plot (angle,ydif,'k');
grid;
xlabel ('Angle - radians')
ylabel ('Difference pattern')
angle = -pi/4:0.01:pi/4;
y1 = sinc(angle + phi0);
y2 = sinc((angle - phi0));
ydif = -y1 + y2;
ysum = y1 + y2;
dovrs = ydif ./ ysum;
figure(4)
plot (angle,dovrs,'k');
grid;
xlabel ('Angle - radians')
ylabel ('voltage gain')
```

Listing 9.2. MATLAB Function "ghk_tracker.m"

```
function [residual, estimate] = ghk_tracker (X0, smoocof, inp, npts, T, nvar)
rn = 1.;
% read the initial estimate for the state vector
X = X0;
theta = smoocof;
%compute values for alpha, beta, gamma
w1 = 1. - (theta^3);
w2 = 1.5 * (1. + theta) * ((1. - theta)^2) / T;
w3 = ((1. - theta)^3) / (T^2);
% setup the transition matrix PHI
PHI = [1. T (T^2)/2.;0. 1. T;0. 0. 1.];
while rn < npts ;
  %use the transition matrix to predict the next state
  XN = PHI * X;
  error = (inp(rn) + normrnd(0,nvar)) - XN(1);
  residual(rn) = error;
  tmp1 = w1 * error;
  tmp2 = w2 * error;
  tmp3 = w3 * error;
  % compute the next state
```

```
    X(1) = XN(1) + tmp1;
    X(2) = XN(2) + tmp2;
    X(3) = XN(3) + tmp3;
    estimate(rn) = X(1);
    rn = rn + 1.;
end
return
```

MATLAB Function "ghk_tracker1.m"

```
function [residual, estimate] = ghk_tracker1 (X0, smoocof, inp, npts, T)
rn = 1.;
% read the initial estimate for the state vector
X = X0;
theta = smoocof;
%compute values for alpha, beta, gamma
w1 = 1. - (theta^3);
w2 = 1.5 * (1. + theta) * ((1. - theta)^2) / T;
w3 = ((1. - theta)^3) / (T^2);
% setup the transition matrix PHI
PHI = [1. T (T^2)/2.;0. 1. T;0. 0. 1.];
while rn < npts ;
   %use the transition matrix to predict the next state
   XN = PHI * X;
   error = inp(rn) - XN(1);
   residual(rn) = error;
   tmp1 = w1 * error;
   tmp2 = w2 * error;
   tmp3 = w3 * error;
   % compute the next state
   X(1) = XN(1) + tmp1;
   X(2) = XN(2) + tmp2;
   X(3) = XN(3) + tmp3;
   estimate(rn) = X(1);
   rn = rn + 1.;
end
return
```

Listing 9.3. MATLAB Program "fig9_21.m"

```
clear all
eps = 0.0000001;
npts = 5000;
del = 1./ 5000.;
t = 0. : del : 1.;
```

```matlab
% generate input sequence
inp = 1.+ t.^3 + .5 .*t.^2 + cos(2.*pi*10 .* t) ;
% read the initial estimate for the state vector
X0 = [2,.1,.01]';
% this is the update interval in seconds
T = 100. * del;
% this is the value of the smoothing coefficient
xi = .91;
[residual, estimate] = ghk_tracker (X0, xi, inp, npts, T, .01);
figure(1)
plot (residual(1:500))
xlabel ('Sample number')
ylabel ('Residual error')
grid
figure(2)
NN = 4999.;
n = 1:NN;
plot (n,estimate(1:NN),'b',n,inp(1:NN),'r')
xlabel ('Sample number')
ylabel ('Position')
legend ('Estimated','Input')
```

Listing 9.4. MATLAB Function "kalman_filter.m"

```matlab
function [residual, estimate] = kalman_filter(npts, T, X0, inp, R, nvar)
N = npts;
rn=1;
% read the initial estimate for the state vector
X = X0;
% it is assumed that the measurement vector H=[1,0,0]
% this is the state noise variance
VAR = nvar;
% setup the initial value for the prediction covariance.
S = [1. 1. 1.; 1. 1. 1.; 1. 1. 1.];
% setup the transition matrix PHI
PHI = [1. T (T^2)/2.; 0. 1. T; 0. 0. 1.];
% setup the state noise covariance matrix
Q(1,1) = (VAR * (T^5)) / 20.;
Q(1,2) = (VAR * (T^4)) / 8.;
Q(1,3) = (VAR * (T^3)) / 6.;
Q(2,1) = Q(1,2);
Q(2,2) = (VAR * (T^3)) / 3.;
Q(2,3) = (VAR * (T^2)) / 2.;
Q(3,1) = Q(1,3);
```

```
Q(3,2) = Q(2,3);
Q(3,3) = VAR * T;
while rn < N ;
   %use the transition matrix to predict the next state
   XN = PHI * X;
   % Perform error covariance extrapolation
   S = PHI * S * PHI' + Q;
   % compute the Kalman gains
   ak(1) = S(1,1) / (S(1,1) + R);
   ak(2) = S(1,2) / (S(1,1) + R);
   ak(3) = S(1,3) / (S(1,1) + R);
   %perform state estimate update:
   error = inp(rn) + normrnd(0,R) - XN(1);
   residual(rn) = error;
   tmp1 = ak(1) * error;
   tmp2 = ak(2) * error;
   tmp3 = ak(3) * error;
   X(1) = XN(1) + tmp1;
   X(2) = XN(2) + tmp2;
   X(3) = XN(3) + tmp3;
   estimate(rn) = X(1);
   % update the error covariance
   S(1,1) = S(1,1) * (1. -ak(1));
   S(1,2) = S(1,2) * (1. -ak(1));
   S(1,3) = S(1,3) * (1. -ak(1));
   S(2,1) = S(1,2);
   S(2,2) = -ak(2) * S(1,2) + S(2,2);
   S(2,3) = -ak(2) * S(1,3) + S(2,3);
   S(3,1) = S(1,3);
   S(3,3) = -ak(3) * S(1,3) + S(3,3);
   rn = rn + 1.;
end
```

Listing 9.5. MATLAB Program "fig9_28.m"

```
clear all
npts = 2000;
del = 1/2000;
t = 0:del:1;
inp = (1+.2 .* t + .1 .*t.^2) + cos(2. * pi * 2.5 .* t);
X0 = [1,.1,.01]';
% it is assumed that the measurement vector H=[1,0,0]
% this is the update interval in seconds
T = 1.;
```

```
% enter the measurement noise variance
R = .035;
% this is the state noise variance
nvar = .5;
[residual, estimate] = kalman_filter(npts, T, X0, inp, R, nvar);
figure(1)
plot(residual)
xlabel ('Sample number')
ylabel ('Residual')
figure(2)
subplot(2,1,1)
plot(inp)
axis tight
ylabel ('position - truth')
subplot(2,1,2)
plot(estimate)
axis tight
xlabel ('Sample number')
ylabel ('Predicted position')
```

Listing 9.6. MATLAB Function "maketraj.m"

```
function [times , trajectory] = maketraj(start_loc, xvelocity, yamp, yperiod,
zamp, zperiod, samplingtime, deltat)
% maketraj.m
% by David J. Hall
% for Bassem Mahafza
% 17 June 2003
% 17:01
% USAGE: [times , trajectory] = maketraj(start_loc, xvelocity, yamp, yperiod,
zamp, zperiod, samplingtime, deltat)
% NOTE: all coordinates are in radar reference coordinates.
% INPUTS
% name         dimension explanation                            units
%------       ------    ---------------                         -------
% start_loc   3 X 1     starting location of target             m
% xvelocity    1        velocity of target                      m/s
% yamp         1        amplitude of oscillation y direction    m
% yperiod      1        period of oscillation y direction       m
% zamp         1        amplitude of oscillation z direction    m
% zperiod      1        period of oscillation z direction       m
% samplingtime 1        length of interval of trajectory        sec
% deltat       1        time between samples                    sec
%
```

```
% OUTPUTS
%
% name      dimension            explanation           units
%------     ----------           ---------------       ------
% times     1 X samplingtime/deltat vector of times
%                               corresponding to samples sec
% trajectory  3 X samplingtime/deltat trajectory x,y,z     m
%
times = 0: deltat: samplingtime ;
x = start_loc(1)+xvelocity.*times ;
if yperiod~=0
  y = start_loc(2)+yamp*cos(2*pi*(1/yperiod).*times) ;
else
  y = ones(1, length(times))*start_loc(2) ;
end
if zperiod~=0
  z = start_loc(3)+zamp*cos(2*pi*(1/zperiod).*times)  ;
else
  z = ones(1, length(times))*start_loc(3) ;
end
trajectory = [x ; y  ; z] ;
```

Listing 9.7. MATLAB Function "addnoise.m"

```
function [noisytraj ] = addnoise(trajectory, sigmaaz, sigmael, sigmarange )
% addnoise.m
% by David J. Hall
% for Bassem Mahafza
% 10 June 2003
% 11:46
% USAGE: [noisytraj ] = addnoise(trajectory, sigmaaz, sigmael, sigmarange )
% INPUTS
% name       dimension  explanation                 units
%------      ------    ---------------              -------
% trajectory  3 X POINTS trajectory in radar reference coords   [m;m;m]
% sigmaaz    1      standard deviation of azimuth error     radians
% sigmael    1      standard deviation of elevation error   radians
% sigmarange 1       standard deviation of range error       m
%
% OUTPUTS
% name       dimension  explanation                 units
%------      ------    ---------------              -------
% noisytraj   3 X POINTS  noisy trajectory            [m;m;m]
noisytraj = zeros(3, size(trajectory,2)) ;
```

```
for loop = 1 : size(trajectory,2)
  x = trajectory(1,loop);
  y = trajectory(2,loop);
  z = trajectory(3,loop);
  azimuth_corrupted =  atan2(y,x) + sigmaaz*randn(1) ;
  elevation_corrupted = atan2(z, sqrt(x^2+y^2)) + sigmael*randn(1) ;
  range_corrupted = sqrt(x^2+y^2+z^2)  + sigmarange*randn(1) ;
  x_corrupted =
range_corrupted*cos(elevation_corrupted)*cos(azimuth_corrupted) ;
  y_corrupted =
range_corrupted*cos(elevation_corrupted)*sin(azimuth_corrupted) ;
  z_corrupted = range_corrupted*sin(elevation_corrupted) ;
  noisytraj(:,loop) = [x_corrupted ; y_corrupted; z_corrupted ] ;
end % next loop
```

Listing 9.8. MATLAB Function "kalfilt.m"

```
function [filtered, residuals , covariances, kalmgains] = kalfilt(trajectory, x0,
P0, phi, R, Q )
% kalfilt.m
% by David J. Hall
% for Bassem Mahafza
% 10 June 2003
% 11:46
% USAGE: [filtered, residuals , covariances, kalmgains] = kalfilt(trajectory,
x0, P0, phi, R, Q )
%
% INPUTS
% name       dimension              explanation                   units
%------     ------                 ---------------               -------
% trajectory   NUMMEASUREMENTS X NUMPOINTS  trajectory in radar
reference coords      [m;m;m]
% x0       NUMSTATES X 1          initial estimate of state vector       m,
m/s
% P0      NUMSTATES X NUMSTATES     initial estimate of covariance
matrix    m, m/s
% phi      NUMSTATES X NUMSTATES      state transition matrix
-
% R         NUMMEASUREMENTS X NUMMEASUREMENTS   measurement
error covariance matrix   m
% Q       NUMSTATES X NUMSTATES    state error covariance matrix
m, m/s
%
```

```matlab
% OUTPUTS
% name       dimension                 explanation                      units
%------      ------          ---------------                  -------
% filtered   NUMSTATES X NUMPOINTS     filtered trajectory x,y,z pos, vel
[m; m/s; m; m/s; m; m/s]
% residuals  NUMSTATES X NUMPOINTS     residuals of filtering
[m;m;m]
% covariances  NUMSTATES X NUMPOINTS   diagonal of covariance
matrix        [m;m;m]
% kalmgains   (NUMSTATES X NUMMEASUREMENTS)
%             X NUMPOINTS              Kalman gain matrix               -
NUMSTATES = 6 ;
NUMMEASUREMENTS = 3 ;
NUMPOINTS = size(trajectory, 2) ;
% initialize output matrices
filtered = zeros(NUMSTATES, NUMPOINTS) ;
residuals = zeros(NUMSTATES, NUMPOINTS) ;
covariances = zeros(NUMSTATES, NUMPOINTS) ;
kalmgains = zeros(NUMSTATES*NUMMEASUREMENTS, NUMPOINTS) ;
% set matrix relating measurements to states
H = [1 0 0 0 0 0 ; 0 0 1 0 0 0 ; 0 0 0 0 1 0];
xhatminus = x0 ;
Pminus = P0 ;
for loop = 1: NUMPOINTS
   % compute the Kalman gain
   K = Pminus*H'*inv(H*Pminus*H' + R) ;
   kalmgains(:,loop) = reshape(K, NUMSTATES*NUMMEASUREMENTS, 1) ;
   % update the estimate with the measurement z
   z = trajectory(:,loop) ;
   xhat = xhatminus + K*(z - H*xhatminus) ;
   filtered(:,loop) = xhat ;
   residuals(:,loop) = xhat - xhatminus ;
   % update the error covariance for the updated estimate
   P = ( eye(NUMSTATES, NUMSTATES) - K*H)*Pminus ;
   covariances(:,loop) = diag(P) ;  % only save diagonal of covariance matrix
   % project ahead
   xhatminus_next = phi*xhat ;
   Pminus_next = phi*P*phi' + Q ;
   xhatminus = xhatminus_next ;
   Pminus = Pminus_next ;
end
```

Chapter 10   *Electronic Countermeasures (ECM)*

This chapter is coauthored with J. Michael Madewell[1]

10.1. Introduction

Any deliberate electronic effort intended to disturb normal radar operation is usually referred to as an Electronic Countermeasure (ECM). This may also include chaff, radar decoys, radar RCS alterations (e.g., radio frequency absorbing materials), and, of course, radar jamming.

In general, ECM is used by the offense to accomplish one, several, or possibly all of the following objectives: (1) deny proper target detection; (2) generate operator confusion and / or deception; (3) force delays in detection and tracking initiation; (4) generate false tracks of non-real targets; (5) overload the radar computer with an excessive number of targets; (6) deny accurate measurements of the target range and range rate; (7) force dropped tracks; and (8) introduce errors in target position and range rate. Alternatively, the defense may utilize Electronic counter-countermeasures (ECCM) to overcome and mitigate the effects of ECM on the radar. When deployed properly, ECCM techniques and / or hardware can have the following effects: (1) prevent receiver saturation; (2) maintain a reasonable CFAR rate; (3) enhance the signal to jammer ratio; (4) properly identify and discriminate directional interference; (5) reject invalid targets; and (6) maintain true target tracks.

ECM techniques can be exploited by a radar system in many different ways and can be categorized into two classes:

---

1. Mr. J. Michael Madewell is with the US Army Space and Missile Defense Command in Huntsville, Alabama.

1. Denial ECM techniques: Denial ECM techniques can be either active or passive. Active denial ECM techniques include: CW, short pulse, long pulse, spot noise, barrage noise, and sidelobe repeaters. Passive ECM techniques include chaff and Radar Absorbing Material (RAM).
2. Deception ECM techniques: Deception ECM techniques are also broken down into active and passive techniques. Active deception ECM techniques include repeater jammers and false target generators. Passive deception ECM include chaff and RAM.

10.2. Jammers

Jammers can be categorized into two general types: (1) barrage jammers and (2) deceptive jammers (repeaters). When strong jamming is present, detection capability is determined by receiver signal-to-noise plus interference ratio rather than SNR. In fact, in most cases, detection is established based on the signal-to-interference ratio alone.

Barrage jammers attempt to increase the noise level across the entire radar operating bandwidth. Consequently, this lowers the receiver SNR, and, in turn, makes it difficult to detect the desired targets. This is the reason why barrage jammers are often called maskers (since they mask the target returns). Barrage jammers can be deployed in the main beam or in the sidelobes of the radar antenna. If a barrage jammer is located in the radar main beam, it can take advantage of the antenna maximum gain to amplify the broadcasted noise signal. Alternatively, sidelobe barrage jammers must either use more power, or operate at a much shorter range than main beam jammers. Main beam barrage jammers can be deployed either on-board the attacking vehicle, or act as an escort to the target. Sidelobe jammers are often deployed to interfere with a specific radar, and since they do not stay close to the target, they have a wide variety of stand-off deployment options.

Repeater jammers carry receiving devices on board in order to analyze the radar's transmission, and then send back false target-like signals in order to confuse the radar. There are two common types of repeater jammers: spot noise repeaters and deceptive repeaters. The spot noise repeater measures the transmitted radar signal bandwidth and then jams only a specific range of frequencies. The deceptive repeater sends back altered signals that make the target appear in some false position (ghosts). These ghosts may appear at different ranges or angles than the actual target. Furthermore, there may be several ghosts created by a single jammer. By not having to jam the entire radar bandwidth, repeater jammers are able to make more efficient use of their jamming power. Radar frequency agility may be the only way possible to defeat spot noise repeaters.

In general a jammer can be identified by its effective operating bandwidth B_J and by its Effective Radiated Power (ERP), which is proportional to the jammer transmitter power P_J. More precisely,

$$ERP = \frac{P_J G_J}{L_J} \tag{10.1}$$

where G_J is the jammer antenna gain and L_J is the total jammer losses. The effect of a jammer on a radar is measured by the Signal-to-Jammer ratio (S/J).

10.2.1. Self-Screening Jammers (SSJ)

Self-screening jammers, also known as self-protecting jammers and as main beam jammers, are a class of ECM systems carried on the vehicle they are protecting. Escort jammers (carried on vehicles that accompany the attacking vehicles) can also be treated as SSJs if they appear at the same range as that of the target(s).

Assume a radar with an antenna gain G, wavelength λ, aperture A_r, bandwidth B_r, receiver losses L, and peak power P_t. The single pulse power received by the radar from a target of RCS σ, at range R, is

$$S = \frac{P_t G^2 \lambda^2 \sigma \tau}{(4\pi)^3 R^4 L} \tag{10.2}$$

τ is the radar pulsewidth. The power received by the radar from an SSJ jammer at the same range is

$$J = \frac{P_J G_J}{4\pi R^2} \frac{A_r}{B_J L_J} \tag{10.3}$$

where P_J, G_J, B_J, L_J are, respectively, the jammer's peak power, antenna gain, operating bandwidth, and losses. Using the relation

$$A_r = \frac{\lambda^2 G}{4\pi} \tag{10.4}$$

then Eq. (10.3) can be written as

$$J = \frac{P_J G_J}{4\pi R^2} \frac{\lambda^2 G}{4\pi} \frac{1}{B_J L_J} \tag{10.5}$$

Note that $B_J > B_r$. This is needed in order to compensate for the fact that the jammer bandwidth is usually larger than the operating bandwidth of the radar. Jammers are normally designed to operate against a wide variety of radar systems with different bandwidths.

Substituting Eq. (10.1) into Eq. (10.5) yields,

$$J = ERP \frac{\lambda^2 G}{(4\pi)^2 R^2} \frac{1}{B_J} \quad (10.6)$$

Thus, S/J ratio for a SSJ case is obtained from Eqs. (10.6) and (10.2),

$$\frac{S}{J} = \frac{P_t \tau G \sigma B_J}{(ERP)(4\pi)R^2 L} \quad (10.7)$$

and when pulse compression is used, with time-bandwidth-product G_{PC}, then Eq. (10.7) can be written as

$$\frac{S}{J} = \frac{P_t G \sigma B_J G_{PC}}{(ERP)(4\pi)R^2 B_r L} \quad (10.8)$$

Note that to obtain Eq. (10.8), one must multiply Eq. (10.7) by the factor B_r/B_r and use the fact that $G_{PC} = B_r \tau$.

The jamming power reaches the radar on a one-way transmission basis, whereas the target echoes involve two-way transmission. Thus, the jamming power is generally greater than the target signal power. In other words, the ratio S/J is less than unity. However, as the target becomes closer to the radar, there will be a certain range such that the ratio S/J is equal to unity. This range is known as the cross-over range. The range window where the ratio S/J is sufficiently larger than unity is denoted as the detection range. In order to compute the crossover range R_{co}, set S/J to unity in Eq. (10.8) and solve for range. It follows that

$$(R_{CO})_{SSJ} = \left(\frac{P_t G \sigma B_J}{4\pi B_r L(ERP)} \right)^{1/2} \quad (10.9)$$

MATLAB Program "ssj_req.m"

The program *"ssj_req.m"* implements Eqs. (10.9); it is given in Listing 10.1 in Section 10.5. This program calculates the cross-over range and generates plots of relative S and J versus range normalized to the cross-over range, as illustrated in Fig. 10.1a.

In this example, the following parameters were utilized: radar peak power $P_t = 50KW$, jammer peak power $P_J = 200W$, radar operating bandwidth $B_r = 667KHz$, jammer bandwidth $B_J = 50MHz$, radar and jammer losses $L = L_J = 0.10dB$, target cross section $\sigma = 10.m^2$, radar antenna gain $G = 35dB$, jammer antenna gain $G_J = 10dB$, the radar operating frequency is $f = 5.6GHz$. The syntax is as follows:

[BR_range] = ssj_req (pt, g, freq, sigma, br, loss, pj, bj, gj, lossj)

where

Symbol	Description	Units	Status
pt	radar peak power	W	input
g	radar antenna gain	dB	input
freq	radar operating frequency	Hz	input
sigma	target cross section	m^2	input
br	radar operating bandwidth	Hz	input
loss	radar losses	dB	input
pj	jammer peak power	W	input
bj	jammer bandwidth	Hz	input
gj	jammer antenna gain	dB	input
lossj	jammer losses	dB	input
BR_range	cross-over range	Km	output

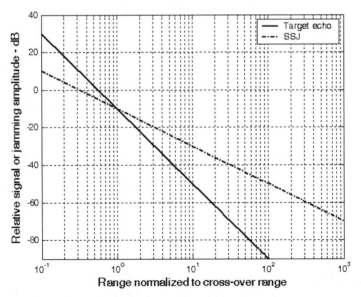

Figure 10.1a. Target and jammer echo signals. Plots were generated using the program *"ssj_req.m"* and using the input parameters defined on the previous page.

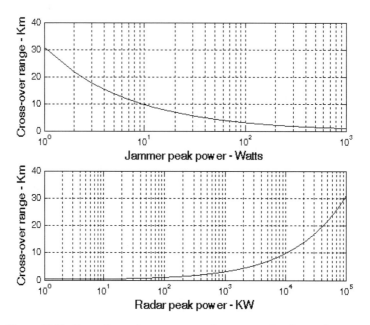

Figure 10.1b. Burn-through range versus jammer and radar peak powers corresponding to example used in generating Fig. 10.1a.

Burn-through Range

If jamming is employed in the form of Gaussian noise, then the radar receiver has to deal with the jamming signal the same way it deals with noise power in the radar. Thus, detection, tracking, and other functions of the radar signal and data processors are no longer dependent on the SNR. In this case, the S/(J+N) ratio must be calculated. More precisely,

$$\frac{S}{J+N} = \frac{\left(\dfrac{P_t G \sigma A_r \tau}{(4\pi)^2 R^4 L}\right)}{\left(\dfrac{(ERP)A_r}{4\pi R^2 B_J} + kT_0\right)} \qquad (10.10)$$

where k is Boltzman's constant and T_0 is the effective noise temperature.

The S/(J+N) ratio should be used in place of the SNR when calculating the the radar equation and when computing the probability of detection. Furthermore, S/(J+N) must also be used in place of the SNR when using coherent or non-coherent pulse integration.

The range at which the radar can detect and perform proper measurements for a given S/(J+N) value is defined as the burn-through range. It is given by

$$R_{BT} = \left\{ \sqrt{\left(\frac{(ERP)A_r}{8\pi B_j kT_0}\right)^2 + \frac{P_t G \sigma A_r \tau}{(4\pi)^2 L \frac{S}{(J+N)} kT_0}} - \frac{(ERP)A_r}{8\pi B_j kT_0} \right\}^{\frac{1}{2}} \quad (10.11)$$

MATLAB Function "sir.m"

The MATLAB function *"sir.m"* implements Eq. (10.10). It generates plots of the S/(J+N) versus detection range and plots of the burn-through range versus the jammer ERP. It is given in Listing 10.2 in Section 10.5. The syntax is as follows:

[SIR] = sir (pt, g, sigma, freq, tau, T0, loss, R, pj, bj, gj, lossj)

where

Symbol	Description	Units	Status
pt	radar peak power	W	input
g	radar antenna gain	dB	input
sigma	target cross section	m^2	input
freq	radar operating frequency	Hz	input
tau	radar pulsewidth	seconds	input
T0	effective noise temperature	Kelvin	input
loss	radar losses	dB	input
R	range. can be single value or a vector	Km	input
pj	jammer peak power	W	input
bj	jammer bandwidth	Hz	input
gj	jammer antenna gain	dB	input
lossj	jammer losses	dB	input
SIR	S/(J+N)	dB	output

Fig. 10.2 shows some typical outputs generated by this function when the inputs are as follows:

Input Parameter	Value
pt	50KW
g	35 dB
sigma	10 square meters
freq	5.6 GHz

Input Parameter	Value
tau	50 micro-seconds
T0	290
loss	5 dB
R	linspace(10,400,5000) Km
pj	200 Watts
bj	50 MHz
gj	10 dB
lossj	0.3 dB

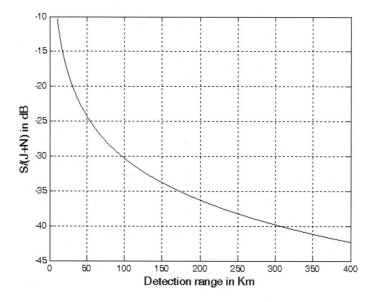

Figure 10.2. S/(J+N) versus detection range.

MATLAB Function "burn_thru.m"

The MATLAB function *"burn_thru.m"* implements Eq. (10.10) and (10.11). It generates plots of the S/(J+N) versus detection range and plots of the burn-through range versus the jammer ERP. It is given in Listing 10.3 in Section 10.5. The syntax is as follows:

[Range] = burn_thru (pt, g, sigma, freq, tau, T0, loss, pj, bj, gj, lossj, sir0, ERP)

where

Symbol	Description	Units	Status
pt	radar peak power	W	input
g	radar antenna gain	dB	input
sigma	target cross section	m^2	input
freq	radar operating frequency	Hz	input
tau	radar pulsewidth	seconds	input
T0	effective noise temperature	Kelvin	input
loss	radar losses	dB	input
pj	jammer peak power	W	input
bj	jammer bandwidth	Hz	input
gj	jammer antenna gain	dB	input
lossj	jammer losses	dB	input
sir0	desired SIR	dB	input
ERP	desired ERP. can be a vector	Watts	input
Range	burn-through range	Km	output

Fig. 10.3 shows some typical outputs generated by this function when the inputs are as follows:

Input Parameter	Value
pt	50KW
g	35 dB
sigma	10 square meters
freq	5.6 GHz
tau	0.5 milli-seconds
T0	290
loss	5 dB
pj	200 Watts
bj	500 MHz
gj	10 dB
lossj	0.3 dB
sir0	15dB
ERP	linspace(1, 1000, 1000) W

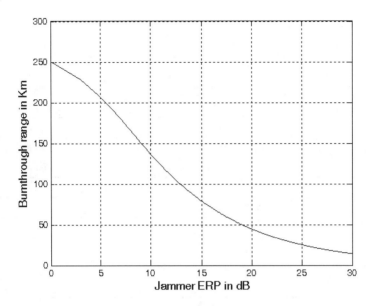

Figure 10.3. Burn-through range versus ERP. (S/(J+N) = 15 dB.

10.2.2. Stand-Off Jammers (SOJ)

Stand-off jammers (SOJ) emit ECM signals from long ranges which are beyond the defense's lethal capability. The power received by the radar from an SOJ jammer at range R_J is

$$J = \frac{P_J G_J}{4\pi R_J^2} \frac{\lambda^2 G'}{4\pi} \frac{1}{B_J L_J} = \frac{ERP}{4\pi R_J^2} \frac{\lambda^2 G'}{4\pi} \frac{1}{B_J} \qquad (10.12)$$

where all terms in Eq. (10.12) are the same as those for the SSJ case except for G'. The gain term G' represents the radar antenna gain in the direction of the jammer and is normally considered to be the sidelobe gain.

The SOJ radar equation is then computed as

$$\frac{S}{J} = \frac{P_t \tau G^2 R_J^2 \sigma B_J}{4\pi (ERP) G' R^4 L} \qquad (10.13)$$

and when pulse compression is used, with time-bandwidth-product G_{PC} then Eq. (10.13) can be written as

$$\frac{S}{J} = \frac{P_t G^2 R_J^2 \sigma B_J P_{PC}}{4\pi (ERP) G' R^4 B_r L} \qquad (10.14)$$

Again, the cross-over range is that corresponding to $S = J$; it is given by

$$(R_{CO})_{SOJ} = \left(\frac{P_t G^2 R_J^2 \sigma B_J P_{PC}}{4\pi (ERP) G' B_r L} \right)^{1/4} \qquad (10.15)$$

MATLAB Program "soj_req.m"

The program "soj_req.m" implements Eqs. (10.15); it is given in Listing 10.4 in Section 10.5. The inputs to the program "soj_req.m" are the same as in the SSJ case, with two additional inputs: the radar antenna gain on the jammer G' and radar-to-jammer range R_J. This program generates the same types of plots as in the case of the SSJ. Typical output is in Fig. 10.4 utilizing the same parameters as those in the SSJ case, with jammer peak power $P_J = 5000W$, jammer antenna gain $G_J = 30dB$, radar antenna gain on the jammer $G' = 10dB$, and radar to jammer range $R_J = 22.2Km$.

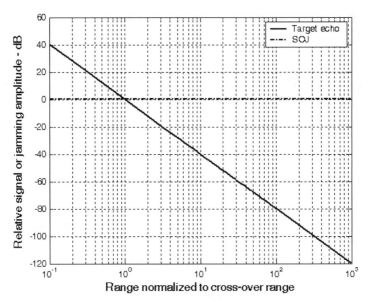

Figure 10.4. Target and jammer echo signals. Plots were generated using the program "soj_req.m".

Again if the jamming is employed in the form of Gaussian noise, then the radar receiver has to deal with the jamming signal the same way it deals with noise power in the radar. In this case, the S/(J+N) is

$$\frac{S}{J+N} = \frac{\left(\dfrac{P_t G \sigma A_r \tau}{(4\pi)^2 R^4 L}\right)}{\left(\dfrac{(ERP)A_r G'}{4\pi R_J^2 B_J} + kT_0\right)} \tag{10.16}$$

10.3. Range Reduction Factor

Consider a radar system whose detection range R in the absence of jamming is governed by

$$(SNR)_o = \frac{P_t G^2 \lambda^2 \sigma}{(4\pi)^3 kT_e B_r FLR^4} \tag{10.17}$$

The term Range Reduction Factor (RRF) refers to the reduction in the radar detection range due to jamming. More precisely, in the presence of jamming the effective radar detection range is

$$R_{dj} = R \times RRF \tag{10.18}$$

In order to compute RRF, consider a radar characterized by Eq. (10.17), and a barrage jammer whose output power spectral density is J_o (i.e., Gaussian-like). Then the amount of jammer power in the radar receiver is

$$J = kT_J B_r \tag{10.19}$$

where T_J is the jammer effective temperature. It follows that the total jammer plus noise power in the radar receiver is given by

$$N_i + J = kT_e B_r + kT_J B_r \tag{10.20}$$

In this case, the radar detection range is now limited by the receiver signal-to-noise plus interference ratio rather than SNR. More precisely,

$$\left(\frac{S}{J+N}\right) = \frac{P_t G^2 \lambda^2 \sigma}{(4\pi)^3 k(T_e + T_J) B_r FLR^4} \tag{10.21}$$

The amount of reduction in the signal-to-noise plus interference ratio because of the jammer effect can be computed from the difference between Eqs. (10.17) and (10.21). It is expressed (in dB) by

Range Reduction Factor

$$\Upsilon = 10.0 \times \log\left(1 + \frac{T_J}{T_e}\right) \tag{10.22}$$

Consequently, the RRF is

$$RRF = 10^{\frac{-\Upsilon}{40}} \tag{10.23}$$

MATLAB Function "range_red_factor.m"

The function *"range_red_factor.m"* implements Eqs. (10.22) and (10.23); it is given in Listing 10.5 in Section 10.5. This function generates plots of RRF versus: (1) the radar operating frequency; (2) radar to jammer range; and (3) jammer power. Its syntax is as follows:

[RRF] = range_red_factor (te, pj, gj, g, freq, bj, rangej, lossj)

where

Symbol	Description	Units	Status
te	radar effective temperature	K	input
pj	jammer peak power	W	input
gj	jammer antenna gain	dB	input
g	radar antenna gain on jammer	dB	input
freq	radar operating frequency	Hz	input
bj	jammer bandwidth	Hz	input
rangej	radar to jammer range	Km	input
lossj	jammer losses	dB	input

The following values were used to produce Figs. 10.5 through 10.7.

Symbol	Value
te	500 kelvin
pj	500 KW
gj	3 dB
g	45 dB
freq	10 GHz
bj	10 MHZ
rangej	750 Km
lossj	1 dB

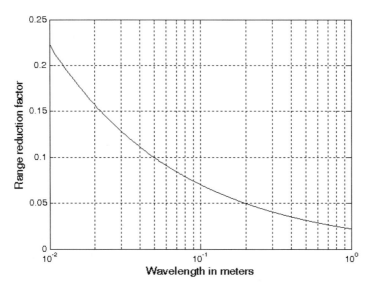

Figure 10.5. Range reduction factor versus radar operating wavelength. This plot was generated using the function *"range_red_factor.m"*.

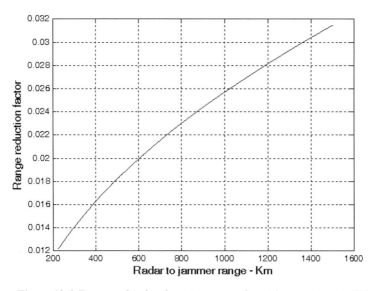

Figure 10.6. Range reduction factor versus radar to jammer range. This plot was generated using the function *"range_red_factor.m"*.

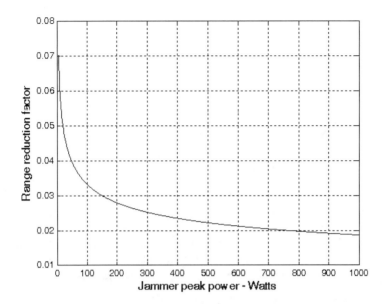

Figure 10.7. Range reduction factor versus jammer peak power. This plot was generated using the function *"range_red_factor.m"*.

10.4. Chaff

In principle, chaff is composed of a large number of small RF reflectors that have large RCS values. Chaff is usually deployed around the target as means of ECM. Historically, chaff was made of aluminum foil; however, in recent years most chaff is made of the more rigid fiber glass with conductive coating.

Chaff can be categorized into two types: (1) denial chaff and (2) deceptive chaff. In the first case, the chaff is deployed in order to screen targets that reside within or near the deployed chaff cloud. In the second case, the chaff cloud is dispersed to complicate and/or overwhelm the tracking and processing functions of the radar by luring the tracker away from the target and/or creating multiple false targets.

The maximum chaff RCS occurs when the individual chaff-dipole length L is one half the radar wavelength. The average RCS for a single dipole when viewed broadside is

$$\sigma_{chaff1} \approx 0.88\lambda^2 \qquad (10.24)$$

and for an average aspect angle, it drops to

$$\sigma_{chaff1} \approx 0.18\lambda^2 \tag{10.25}$$

where the subscript $chaff1$ is used to indicate a single dipole, and λ is the radar wavelength. The total chaff RCS within a radar resolution volume is

$$\sigma_c \approx \frac{0.18\lambda^2 N_D V_{CS}}{L_{beam} V_R} \tag{10.26}$$

where N_D is the total number of dipoles, V_R is the radar resolution cell volume, V_{CS} is the chaff scattering volume, and L_{beam} is the radar antenna beam shape loss for the chaff cloud.

Echoes from a chaff cloud are typically random and have thermal noise-like characteristics because the individual clutter components (scatterers) have random phases and amplitudes. Due to these characteristics, chaff is often statistically described by a probability distribution function. The type of distribution depends on the nature of the chaff cloud itself, radar operating parameters, and the viewing angle of the radar. Thus, the signal-to-chaff ratio is given by

$$\frac{S}{C_{chaff}} = \frac{\sigma}{\sigma_c} CCR \tag{10.27}$$

where σ is the target RCS and CCR is the chaff-cancellation-ratio. The value of CCR depends on the type of chaff mitigation techniques adopted by the radar signal and data processors. Since chaff is a form of volumetric clutter, signal processing and MTI techniques developed for rain and other forms of volumetric clutter can be applied to mitigate many of the effects of chaff. The next section provides an example of one such chaff mitigation technique.

10.4.1. Multiple MTI Chaff Mitigation Technique[1]

In this section, an algorithmic (schema) approach for detecting and tracking targets in highly cluttered environments is presented. The approach is to accurately track the centroid of the chaff cloud using a combination of medium band (MB) and wide-band (WB) range resolution radar waveforms.

At moderate Pulse Repetition Frequencies (PRFs), differential target velocities (about the centroid of the chaff cloud) are detected and tracked via Doppler banks of transversal filters that are tuned to detect the target velocity differ-

---

1. This section is extracted from the paper: J. Michael Madewell, *Mitigating the Effects of Chaff in Ballistic Missile Defense,* 2003 IEEE Radar Conference, Huntsville, AL, May 2003.

Chaff

ences. Through sensitivity analysis models, the theoretical lower bound on detectable differential target velocity as a function of the chaff cloud composition (e.g., clutter cross section, clutter spectral width, number of dipoles, and clutter velocity standard deviation) and radar related parameters (e.g., waveform frequency, bandwidth, integration times, PRFs, and signal-to-clutter ratio) are analyzed.

Overview

A five-step approach for detecting and tracking targets in highly cluttered environments has been developed. The five steps are:

1. Utilize a 1 to 5 percent MB bandwidth, high PRF radar waveform, to measure the chaff cloud range extent, centroid, and velocity growth rate.
2. Establish track on the centroid of the chaff cloud with the MB waveform.
3. Based on course track information obtained in steps 1) and 2), implement WB track (10% or greater bandwidth waveform) on the cloud centroid.
4. Design a doppler bank of Moving Target Indicator (MTI) transversal filters to provide adequate gain at specific velocity increments about the WB centroid track.
5. Process the Multiple MTI (M^2) doppler filters in parallel to detect differences in target Doppler (with respect to the cloud centroid track velocity). Targets are detected when integration at the correct Doppler difference occurs.

Operational concerns that have been identified for implementation of this approach include: (1) the ability of a radar to adequately track the centroid of the chaff cloud (i.e., track precision); (2) the ability of a radar to detect small differences in target Doppler relative to the chaff cloud centroid (i.e., Doppler precision); and (3) the ability of a filter (in this case, a bank of MTI's) to achieve the necessary processing gain to detect the target

Theoretical tracking accuracy of a chaff cloud

The single pulse thermal-noise error σ_f in a velocity tracking measurement for optimum processing can be described by

$$\sigma_f = \frac{1}{1.81\tau\sqrt{2 \times SNR}} \qquad (10.28)$$

where τ is the pulsewidth and SNR is that for the target in track. To detect targets in clutter, substitute the difference-channel chaff-to-signal ratio for SNR. More precisely,

$$\sigma_f = \frac{1}{1.81\tau\sqrt{2 \times C_{chaff}/S}} \qquad (10.29)$$

Fig. 10.8 shows a graph for σ_f versus C_{chaff}/S and τ. This figure can be reproduced using MATLAB program *"fig10_8.m"* given in Listing 10.6 in Section 10.5. This graph will be utilized in the analysis and of the expected M^2 signal processing performance.

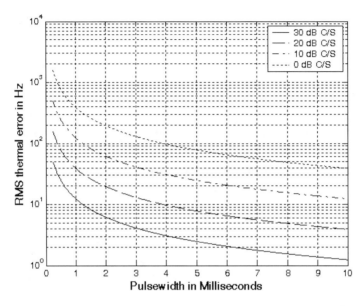

Figure 10.8. Single pulse thermal noise error versus C_{chaff}/S and τ.

Multiple MTI (M^2) Doppler Filter Design

The M^2 Doppler filter design is derived from the theoretical N-tap delay line MTI canceller. The general formula for the improvement factor was derived in Chapter 7 (Section 7.7.2). A bank of N MTI Doppler filters that cover the frequency range from 0 to the PRF will achieve performance beyond that of a conventional MTI. The weights are given by:

$$W_{i,k} = e^{\frac{j2\pi(i-1)}{k/N}} \qquad (10.30)$$

where the index k is between 0 to N-1 and corresponds to the N MTI Doppler filter bank. In this design, a 5-tap delay line MTI filter is considered. The transfer function for the overall Doppler bank is

$$H^k(f) = \sum_{i-1}^{N} e^{(-j2\pi)(i-1)(fT-k/N)} \qquad (10.31)$$

where

$$T = 1/PRF \tag{10.32}$$

It follows that the magnitude of the frequency response is

$$|H^k(f)| = \left|\frac{\sin(\pi N(fT - k/N))}{\sin(\pi(fT - k/N))}\right| \tag{10.33}$$

The impulse response for a kth 5-tap MTI filter is

$$y^k(t) = v_k(t) - 5v_k(t-T) + 10v_k(t-2T) - \\ 10v_k(t-3T) + 5v_k(t-4T) - v_k(t-5T) \tag{10.34}$$

v_k is the input signal. The corresponding transfer function is

$$Y^k(f) = 25(\sin(\pi fT))^5 \tag{10.35}$$

Fig. 10.9 shows a block diagram for the M^2 filter. Since each filter occupies approximately $(1/N)th$ the clutter and signal bandwidth, the combined performance of the M^2 Doppler filter performance is greater than that of a single delay-line canceller that does not utilize Doppler information. The clutter mitigation performance of the M^2 Doppler filter, however, will likely be determined by the coherence times of the target and/or the clutter.

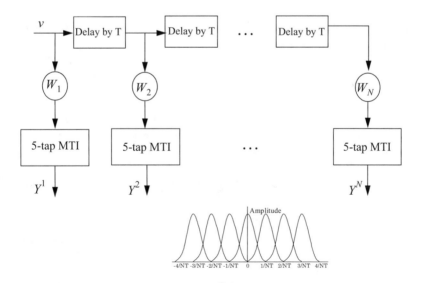

Figure 10.9. Block diagram for the M^2 algorithm, and corresponding frequency response of the MTI filters (N=8).

Processor Implementation And Simulated Results

The M^2 filter approach outlined in this section requires a very accurate track of the centroid of the chaff cloud being probed. As described earlier, initiation of track on the chaff cloud centroid is achieved with a MB range resolution waveform (step 1). As an example, assume that an X-band radar (10 GHz) is engaging one or more ballistic targets enveloped in a chaff cloud that contains 1 million dipoles occupying a 1-kilometer range extent. Assuming that the chaff cloud velocity distribution can be accurately modeled by Gaussian statistics, approximately 67% of these dipoles will reside in 333 meters of range extent. With these assumptions, the combined average RCS of the dipoles (RCS_D) contained within a radar range resolution cell of this length (333 m) can be approximated by

$$RCS_D = 0.18 N_D \lambda^2 = 0.18 \times 670,000 \times 0.03^2 = 108.54 \Rightarrow 20.4 dBsm \quad \textbf{(10.36)}$$

The RCS of a typical ballistic Reentry Vehicle (RV) at forward aspect viewing angles can be $-20 dBsm$ or smaller. Therefore, the MB C_{chaff}/S for a typical RV enveloped by the chaff cloud assumed above can approach $40 dB$ or greater. Using an 8-msec pulsewidth and $30 dB$ C_{chaff}/S, the theoretical, single pulse, minimum rms track error is approximately $f_e = 1 Hz$. At X-band frequencies, this translates to a single pulse velocity error of

$$v_e = \frac{f_e \lambda}{2} = 0.015 m/s \quad \textbf{(10.37)}$$

Note that for a train of pulses, this velocity error can be reduced by a factor of 10 or more. Thus, for a typical X-band radar, theory suggests that the track precision of the chaff cloud centroid can approach 0.0015 m/s or better. This track precision is much less than the WB range resolution capability of the radar and therefore can be utilized to bootstrap the WB tracker (steps 2 and 3).

Assume a Gaussian chaff clutter velocity distribution and denote it v_g. If $v_g = 1.8 m/s$ ($\pm 0.9 m/s$ relative to the cloud centroid velocity), the minimum PRF required to meet the Nyquist sampling criterion is

$$PRF = f_r \geq 2 \times \frac{2 v_g}{\lambda} = 240 Hz \quad \textbf{(10.38)}$$

Also, assume that a bank of Doppler MTI's (step 4) can be formed to cover this frequency range. Note that 256 is the closest 2^N multiple for implementation with the Fast Fourier Transform (FFT). Using a 256 point FFT design, each filter will contain approximately 1/256 of the total clutter velocities (about 0.03 m/s of velocity clutter per MTI Doppler filter). In addition, by utilizing the WB track waveform, a very precise range-Doppler image can be formed (with each range-Doppler resolution cell containing approximately 15 cm by 0.03 m/s of clutter). This design effectively reduces the amount of clutter that competes with an individual target scatter by a factor of more than 40 dB, thus reducing the C_{chaff}/S by this same amount.

For extreme chaff cases where the initial WB range-Doppler image S/C is negative, an N-pulse coherent sliding window routine can be applied to the data prior to implementing the M^2 algorithm. For example, a 16 pulse coherent sliding window can provide up to 12 dB of S/C_{chaff} improvement. One should ensure that the number of pulses integrated is less than the coherency time of the target and clutter. Other constraints in implementing this approach are to ensure that the target phase does not deviate very much during the integration period (to ensure optimum coherent processing gain) and the target position does not migrate to another range and/or Doppler cell (often referred to as range-Doppler walk). The zero Doppler filter (and/or near zero Doppler filters) can be used to perform statistics on the clutter and to adaptively adjust the optimal threshold setting to obtain low false alarms and high probabilities of detection over time.

A model for the M^2 signal processor has been developed using MATLAB. Fig. 10.10 shows a plot of the amplitude versus range and Doppler (256x256 range-Doppler image) of three constant -20 dBsm target scatterers that are embedded in approximately -15 dBsm Gaussian white noise. In this figure, the noise completely envelops the signal. These modeling results are comparable to the output of a typical range Doppler imaging radar. Fig. 10.11 shows the results obtained by executing the first two blocks of the M^2 signal processor. As expected, the three scatterers rise from above the noise and now have an S/C_{chaff} ratio of approximately 7 dB.

Finally, Fig. 10.12 shows the results obtained by implementing the entire top portion of the M^2 signal processing chain. No attempt was made to optimize the threshold level. Instead, the threshold was manually set to -43 dB to allow for some of the higher false alarms to be seen in the figure. The largest amplitude false alarms are approximately -34 dB. Meanwhile, the amplitudes of the target returns have been reduced (less than 1 dB) from that of Fig. 10.11. Therefore, the S/C_{chaff} improvement in Fig. 10.12 over that shown in Fig. 10.11 is approximately 8 to 9 dB. Hence, the processing gain attributed to the M^2 signal processor is more than 20 dB above that of traditional range Doppler processing.

In summary, one concludes that the M^2 signal processing algorithm for detecting and tracking ballistic missile targets in highly cluttered environments can provide better than 20 dB S/C_{chaff} improvement over that of traditional range Doppler processing techniques alone.

492 *MATLAB Simulations for Radar Systems Design*

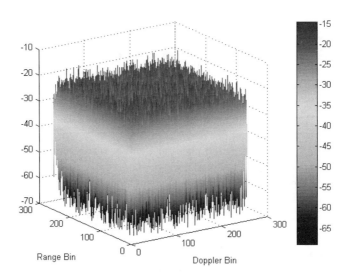

Figure 10.10. Range -Doppler image for three targets embedded in chaff.

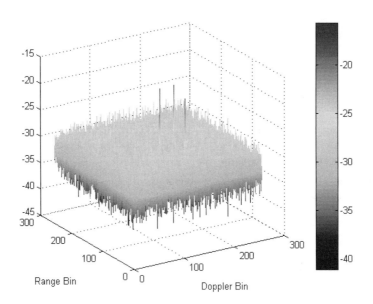

Figure 10.11. Image from Fig. 10.10 after a 16-point sliding window coherent integration process.

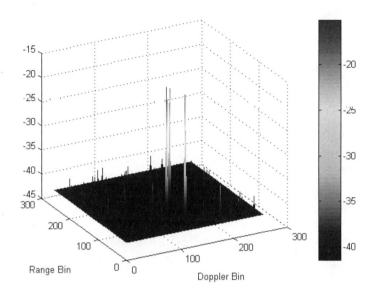

Figure 10.12. Image from Fig. 10.11 after applying the M^2 algorithm.

10.5. MATLAB Program and Function Listings

This section presents listings for all MATLAB programs/functions used in this chapter. The user is advised to rerun these programs with different input parameters.

Listing 10.1. MATLAB Function "ssj_req.m"

```
function [BR_range] = ssj_req (pt, g, freq, sigma, b, loss, ...
   pj, bj, gj, lossj)
% This function implements Eq. (10.9)
c = 3.0e+8;
lambda = c / freq;
lambda_db = 10*log10(lambda^2);
if (loss == 0.0)
   loss = 0.000001;
end
if (lossj == 0.0)
   lossj =0.000001;
end
sigmadb =10*log10(sigma);
```

```
pt_db = 10*log10(pt);
b_db = 10*log10(b);
bj_db = 10*log10(bj);
pj_db = 10*log10(pj);
factor = 10*log10(4.0 *pi);
BR_range = sqrt((pt * (10^(g/10)) * sigma * bj * (10^(lossj/10))) / ...
   (4.0 * pi * pj * (10^(gj/10)) * b * ...
   (10^(loss/10)))) / 1000.0
s_at_br = pt_db + 2.0 * g + lambda_db + sigmadb - ...
      3.0 * factor - 4.* 10*log10(BR_range) - loss
index =0;
for ran_var = .1:10:10000
   index = index + 1;
   ran_db = 10*log10(ran_var * 1000.0);
   ssj(index) = pj_db + gj + lambda_db + g + b_db - 2.0 * factor - ...
      2.0 * ran_db - bj_db - lossj + s_at_br ;
   s(index) = pt_db + 2.0 * g + lambda_db + sigmadb - ...
      3.0 * factor - 4.* ran_db - loss + s_at_br ;
end
ranvar = .1:10:10000;
ranvar = ranvar ./ BR_range;
semilogx (ranvar,s,'k',ranvar,ssj,'k-.');
axis([.1 1000 -90 40])
xlabel ('Range normalized to cross-over range');
legend('Target echo','SSJ')
ylabel ('Relative signal or jamming amplitude - dB');
grid
pj_var = 1:1:1000;
BR_pj = sqrt((pt * (10^(g/10)) * sigma * bj * (10^(lossj/10))) ...
   ./ (4.0 * pi .* pj_var * (10^(gj/10)) * b * (10^(loss/10)))) ./ 1000;
pt_var = 1000:100:10e6;
BR_pt = sqrt((pt_var * (10^(g/10)) * sigma * bj * (10^(lossj/10))) ...
   ./ (4.0 * pi .* pj * (10^(gj/10)) * b * (10^(loss/10)))) ./ 1000;
figure (2)
subplot (2,1,1)
semilogx (BR_pj,'k')
xlabel ('Jammer peak power - Watts');
ylabel ('Burn-through range - Km')
grid
subplot (2,1,2)
semilogx (BR_pt,'k')
xlabel ('Radar peak power - KW')
ylabel ('Burn-through range - Km')
grid
```

Listing 10.2. MATLAB Function "sir.m"

```
function [SIR] = sir (pt, g, freq, sigma, tau,T0,  loss, R, pj, bj, gj, lossj);
c = 3.0e+8;
k = 1.38e-23;
%R = linspace(rmin, rmax, 1000);
range = R .* 1000;
lambda = c / freq;
gj = 10^(gj/10);
G = 10^(g/10);
ERP1 = pj * gj / lossj;
ERP_db = 10*log10(ERP1);
% Calculate Eq. (10.10)
Ar = lambda *lambda * G / 4 /pi;
num1 = pt * tau * G * sigma * Ar;
demo1 = 4^2 * pi^2 * loss .* range.^4;
demo2 = 4 * pi * bj .* range.^2;
num2 = ERP1 * Ar;
val11 = num1 ./ demo1;
val21 = num2 ./demo2;
sir = val11 ./ (val21 + k * T0);
SIR = 10*log10(sir);
figure (1)
plot (R, SIR,'k')
xlabel ('Detection range in Km');
ylabel ('S/(J+N) in dB')
grid
```

Listing 10.3. MATLAB Function "burn_thru.m"

```
function [Range] = burn_thru (pt, g, freq, sigma, tau, T0, loss, pj, bj, gj,
lossj,sir0,ERP);
c = 3.0e+8;
k = 1.38e-23;
%R = linspace(rmin, rmax, 1000);
sir0 = 10^(sir0/10);
lambda = c / freq;
gj = 10^(gj/10);
G = 10^(g/10);
 Ar = lambda *lambda * G / 4 /pi;
%ERP = linspace(1,1000,5001);
num32 = ERP .* Ar;
demo3 = 8 *pi * bj * k * T0;
demo4 = 4^2 * pi^2 * k * T0 * sir0;
```

```
val1 = (num32 ./ demo3).^2;
val2 = (pt * tau * G * sigma * Ar)/(4^2 * pi^2 * loss * sir0 * k *T0);
val3 = sqrt(val1 + val2);
val4 = (ERP .* Ar) ./ demo3;
Range = sqrt(val3 - val4) ./ 1000;
figure (1)
plot (10*log10(ERP), Range,'k')
xlabel (' Jammer ERP in dB')
ylabel ('Burnthrough range in Km')
grid
```

Listing 10.4. MATLAB Function "soj_req.m"

```
function [BR_range] = soj_req (pt, g, sigma, b, freq, loss, range, ...
   pj, bj,gj, lossj, gprime, rangej)
% This function implements equations for SOJs
c = 3.0e+8;
lambda = c / freq;
lambda_db = 10*log10(lambda^2)
if (loss == 0.0)
   loss = 0.000001;
end
if (lossj == 0.0)
   lossj =0.000001;
end
sigmadb = 10*log10(sigma);
range_db = 10*log10(range * 1000.);
rangej_db = 10*log10(rangej * 1000.)
pt_db = 10*log10(pt);
b_db = 10*log10(b);
bj_db = 10*log10(bj);
pj_db = 10*log10(pj);
factor = 10*log10(4.0 *pi);
BR_range = ((pt * 10^(2.0*g/10) * sigma * bj * 10^(lossj/10) * ...
   (rangej)^2) / (4.0 * pi * pj * 10^(gj/10) * 10^(gprime/10) * ...
   b * 10^(loss/10)))^.25 / 1000.
s_at_br = pt_db + 2.0 * g + lambda_db + sigmadb - ...
   3.0 * factor - 4.0 * 10*log10(BR_range) - loss
index =0;
for ran_var = .1:1:1000;
   index = index + 1;
   ran_db = 10*log10(ran_var * 1000.0);
   s(index) = pt_db + 2.0 * g + lambda_db + sigmadb - ...
      3.0 * factor - 4.0 * ran_db - loss + s_at_br;
```

```
    soj(index) = s_at_br - s_at_br;
end
ranvar = .1:1:1000;
%ranvar = ranvar ./BR_range;
semilogx (ranvar,s,'k',ranvar,soj,'k-.');
xlabel ('Range normalized to cross-over range');
legend('Target echo','SOJ')
ylabel ('Relative signal or jamming amplitude - dB');
```

Listing 10.5. MATLAB Function "range_red_factor.m"

```
function RRF = range_red_factor (te, pj, gj, g, freq, bj, rangej, lossj)
% This function computes the range reduction factor and produces
% plots of RRF versus wavelength, radar to jammer range, and jammer power
c = 3.0e+8;
k = 1.38e-23;
lambda = c / freq;
gj_10 = 10^( gj/10);
g_10 = 10^( g/10);
lossj_10 = 10^(lossj/10);
index = 0;
for wavelength = .01:.001:1
   index = index +1;
   jamer_temp = (pj * gj_10 * g_10 *wavelength^2) / ...
      (4.0^2 * pi^2 * k * bj * lossj_10 * (rangej * 1000.0)^2);
   delta = 10.0 * log10(1.0 + (jamer_temp / te));
   rrf(index) = 10^(-delta /40.0);
end
w = 0.01:.001:1;
figure (1)
semilogx(w,rrf,'k')
grid
xlabel ('Wavelength in meters')
ylabel ('Range reduction factor')
index = 0;
for ran =rangej*.3:1:rangej*2
   index = index + 1;
   jamer_temp = (pj * gj_10 * g_10 *wavelength^2) / ...
      (4.0^2 * pi^2 * k * bj * lossj_10 * (ran * 1000.0)^2);
   delta = 10.0 * log10(1.0 + (jamer_temp / te));
   rrf1(index) = 10^(-delta /40.0);
end
figure(2)
ranvar = rangej*.3:1:rangej*2 ;
```

```
plot(ranvar,rrf1,'k')
grid
xlabel ('Radar to jammer range - Km')
ylabel ('Range reduction factor')
index = 0;
for pjvar = pj*.01:1:pj*2
   index = index + 1;
   jamer_temp = (pjvar * gj_10 * g_10 *wavelength^2) / ...
      (4.0^2 * pi^2 * k * bj * lossj_10 * (rangej * 1000.0)^2);
   delta = 10.0 * log10(1.0 + (jamer_temp / te));
   rrf2(index) = 10^(-delta /40.0);
end
figure(3)
pjvar = pj*.01:1:pj*2;
plot(pjvar,rrf2,'k')
grid
xlabel ('Jammer peak power - Watts')
ylabel ('Range reduction factor')
%%%%%%%%%%%%%%%%%%%%%%%%%%%%%%%%%%%%%
% Use this input file to reproduce Figs. 10.5 through 10.7
clear all
te = 500.0;    % radar effective temp. in Kelvin
pj= 500;  % jammer peak power in W
gj = 3.0;     % jammer antenna gain in dB
g = 45.0;     % radar antenna gain
freq = 10.0e+9;% radar operating frequency in Hz
bj= 10.0e+6;   % radar operating bandwidth in Hz
rangej = 750.0;% radar to jammer range in Km
lossj = 1.0;   % jammer losses in dB
```

Listing 10.6. MATLAB Program "fig10_8.m"

```
% Use this program to reproduce Fig. 10.8 in the text
clear all
close all
tau = linspace(.25,10,500);
taum = tau .* 1e-3;
C_S = [-20 -10 0 10];
c_s = 10.^(C_S./10);
for n = 1:size(C_S,2)
   val1 = 1 / (1.81*sqrt(2*c_s(n)));
   sigma(n,:) = val1 ./ taum;
end
figure (1)
```

```
semilogy(tau,sigma(1,:),'k',tau,sigma(2,:),'k-- ',tau,sigma(3,:),'k-.', ...
    tau,sigma(4,:),'k:');
xlabel('Pulsewidth in Milliseconds')
ylabel('RMS thermal error in Hz')
legend('-20 dB C/S','-10 dB C/S','0 dB C/S','10 dB C/S')
grid
```

Chapter 11

Radar Cross Section (RCS)

In this chapter, the phenomenon of target scattering and methods of RCS calculation are examined. Target RCS fluctuations due to aspect angle, frequency, and polarization are presented. Radar cross section characteristics of some simple and complex targets are also introduced.

11.1. RCS Definition

Electromagnetic waves, with any specified polarization, are normally diffracted or scattered in all directions when incident on a target. These scattered waves are broken down into two parts. The first part is made of waves that have the same polarization as the receiving antenna. The other portion of the scattered waves will have a different polarization to which the receiving antenna does not respond. The two polarizations are orthogonal and are referred to as the Principal Polarization (PP) and Orthogonal Polarization (OP), respectively. The intensity of the *backscattered* energy that has the same polarization as the radar's receiving antenna is used to define the target RCS. When a target is illuminated by RF energy, it acts like an antenna, and will have near and far fields. Waves reflected and measured in the near field are, in general, spherical. Alternatively, in the far field the wavefronts are decomposed into a linear combination of plane waves.

Assume the power density of a wave incident on a target located at range R away from the radar is P_{Di}, as illustrated in Fig. 11.1. The amount of reflected power from the target is

$$P_r = \sigma P_{Di} \tag{11.1}$$

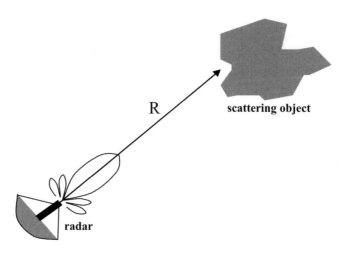

Figure 11.1. Scattering object located at range R.

σ denotes the target cross section. Define P_{Dr} as the power density of the scattered waves at the receiving antenna. It follows that

$$P_{Dr} = P_r/(4\pi R^2) \tag{11.2}$$

Equating Eqs. (11.1) and (11.2) yields

$$\sigma = 4\pi R^2 \left(\frac{P_{Dr}}{P_{Di}}\right) \tag{11.3}$$

and in order to ensure that the radar receiving antenna is in the far field (i.e., scattered waves received by the antenna are planar), Eq. (11.3) is modified

$$\sigma = 4\pi R^2 \lim_{R \to \infty} \left(\frac{P_{Dr}}{P_{Di}}\right) \tag{11.4}$$

The RCS defined by Eq. (11.4) is often referred to as either the monostatic RCS, the backscattered RCS, or simply target RCS.

The backscattered RCS is measured from all waves scattered in the direction of the radar and has the same polarization as the receiving antenna. It represents a portion of the total scattered target RCS σ_t, where $\sigma_t > \sigma$. Assuming a spherical coordinate system defined by (ρ, θ, φ), then at range ρ the target scattered cross section is a function of (θ, φ). Let the angles (θ_i, φ_i) define the direction of propagation of the incident waves. Also, let the angles (θ_s, φ_s) define the direction of propagation of the scattered waves. The special case, when $\theta_s = \theta_i$ and $\varphi_s = \varphi_i$, defines the monostatic RCS. The RCS measured by the radar at angles $\theta_s \neq \theta_i$ and $\varphi_s \neq \varphi_i$ is called the bistatic RCS.

The total target scattered RCS is given by

$$\sigma_t = \frac{1}{4\pi} \int_{\varphi_s = 0}^{2\pi} \int_{\theta_s = 0}^{\pi} \sigma(\theta_s, \varphi_s) \sin\theta_s \, d\theta \, d\varphi_s \tag{11.5}$$

The amount of backscattered waves from a target is proportional to the ratio of the target extent (size) to the wavelength, λ, of the incident waves. In fact, a radar will not be able to detect targets much smaller than its operating wavelength. For example, if weather radars use L-band frequency, rain drops become nearly invisible to the radar since they are much smaller than the wavelength. RCS measurements in the frequency region, where the target extent and the wavelength are comparable, are referred to as the Rayleigh region. Alternatively, the frequency region where the target extent is much larger than the radar operating wavelength is referred to as the optical region. In practice, the majority of radar applications fall within the optical region.

The analysis presented in this book mainly assumes far field monostatic RCS measurements in the optical region. Near field RCS, bistatic RCS, and RCS measurements in the Rayleigh region will not be considered since their treatment falls beyond this book's intended scope. Additionally, RCS treatment in this chapter is mainly concerned with Narrow Band (NB) cases. In other words, the extent of the target under consideration falls within a single range bin of the radar. Wide Band (WB) RCS measurements will be briefly addressed in a later section. Wide band radar range bins are small (typically 10 - 50 cm); hence, the target under consideration may cover many range bins. The RCS value in an individual range bin corresponds to the portion of the target falling within that bin.

11.2. RCS Prediction Methods

Before presenting the different RCS calculation methods, it is important to understand the significance of RCS prediction. Most radar systems use RCS as a means of discrimination. Therefore, accurate prediction of target RCS is critical in order to design and develop robust discrimination algorithms. Additionally, measuring and identifying the scattering centers (sources) for a given target aid in developing RCS reduction techniques. Another reason of lesser importance is that RCS calculations require broad and extensive technical knowledge; thus, many scientists and scholars find the subject challenging and intellectually motivating. Two categories of RCS prediction methods are available: exact and approximate.

Exact methods of RCS prediction are very complex even for simple shape objects. This is because they require solving either differential or integral equations that describe the scattered waves from an object under the proper set of boundary conditions. Such boundary conditions are governed by Maxwell's equations. Even when exact solutions are achievable, they are often difficult to interpret and to program using digital computers.

Due to the difficulties associated with the exact RCS prediction, approximate methods become the viable alternative. The majority of the approximate methods are valid in the optical region, and each has its own strengths and limitations. Most approximate methods can predict RCS within few dBs of the truth. In general, such a variation is quite acceptable by radar engineers and designers. Approximate methods are usually the main source for predicting RCS of complex and extended targets such as aircrafts, ships, and missiles. When experimental results are available, they can be used to validate and verify the approximations.

Some of the most commonly used approximate methods are Geometrical Optics (GO), Physical Optics (PO), Geometrical Theory of Diffraction (GTD), Physical Theory of Diffraction (PTD), and Method of Equivalent Currents (MEC). Interested readers may consult Knott or Ruck (see bibliography) for more details on these and other approximate methods.

11.3. Dependency on Aspect Angle and Frequency

Radar cross section fluctuates as a function of radar aspect angle and frequency. For the purpose of illustration, isotropic point scatterers are considered. An isotropic scatterer is one that scatters incident waves equally in all directions. Consider the geometry shown in Fig. 11.2. In this case, two unity ($1m^2$) isotropic scatterers are aligned and placed along the radar line of sight (zero aspect angle) at a far field range R. The spacing between the two scatterers is 1 meter. The radar aspect angle is then changed from zero to 180 degrees, and the composite RCS of the two scatterers measured by the radar is computed.

This composite RCS consists of the superposition of the two individual radar cross sections. At zero aspect angle, the composite RCS is $2m^2$. Taking scatterer-1 as a phase reference, when the aspect angle is varied, the composite RCS is modified by the phase that corresponds to the electrical spacing between the two scatterers. For example, at aspect angle $10°$, the electrical spacing between the two scatterers is

$$elec\text{-}spacing = \frac{2 \times (1.0 \times \cos(10°))}{\lambda} \quad \text{(11.6)}$$

λ is the radar operating wavelength.

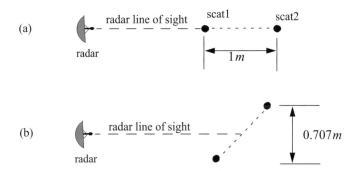

Figure 11.2. RCS dependency on aspect angle. (a) Zero aspect angle, zero electrical spacing. (b) $45°$ aspect angle, 1.414λ electrical spacing.

Fig. 11.3 shows the composite RCS corresponding to this experiment. This plot can be reproduced using MATLAB function *"rcs_aspect.m"* given in Listing 11.1 in Section 11.9. As clearly indicated by Fig. 11.3, RCS is dependent on the radar aspect angle; thus, knowledge of this constructive and destructive interference between the individual scatterers can be very critical when a radar tries to extract the RCS of complex or maneuvering targets. This is true because of two reasons. First, the aspect angle may be continuously changing. Second, complex target RCS can be viewed to be made up from contributions of many individual scattering points distributed on the target surface. These scattering points are often called scattering centers. Many approximate RCS prediction methods generate a set of scattering centers that define the backscattering characteristics of such complex targets.

MATLAB Function *"rcs_aspect.m"*

The function *"rcs_aspect.m"* computes and plots the RCS dependency on aspect angle. Its syntax is as follows:

[rcs] = rcs_aspect (scat_spacing, freq)

where

Symbol	Description	Units	Status
scat_spacing	scatterer spacing	meters	input
freq	radar frequency	Hz	input
rcs	array of RCS versus aspect angle	dBsm	output

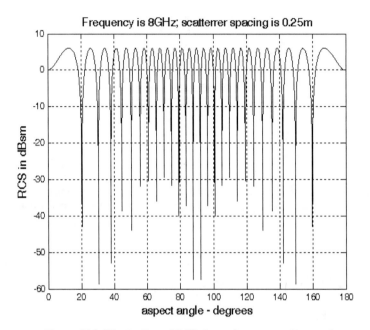

Figure 11.3. Illustration of RCS dependency on aspect angle.

Next, to demonstrate RCS dependency on frequency, consider the experiment shown in Fig. 11.4. In this case, two far field unity isotropic scatterers are aligned with radar line of sight, and the composite RCS is measured by the radar as the frequency is varied from 8 GHz to 12.5 GHz (X-band). Figs. 11.5 and 11.6 show the composite RCS versus frequency for scatterer spacing of 0.25 and 0.75 meters.

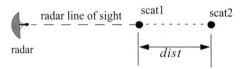

Figure 11.4. Experiment setup which demonstrates RCS dependency on frequency; dist = 0.1, or 0.7 m.

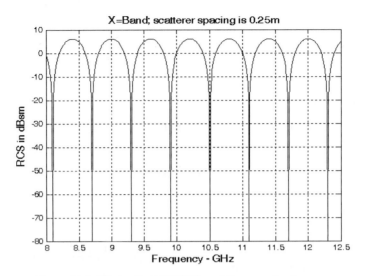

Figure 11.5. Illustration of RCS dependency on frequency.

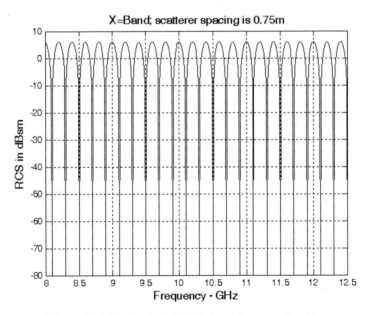

Figure 11.6. Illustration of RCS dependency on frequency.

The plots shown in Figs. 11.5 and 11.6 can be reproduced using MATLAB function *"rcs_frequency.m"* given in Listing 11.2 in Section 11.9. From those two figures, RCS fluctuation as a function of frequency is evident. Little frequency change can cause serious RCS fluctuation when the scatterer spacing is large. Alternatively, when scattering centers are relatively close, it requires more frequency variation to produce significant RCS fluctuation.

MATLAB Function *"rcs_frequency.m"*

The function *"rcs_frequency.m"* computes and plots the RCS dependency on frequency. Its syntax is as follows:

$$[rcs] = rcs\_frequency\ (scat\_spacing, frequ, freql)$$

where

Symbol	Description	Units	Status
scat_spacing	scatterer spacing	meters	input
freql	start of frequency band	Hz	input
frequ	end of frequency band	Hz	input
rcs	array of RCS versus aspect angle	dBsm	output

Referring to Fig. 11.2, assume that the two scatterers complete a full revolution about the radar line of sight in $T_{rev} = 3 \sec$. Furthermore, assume that an X-band radar ($f_0 = 9GHz$) is used to detect (observe) those two scatterers using a PRF $f_r = 300Hz$ for a period of 3 seconds. Finally, assume a NB bandwidth $B_{NB} = 1MHz$ and a WB bandwidth $B_{WB} = 2GHz$. It follows that the radar's NB and WB range resolutions are respectively equal to $\Delta R_{NB} = 150m$ and $\Delta R_{WB} = 7.5cm$.

Fig. 11.7 shows a plot of the detected range history for the two scatterers using NB detection. Clearly, the two scatterers are completely contained within one range bin. Fig. 11.8 shows the same; however, in this case WB detection is utilized. The two scatterers are now completely resolved as two distinct scatterers, except during the times where both point scatterers fall within the same range bin.

11.4. RCS Dependency on Polarization

The material in this section covers two topics. First, a review of polarization fundamentals is presented. Second, the concept of the target scattering matrix is introduced.

RCS Dependency on Polarization 509

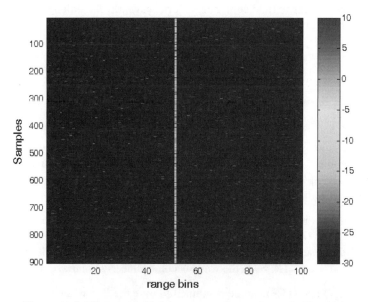

Figure 11.7. NB detection of the two scatterers shown in Fig. 11.2.

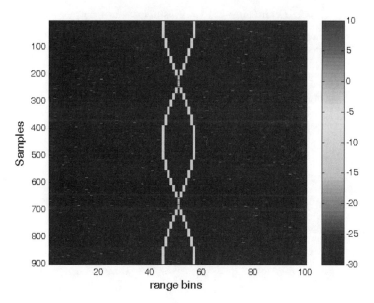

Figure 11.8. WB detection of the two scatterers shown in Fig. 11.2.

11.4.1. Polarization

The x and y electric field components for a wave traveling along the positive z direction are given by

$$E_x = E_1 \sin(\omega t - kz) \tag{11.7}$$

$$E_y = E_2 \sin(\omega t - kz + \delta) \tag{11.8}$$

where $k = 2\pi/\lambda$, ω is the wave frequency, the angle δ is the time phase angle which E_y leads E_x, and, finally, E_1 and E_2 are, respectively, the wave amplitudes along the x and y directions. When two or more electromagnetic waves combine, their electric fields are integrated vectorially at each point in space for any specified time. In general, the combined vector traces an ellipse when observed in the x-y plane. This is illustrated in Fig. 11.9.

The ratio of the major to the minor axes of the polarization ellipse is called the Axial Ratio (AR). When AR is unity, the polarization ellipse becomes a circle, and the resultant wave is then called circularly polarized. Alternatively, when $E_1 = 0$ and $AR = \infty$ the wave becomes linearly polarized.

Eqs. (11.7) and (11.8) can be combined to give the instantaneous total electric field,

$$\vec{E} = \hat{a}_x E_1 \sin(\omega t - kz) + \hat{a}_y E_2 \sin(\omega t - kz + \delta) \tag{11.9}$$

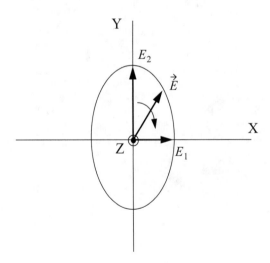

Figure 11.9. Electric field components along the x and y directions. The positive z direction is out of the page.

where $\hat{a}_x$ and $\hat{a}_y$ are unit vectors along the x and y directions, respectively. At $z = 0$, $E_x = E_1 \sin(\omega t)$ and $E_y = E_2 \sin(\omega t + \delta)$, then by replacing $\sin(\omega t)$ by the ratio E_x/E_1 and by using trigonometry properties Eq. (11.9) can be rewritten as

$$\frac{E_x^2}{E_1^2} - \frac{2 E_x E_y \cos \delta}{E_1 E_2} + \frac{E_y^2}{E_2^2} = (\sin \delta)^2 \qquad (11.10)$$

Note that Eq. (11.10) has no dependency on ωt.

In the most general case, the polarization ellipse may have any orientation, as illustrated in Fig. 11.10. The angle ξ is called the tilt angle of the ellipse. In this case, AR is given by

$$AR = \frac{OA}{OB} \qquad (1 \leq AR \leq \infty) \qquad (11.11)$$

When $E_1 = 0$, the wave is said to be linearly polarized in the y direction, while if $E_2 = 0$ the wave is said to be linearly polarized in the x direction. Polarization can also be linear at an angle of 45° when $E_1 = E_2$ and $\xi = 45°$. When $E_1 = E_2$ and $\delta = 90°$, the wave is said to be Left Circularly Polarized (LCP), while if $\delta = -90°$ the wave is said to Right Circularly Polarized (RCP). It is a common notation to call the linear polarizations along the x and y directions by the names horizontal and vertical polarizations, respectively.

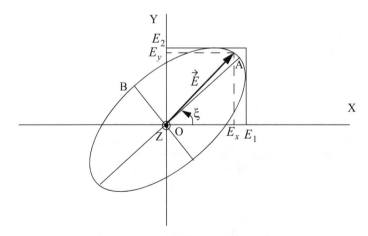

Figure 11.10. Polarization ellipse in the general case.

In general, an arbitrarily polarized electric field may be written as the sum of two circularly polarized fields. More precisely,

$$\vec{E} = \vec{E_R} + \vec{E_L} \tag{11.12}$$

where $\vec{E_R}$ and $\vec{E_L}$ are the RCP and LCP fields, respectively. Similarly, the RCP and LCP waves can be written as

$$\vec{E_R} = \vec{E_V} + j\vec{E_H} \tag{11.13}$$

$$\vec{E_L}' = \vec{E_V} - j\vec{E_H} \tag{11.14}$$

where $\vec{E_V}$ and $\vec{E_H}$ are the fields with vertical and horizontal polarizations, respectively. Combining Eqs. (11.13) and (11.14) yields

$$E_R = \frac{E_H - jE_V}{\sqrt{2}} \tag{11.15}$$

$$E_L = \frac{E_H + jE_V}{\sqrt{2}} \tag{11.16}$$

Using matrix notation Eqs. (11.15) and (11.16) can be rewritten as

$$\begin{bmatrix} E_R \\ E_L \end{bmatrix} = \frac{1}{\sqrt{2}} \begin{bmatrix} 1 & -j \\ 1 & j \end{bmatrix} \begin{bmatrix} E_H \\ E_V \end{bmatrix} = [T] \begin{bmatrix} E_H \\ E_V \end{bmatrix} \tag{11.17}$$

$$\begin{bmatrix} E_H \\ E_V \end{bmatrix} = \frac{1}{\sqrt{2}} \begin{bmatrix} 1 & 1 \\ j & -j \end{bmatrix} \begin{bmatrix} E_R \\ E_L \end{bmatrix} = [T]^{-1} \begin{bmatrix} E_H \\ E_V \end{bmatrix} \tag{11.18}$$

For many targets the scattered waves will have different polarization than the incident waves. This phenomenon is known as depolarization or cross-polarization. However, perfect reflectors reflect waves in such a fashion that an incident wave with horizontal polarization remains horizontal, and an incident wave with vertical polarization remains vertical but is phase shifted $180°$. Additionally, an incident wave which is RCP becomes LCP when reflected, and a wave which is LCP becomes RCP after reflection from a perfect reflector. Therefore, when a radar uses LCP waves for transmission, the receiving antenna needs to be RCP polarized in order to capture the PP RCS, and LCR to measure the OP RCS.

Example:

Plot the locus of the electric field vector for the following cases:

case 1: $\vec{E}(t,z) = \hat{a}_x \cos\left(\omega_0 t + \frac{2\pi z}{\lambda}\right) + \hat{a}_y \sqrt{3} \cos\left(\omega_0 t + \frac{2\pi z}{\lambda}\right)$

case 2: $\vec{E}(t,z) = \hat{a}_x \cos\left(\omega_0 t + \frac{2\pi z}{\lambda}\right) + \hat{a}_y \sin\left(\omega_0 t + \frac{2\pi z}{\lambda}\right)$

case 3: $\vec{E}(t,z) = \hat{a}_x \cos\left(\omega_0 t + \frac{2\pi z}{\lambda}\right) + \hat{a}_y \cos\left(\omega_0 t + \frac{2\pi z}{\lambda} + \frac{\pi}{6}\right)$

case 4: $\vec{E}(t,z) = \hat{a}_x \cos\left(\omega_0 t + \frac{2\pi z}{\lambda}\right) + \hat{a}_y \sqrt{3} \cos\left(\omega_0 t + \frac{2\pi z}{\lambda} + \frac{\pi}{3}\right)$

Solution:

The MATLAB program "example11_1.m" was developed to calculate and plot the loci of the electric fields. Figs. 11.11 through 11.14 show the desired electric fields' loci. See listing 11.3 in Section 11.9.

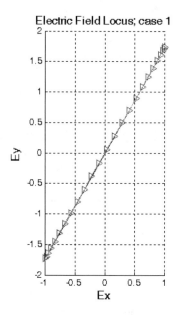

Figure 11.11. Linearly polarized electric field.

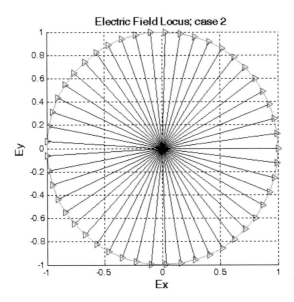

Figure 11.12. Circularly polarized electric field.

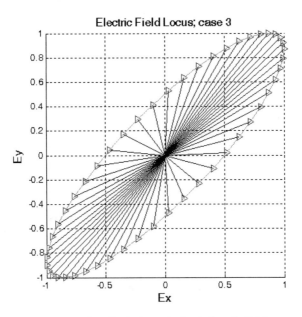

Figure 11.13. Elliptically polarized electric field.

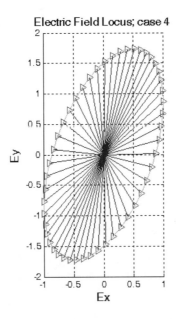

Figure 11.14. Elliptically polarized electric field.

11.4.2. Target Scattering Matrix

Target backscattered RCS is commonly described by a matrix known as the scattering matrix, and is denoted by $[S]$. When an arbitrarily linearly polarized wave is incident on a target, the backscattered field is then given by

$$\begin{bmatrix} E_1^s \\ E_2^s \end{bmatrix} = [S] \begin{bmatrix} E_1^i \\ E_2^i \end{bmatrix} = \begin{bmatrix} s_{11} & s_{12} \\ s_{21} & s_{22} \end{bmatrix} \begin{bmatrix} E_1^i \\ E_2^i \end{bmatrix} \qquad (11.19)$$

The superscripts i and s denote incident and scattered fields. The quantities s_{ij} are in general complex and the subscripts 1 and 2 represent any combination of orthogonal polarizations. More precisely, $1 = H, R$, and $2 = V, L$. From Eq. (11.3), the backscattered RCS is related to the scattering matrix components by the following relation:

$$\begin{bmatrix} \sigma_{11} & \sigma_{12} \\ \sigma_{21} & \sigma_{22} \end{bmatrix} = 4\pi R^2 \begin{bmatrix} |s_{11}|^2 & |s_{12}|^2 \\ |s_{21}|^2 & |s_{22}|^2 \end{bmatrix} \qquad (11.20)$$

It follows that once a scattering matrix is specified, the target backscattered RCS can be computed for any combination of transmitting and receiving polarizations. The reader is advised to see Ruck for ways to calculate the scattering matrix $[S]$.

Rewriting Eq. (11.20) in terms of the different possible orthogonal polarizations yields

$$\begin{bmatrix} E_H^s \\ E_V^s \end{bmatrix} = \begin{bmatrix} s_{HH} & s_{HV} \\ s_{VH} & s_{VV} \end{bmatrix} \begin{bmatrix} E_H^i \\ E_V^i \end{bmatrix} \quad (11.21)$$

$$\begin{bmatrix} E_R^s \\ E_L^s \end{bmatrix} = \begin{bmatrix} s_{RR} & s_{RL} \\ s_{LR} & s_{LL} \end{bmatrix} \begin{bmatrix} E_R^i \\ E_L^i \end{bmatrix} \quad (11.22)$$

By using the transformation matrix $[T]$ in Eq. (11.17), the circular scattering elements can be computed from the linear scattering elements

$$\begin{bmatrix} s_{RR} & s_{RL} \\ s_{LR} & s_{LL} \end{bmatrix} = [T] \begin{bmatrix} s_{HH} & s_{HV} \\ s_{VH} & s_{VV} \end{bmatrix} \begin{bmatrix} 1 & 0 \\ 0 & -1 \end{bmatrix} [T]^{-1} \quad (11.23)$$

and the individual components are

$$s_{RR} = \frac{-s_{VV} + s_{HH} - j(s_{HV} + s_{VH})}{2}$$

$$s_{RL} = \frac{s_{VV} + s_{HH} + j(s_{HV} - s_{VH})}{2}$$

$$s_{LR} = \frac{s_{VV} + s_{HH} - j(s_{HV} - s_{VH})}{2} \quad (11.24)$$

$$s_{LL} = \frac{-s_{VV} + s_{HH} + j(s_{HV} + s_{VH})}{2}$$

Similarly, the linear scattering elements are given by

$$\begin{bmatrix} s_{HH} & s_{HV} \\ s_{VH} & s_{VV} \end{bmatrix} = [T]^{-1} \begin{bmatrix} s_{RR} & s_{RL} \\ s_{LR} & s_{LL} \end{bmatrix} \begin{bmatrix} 1 & 0 \\ 0 & -1 \end{bmatrix} [T] \quad (11.25)$$

and the individual components are

$$s_{HH} = \frac{-s_{RR} + s_{RL} + s_{LR} - s_{LL}}{2}$$

$$s_{VH} = \frac{j(s_{RR} - s_{LR} + s_{RL} - s_{LL})}{2}$$

$$s_{HV} = \frac{-j(s_{RR} + s_{LR} - s_{RL} - s_{LL})}{2} \quad (11.26)$$

$$s_{VV} = \frac{s_{RR} + s_{LL} + js_{RL} + s_{LR}}{2}$$

11.5. RCS of Simple Objects

This section presents examples of backscattered radar cross section for a number of simple shape objects. In all cases, except for the perfectly conducting sphere, only optical region approximations are presented. Radar designers and RCS engineers consider the perfectly conducting sphere to be the simplest target to examine. Even in this case, the complexity of the exact solution, when compared to the optical region approximation, is overwhelming. Most formulas presented are Physical Optics (PO) approximation for the backscattered RCS measured by a far field radar in the direction (θ, φ), as illustrated in Fig. 11.15.

In this section, it is assumed that the radar is always illuminating an object from the positive z-direction.

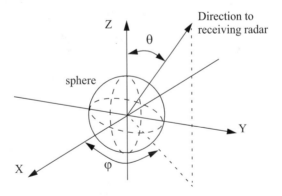

Figure 11.15. Direction of antenna receiving backscattered waves.

11.5.1. Sphere

Due to symmetry, waves scattered from a perfectly conducting sphere are co-polarized (have the same polarization) with the incident waves. This means that the cross-polarized backscattered waves are practically zero. For example, if the incident waves were Left Circularly Polarized (LCP), then the backscattered waves will also be LCP. However, because of the opposite direction of propagation of the backscattered waves, they are considered to be Right Circularly Polarized (RCP) by the receiving antenna. Therefore, the PP backscattered waves from a sphere are LCP, while the OP backscattered waves are negligible.

The normalized exact backscattered RCS for a perfectly conducting sphere is a Mie series given by

$$\frac{\sigma}{\pi r^2} = \left(\frac{j}{kr}\right) \sum_{n=1}^{\infty} (-1)^n (2n+1) \left[\left(\frac{kr J_{n-1}(kr) - n J_n(kr)}{kr H^{(1)}_{n-1}(kr) - n H^{(1)}_n(kr)} \right) - \left(\frac{J_n(kr)}{H^{(1)}_n(kr)} \right) \right] \quad (11.27)$$

where r is the radius of the sphere, $k = 2\pi/\lambda$, λ is the wavelength, J_n is the spherical Bessel of the first kind of order n, and $H^{(1)}_n$ is the Hankel function of order n, and is given by

$$H^{(1)}_n(kr) = J_n(kr) + j Y_n(kr) \quad (11.28)$$

Y_n is the spherical Bessel function of the second kind of order n. Plots of the normalized perfectly conducting sphere RCS as a function of its circumference in wavelength units are shown in Figs. 11.16a and 11.16b. These plots can be reproduced using the function *"rcs_sphere.m"* given in Listing 11.4 in Section 11.9.

In Fig. 11.16, three regions are identified. First is the optical region (corresponds to a large sphere). In this case,

$$\sigma = \pi r^2 \qquad r \gg \lambda \quad (11.29)$$

Second is the Rayleigh region (small sphere). In this case,

$$\sigma \approx 9\pi r^2 (kr)^4 \qquad r \ll \lambda \quad (11.30)$$

The region between the optical and Rayleigh regions is oscillatory in nature and is called the Mie or resonance region.

RCS of Simple Objects

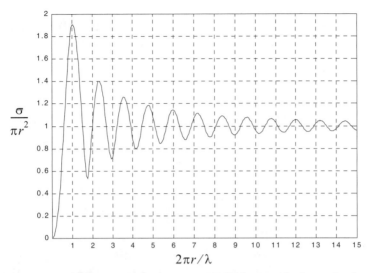

Figure 11.16a. Normalized backscattered RCS for a perfectly conducting sphere.

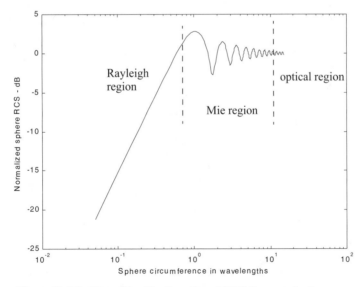

Figure 11.16b. Normalized backscattered RCS for a perfectly conducting sphere using semi-log scale.

The backscattered RCS for a perfectly conducting sphere is constant in the optical region. For this reason, radar designers typically use spheres of known cross sections to experimentally calibrate radar systems. For this purpose, spheres are flown attached to balloons. In order to obtain Doppler shift, spheres of known RCS are dropped out of an airplane and towed behind the airplane whose velocity is known to the radar.

11.5.2. Ellipsoid

An ellipsoid centered at (0,0,0) is shown in Fig. 11.17. It is defined by the following equation:

$$\left(\frac{x}{a}\right)^2 + \left(\frac{y}{b}\right)^2 + \left(\frac{z}{c}\right)^2 = 1 \tag{11.31}$$

One widely accepted approximation for the ellipsoid backscattered RCS is given by

$$\sigma = \frac{\pi a^2 b^2 c^2}{\left(a^2(\sin\theta)^2(\cos\varphi)^2 + b^2(\sin\theta)^2(\sin\varphi)^2 + c^2(\cos\theta)^2\right)^2} \tag{11.32}$$

When $a = b$, the ellipsoid becomes roll symmetric. Thus, the RCS is independent of φ, and Eq. (11.32) is reduced to

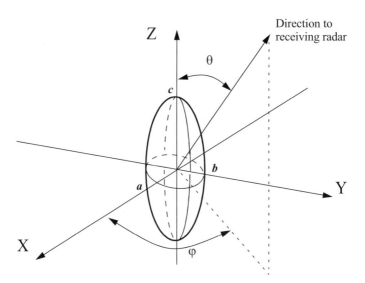

Figure 11.17. Ellipsoid.

$$\sigma = \frac{\pi b^4 c^2}{(a^2(\sin\theta)^2 + c^2(\cos\theta)^2)^2} \tag{11.33}$$

and for the case when $a = b = c$,

$$\sigma = \pi c^2 \tag{11.34}$$

Note that Eq. (11.34) defines the backscattered RCS of a sphere. This should be expected, since under the condition $a = b = c$ the ellipsoid becomes a sphere. Fig. 11.18a shows the backscattered RCS for an ellipsoid versus θ for $\varphi = 45°$. This plot can be generated using MATLAB program *"fig11_18a.m"* given in Listing 11.5 in Section 11.9. Note that at normal incidence ($\theta = 90°$) the RCS corresponds to that of a sphere of radius c, and is often referred to as the broadside specular RCS value.

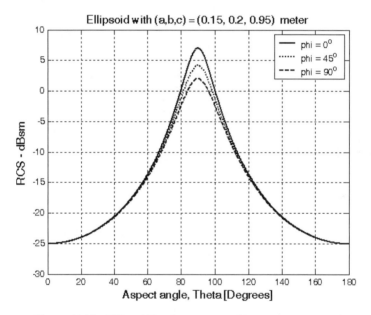

Figure 11.18a. Ellipsoid backscattered RCS versus aspect angle.

MATLAB Function *"rcs_ellipsoid.m"*

The function *"rcs_ellipsoid.m"* computes and plots the RCS of an ellipsoid versus aspect angle. It is given in Listing 11.6 in Section 11.9, and its syntax is as follows:

[rcs] = rcs_ellipsoid (a, b, c, phi)

where

Symbol	Description	Units	Status
a	ellipsoid a-radius	meters	input
b	ellipsoid b-radius	meters	input
c	ellipsoid c-radius	meters	input
phi	ellipsoid roll angle	degrees	input
rcs	array of RCS versus aspect angle	dBsm	output

Fig. 11.18b shows the GUI workspace associated with function. To execute this GUI type *"rcs_ellipsoid_gui"* from the MATLAB Command window.

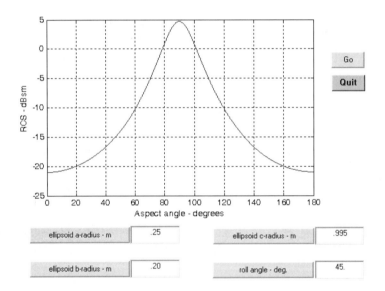

Figure 11.18b. GUI workspace associated with the function *"rcs_ellipsoid.m"*.

11.5.3. Circular Flat Plate

Fig. 11.19 shows a circular flat plate of radius r, centered at the origin. Due to the circular symmetry, the backscattered RCS of a circular flat plate has no dependency on φ. The RCS is only aspect angle dependent. For normal incidence (i.e., zero aspect angle) the backscattered RCS for a circular flat plate is

$$\sigma = \frac{4\pi^3 r^4}{\lambda^2} \qquad \theta = 0° \qquad (11.35)$$

For non-normal incidence, two approximations for the circular flat plate backscattered RCS for any linearly polarized incident wave are

$$\sigma = \frac{\lambda r}{8\pi \sin\theta(\tan(\theta))^2} \qquad (11.36)$$

$$\sigma = \pi k^2 r^4 \left(\frac{2J_1(2kr\sin\theta)}{2kr\sin\theta}\right)^2 (\cos\theta)^2 \qquad (11.37)$$

where $k = 2\pi/\lambda$, and $J_1(\beta)$ is the first order spherical Bessel function evaluated at β. The RCS corresponding to Eqs. (11.35) through (11.37) is shown in Fig. 11.20. These plots can be reproduced using MATLAB function *"rcs_circ_gui.m"*.

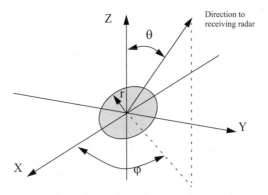

Figure 11.19. Circular flat plate.

MATLAB Function *"rcs_circ_plate.m"*

The function *"rcs_circ_plate.m"* calculates and plots the backscattered RCS from a circular plate. It is given in Listing 11.7 in Section 11.9; its syntax is as follows:

$$[rcs] = rcs\_circ\_plate\ (r, freq)$$

where

Symbol	Description	Units	Status
r	radius of circular plate	meters	input
freq	frequency	Hz	input
rcs	array of RCS versus aspect angle	dBsm	output

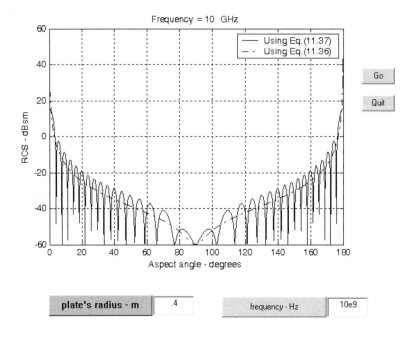

Figure 11.20. Backscattered RCS for a circular flat plate.

11.5.4. Truncated Cone (Frustum)

Figs. 11.21 and 11.22 show the geometry associated with a frustum. The half cone angle α is given by

$$\tan\alpha = \frac{(r_2 - r_1)}{H} = \frac{r_2}{L} \tag{11.38}$$

Define the aspect angle at normal incidence with respect to the frustum's surface (broadside) as θ_n. Thus, when a frustum is illuminated by a radar located at the same side as the cone's small end, the angle θ_n is

$$\theta_n = 90° - \alpha \tag{11.39}$$

Alternatively, normal incidence occurs at

$$\theta_n = 90° + \alpha \tag{11.40}$$

At normal incidence, one approximation for the backscattered RCS of a truncated cone due to a linearly polarized incident wave is

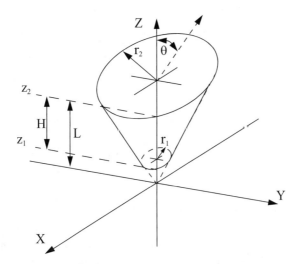

Figure 11.21. Truncated cone (frustum).

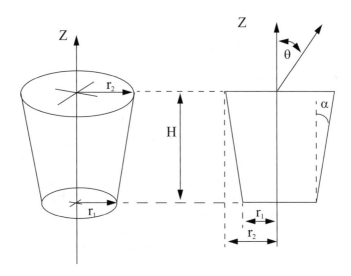

Figure 11.22. Definition of half cone angle.

$$\sigma_{\theta_n} = \frac{8\pi(z_2^{3/2} - z_1^{3/2})^2}{9\lambda \sin\theta_n} \tan\alpha (\sin\theta_n - \cos\theta_n \tan\alpha)^2 \quad (11.41)$$

where λ is the wavelength, and z_1, z_2 are defined in Fig. 11.21. Using trigonometric identities, Eq. (11.41) can be reduced to

$$\sigma_{\theta_n} = \frac{8\pi(z_2^{3/2} - z_1^{3/2})^2}{9\lambda} \frac{\sin\alpha}{(\cos\alpha)^4} \quad (11.42)$$

For non-normal incidence, the backscattered RCS due to a linearly polarized incident wave is

$$\sigma = \frac{\lambda z \tan\alpha}{8\pi \sin\theta} \left(\frac{\sin\theta - \cos\theta \tan\alpha}{\sin\theta \tan\alpha + \cos\theta} \right)^2 \quad (11.43)$$

where z is equal to either z_1 or z_2 depending on whether the RCS contribution is from the small or the large end of the cone. Again, using trigonometric identities Eq. (11.43) (assuming the radar illuminates the frustum starting from the large end) is reduced to

$$\sigma = \frac{\lambda z \tan\alpha}{8\pi \sin\theta} (\tan(\theta - \alpha))^2 \quad (11.44)$$

When the radar illuminates the frustum starting from the small end (i.e., the radar is in the negative z direction in Fig. 11.21), Eq. (11.44) should be modified to

$$\sigma = \frac{\lambda z \tan\alpha}{8\pi \sin\theta} (\tan(\theta + \alpha))^2 \quad (11.45)$$

For example, consider a frustum defined by $H = 20.945 cm$, $r_1 = 2.057 cm$, $r_2 = 5.753 cm$. It follows that the half cone angle is $10°$. Fig. 11.23a shows a plot of its RCS when illuminated by a radar in the positive z direction. Fig. 11.23b shows the same thing, except in this case, the radar is in the negative z direction. Note that for the first case, normal incidence occur at $100°$, while for the second case it occurs at $80°$. These plots can be reproduced using MATLAB function *"rcs_frustum_gui.m"* given in Listing 11.8 in Section 11.9.

MATLAB Function "rcs_frustum.m"

The function *"rcs_frustum.m"* computes and plots the backscattered RCS of a truncated conic section. The syntax is as follows:

[rcs] = rcs_frustum (r1, r2, freq, indicator)

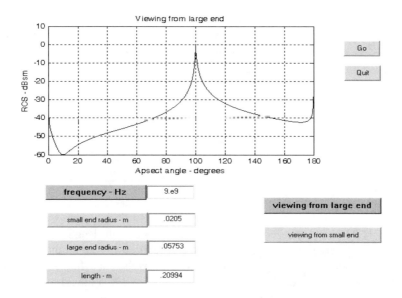

Figure 11.23a. Backscattered RCS for a frustum.

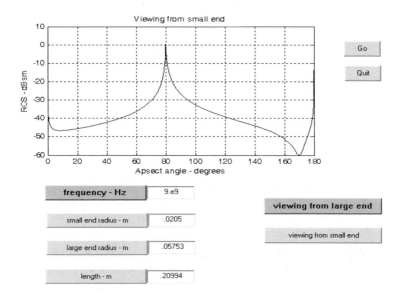

Figure 11.23b. Backscattered RCS for a frustum.

where

Symbol	Description	Units	Status
r1	small end radius	meters	input
r2	large end radius	meters	input
freq	frequency	Hz	input
indicator	indicator = 1 when viewing from large end	none	input
	indicator = 0 when viewing from small end		
rcs	array of RCS versus aspect angle	dBsm	output

11.5.5. Cylinder

Fig. 11.24 shows the geometry associated with a finite length conducting cylinder. Two cases are presented: first, the general case of an elliptical cross section cylinder; second, the case of a circular cross section cylinder. The normal and non-normal incidence backscattered RCS due to a linearly polarized incident wave from an elliptical cylinder with minor and major radii being r_1 and r_2 are, respectively, given by

$$\sigma_{\theta_n} = \frac{2\pi H^2 r_2^2 r_1^2}{\lambda [r_1^2(\cos\varphi)^2 + r_2^2(\sin\varphi)^2]^{1.5}} \quad (11.46)$$

$$\sigma = \frac{\lambda r_2^2 r_1^2 \sin\theta}{8\pi \ (\cos\theta)^2 [r_1^2(\cos\varphi)^2 + r_2^2(\sin\varphi)^2]^{1.5}} \quad (11.47)$$

For a circular cylinder of radius r, then due to roll symmetry, Eqs. (11.46) and (11.47), respectively, reduce to

$$\sigma_{\theta_n} = \frac{2\pi H^2 r}{\lambda} \quad (11.48)$$

$$\sigma = \frac{\lambda r \sin\theta}{8\pi(\cos\theta)^2} \quad (11.49)$$

Fig. 11.25a shows a plot of the cylinder backscattered RCS for a symmetrical cylinder. Fig. 11.25b shows the backscattered RCS for an elliptical cylinder. These plots can be reproduced using MATLAB function *"rcs_cylinder.m"* given in Listing 11.9 in Section 11.9.

RCS of Simple Objects

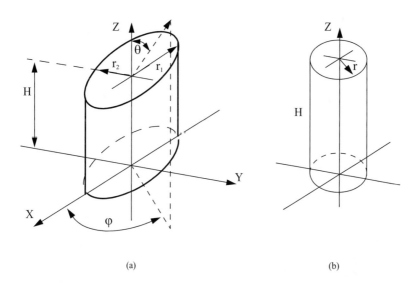

Figure 11.24. (a) Elliptical cylinder; (b) circular cylinder.

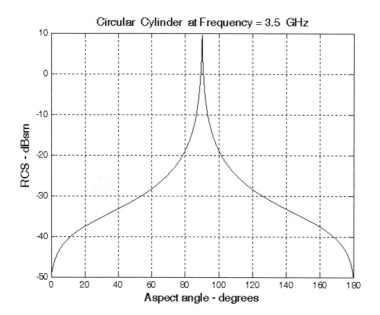

Figure 11.25a. Backscattered RCS for a symmetrical cylinder, $r = 0.125m$ and $H = 1m$.

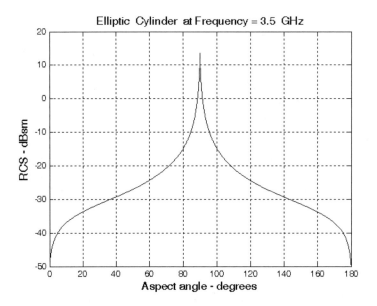

Figure 11.25b. Backscattered RCS for an elliptical cylinder, $r_1 = 0.125m$, $r_2 = 0.05m$, and $H = 1m$.

MATLAB Function "rcs_cylinder.m"

The function *"rcs_cylinder.m"* computes and plots the backscattered RCS of a cylinder. The syntax is as follows:

[rcs] = rcs_cylinder(r1, r2, h, freq, phi, CylinderType)

where

Symbol	Description	Units	Status
r1	radius r1	meters	input
r2	radius r2	meters	input
h	length of cylinder	meters	input
freq	frequency	Hz	input
phi	roll viewing angle	degrees	input
CylinderType	'Circular,' i.e., $r_1 = r_2$ 'Elliptic,' i.e., $r_1 \neq r_2$	none	input
rcs	array of RCS versus aspect angle	dBsm	output

11.5.6. Rectangular Flat Plate

Consider a perfectly conducting rectangular thin flat plate in the x-y plane as shown in Fig. 11.26. The two sides of the plate are denoted by $2a$ and $2b$. For a linearly polarized incident wave in the x-z plane, the horizontal and vertical backscattered RCS are, respectively, given by

$$\sigma_V = \frac{b^2}{\pi}\left|\sigma_{1V} - \sigma_{2V}\left[\frac{1}{\cos\theta} + \frac{\sigma_{2V}}{4}(\sigma_{3V} + \sigma_{4V})\right]\sigma_{5V}^{-1}\right|^2 \quad (11.50)$$

$$\sigma_H = \frac{b^2}{\pi}\left|\sigma_{1H} - \sigma_{2H}\left[\frac{1}{\cos\theta} - \frac{\sigma_{2H}}{4}(\sigma_{3H} + \sigma_{4H})\right]\sigma_{5H}^{-1}\right|^2 \quad (11.51)$$

where $k = 2\pi/\lambda$ and

$$\sigma_{1V} = \cos(ka\sin\theta) - j\frac{\sin(ka\sin\theta)}{\sin\theta} = (\sigma_{1H})^* \quad (11.52)$$

$$\sigma_{2V} = \frac{e^{j(ka - \pi/4)}}{\sqrt{2\pi}(ka)^{3/2}} \quad (11.53)$$

$$\sigma_{3V} = \frac{(1 + \sin\theta)e^{-jka\sin\theta}}{(1 - \sin\theta)^2} \quad (11.54)$$

$$\sigma_{4V} = \frac{(1 - \sin\theta)e^{jka\sin\theta}}{(1 + \sin\theta)^2} \quad (11.55)$$

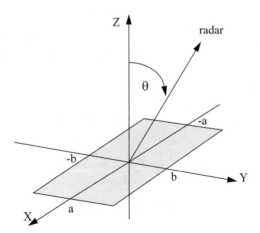

Figure 11.26. Rectangular flat plate.

$$\sigma_{5V} = 1 - \frac{e^{j(2ka - \pi/2)}}{8\pi(ka)^3} \qquad (11.56)$$

$$\sigma_{2H} = \frac{4e^{j(ka + \pi/4)}}{\sqrt{2\pi}(ka)^{1/2}} \qquad (11.57)$$

$$\sigma_{3H} = \frac{e^{-jka\sin\theta}}{1 - \sin\theta} \qquad (11.58)$$

$$\sigma_{4H} = \frac{e^{jka\sin\theta}}{1 + \sin\theta} \qquad (11.59)$$

$$\sigma_{5H} = 1 - \frac{e^{j(2ka + (\pi/2))}}{2\pi(ka)} \qquad (11.60)$$

Eqs. (11.50) and (11.51) are valid and quite accurate for aspect angles $0° \le \theta \le 80$. For aspect angles near $90°$, Ross[1] obtained by extensive fitting of measured data an empirical expression for the RCS. It is given by

$$\sigma_H \to 0$$

$$\sigma_V = \frac{ab^2}{\lambda}\left\{\left[1 + \frac{\pi}{2(2a/\lambda)^2}\right] + \left[1 - \frac{\pi}{2(2a/\lambda)^2}\right]\cos\left(2ka - \frac{3\pi}{5}\right)\right\} \qquad (11.61)$$

The backscattered RCS for a perfectly conducting thin rectangular plate for incident waves at any θ, φ can be approximated by

$$\sigma = \frac{4\pi a^2 b^2}{\lambda^2}\left(\frac{\sin(ak\sin\theta\cos\varphi)}{ak\sin\theta\cos\varphi} \frac{\sin(bk\sin\theta\sin\varphi)}{bk\sin\theta\sin\varphi}\right)^2 (\cos\theta)^2 \qquad (11.62)$$

Eq. (11.62) is independent of the polarization, and is only valid for aspect angles $\theta \le 20°$. Fig. 11.27 shows an example for the backscattered RCS of a rectangular flat plate, for both vertical (Fig. 11.27a) and horizontal (Fig. 11.27b) polarizations, using Eqs. (11.50), (11.51), and (11.62). In this example, $a = b = 10.16 cm$ and wavelength $\lambda = 3.33 cm$. This plot can be reproduced using MATLAB function *"rcs_rect_plate"* given in Listing 11.10.

MATLAB Function *"rcs_rect_plate.m"*

The function *"rcs_rect_plate.m"* calculates and plots the backscattered RCS of a rectangular flat plate. Its syntax is as follows:

$$[rcs] = rcs\_rect\_plate\ (a, b, freq)$$

---

1. Ross, R. A., Radar Cross Section of Rectangular Flat Plate as a Function of Aspect Angle, *IEEE Trans.*, AP-14,320, 1966.

RCS of Simple Objects 533

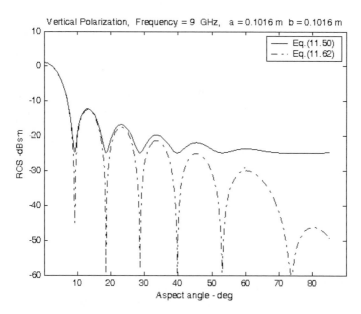

Figure 11.27a. Backscattered RCS for a rectangular flat plate.

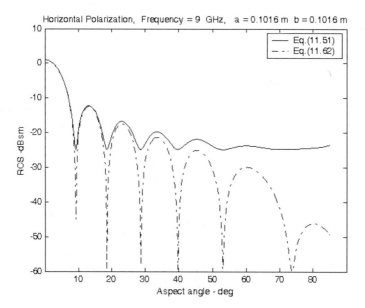

Figure 11.27b. Backscattered RCS for a rectangular flat plate.

where

Symbol	Description	Units	Status
a	short side of plate	meters	input
b	long side of plate	meters	input
freq	frequency	Hz	input
rcs	array of RCS versus aspect angle	dBsm	output

Fig. 11.27c shows the GUI workspace associated with this function.

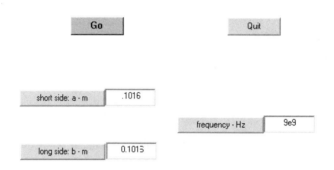

Figure 11.27c. GUI workspace associated with the function *"rcs_rect_plate.m"*.

11.5.7. Triangular Flat Plate

Consider the triangular flat plate defined by the isosceles triangle as oriented in Fig. 11.28. The backscattered RCS can be approximated for small aspect angles ($\theta \leq 30°$) by

$$\sigma = \frac{4\pi A^2}{\lambda^2}(\cos\theta)^2 \sigma_0 \qquad (11.63)$$

$$\sigma_0 = \frac{[(\sin\alpha)^2 - (\sin(\beta/2))^2]^2 + \sigma_{01}}{\alpha^2 - (\beta/2)^2} \qquad (11.64)$$

$$\sigma_{01} = 0.25(\sin\varphi)^2[(2a/b)\cos\varphi\sin\beta - \sin\varphi\sin 2\alpha]^2 \qquad (11.65)$$

where $\alpha = ka\sin\theta\cos\varphi$, $\beta = kb\sin\theta\sin\varphi$, and $A = ab/2$. For waves incident in the plane $\varphi = 0$, the RCS reduces to

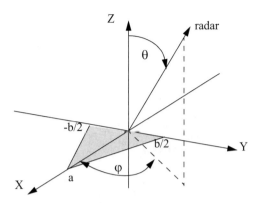

Figure 11.28. Coordinates for a perfectly conducting isosceles triangular plate.

$$\sigma = \frac{4\pi A^2}{\lambda^2}(\cos\theta)^2\left[\frac{(\sin\alpha)^4}{\alpha^4} + \frac{(\sin 2\alpha - 2\alpha)^2}{4\alpha^4}\right] \quad (11.66)$$

and for incidence in the plane $\varphi = \pi/2$

$$\sigma = \frac{4\pi A^2}{\lambda^2}(\cos\theta)^2\left[\frac{(\sin(\beta/2))^4}{(\beta/2)^4}\right] \quad (11.67)$$

Fig. 11.29 shows a plot for the normalized backscattered RCS from a perfectly conducting isosceles triangular flat plate. In this example $a = 0.2m$, $b = 0.75m$. This plot can be reproduced using MATLAB function "rcs_isosceles.m" given in Listing 11.11 in Section 11.9.

MATLAB Function "rcs_isosceles.m"

The function "rcs_isosceles.m" calculates and plots the backscattered RCS of a triangular flat plate. Its syntax is as follows:

$$[rcs] = rcs\_isosceles\ (a,\ b,\ freq,\ phi)$$

where

Symbol	Description	Units	Status
a	height of plate	meters	input
b	base of plate	meters	input
freq	frequency	Hz	input
phi	roll angle	degrees	input
rcs	array of RCS versus aspect angle	dBsm	output

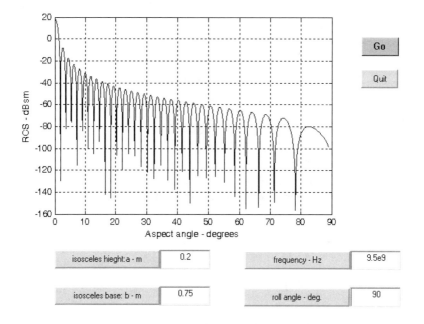

Figure 11.29. Backscattered RCS for a perfectly conducting triangular flat plate, $a = 20cm$ **and** $b = 75cm$.

11.6. Scattering From a Dielectric-Capped Wedge

The geometry of a dielectric-capped wedge is shown in Fig. 11.30. It is required to find to the field expressions for the problem of scattering by a 2-D perfect electric conducting (PEC) wedge capped with a dielectric cylinder. Using the cylindrical coordinates system, the excitation due to an electric line current of complex amplitude I_0 located at (ρ_0, φ_0) results in TM^z incident field with the electric field expression given by

$$E_z^i = -I_e \frac{\omega \mu_0}{4} H_0^{(2)}\left(k|\boldsymbol{\rho}-\boldsymbol{\rho}_0|\right) \tag{11.68}$$

The problem is divided into three regions, I, II, and III shown in Fig. 11.30. The field expressions may be assumed to take the following forms:

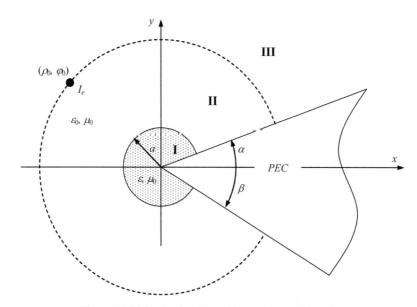

Figure 11.30. Scattering from dielectric-capped wedge.

$$E_z^I = \sum_{n=0}^{\infty} a_n J_\nu(k_1\rho) \sin\nu(\varphi-\alpha)\sin\nu(\varphi_0-\alpha)$$

$$E_z^{II} = \sum_{n=0}^{\infty} \left(b_n J_\nu(k\rho) + c_n H_\nu^{(2)}(k\rho)\right) \sin\nu(\varphi-\alpha)\sin\nu(\varphi_0-\alpha) \quad \text{(11.69)}$$

$$E_z^{III} = \sum_{n=0}^{\infty} d_n H_\nu^{(2)}(k\rho) \sin\nu(\varphi-\alpha)\sin\nu(\varphi_0-\alpha)$$

where

$$\nu = \frac{n\pi}{2\pi - \alpha - \beta} \quad \text{(11.70)}$$

while $J_\nu(x)$ is the Bessel function of order ν and argument x and $H_\nu^{(2)}$ is the Hankel function of the second kind of order ν and argument x. From Maxwell's equations, the magnetic field component H_φ is related to the electric field component E_z for a TM^z wave by

$$H_\varphi = \frac{1}{j\omega\mu} \frac{\partial E_z}{\partial \rho} \quad \text{(11.71)}$$

Thus, the magnetic field component H_φ in the various regions may be written as

$$H_\varphi^I = \frac{k_1}{j\omega\mu_0} \sum_{n=0}^{\infty} a_n J_v'(k_1\rho) \sin v(\varphi-\alpha) \sin v(\varphi_0-\alpha)$$

$$H_\varphi^{II} = \frac{k}{j\omega\mu_0} \sum_{n=0}^{\infty} \left(b_n J_v'(k\rho) + c_n H_v^{(2)\prime}(k\rho)\right) \sin v(\varphi-\alpha) \sin v(\varphi_0-\alpha) \quad (11.72)$$

$$H_\varphi^{III} = \frac{k}{j\omega\mu_0} \sum_{n=0}^{\infty} d_n H_v^{(2)\prime}(k\rho) \sin v(\varphi-\alpha) \sin v(\varphi_0-\alpha)$$

Where the prime indicated derivatives with respect to the full argument of the function. The boundary conditions require that the tangential electric field components vanish at the PEC surface. Also, the tangential field components should be continuous across the air-dielectric interface and the virtual boundary between region II and III, except for the discontinuity of the magnetic field at the source point. Thus,

$$E_z = 0 \text{ at } \quad \varphi = \alpha, 2\pi - \beta \quad (11.73)$$

$$\left.\begin{array}{l} E_z^I = E_z^{II} \\ H_\varphi^I = H_\varphi^{II} \end{array}\right\} \text{ at } \quad \rho = a \quad (11.74)$$

$$\left.\begin{array}{l} E_z^{II} = E_z^{III} \\ H_\varphi^{II} - H_\varphi^{III} = -J_e \end{array}\right\} \text{ at } \quad \rho = \rho_0 \quad (11.75)$$

The current density J_e may be given in Fourier series expansion as

$$J_e = \frac{I_e}{\rho_0} \delta(\varphi-\varphi_0) = \frac{2}{2\pi-\alpha-\beta} \frac{I_e}{\rho_0} \sum_{n=0}^{\infty} \sin v(\varphi-\alpha) \sin v(\varphi_0-\alpha) \quad (11.76)$$

The boundary condition on the PEC surface is automatically satisfied by the φ dependence of the electric field Eq. (11.72). From the boundary conditions in Eq. (11.73)

$$\sum_{n=0}^{\infty} a_n J_v(k_1 a) \sin v(\varphi-\alpha) \sin v(\varphi_0-\alpha) = \\ \sum_{n=0}^{\infty} \left(b_n J_v(ka) + c_n H_v^{(2)}(ka)\right) \sin v(\varphi-\alpha) \sin v(\varphi_0-\alpha) \quad (11.77)$$

$$\frac{k_1}{j\omega\mu_0}\sum_{n=0}^{\infty}a_n J_v'(k_1 a)\sin v(\varphi-\alpha)\sin v(\varphi_0-\alpha) =$$
$$\frac{k}{j\omega\mu_0}\sum_{n=0}^{\infty}\left(b_n J_v'(ka) + c_n H_v^{(2)'}(ka)\right)\sin v(\varphi-\alpha)\sin v(\varphi_0-\alpha) \quad (11.78)$$

From the boundary conditions in Eq. (11.75), we have

$$\sum_{n=0}^{\infty}\left(b_n J_v(k\rho_0) + c_n H_v^{(2)}(k\rho_0)\right)\sin v(\varphi-\alpha)\sin v(\varphi_0-\alpha) -$$
$$\sum_{n=0}^{\infty}d_n H_v^{(2)}(k\rho_0)\sin v(\varphi-\alpha)\sin v(\varphi_0-\alpha) \quad (11.79)$$

$$\frac{k}{j\omega\mu_0}\sum_{n=0}^{\infty}\left(b_n J_v'(k\rho_0) + c_n H_v^{(2)'}(k\rho_0)\right)\sin v(\varphi-\alpha)\sin v(\varphi_0-\alpha) =$$
$$\frac{k}{j\omega\mu_0}\sum_{n=0}^{\infty}d_n H_v^{(2)'}(k\rho_0)\sin v(\varphi-\alpha)\sin v(\varphi_0-\alpha) \quad (11.80)$$
$$-\frac{2}{2\pi-\alpha-\beta}\frac{I_e}{\rho_0}\sum_{n=0}^{\infty}\sin v(\varphi-\alpha)\sin v(\varphi_0-\alpha)$$

Since Eqs. (11.77) and (11.80) hold for all φ, the series on the left and right hand sides should be equal term by term. More precisely,

$$a_n J_v(k_1 a) = b_n J_v(ka) + c_n H_v^{(2)}(ka) \quad (11.81)$$

$$\frac{k_1}{\mu_0}a_n J_v'(k_1 a) = \frac{k}{\mu_0}\left(b_n J_v'(ka) + c_n H_v^{(2)'}(ka)\right) \quad (11.82)$$

$$b_n J_v(k\rho_0) + c_n H_v^{(2)}(k\rho_0) = d_n H_v^{(2)}(k\rho_0) \quad (11.83)$$

$$b_n J_v'(k\rho_0) + c_n H_v^{(2)'}(k\rho_0) = d_n H_v^{(2)'}(k\rho_0) - \frac{2j\eta_0}{2\pi-\alpha-\beta}\frac{I_e}{\rho_0} \quad (11.84)$$

From Eqs. (11.81) and (11.83), we have

$$a_n = \frac{1}{J_v(k_1 a)}\left[b_n J_v(ka) + c_n H_v^{(2)}(ka)\right] \quad (11.85)$$

$$d_n = c_n + b_n \frac{J_v(k\rho_0)}{H_v^{(2)}(k\rho_0)} \quad (11.86)$$

Multiplying Eq. (11.83) by $H_v^{(2)'}$ and Eq. (11.84) by $H_v^{(2)}$, and by subtraction and using the Wronskian of the Bessel and Hankel functions, we get

$$b_n = -\frac{\pi\omega\mu_0 I_e}{2\pi - \alpha - \beta} H_v^{(2)}(k\rho_0) \tag{11.87}$$

Substituting b_n in Eqs. (11.81) and (11.82) and solving for c_n yield

$$c_n = \frac{\pi\omega\mu_0 I_e}{2\pi - \alpha - \beta}\left[H_v^{(2)}(k\rho_0)\frac{kJ_v'(ka)J_v(k_1a) - k_1J_v(ka)J_v'(k_1a)}{kH_v^{(2)'}(ka)J_v(k_1a) - k_1H_v^{(2)}(ka)J_v'(k_1a)}\right] \tag{11.88}$$

From Eqs. (11.86) through (11.88), d_n may be given by

$$d_n = \frac{\pi\omega\mu_0 I_e}{2\pi - \alpha - \beta}\left[H_v^{(2)}(k\mathfrak{I}_0)\frac{kJ_v'(ka)J_v(k_1a) - k_1J_v(ka)J_v'(k_1a)}{kH_v^{(2)'}(ka)J_v(k_1a) - k_1H_v^{(2)}(ka)J_v'(k_1a)} - J_v(k\rho_0)\right] \tag{11.89}$$

which can be written as

$$d_n = \frac{\pi\omega\mu_0 I_e}{2\pi - \alpha - \beta}\left\{\frac{\begin{aligned}&kJ_v(k_1a)\left[J_v'(ka)H_v^{(2)}(k\rho_0) - H_v^{(2)'}(ka)J_v(k\rho_0)\right] + K\\&k_1J_v'(k_1a)\left[H_v^{(2)}(ka)J_v(k\rho_0) - J_v(ka)H_v^{(2)}(k\rho_0)\right]\end{aligned}}{kH_v^{(2)'}(ka)J_v(k_1a) - k_1H_v^{(2)}(ka)J_v'(k_1a)}\right\} \tag{11.90}$$

Substituting for the Hankel function in terms of Bessel and Neumann functions, Eq. (11.90) reduces to

$$d_n = -j\frac{\pi\omega\mu_0 I_e}{2\pi - \alpha - \beta}\left\{\frac{\begin{aligned}&kJ_v(k_1a)\left[J_v'(ka)Y_v(k\rho_0) - Y_v'(ka)J_v(k\rho_0)\right] + K\\&k_1J_v'(k_1a)\left[Y_v(ka)J_v(k\rho_0) - J_v(ka)Y_v(k\rho_0)\right]\end{aligned}}{kH_v^{(2)'}(ka)J_v(k_1a) - k_1H_v^{(2)}(ka)J_v'(k_1a)}\right\} \tag{11.91}$$

With these closed form expressions for the expansion coeffiecients a_n, b_n, c_n and d_n, the field components E_z and H_φ can be determined from Eq. (11.69) and Eq. (11.72), respectively. Alternatively, the magnetic field component H_ρ can be computed from

$$H_\rho = -\frac{1}{j\omega\mu}\frac{1}{\rho}\frac{\partial E_z}{\partial \phi} \tag{11.92}$$

Thus, the H_ρ expressions for the three regions defined in Fig. 11.30 become

$$H_\rho^I = -\frac{1}{j\omega\mu\rho}\sum_{n=0}^{\infty} a_n v J_v(k_1\rho)\cos v(\varphi-\alpha)\sin v(\varphi_0-\alpha)$$

$$H_\rho^{II} = -\frac{1}{j\omega\mu\rho}\sum_{n=0}^{\infty} v\left(b_n J_v(k\rho) + c_n H_v^{(2)}(k\rho)\right)\cos v(\varphi-\alpha)\sin v(\varphi_0-\alpha) \quad (11.93)$$

$$H_\rho^{III} = -\frac{1}{j\omega\mu\rho}\sum_{n=0}^{\infty} d_n v H_v^{(2)}(k\rho)\cos v(\varphi-\alpha)\sin v(\varphi_0-\alpha)$$

11.6.1. Far Scattered Field

In region III, the scattered field may be found as the difference between the total and incident fields. Thus, using Eqs. (11.68) and (11.69) and considering the far field condition ($\rho \to \infty$) we get

$$E_z^{III} = E_z^i + E_z^s = \sqrt{\frac{2j}{\pi k\rho}}e^{-jk\rho}\sum_{n=0}^{\infty} d_n j^v \sin v(\varphi-\alpha)\sin v(\varphi_0-\alpha)$$

$$E_z^i = -I_e\frac{\omega\mu_0}{4}\sqrt{\frac{2j}{\pi k\rho}}e H_\rho = -\frac{1}{j\omega\mu}\frac{1}{\rho}\frac{\partial E_z}{\partial \phi} \quad (11.94)$$

Note that d_n can be written as

$$d_n = -\frac{\omega\mu_0 I_e}{4}\tilde{d}_n \quad (11.95)$$

where

$$\tilde{d}_n = j\frac{4\pi}{2\pi-\alpha-\beta}\left\{\frac{\begin{array}{c}kJ_v(k_1 a)\left[J_v'(ka)Y_v(k\rho_0)-Y_v'(ka)J_v(k\rho_0)\right]+\\ k_1 J_v'(k_1 a)\left[Y_v(ka)J_v(k\rho_0)-J_v(ka)Y_v(k\rho_0)\right]\end{array}}{kH_v^{(2)'}(ka)J_v(k_1 a)-k_1 H_v^{(2)}(ka)J_v'(k_1 a)}\right\} \quad (11.96)$$

Substituting Eq. (11.95) into Eq. (11.94), the scattered field $f(\varphi)$ is

$$E_z^s = \frac{-\omega\mu_0 I_e}{4}\sqrt{\frac{2j}{\pi k\rho}}e^{-jk\rho} \qquad (11.97)$$

$$\left(\sum_{n=0}^{\infty} d_n j^\nu \sin\nu(\varphi-\alpha)\sin\nu(\varphi_0-\alpha) - e^{jk\rho_0\cos(\varphi-\varphi_0)}\right)$$

11.6.2. Plane Wave Excitation

For plane wave excitation ($\rho_0 \to \infty$), the expression in Eqs. (11.87) and (11.88) reduce to

$$b_n = -\frac{\pi\omega\mu_0 I_e}{2\pi-\alpha-\beta}j^\nu\sqrt{\frac{2j}{\pi k\rho_0}}e^{-jk\rho_0}$$

$$c_n = \frac{\pi\omega\mu_0 I_e}{2\pi-\alpha-\beta}j^\nu\sqrt{\frac{2j}{\pi k\rho_0}}e^{-jk\rho_0}\frac{kJ_\nu'(ka)J_\nu(k_1a)-k_1J_\nu(ka)J_\nu'(k_1a)}{kH_\nu^{(2)'}(ka)J_\nu(k_1a)-k_1H_\nu^{(2)}(ka)J_\nu'(k_1a)} \qquad (11.98)$$

where the complex amplitude of the incident plane wave, E_0, can be given by

$$E_0 = -I_e\frac{\omega\mu_0}{4}\sqrt{\frac{2j}{\pi k\rho_0}}e^{-jk\rho_0} \qquad (11.99)$$

In this case, the field components can be evaluated in regions I and II only.

11.6.3. Special Cases

Case I: $\alpha = \beta$ (reference at bisector); The definition of ν reduces to

$$\nu = \frac{n\pi}{2(\pi-\beta)} \qquad (11.100)$$

and the same expression will hold for the coefficients (with $\alpha = \beta$).

Case II: $\alpha = 0$ (reference at face); the definition of ν takes on the form

$$\nu = \frac{n\pi}{2\pi-\beta} \qquad (11.101)$$

and the same expression will hold for the coefficients (with $\alpha = 0$).

Case III: $k_1 \to \infty$ (PEC cap); Fields at region I will vanish, and the coefficients will be given by

$$b_n = -\frac{\pi\omega\mu_0 I_e}{2\pi - \alpha - \beta} H_v^{(2)}(k\rho_0)$$

$$c_n = \frac{\pi\omega\mu_0 I_e}{2\pi - \alpha - \beta} H_v^{(2)}(k\rho_0) \frac{J_v(ka)}{H_v^{(2)}(ka)}$$

$$d_n = j\frac{\pi\omega\mu_0 I_e}{2\pi - \alpha - \beta} \frac{Y_v(ka) J_v(k\rho_0) - J_v(ka) Y_v(k\rho_0)}{H_v^{(2)}(ka)} \quad (11.102)$$

$$a_n = \frac{1}{J_v(k_1 a)} \left[b_n J_v(ka) + c_n H_v^{(2)}(ka) \right] = 0$$

Note that the expressions of b_n and c_n will yield zero tangential electric field at $\rho = a$ when substituted in Eq.(11.69).

Case IV: $a \to 0$ (no cap); The expressions of the coefficients in this case may be obtained by setting $k_1 = k$, or by taking the limit as a approaches zero. Thus,

$$c_n = \frac{\pi\omega\mu_0 I_e}{2\pi - \alpha - \beta} \left[H_v^{(2)}(k\rho_0) \frac{kJ_v'(ka) J_v(ka) - kJ_v(ka) J_v'(ka)}{kH_v^{(2)'}(ka) J_v(ka) - kH_v^{(2)}(ka) J_v'(ka)} \right] = 0$$

$$b_n = -\frac{\pi\omega\mu_0 I_e}{2\pi - \alpha - \beta} H_v^{(2)}(k\rho_0)$$

$$a_n = \frac{1}{J_v(ka)} \left[b_n J_v(ka) + c_n H_v^{(2)}(ka) \right] = b_n$$

$$d_n = \frac{\pi\omega\mu_0 I_e}{2\pi - \alpha - \beta} \left\{ \frac{kJ_v(k_1 a)\left[J_v'(ka) H_v^{(2)}(k\rho_0) - H_v^{(2)'}(ka) J_v(k\rho_0) \right] + K}{kH_v^{(2)'}(ka) J_v(k_1 a) - k_1 H_v^{(2)}(ka) J_v'(k_1 a)} \right\} \quad (11.103)$$

$$= -\frac{\pi\omega\mu_0 I_e}{2\pi - \alpha - \beta} J_v(k\rho_0)$$

Case V: $a \to 0$ and $\alpha = \beta = 0$ (semi-infinite PEC plane); In this case, the coefficients in Eq. (11.103) become valid with the exception that the values of v reduce to $n/2$. Once, the electric field component E_z in the different regions is computed, the corresponding magnetic field component H_φ can be computed using Eq. (11.71) and the magnetic field component H_ρ may be computed as

$$H_\rho = -\frac{1}{j\omega\mu} \frac{1}{\rho} \frac{\partial E_z}{\partial \phi} \quad (11.104)$$

MATLAB Program "Capped_WedgeTM.m"

The MATLAB program *"Capped_WedgeTM.m"* given in listing 11.12, along with the following associated functions *"DielCappedWedgeTMFields_Ls.m"*, *"DielCappedWedgeTMFields_PW"*, *"polardb.m"*, *"dbesselj.m"*, *"dbesselh.m"*, and *"dbessely.m"* given in the following listings, calculates and plots the far field of a capped wedge in the presence of an electric line source field. The near field distribution is also computed for both line source or plane wave excitation. All near field components are computed and displayed, in separate windows, using 3-D output format. The program is also capable of analyzing the field variations due to the cap parameters. The user can execute this MATLAB program from the MATLAB command window and manually change the input parameters in the designated section in the program in order to perform the desired analysis. Alternatively, the *"Capped_Wedge_GUI.m"* function along with the *"Capped_Wedge_GUI.fig"* file can be used to simplify the data entry procedure.

A sample of the data entry screen of the *"Capped_Wedge_GUI"* program is shown in Fig. 11.31 for the case of a line source exciting a sharp conducting wedge. The corresponding far field pattern is shown in Fig. 11.32. When keeping all the parameters in Fig. 11.31 the same except that selecting a dielectric or conducting cap, one obtains the far field patterns in Figs. 11.33 and 11.34, respectively. It is clear from these figures how the cap parameters affect the direction of the maximum radiation of the line source in the presence of the wedge. The distribution of the components of the fields in the near field for these three cases (sharp edge, dielectric capped edge, and conducting capped edge) is computed and shown in Figs. 11.35 to 11.43. The near field distribution for an incident plane wave field on these three types of wedges is also computed and shown in Figs. 11.44 to 11.52. These near field distributions clearly demonstrated the effect or cap parameters in altering the sharp edge singular behavior. To further illustrate this effect, the following set of figures (Figs. (11.53) to (11.55)) presents the near field of the electric component of plane wave incident on a half plane with a sharp edge, dielectric capped edge, and conducting capped edge.

The user is encouraged to experiment with this program as there are many parameters that can be altered to change the near and far field characteristic due to the scattering from a wedge structure.

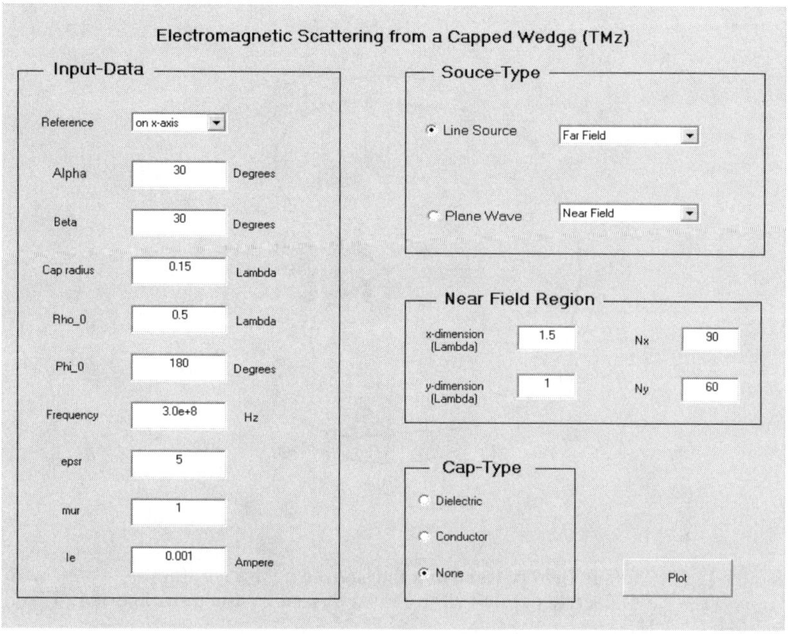

Scattering From a Dielectric-Capped Wedge

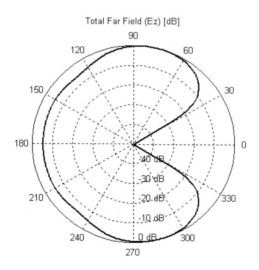

Figure 11.31. The parameters for computing the far field pattern of a 60 degrees wedge excited by a line source

Figure 11.32. The far field pattern of a line source near a conducting wedge with sharp edge characterized by the parameters in Fig. 11.31.

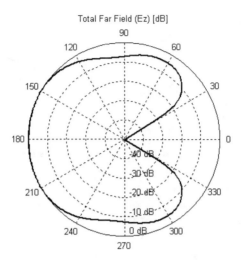

Figure 11.33. The far field pattern of a line source near a conducting wedge with a dielectric capped edge characterized by the parameters in Fig. 11.31.

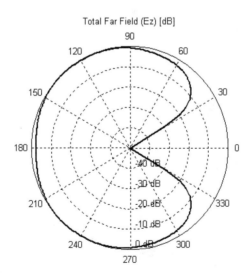

Figure 11.34. The far field pattern of a line source near a conducting wedge with a conducting capped edge characterized by the parameters in Fig. 11.31.

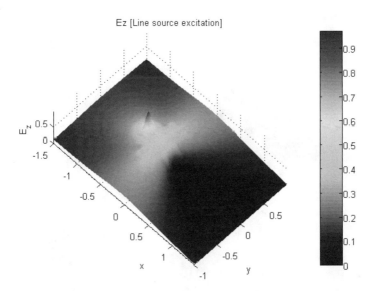

Figure 11.35. The E_z **near field pattern of a line source near a conducting wedge with a sharp edge characterized by the parameters in Fig. 11.31.**

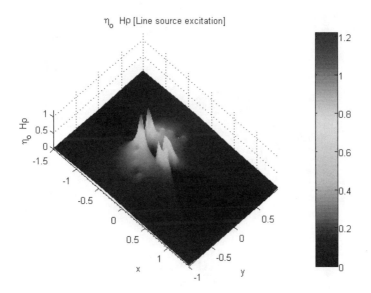

Figure 11.36. The H_ρ **near field pattern of a line source near a conducting wedge with a sharp edge characterized by the parameters in Fig. 11.31.**

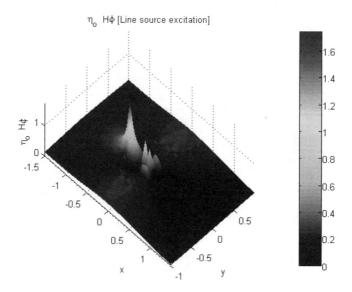

Figure 11.37. The H_φ **near field pattern of a line source near a conducting wedge with a sharp edge characterized by the parameters in Fig. 11.31.**

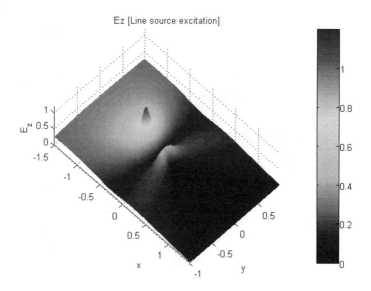

Figure 11.38. The E_z **near field pattern of a line source near a conducting wedge with a dielectric cap edge characterized by Fig. 11.31.**

Scattering From a Dielectric-Capped Wedge 549

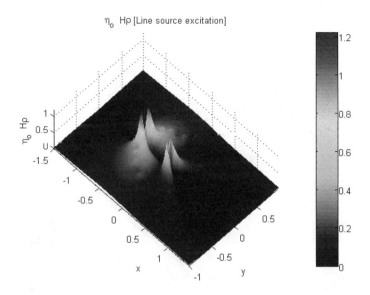

Figure 11.39. The H_ρ near field pattern of a line source near a conducting wedge with a dielectric cap edge characterized by Fig. 11.31.

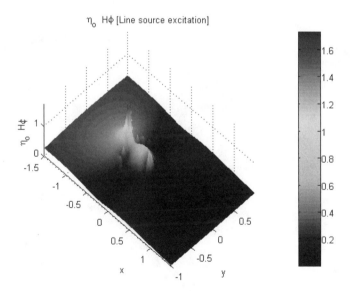

Figure 11.40. The H_ϕ near field pattern of a line source near a conducting wedge with a dielectric cap edge characterized by Fig. 11.31.

550   *MATLAB Simulations for Radar Systems Design*

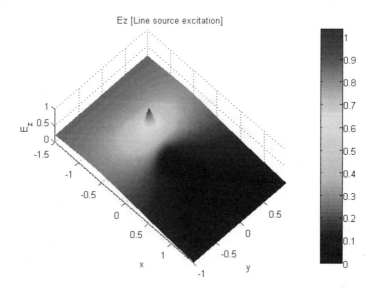

Figure 11.41. The E_z near field pattern of a line source near a conducting wedge with a conducting capped edge characterized by Fig. 11.31.

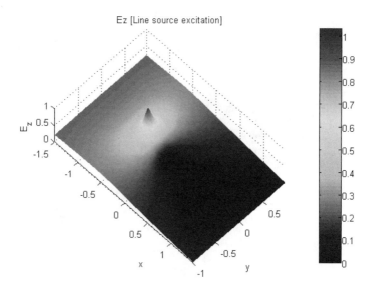

Figure 11.42. The H_ρ near field pattern of a line source near a conducting wedge with a conducting capped edge characterized by Fig. 11.31.

Scattering From a Dielectric-Capped Wedge 551

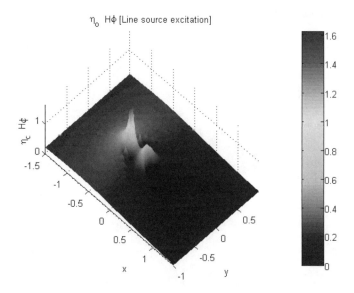

Figure 11.43. The H_φ near field pattern of a line source near a conducting wedge with a conducting capped edge characterized by Fig. 11.31.

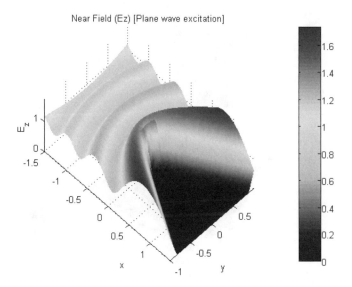

Figure 11.44. The E_z near field pattern of a plane wave incident on a conducting wedge with a sharp edge characterized by Fig. 11.31.

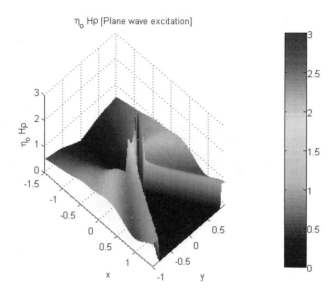

Figure 11.45. The H_ρ near field pattern of a plane wave incident on a conducting wedge with a sharp edge characterized by Fig. 11.31.

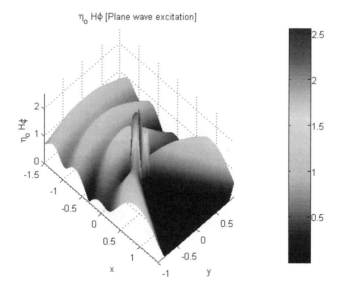

Figure 11.46. The H_φ near field pattern of a plane wave incident on a conducting wedge with a sharp edge characterized by Fig. 11.31.

Scattering From a Dielectric-Capped Wedge 553

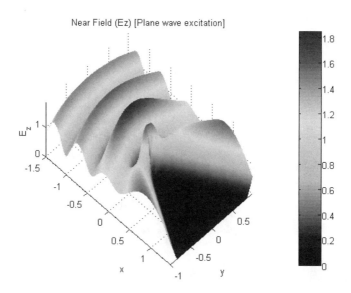

Figure 11.47. The E_z near field pattern of a plane wave incident on a conducting wedge with a dielectric edge characterized by Fig. 11.31.

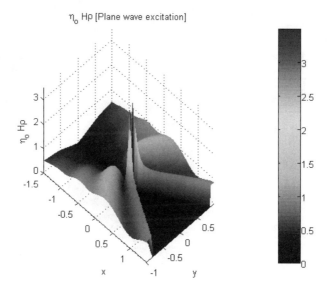

Figure 11.48. The H_ρ near field pattern of a plane wave incident on a conducting wedge with a dielectric edge characterized by Fig. 11.31.

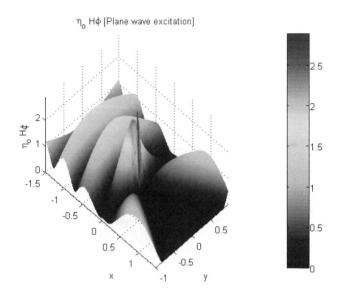

Figure 11.49. The H_φ **near field pattern of a plane wave incident on a conducting wedge with dielectric capped edge characterized by Fig. 11.31.**

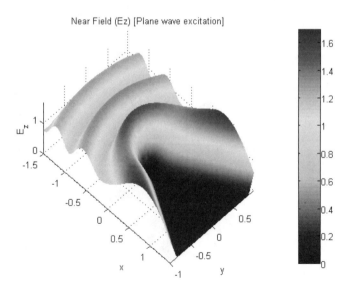

Figure 11.50. The E_z **near field pattern of a plane wave incident on a conducting wedge with a conducting capped edge characterized by Fig. 11.31.**

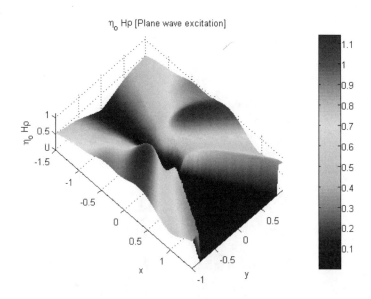

Figure 11.51. The H_ρ near field pattern of a plane wave incident on a conducting wedge with a conducting capped edge characterized by Fig. 11.31.

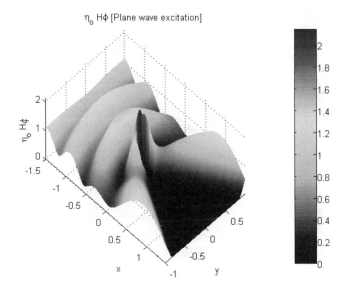

Figure 11.52. The H_φ near field pattern of a plane wave incident on a conducting wedge with a conducting capped edge characterized by Fig. 11.31.

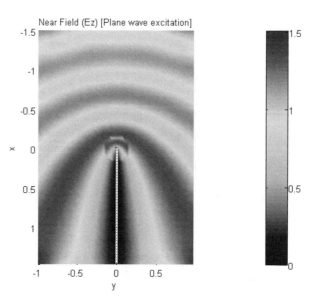

Figure 11.53. The E_z near field pattern of a plane wave incident on a half plane with sharp edge. All other parameters are as in Fig. 11.31.

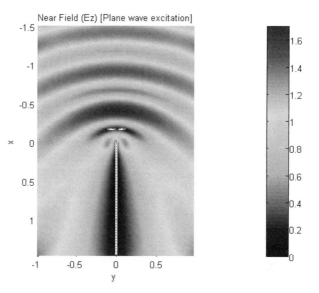

Figure 11.54. E_z near field pattern of a plane wave incident on a half plane with a dielectric capped edge. All other parameters are as in Fig. 11.31.

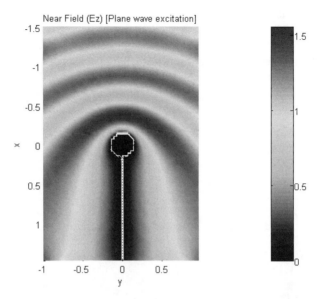

Figure 11.55. E_z near field pattern of a plane wave incident on a half plane with a conducting capped edge. All other parameters are as in Fig. 11.31.

11.7. RCS of Complex Objects

A complex target RCS is normally computed by coherently combining the cross sections of the simple shapes that make that target. In general, a complex target RCS can be modeled as a group of individual scattering centers distributed over the target. The scattering centers can be modeled as isotropic point scatterers (N-point model) or as simple shape scatterers (N-shape model). In any case, knowledge of the scattering centers' locations and strengths is critical in determining complex target RCS. This is true, because as seen in Section 11.3, relative spacing and aspect angles of the individual scattering centers drastically influence the overall target RCS. Complex targets that can be modeled by many equal scattering centers are often called Swerling 1 or 2 targets. Alternatively, targets that have one dominant scattering center and many other smaller scattering centers are known as Swerling 3 or 4 targets.

In NB radar applications, contributions from all scattering centers combine coherently to produce a single value for the target RCS at every aspect angle. However, in WB applications, a target may straddle many range bins. For each range bin, the average RCS extracted by the radar represents the contributions from all scattering centers that fall within that bin.

As an example, consider a circular cylinder with two perfectly conducting circular flat plates on both ends. Assume linear polarization and let $H = 1m$ and $r = 0.125m$. The backscattered RCS for this object versus aspect angle is shown in Fig. 11.56. Note that at aspect angles close to $0°$ and $180°$ the RCS is mainly dominated by the circular plate, while at aspect angles close to normal incidence, the RCS is dominated by the cylinder broadside specular return. The reader can reproduced this plot using the MATLAB program *"rcs_cyliner_complex.m"* given in Listing 11.19 in Section 11.9.

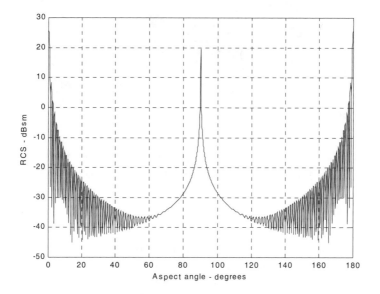

Figure 11.56. Backscattered RCS for a cylinder with flat plates.

11.8. RCS Fluctuations and Statistical Models

In most practical radar systems there is relative motion between the radar and an observed target. Therefore, the RCS measured by the radar fluctuates over a period of time as a function of frequency and the target aspect angle. This observed RCS is referred to as the radar dynamic cross section. Up to this point, all RCS formulas discussed in this chapter assumed a stationary target, where in this case, the backscattered RCS is often called static RCS.

Dynamic RCS may fluctuate in amplitude and/or in phase. Phase fluctuation is called glint, while amplitude fluctuation is called scintillation. Glint causes the far field backscattered wavefronts from a target to be non-planar. For most radar applications, glint introduces linear errors in the radar measurements, and thus it is not of a major concern. However, in cases where high precision and accuracy are required, glint can be detrimental. Examples include precision instrumentation tracking radar systems, missile seekers, and automated aircraft landing systems. For more details on glint, the reader is advised to visit cited references listed in the bibliography.

Radar cross-section scintillation can vary slowly or rapidly depending on the target size, shape, dynamics, and its relative motion with respect to the radar. Thus, due to the wide variety of RCS scintillation sources, changes in the radar cross section are modeled statistically as random processes. The value of an RCS random process at any given time defines a random variable at that time. Many of the RCS scintillation models were developed and verified by experimental measurements.

11.8.1. RCS Statistical Models - Scintillation Models

This section presents the most commonly used RCS statistical models. Statistical models that apply to sea, land, and volume clutter, such as the Weibull and Log-normal distributions, will be discussed in a later chapter. The choice of a particular model depends heavily on the nature of the target under examination.

Chi-Square of Degree $2m$

The Chi-square distribution applies to a wide range of targets; its *pdf* is given by

$$f(\sigma) = \frac{m}{\Gamma(m)\sigma_{av}} \left(\frac{m\sigma}{\sigma_{av}}\right)^{m-1} e^{-m\sigma/\sigma_{av}} \qquad \sigma \geq 0 \qquad (11.105)$$

where $\Gamma(m)$ is the gamma function with argument m, and σ_{av} is the average value. As the degree gets larger the distribution corresponds to constrained RCS values (narrow range of values). The limit $m \to \infty$ corresponds to a constant RCS target (steady-target case).

Swerling I and II (Chi-Square of Degree 2)

In Swerling I, the RCS samples measured by the radar are correlated throughout an entire scan, but are uncorrelated from scan to scan (slow fluctuation). In this case, the *pdf* is

$$f(\sigma) = \frac{1}{\sigma_{av}} \exp\left(-\frac{\sigma}{\sigma_{av}}\right) \qquad \sigma \geq 0 \qquad (11.106)$$

where σ_{av} denotes the average RCS overall target fluctuation. Swerling II target fluctuation is more rapid than Swerling I, but the measurements are pulse to pulse uncorrelated. Swerlings I and II apply to targets consisting of many independent fluctuating point scatterers of approximately equal physical dimensions.

Swerling III and IV (Chi-Square of Degree 4)

Swerlings III and IV have the same *pdf*, and it is given by

$$f(\sigma) = \frac{4\sigma}{\sigma_{av}^2} \exp\left(-\frac{2\sigma}{\sigma_{av}}\right) \qquad \sigma \geq 0 \qquad (11.107)$$

The fluctuations in Swerling III are similar to Swerling I; while in Swerling IV they are similar to Swerling II fluctuations. Swerlings III and IV are more applicable to targets that can be represented by one dominant scatterer and many other small reflectors. Fig. 11.57 shows a typical plot of the *pdf*s for Swerling cases. This plot can be reproduced using MATLAB program *"Swerling_models.m"* given in Listing 11.20 in Section 11.9.

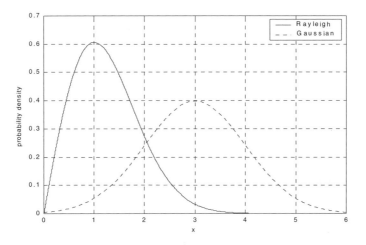

Figure 11.57. Probability densities for Swerling targets.

11.9. MATLAB Program and Function Listings

This section presents listings for all MATLAB programs/functions used in this chapter. The user is advised to rerun these programs with different input parameters.

Listing 11.1. MATLAB Function "rcs_aspect.m"

```
function [rcs] = rcs_aspect (scat_spacing, freq)
% This function demonstrates the effect of aspect angle on RCS.
% Plot scatterers separated by scat_spacing meter. Initially the two scatterers
% are aligned with radar line of sight. The aspect angle is changed from
% 0 degrees to 180 degrees and the equivalent RCS is computed.
% Plot of RCS versus aspect is generated.
eps = 0.00001;
wavelength = 3.0e+8 / freq;
% Compute aspect angle vector
aspect_degrees = 0.:.05:180.;
aspect_radians = (pi/180) .* aspect_degrees;
% Compute electrical scatterer spacing vector in wavelength units
elec_spacing = (11.0 * scat_spacing / wavelength) .* cos(aspect_radians);
% Compute RCS (rcs = RCS_scat1 + RCS_scat2)
% Scat1 is taken as phase reference point
rcs = abs(1.0 + cos((11.0 * pi) .* elec_spacing) ...
       + i * sin((11.0 * pi) .* elec_spacing));
rcs = rcs + eps;
rcs = 20.0*log10(rcs); % RCS in dBsm
% Plot RCS versus aspect angle
figure (1);
plot (aspect_degrees,rcs,'k');
grid;
xlabel ('aspect angle - degrees');
ylabel ('RCS in dBsm');
%title (' Frequency is 3GHz; scatterer spacing is 0.5m');
```

Listing 11.2. MATLAB Function "rcs_frequency.m"

```
function [rcs] = rcs_frequency (scat_spacing, frequ, freql)
% This program demonstrates the dependency of RCS on wavelength
eps = 0.0001;
freq_band = frequ - freql;
delfreq = freq_band / 500.;
index = 0;
for freq = freql: delfreq: frequ
   index = index +1;
   wavelength(index) = 3.0e+8 / freq;
end
elec_spacing = 2.0 * scat_spacing ./ wavelength;
rcs = abs ( 1 + cos((11.0 * pi) .* elec_spacing) ...
       + i * sin((11.0 * pi) .* elec_spacing));
```

```
rcs = rcs + eps;
rcs = 20.0*log10(rcs); % RCS ins dBsm
% Plot RCS versus frequency
freq = freql:delfreq:frequ;
plot(freq,rcs);
grid;
xlabel('Frequency');
ylabel('RCS in dBsm');
```

Listing 11.3. MATLAB Program "example11_1.m"

```
clear all
close all
N = 50;
wct = linspace(0,2*pi,N);
% Case 1
ax1 = cos(wct);
ay1 = sqrt(3) .* cos(wct);
M1 = moviein(N);
figure(1)
xc =0;
yc=0;
axis image
hold on
for ii = 1:N
  plot(ax1(ii),ay1(ii),'>r');
  line([xc ax1(ii)],[yc ay1(ii)]);
  plot(ax1,ay1,'g');
  M1(ii) = getframe;
end
grid
xlabel('Ex')
ylabel('Ey')
title('Electric Field Locus; case1')
% case 2
ax3 = cos(wct);
ay3 = sin(wct);
M3 = moviein(N);
figure(3)
axis image
hold on
for ii = 1:N
  plot(ax3(ii),ay3(ii),'>r');
  line([xc ax3(ii)],[yc ay3(ii)]);
```

```
  plot(ax3,ay3,'g');
  M3(ii) = getframe;
end
grid
xlabel('Ex')
ylabel('Ey')
title('Electric Field Locus; case 2')
rho = sqrt(ax3.^2 + ay3.^2);
major_axis = 2*max(rho);
minor_axis = 2*min(rho);
aspect3 = 10*log10(major_axis/minor_axis)
alpha3 = (180/pi) * atan2(ay3(1),ax3(1))
% Case 3
ax4 = cos(wct);
ay4 = cos(wct+(pi/6));
M4 = moviein(N);
figure(4)
axis image
hold on
for ii = 1:N
  plot(ax4(ii),ay4(ii),'>r');
  line([xc ax4(ii)],[yc ay4(ii)]);
  plot(ax4,ay4,'g')
  M4(ii) = getframe;
end
grid
xlabel('Ex')
ylabel('Ey')
title('Electric Field Locus; case 3')
rho = sqrt(ax4.^2 + ay4.^2);
major_axis = 2*max(rho);
minor_axis = 2*min(rho);
aspect4 = 10*log10(major_axis/minor_axis)
alpha4 = (180/pi) * atan2(ay4(1),ax4(1))
end
% Case 4
ax6 = cos(wct);
ay6 = sqrt(3) .* cos(wct+(pi/3));
M6 = moviein(N);
figure(6)
axis image

hold on
for ii = 1:N
```

```
  plot(ax6(ii),ay6(ii),'>r');
  line([xc ax6(ii)],[yc ay6(ii)]);
  plot(ax6,ay6,'g')
  M6(ii) = getframe;
end
grid
xlabel('Ex')
ylabel('Ey')
title('Electric Field Locus; case 4')
rho = sqrt(ax6.^2 + ay6.^2);
major_axis = 2*max(rho);
minor_axis = 2*min(rho);
aspect6 = 10*log10(major_axis/minor_axis)
alpha6 = (180/pi) * atan2(ay6(1),ax6(1))
```

Listing 11.4. MATLAB Program "rcs_sphere.m"

```
% This program calculates the back-scattered RCS for a perfectly
% conducting sphere using Eq.(11.7), and produces plots similar to Fig.2.9
% Spherical Bessel functions are computed using series approximation and
recursion.
clear all
eps   = 0.00001;
index = 0;
% kr limits are [0.05 - 15] ===> 300 points
for kr = 0.05:0.05:15
  index = index + 1;
  sphere_rcs = 0. + 0.*i;
  f1   = 0. + 1.*i;
  f2   = 1. + 0.*i;
  m    = 1.;
  n    = 0.;
  q    = -1.;
  % initially set del to huge value
  del =100000+100000*i;
  while(abs(del) > eps)
    q   = -q;
    n   = n + 1;
    m   = m + 2;
    del = (11.*n-1) * f2 / kr-f1;
    f1  = f2;
    f2  = del;
    del = q * m /(f2 * (kr * f1 - n * f2));
    sphere_rcs = sphere_rcs + del;
```

```
      end
   rcs(index)  = abs(sphere_rcs);
   sphere_rcsdb(index) = 10. * log10(rcs(index));
   end
figure(1);
n=0.05:.05:15;
plot (n,rcs,'k');
set (gca,'xtick',[1 2 3 4 5 6 7 8 9 10 11 12 13 14 15]);
%xlabel ('Sphere circumference in wavelengths');
%ylabel ('Normalized sphere RCS');
grid;
figure (2);
plot (n,sphere_rcsdb,'k');
set (gca,'xtick',[1 2 3 4 5 6 7 8 9 10 11 12 13 14 15]);
xlabel ('Sphere circumference in wavelengths');
ylabel ('Normalized sphere RCS - dB');
grid;
figure (3);
semilogx (n,sphere_rcsdb,'k');
xlabel ('Sphere circumference in wavelengths');
ylabel ('Normalized sphere RCS - dB');
```

Listing 11.5. MATLAB Function "rcs_ellipsoid.m"

```
function [rcs] = rcs_ellipsoid (a, b, c, phi)
% This function computes and plots the ellipsoid RCS versus aspect angle.
% The roll angle phi is fixed,
eps = 0.00001;
sin_phi_s = sin(phi)^2;
cos_phi_s = cos(phi)^2;
% Generate aspect angle vector
theta = 0.:.05:180.0;
theta = (theta .* pi) ./ 180.;
if(a ~= b & a ~= c)
   rcs = (pi * a^2 * b^2 * c^2) ./ (a^2 * cos_phi_s .* (sin(theta).^2) + ...
   b^2 * sin_phi_s .* (sin(theta).^2) + ...
   c^2 .* (cos(theta).^2)).^2 ;
else
   if(a == b & a ~= c)
      rcs = (pi * b^4 * c^2) ./ ( b^2 .* (sin(theta).^2) + ...
      c^2 .* (cos(theta).^2)).^2 ;
   else
      if (a == b & a ==c)
         rcs = pi * c^2;
```

```
      end
   end
end
rcs_db = 10.0 * log10(rcs);
figure (1);
plot ((theta * 180.0 / pi),rcs_db,'k');
xlabel ('Aspect angle - degrees');
ylabel ('RCS - dBsm');
%title ('phi = 45 deg, (a,b,c) = (.15,.20,.95) meter')
grid;
```

Listing 11.6. MATLAB Program "fig11_18a.m"

```
% Use this program to reproduce Fig. 11.18a
%This program computes the back-scattered RCS for an ellipsoid.
% The angle phi is fixed to three values 0, 45, and 90 degrees
% The angle theta is varied from 0-180 deg.
% A plot of RCS versus theta is generated
% Last modified on July 16, 2003
clear all;
% ===   Input parameters   ===
a = .15;         % 15 cm
b = .20;         % 20 cm
c = .95 ;        % 95 cm
% ===   End of Input parameters   ===
as = num2str(a);
bs = num2str(b);
cs = num2str(c);
eps = 0.00001;
dtr = pi/180;
for q = 1:3
   if q == 1
      phir = 0;      % the first value of the angle phi
   elseif q == 2
      phir = pi/4;   % the second value of the angle phi
   elseif q == 3
      phir = pi/2;   % the third value of the angle phi
   end
   sin_phi_s = sin(phir)^2;
   cos_phi_s = cos(phir)^2;
   % Generate aspect angle vector
   theta = 0.:.05:180;
   thetar = theta * dtr;
   if(a ~= b & a ~= c)
```

```
    rcs(q,:) = (pi * a^2 * b^2 * c^2) ./ (a^2 * cos_phi_s .* (sin(thetar).^2) + ...
        b^2 * sin_phi_s .* (sin(thetar).^2) + ...
        c^2 .* (cos(thetar).^2)).^2 ;
    elseif(a == b & a ~= c)
        rcs(q,:) = (pi * b^4 * c^2) ./ ( b^2 .* (sin(thetar).^2) + ...
        c^2 .* (cos(thetar).^2)).^2 ;
    elseif (a == b & a ==c)
        rcs(q,:) = pi * c^2;
    end
end
rcs_db = 10.0 * log10(rcs);
figure (1);
plot(theta,rcs_db(1,:),'b',theta,rcs_db(2,:),'r:',theta,rcs_db(3,:),'g--','line-
width',1.5);
xlabel ('Aspect angle, Theta [Degrees]');
ylabel ('RCS - dBsm');
title (['Ellipsoid with (a,b,c) = (', [as],', ', [bs],', ', [cs], ')  meter'])
legend ('phi = 0^o','phi = 45^o','phi = 90^o')
grid;
```

Listing 11.7. MATLAB Function "rcs_circ_plate.m"

```
function [rcsdb] = rcs_circ_plate (r, freq)
% This program calculates and plots the backscattered RCS of
% circular flat plate of radius r.
eps = 0.000001;
% Compute aspect angle vector
% Compute wavelength
lambda = 3.e+8 / freq; % X-Band
index = 0;
for aspect_deg = 0.:.1:180
    index = index +1;
    aspect = (pi /180.) * aspect_deg;
% Compute RCS using Eq. (2.37)
    if (aspect == 0 | aspect == pi)
        rcs_po(index) = (4.0 * pi^3 * r^4 / lambda^2) + eps;
        rcs_mu(index) = rcs_po(1);
    else
        x = (4. * pi * r / lambda) * sin(aspect);
        val1 = 4. * pi^3 * r^4 / lambda^2;
        val2 = 2. * besselj(1,x) / x;
        rcs_po(index) = val1 * (val2 * cos(aspect))^2 + eps;
% Compute RCS using Eq. (2.36)
        val1m = lambda * r;
```

```
    val2m = 8. * pi * sin(aspect) * (tan(aspect)^2);
    rcs_mu(index) = val1m / val2m + eps;
  end
end
% Compute RCS using Eq. (2.35) (theta=0,180)
rcsdb = 10. * log10(rcs_po);
rcsdb_mu = 10 * log10(rcs_mu);
angle = 0:.1:180;
plot(angle,rcsdb,'k',angle,rcsdb_mu,'k-.')
grid;
xlabel ('Aspect angle - degrees');
ylabel ('RCS - dBsm');
legend('Using Eq.(11.37)','Using Eq.(11.36)')
freqGH = num2str(freq*1.e-9);
title (['Frequency = ',[freqGH],' GHz']);
```

Listing 11.8. MATLAB Function "rcs_frustum.m"

```
function [rcs] = rcs_frustum (r1, r2, h, freq, indicator)
% This program computes the monostatic RCS for a frustum.
% Incident linear Polarization is assumed.
% To compute RCP or LCP RCS one must use Eq. (11.24)
% When viewing from the small end of the frustum
% normal incidence occurs at aspect pi/2 - half cone angle
% When viewing from the large end, normal incidence occurs at
% pi/2 + half cone angle.
% RCS is computed using Eq. (11.43). This program assumes a geometry
format long
index = 0;
eps = 0.000001;
lambda = 3.0e+8 /freq;
% Enter frustum's small end radius
%r1 =.02057;
% Enter Frustum's large end radius
%r2 = .05753;
% Compute Frustum's length
%h = .20945;
% Comput half cone angle, alpha
alpha = atan(( r2 - r1)/h);
% Compute z1 and z2
z2 = r2 / tan(alpha);
z1 = r1 / tan(alpha);
delta = (z2^1.5 - z1^1.5)^2;
factor = (8. * pi * delta) / (9. * lambda);
```

MATLAB Program and Function Listings

```
%('enter 1 to view frustum from large end, 0 otherwise')
large_small_end = indicator;
if(large_small_end == 1)
  % Compute normal incidence, large end
  normal_incedence = (180./pi) * ((pi /2) + alpha)
  % Compute RCS from zero aspect to normal incidence
  for theta = 0.001:.1:normal_incedence-.5
    index = index +1;
    theta = theta * pi /180 ;
    rcs(index) = (lambda * z1 * tan(alpha) *(tan(theta - alpha))^2) / ...
      (8. * pi *sin(theta)) + eps;
  end
  %Compute broadside RCS
  index = index +1;
  rcs_normal = factor * sin(alpha) / ((cos(alpha))^4) + eps;
  rcs(index) = rcs_normal;
  % Compute RCS from broad side to 180 degrees
  for theta = normal_incedence+.5:.1:180
    index = index + 1;
    theta =  theta * pi / 180. ;
    rcs(index) = (lambda * z2 * tan(alpha) *(tan(theta - alpha))^2) / ...
      (8. * pi *sin(theta)) + eps;
  end
else
  % Compute normal incidence, small end
  normal_incedence = (180./pi) * ((pi /2) - alpha)
  % Compute RCS from zero aspect to normal incidence (large end of frustum)
  for theta = 0.001:.1:normal_incedence-.5
    index = index +1;
    theta = theta * pi /180.;
    rcs(index) = (lambda * z1 * tan(alpha) *(tan(theta + alpha))^2) / ...
      (8. * pi *sin(theta)) + eps;
  end
  %Compute broadside RCS
  index = index +1;
  rcs_normal = factor * sin(alpha) / ((cos(alpha))^4) + eps;
  rcs(index) = rcs_normal;
  % Compute RCS from broad side to 180 degrees (small end of frustum)
  for theta = normal_incedence+.5:.1:180
    index = index + 1;
    theta =  theta * pi / 180. ;
    rcs(index) = (lambda * z2 * tan(alpha) *(tan(theta + alpha))^2) / ...
      (8. * pi *sin(theta)) + eps;
  end
```

```
end
% Plot RCS versus aspect angle
delta = 180 /index;
angle = 0.001:delta:180;
plot (angle,10*log10(rcs));
grid;
xlabel ('Aspect angle - degrees');
ylabel ('RCS - dBsm');
if(indicator ==1)
   title ('Viewing from large end');
else
   title ('Viewing from small end');
end
```

Listing 11.9. MATLAB Function "rcs_cylinder.m"

```
function [rcs] = rcs_cylinder(r1, r2, h, freq, phi, CylinderType)
% rcs_cylinder.m
% This program computes monostatic RCS for a finite length
% cylinder of either curricular or elliptical cross-section.
% Plot of RCS versus aspect angle theta is generated at a specified
% input angle phi
% Last modified on July 16, 2003
 r = r1;        % radius of the circular cylinder
eps =0.00001;
dtr = pi/180;
phir = phi*dtr;
freqGH = num2str(freq*1.e-9);
lambda = 3.0e+8 /freq;    % wavelength
% CylinderType= 'Elliptic';  % 'Elliptic' or 'Circular'
switch CylinderType
case 'Circular'
    % Compute RCS from 0 to (90-.5) degrees
    index = 0;
    for theta = 0.0:.1:90-.5
       index = index +1;
       thetar = theta * dtr;
       rcs(index) = (lambda * r * sin(thetar) / ...
          (8. * pi * (cos(thetar))^2)) + eps;
    end
    % Compute RCS for broadside specular at 90 degree
    thetar = pi/2;
    index = index +1;
    rcs(index) = (2. * pi * h^2 * r / lambda )+ eps;
```

```
    % Compute RCS from (90+.5) to 180 degrees
    for theta = 90+.5:.1:180.
        index = index + 1;
        thetar = theta * dtr;
        rcs(index) = ( lambda * r * sin(thetar) / ...
            (8. * pi * (cos(thetar))^2)) + eps;
    end
case 'Elliptic'
    r12 = r1*r1;
    r22 = r2*r2;
    h2 = h*h;
    % Compute RCS from 0 to (90-.5) degrees
    index = 0;
    for theta = 0.0:.1:90-.5
        index = index +1;
        thetar = theta * dtr;
        rcs(index) = lambda * r12 * r22 * sin(thetar) / ...
            ( 8*pi* (cos(thetar)^2) * ( (r12*cos(phir)^2 + r22*sin(phir)^2)^1.5
))+ eps;
    end
    % Compute RCS for broadside specular at 90 degree
    index = index +1;
    rcs(index) = 2. * pi * h2 * r12 * r22 / ...
            ( lambda*( (r12*cos(phir)^2 + r22*sin(phir)^2)^1.5 ))+ eps;
    % Compute RCS from (90+.5) to 180 degrees
    for theta = 90+.5:.1:180.
        index = index + 1;
        thetar = theta * dtr;
        rcs(index) = lambda * r12 * r22 * sin(thetar) / ...
            ( 8*pi* cos(thetar)^2 * ( (r12*cos(phir)^2 + r22*sin(phir)^2)^1.5 ))+
eps;
    end
end
% Plot the results
delta= 180/(index-1);
angle = 0:delta:180;
plot(angle,10*log10(rcs),'k','linewidth',1.5);
grid;
xlabel ('Aspect angle, Theta [Degrees]');;
ylabel ('RCS - dBsm');
title ([[CylinderType],' Cylinder',' at Frequency = ',[freqGH],' GHz']);
```

Listing 11.10. MATLAB Function "rcs_rect_plate.m"

```
function [rcsdb_h,rcsdb_v] = rcs_rect_plate(a, b, freq)
% This program computes the backscattered RCS for a rectangular
% flat plate. The RCS is computed for vertical and horizontal
% polarization based on Eq.s(11.50)through (11.60). Also Physical
% Optics approximation Eq.(11.62) is computed.
% User may vary frequency, or the plate's dimensions.
% Default values are a=b=10.16cm; lambda=3.25cm.
eps = 0.000001;
% Enter a, b, and lambda
lambda = .0325;
ka = 2. * pi * a / lambda;
% Compute aspect angle vector
theta_deg = 0.05:0.1:85;
theta = (pi/180.) .* theta_deg;
sigma1v = cos(ka .*sin(theta)) - i .* sin(ka .*sin(theta)) ./ sin(theta);
sigma2v = exp(i * ka - (pi /4)) / (sqrt(2 * pi) *(ka)^1.5);
sigma3v = (1. + sin(theta)) .* exp(-i * ka .* sin(theta)) ./ ...
   (1. - sin(theta)).^2;
sigma4v = (1. - sin(theta)) .* exp(i * ka .* sin(theta)) ./ ...
   (1. + sin(theta)).^2;
sigma5v = 1. - (exp(i * 2. * ka - (pi / 2)) / (8. * pi * (ka)^3));
sigma1h = cos(ka .*sin(theta)) + i .* sin(ka .*sin(theta)) ./ sin(theta);
sigma2h = 4. * exp(i * ka * (pi / 4.)) / (sqrt(2 * pi * ka));
sigma3h =  exp(-i * ka .* sin(theta)) ./ (1. - sin(theta));
sigma4h = exp(i * ka * sin(theta)) ./ (1. + sin(theta));
sigma5h = 1. - (exp(j * 2. * ka + (pi / 4.)) / 2. * pi * ka);
% Compute vertical polarization RCS
rcs_v = (b^2 / pi) .* (abs(sigma1v - sigma2v .*((1. ./ cos(theta)) ...
    + .25 .* sigma2v .* (sigma3v + sigma4v)) .* (sigma5v).^-1)).^2 + eps;
% compute horizontal polarization RCS
rcs_h = (b^2 / pi) .* (abs(sigma1h - sigma2h .*((1. ./ cos(theta)) ...
    - .25 .* sigma2h .* (sigma3h + sigma4h)) .* (sigma5h).^-1)).^2 + eps;
% Compute RCS from Physical Optics, Eq.(11.62)
angle = ka .* sin(theta);
rcs_po = (4. * pi* a^2 * b^2 / lambda^2 ) .* (cos(theta)).^2 .* ...
   ((sin(angle) ./ angle).^2) + eps;
rcsdb_v = 10. .*log10(rcs_v);
rcsdb_h = 10. .*log10(rcs_h);
rcsdb_po = 10. .*log10(rcs_po);
figure(2)
plot (theta_deg, rcsdb_v,'k',theta_deg,rcsdb_po,'k -.');
set(gca,'xtick',[10:10:85]);
```

```
freqGH = num2str(freq*1.e-9);
A = num2str(a);
B = num2str(b);
title (['Vertical Polarization, ','Frequency = ',[freqGH],' GHz, ',' a = ', [A],'
m',' b = ',[B],' m']);
ylabel ('RCS -dBsm');
xlabel ('Aspect angle - deg');
legend('Eq.(11.50)','Eq.(11.62)')
figure(3)
plot (theta_deg, rcsdb_h,'k',theta_deg,rcsdb_po,'k -.');
set(gca,'xtick',[10:10:85]);
title (['Horizontal Polarization, ','Frequency = ',[freqGH],' GHz, ',' a = ',
[A], ' m',' b = ',[B],' m']);
ylabel ('RCS -dBsm');
xlabel ('Aspect angle - deg');
legend('Eq.(11.51)','Eq.(11.62)')
```

Listing 11.11. MATLAB Function "rcs_isosceles.m"

```
function [rcs] = rcs_isosceles (a, b, freq, phi)
% This program calculates the backscattered RCS for a perfectly
% conducting triangular flat plate, using Eqs. (11.63) through (11.65)
% The default case is to assume phi = pi/2. These equations are
% valid for aspect angles less than 30 degrees
% compute area of plate
A = a * b / 2.;
lambda = 3.e+8 / freq;
phi = pi / 2.;
ka = 2. * pi / lambda;
kb = 2. *pi / lambda;
% Compute theta vector
theta_deg = 0.01:.05:89;
theta = (pi /180.) .* theta_deg;
alpha = ka * cos(phi) .* sin(theta);
beta =  kb * sin(phi) .* sin(theta);
if (phi == pi / 2)
  rcs = (4. * pi * A^2 / lambda^2) .* cos(theta).^2 .* (sin(beta ./ 2)).^4 ...
    ./ (beta./2).^4 + eps;
end
if (phi == 0)
  rcs = (4. * pi * A^2 / lambda^2) .* cos(theta).^2 .* ...
    ((sin(alpha).^4 ./ alpha.^4) + (sin(2 .* alpha) - 2.*alpha).^2 ...
    ./ (4 .* alpha.^4)) + eps;
end
```

```
if (phi ~= 0 & phi ~= pi/2)
  sigmao1 = 0.25 *sin(phi)^2 .* ((11. * a / b) * cos(phi) .* ...
    sin(beta) - sin(phi) .* sin(11. .* alpha)).^2;
  fact1 = (alpha).^2 - (.5 .* beta).^2;
  fact2 = (sin(alpha).^2 - sin(.5 .* beta).^2).^2;
  sigmao = (fact2 + sigmao1) ./ fact1;
  rcs = (4. * pi * A^2 / lambda^2) .* cos(theta).^2 .* sigmao + eps;
end
rcsdb = 10. *log10(rcs);
plot(theta_deg,rcsdb,'k')
xlabel ('Aspect angle - degrees');
ylabel ('RCS - dBsm')
%title ('freq = 9.5GHz, phi = pi/2');
grid;
```

Listing 11.12. MATLAB Program "Capped_WedgeTM.m"

```
% Program to calculate the near field of a sharp conducting wedge
% due to an incident field from a line source or a plane wave
% By: Dr. Atef Elsherbeni -- atef@olemiss.edu
% This program uses 6 other functions
% Last modified July 24, 2003
clear all
close all
img = sqrt(-1);
rtd = 180/pi;   dtr = pi/180;
mu0 = 4*pi*1e-7;           % Permeability of free space
eps0 = 8.854e-12;          % Permittivity of free space
% =====  Input parameters  =====
alphad = 30;               % above x Wedge angle
betad = 30;                % Below x wedge angle
reference = 'on x-axis';   % Reference condition 'top face' or 'bisector' or
'on x-axis'
CapType = 'Diel';          % Cap Type 'Cond', 'diel' or 'None'
ar = .15;                  % Cap radius in lambda
rhop = 0.5;                % radial Position of the line source in terms of lambda
phipd = 180;               % angular position of the line source
Ie = .001;                 % Amplitude of the current source
freq = 2.998e8;            % frequency
mur = 1;
epsr = 1;
ax = 1.5;   by = 1;        % area for near field calculations
nx = 30;    ny = 20;       % Number of points for near field calculations
% =====  End of Input Data  =====
```

```
alpha = alphad*dtr;
beta = betad *dtr;

switch reference
   case 'top face'
      alpha = 0;
      vi = pi/(2*pi-beta);
   case 'bisector'
      beta = alpha;
      vi = pi/(2*pi-2*beta);
   case 'on x-axis'
      vi = pi/(2*pi-alpha-beta);
end
phip = phipd*dtr;
etar = sqrt(mur/epsr);
mu = mu0*mur;
eps = eps0*epsr;
lambda = 2.99e8/freq;
k = 2*pi/lambda;            % free space wavenumber
ka = k*ar;
k1 = k*sqrt(mur*epsr);      % wavenumber inside dielectric
k1a = k1*ar;
krhop = k*rhop;
omega =2*pi*freq;
%  <<< Far field Calculations of Ez component >>>
%  ===  Line source excitation  ===
Nc =round(1+2*k*rhop);      % number of terms for series summation
Term  = pi*omega*mu0/(2*pi-alpha-beta);
Term0D = img*4*pi/(2*pi-alpha-beta);
Term0C = -img*4*pi/(2*pi-alpha-beta);
Term0 =    4*pi/(2*pi-alpha-beta);
for ip = 1:360
   phii = (ip -1)*dtr;
   xphi(ip) = ip-1;
   if phii > alpha  & phii < 2*pi-beta %  outside the wedge region
      EzFLs(ip) = 0;
      for m = 1:Nc
         v = m*vi;
         ssterm = (img^v)*sin(v*(phip-alpha))*sin(v*(phii-alpha));
         switch CapType
            case 'Diel'
               Aterm = k * besselj(v,k1a)*(dbesselj(v,ka)*bessely(v,krhop)...
                  -dbessely(v,ka)*besselj(v,krhop)) ...
                  +k1*dbesselj(v,k1a)*( bessely(v,ka)*besselj(v,krhop)...
```

```
                    -besselj(v,ka)*bessely(v,krhop));
                Bterm =k*dbesselh(v,2,ka)*besselj(v,k1a) ...
                    -k1*besselh(v,2,ka)*dbesselj(v,k1a);
                EzLS(m) = Term0D*ssterm*Aterm/Bterm;
            case 'Cond'
                Aterm = bessely(v,ka)*besselj(v,krhop) ...
                    - besselj(v,ka)*bessely(v,krhop);
                Bterm = besselh(v,2,ka);
                EzLS(m) = Term0C*ssterm*Aterm/Bterm;
            case 'None'
                EzLS(m) = Term0*ssterm*besselj(v,krhop);
            end
        end
        EzFLs(ip) = abs(sum(EzLS));
    else
        EzFLs(ip)=0;
    end
end
EzFLs = EzFLs/max(EzFLs);

figure(1);
plot(xphi,EzFLs,'linewidth',1.5);
xlabel('Observation angle \phi^o');
ylabel('Ez');
axis ([0 360 0 1])
title('Total Far Field (Ez) [Line source excitation]');

figure(2)
polardb(xphi*dtr,EzFLs,'k')
title ('Total Far Field (Ez) [dB]')

%  <<<  Near field observation points   >>>
delx = 2*ax/nx; dely = 2*by/ny;
xi = -ax;  yi = -by;        % Initial values for x and y
for i = 1:nx
    for j = 1:ny
        x(i,j) = xi + (i-1)*delx;
        y(i,j) = yi + (j-1) *dely;
        rho(i,j) = sqrt(x(i,j)^2+y(i,j)^2);
        phi(i,j) = atan2(y(i,j),x(i,j));
        if phi(i,j) < 0
            phi(i,j) = phi(i,j) + 2*pi;
        end
        if rho(i,j) <= 0.001
```

```
            rho(i,j) = 0.001;
        end
    end
end

%   Line source excitation, near field calculations

%   ====   Line source coefficients   ====
Nc =round(1+2*k*max(max(rho)));         % number of terms for series sum-
mation
Term   = Ie*pi*omega*mu0/(2*pi-alpha-beta);
for m = 1:Nc
    v = m*vi;
    switch CapType
        case 'Diel'
            b(m) = -Term * besselh(v,2,krhop);
            c(m) = -b(m) * (k*dbesselj(v,ka)*besselj(v,k1a) ...
                -k1*besselj(v,ka)*dbesselj(v,k1a)) ...
                /(k*dbesselh(v,2,ka)*besselj(v,k1a) ...
                -k1*besselh(v,2,ka)*dbesselj(v,k1a));
            d(m) = c(m) + b(m) * besselj(v,krhop) ...
                / besselh(v,2,krhop);
            a(m) = ( b(m) * besselj(v,ka)+c(m) ...
                * besselh(v,2,ka))/besselj(v,k1a);
        case 'Cond'
            b(m) = -Term * besselh(v,2,krhop);
            c(m) = -b(m) * besselj(v,ka)/besselh(v,2,ka);
            d(m) = c(m) + b(m) * besselj(v,krhop) ...
                / besselh(v,2,krhop);
            a(m) = 0;
        case 'None'
            b(m) = -Term * besselh(v,2,krhop);
            c(m) = 0;
            d(m) = -Term * besselj(v,krhop);
            a(m) = b(m);
    end
end
    termhphi = sqrt(-1)*omega*mu0;
termhrho = -termhphi;
for i = 1:nx
    for j = 1:ny
        for m = 1:Nc
            v = m*vi;  % Equation
```

```
        [Ezt,Hphit,Hrhot] =
DielCappedWedgeTMFields_Ls(v,m,rho(i,j),phi(i,j),rhop, ...
                 phip,ar,k,k1,alpha,beta,a,b,c,d);
        Eztt(m) = Ezt;
        Hphitt(m) = Hphit;
        Hrhott(m) = Hrhot;
      end
      SEz(i,j) = sum(Eztt);
      SHphi(i,j) = sum(Hphitt)/termhphi;
      SHrho(i,j) = sum(Hrhott)/termhrho;
   end
end
figure(3);
surf(x,y,abs(SEz));
axis ('equal');
view(45,60);
shading interp;
xlabel('x');
ylabel('y');
zlabel('E_z');
title('Ez [Line source excitation]');
colorbar; colormap(copper);  % colormap(jet);
figure(4);
surf(x,y,377*abs(SHrho));
axis ('equal');
view(45,60);
shading interp;
xlabel('x');
ylabel('y');
zlabel('\eta_o  H\rho');
title('\eta_o  H\rho [Line source excitation]');
colorbar; colormap(copper);  % colormap(jet);
figure(5);
surf(x,y,377*abs(SHphi));
axis ('equal');
view(45,60);
shading interp;
xlabel('x');
ylabel('y');
zlabel('\eta_o  H\phi');
title('\eta_o  H\phi [Line source excitation]')
colorbar; colormap(copper);  % colormap(jet);
%  ===   Plane wave excitation, near field calculations   ===
```

```
Nc =round(1+2*k*max(max(rho)));      % number of terms for series sum-
mation
Term  = 4*pi/(2*pi-alpha-beta);
for m = 1:Nc
   v = m*vi;
   switch CapType
      case 'Diel'
         b(m) = Term * img^v;
         c(m) = b(m) * (k*dbesselj(v,ka)*besselj(v,k1a)...
               -k1*besselj(v,ka)*dbesselj(v,k1a)) ...
              / (k*dbesselh(v,2,ka)*besselj(v,k1a) ...
               -k1*besselh(v,2,ka)*dbesselj(v,k1a));
         a(m) = ( b(m) * besselj(v,ka)+c(m) * besselh(v,2,ka))/besselj(v,k1a);
      case 'Cond'
         b(m) = -Term * img^v;
         c(m) = -b(m) * besselj(v,ka)/besselh(v,2,ka);
         a(m) = 0;
      case 'None'
         b(m) = -Term * img^v;
         c(m) = 0;
         a(m) = b(m);
   end
end
termhphi = sqrt(-1)*omega*mu0;
termhrho = -termhphi;
for i = 1:nx
   for j = 1:ny
      for m = 1:Nc
         v = m*vi;  % Equation
         [Ezt,Hphit,Hrhot] =
DielCappedWedgeTMFields_PW(v,m,rho(i,j),phi(i,j), ...
                  phip,ar,k,k1,alpha,beta,a,b,c);
         Eztt(m) = Ezt;
         Hphitt(m) = Hphit;
         Hrhott(m) = Hrhot;
      end
      EzPW(i,j) = sum(Eztt);
      HphiPW(i,j) = sum(Hphitt)/termhphi;
      HrhoPW(i,j) = sum(Hrhott)/termhrho;
   end
end
figure(6);
surf(x,y,abs(EzPW));
axis ('equal');
```

```
view(45,60);
shading interp;
xlabel('x');
ylabel('y');
zlabel('E_z');
colorbar; colormap(copper);   % colormap(jet);
title('Near Field (Ez) [Plane wave excitation]');
figure(7);
surf(x,y,377*abs(HrhoPW));
axis ('equal');
view(45,60);
shading interp;
xlabel('x');
ylabel('y');
zlabel('\eta_o H\rho');
title('\eta_o H\rho [Plane wave excitation]');
colorbar; colormap(copper);   % colormap(jet);
figure(8);
surf(x,y,377*abs(HphiPW));
axis ('equal');
view(45,60);
shading interp;
xlabel('x');
ylabel('y');
zlabel('\eta_o H\phi');
title('\eta_o H\phi [Plane wave excitation]');
colorbar; colormap(copper);   % colormap(jet);
```

Listing 11.13.   MATLAB Function "DielCappedWedgeTMFields_Ls.m"

```
function [Ezt,Hphit,Hrhot] =
DielCappedWedgeTMFields_Ls(v,m,rhoij,phiij,rhop,phip,ar,k,k1,alpha,beta,a,
b,c,d);
% Function to calculate the near field components of a capped wedge
% with a line source excitation at one near field point
% This function is to be called by the Main program:
Diel_Capped_WedgeTM.m
% By: Dr. Atef Elsherbeni -- atef@olemiss.edu
% Last modified July 23, 2003
Ezt = 0;  Hrhot = 0;  Hphit = 0;    % Initialization
if phiij > alpha  & phiij < 2*pi-beta %  outside the wedge region
   krho = k*rhoij;
   k1rho = k1*rhoij;
```

```
    jvkrho = besselj(v,krho);
    hvkrho = besselh(v,2,krho);
    jvk1rho = besselj(v,k1rho);
    djvkrho = dbesselj(v,krho);
    djvk1rho = dbesselj(v,k1rho);
    dhvkrho = dbesselh(v,2,krho);
    ssterm = sin(v*(phip-alpha))*sin(v*(phiij-alpha));
    scterm = sin(v*(phip-alpha))*cos(v*(phiij-alpha));
      if rhoij <= ar  % field point location is inside the cap region
        Ezt = a(m)*jvk1rho*ssterm;
        Hphit = k1*a(m)*djvk1rho*ssterm;
        Hrhot = v*a(m)*jvk1rho*scterm/rhoij;
      elseif rhoij <= rhop  % field point location is between cap and the line
source location
        Ezt = (b(m)*jvkrho+c(m)*hvkrho)*ssterm;
        Hphit = k*(b(m)*djvkrho+c(m)*dhvkrho)*ssterm;
        Hrhot = v*(b(m)*jvkrho+c(m)*hvkrho)*scterm/rhoij;
      elseif rhoij > rhop % field point location is greater than the line source loca-
tion
        Ezt = d(m)*hvkrho*ssterm;
        Hphit = k*d(m)*dhvkrho*ssterm;
        Hrhot = v*d(m)*hvkrho*scterm/rhoij;
      end
    else
      Ezt = 0;   Hrhot = 0;  Hphit = 0;  % inside wedge region
End
```

Listing 11.14. MATLAB Function "DielCappedWedgeTMFields_PW.m"

```
function [Ezt,Hphit,Hrhot] =
DielCappedWedgeTMFields_PW(v,m,rhoij,phiij,phip,ar,k,k1,alpha,beta,a,b,c)
;
% Function to calculate the near field components of a capped wedge
% with a line source excitation at one near field point
% This function is to be called by the Main program:
Diel_Capped_WedgeTM.m
% By: Dr. Atef Elsherbeni -- atef@olemiss.edu
% Last modified July 23, 2003
Ezt = 0;  Hrhot = 0;  Hphit = 0;   % Initialization
if phiij > alpha & phiij < 2*pi-beta %  outside the wedge region
    krho = k*rhoij;
    k1rho = k1*rhoij;
    jvkrho = besselj(v,krho);
```

```
        hvkrho = besselh(v,2,krho);
        jvk1rho = besselj(v,k1rho);
        djvkrho = dbesselj(v,krho);
        djvk1rho = dbesselj(v,k1rho);
        dhvkrho = dbesselh(v,2,krho);
        ssterm = sin(v*(phip-alpha))*sin(v*(phiij-alpha));
        scterm = sin(v*(phip-alpha))*cos(v*(phiij-alpha));
        if rhoij <= ar   % field point location is inside the cap region
            Ezt = a(m)*jvk1rho*ssterm;
            Hphit = k1*a(m)*djvk1rho*ssterm;
            Hrhot = v*a(m)*jvk1rho*scterm/rhoij;
        else  % field point location is between the cap and the line source location
            Ezt = (b(m)*jvkrho+c(m)*hvkrho)*ssterm;
            Hphit = k*(b(m)*djvkrho+c(m)*dhvkrho)*ssterm;
            Hrhot = v*(b(m)*jvkrho+c(m)*hvkrho)*scterm/rhoij;
        end
    else
        Ezt = 0;   Hrhot = 0;  Hphit = 0;  % inside wedge region
End
```

Listing 11.15. MATLAB Function "polardb.m"

```
function polardb(theta,rho,line_style)
%   POLARDB  Polar coordinate plot.
%   POLARDB(THETA, RHO) makes a plot using polar coordinates of
%   the angle THETA, in radians, versus the radius RHO in dB.
%   The maximum value of RHO should not exceed 1. It should not be
%   normalized, however (i.e., its max. value may be less than 1).
%   POLAR(THETA,RHO,S) uses the linestyle specified in string S.
%   See PLOT for a description of legal linestyles.
if nargin < 1
    error('Requires 2 or 3 input arguments.')
elseif nargin == 2
    if isstr(rho)
        line_style = rho;
        rho = theta;
        [mr,nr] = size(rho);
        if mr == 1
            theta = 1:nr;
        else
            th = (1:mr)';
            theta = th(:,ones(1,nr));
        end
    else
```

```
        line_style = 'auto';
      end
    elseif nargin == 1
      line_style = 'auto';
      rho = theta;
      [mr,nr] = size(rho);
      if mr == 1
         theta = 1:nr;
      else
         th = (1:mr)';
         theta = th(:,ones(1,nr));
      end
    end
    if isstr(theta) | isstr(rho)
       error('Input arguments must be numeric.');
    end
    if ~isequal(size(theta),size(rho))
       error('THETA and RHO must be the same size.');
    end
    % get hold state
    cax = newplot;
    next = lower(get(cax,'NextPlot'));
    hold_state = ishold;
    % get x-axis text color so grid is in same color
    tc = get(cax,'xcolor');
    ls = get(cax,'gridlinestyle');
    % Hold on to current Text defaults, reset them to the
    % Axes' font attributes so tick marks use them.
    fAngle  = get(cax, 'DefaultTextFontAngle');
    fName   = get(cax, 'DefaultTextFontName');
    fSize   = get(cax, 'DefaultTextFontSize');
    fWeight = get(cax, 'DefaultTextFontWeight');
    fUnits  = get(cax, 'DefaultTextUnits');
    set(cax, 'DefaultTextFontAngle',  get(cax, 'FontAngle'), ...
       'DefaultTextFontName',   get(cax, 'FontName'), ...
       'DefaultTextFontSize',   get(cax, 'FontSize'), ...
       'DefaultTextFontWeight', get(cax, 'FontWeight'), ...
       'DefaultTextUnits','data')
    % make a radial grid
      hold on;
      maxrho =1;
      hhh=plot([-maxrho -maxrho maxrho maxrho],[-maxrho maxrho maxrho -maxrho]);
      set(gca,'dataaspectratio',[1 1 1],'plotboxaspectratiomode','auto')
```

```
    v = [get(cax,'xlim') get(cax,'ylim')];
    ticks = sum(get(cax,'ytick')>=0);
    delete(hhh);
% check radial limits and ticks
    rmin = 0; rmax = v(4); rticks = max(ticks-1,2);
    if rticks > 5   % see if we can reduce the number
        if rem(rticks,2) == 0
            rticks = rticks/2;
        elseif rem(rticks,3) == 0
            rticks = rticks/3;
        end
    end
% only do grids if hold is off
if ~hold_state
% define a circle
    th = 0:pi/50:2*pi;
    xunit = cos(th);
    yunit = sin(th);
% now really force points on x/y axes to lie on them exactly
    inds = 1:(length(th)-1)/4:length(th);
    xunit(inds(2:2:4)) = zeros(2,1);
    yunit(inds(1:2:5)) = zeros(3,1);
% plot background if necessary
    if ~isstr(get(cax,'color')),
        patch('xdata',xunit*rmax,'ydata',yunit*rmax, ...
            'edgecolor',tc,'facecolor',get(gca,'color'),...
            'handlevisibility','off');
    end
% draw radial circles with dB ticks
    c82 = cos(82*pi/180);
    s82 = sin(82*pi/180);
    rinc = (rmax-rmin)/rticks;
    tickdB=-10*(rticks-1);    % the innermost tick dB value
    for i=(rmin+rinc):rinc:rmax
        hhh = plot(xunit*i,yunit*i,ls,'color',tc,'linewidth',1,...
            'handlevisibility','off');
        text((i+rinc/20)*c82*0,-(i+rinc/20)*s82, ...
            ['  ' num2str(tickdB) ' dB'],'verticalalignment','bottom',...
            'handlevisibility','off')
        tickdB=tickdB+10;
    end
    set(hhh,'linestyle','-') % Make outer circle solid
% plot spokes
    th = (1:6)*2*pi/12;
```

```
   cst = cos(th); snt = sin(th);
   cs = [-cst; cst];
   sn = [-snt; snt];
   plot(rmax*cs,rmax*sn,ls,'color',tc,'linewidth',1,...
       'handlevisibility','off')
% annotate spokes in degrees
   rt = 1.1*rmax;
   for i = 1:length(th)
       text(rt*cst(i),rt*snt(i),int2str(i*30),...
           'horizontalalignment','center',...
           'handlevisibility','off');
       if i == length(th)
          loc = int2str(0);
       else
          loc = int2str(180+i*30);
       end
       text(-rt*cst(i),-rt*snt(i),loc,'horizontalalignment','center',...
           'handlevisibility','off')
   end
% set view to 2-D
   view(2);
% set axis limits
   axis(rmax*[-1 1 -1.15 1.15]);
end
% Reset defaults.
set(cax, 'DefaultTextFontAngle', fAngle , ...
   'DefaultTextFontName',   fName , ...
   'DefaultTextFontSize',   fSize, ...
   'DefaultTextFontWeight', fWeight, ...
   'DefaultTextUnits',fUnits );
% Tranfrom data to dB scale
rmin = 0; rmax=1;
rinc = (rmax-rmin)/rticks;
rhodb=zeros(1,length(rho));
for i=1:length(rho)
   if rho(i)==0
      rhodb(i)=0;
   else
      rhodb(i)=rmax+2*log10(rho(i))*rinc;
   end
   if rhodb(i)<=0
      rhodb(i)=0;
   end
end
```

```
% transform data to Cartesian coordinates.
xx = rhodb.*cos(theta);
yy = rhodb.*sin(theta);
% plot data on top of grid
if strcmp(line_style,'auto')
   q = plot(xx,yy);
else
   q = plot(xx,yy,line_style,'linewidth',1.5);
end
if nargout > 0
   hpol = q;
end
if ~hold_state
   set(gca,'dataaspectratio',[1 1 1]), axis off; set(cax,'NextPlot',next);
end
set(get(gca,'xlabel'),'visible','on')
set(get(gca,'ylabel'),'visible','on')
```

Listing 11.16. MATLAB Function "dbesselj.m"

```
function [ res ] = dbesselj( nu,z )
res=besselj(nu-1,z)-besselj(nu,z)*nu/z;
```

Listing 11.17. MATLAB Function "dbessely.m"

```
function [ res ] = dbessely( nu,z )
res=bessely(nu-1,z)-bessely(nu,z)*nu/z;
```

Listing 11.18. MATLAB Function "dbesselh.m"

```
function [ res ] = dbesselh(nu,kind,z)
res=besselh(nu-1,kind,z)-besselh(nu,kind,z)*nu/z;
```

Listing 11.19. MATLAB Program "rcs_cylinder_complex.m"

```
% This program computes the backscattered RCS for a cylinder
% with flat plates.
clear all
index = 0;
eps =0.00001;
a1 =.125;
h = 1.;
lambda = 3.0e+8 /9.5e+9;
lambda = 0.00861;
index = 0;
```

```
for theta = 0.0:.1:90-.1
   index = index +1;
   theta = theta * pi /180.;
   rcs(index) = (lambda * a1 * sin(theta) / ...
      (8 * pi * (cos(theta))^2)) + eps;
end
theta*180/pi;
theta = pi/2;
index = index +1;
rcs(index) = (2 * pi * h^2 * a1 / lambda )+ eps;
for theta = 90+.1:.1:180.
   index = index + 1;
   theta = theta * pi / 180.;
   rcs(index) = ( lambda * a1 * sin(theta) / ...
      (8 * pi * (cos(theta))^2)) + eps;
end
r = a1;
index = 0;
for aspect_deg = 0.:.1:180
   index = index +1;
   aspect = (pi /180.) * aspect_deg;
% Compute RCS using Eq. (11.37)
   if (aspect == 0 | aspect == pi)
      rcs_po(index) = (4.0 * pi^3 * r^4 / lambda^2) + eps;
      rcs_mu(index) = rcs_po(1);
   else
      x = (4. * pi * r / lambda) * sin(aspect);
      val1 = 4. * pi^3 * r^4 / lambda^2;
      val2 = 2. * besselj(1,x) / x;
      rcs_po(index) = val1 * (val2 * cos(aspect))^2 + eps;
   end
 end
rcs_t =(rcs_po + rcs);
angle = 0:.1:180;
plot(angle,10*log10(rcs_t(1:1801)),'k');
grid;
xlabel ('Aspect angle -degrees');
ylabel ('RCS -dBsm');
```

Listing 11.20. MATLAB Program "Swerling_models.m"

```
% This program computes and plots Swerling statistical models
% sigma_bar = 1.5;
clear all
```

```
sigma = 0:0.001:6;
sigma_bar = 1.5;
swer_3_4 = (4. / sigma_bar^2) .* sigma .* ...
   exp(-2. * (sigma ./ sigma_bar));
%t.*exp(-(t.^2)./2.
swer_1_2 = (1. /sigma_bar) .* exp( -sigma ./ sigma_bar);
plot(sigma,swer_1_2,'k',sigma,swer_3_4,'k');
grid;
gtext ('Swerling I,II');
gtext ('Swerling III,IV');
xlabel ('sigma');
ylabel ('Probability density');
title ('sigma-bar = 1.5');
```

Chapter 12 *High Resolution Tactical Synthetic Aperture Radar (TSAR)*

This chapter is coauthored with Brian J. Smith[1]

This chapter provides an introduction to Tactical Synthetic Aperture Radar (TSAR). The purpose of this chapter is to further develop the readers' understanding of SAR by taking a closer look at high resolution spotlight SAR image formation algorithms, motion compensation techniques, autofocus algorithms, and performance metrics.

12.1. Introduction

Modern airborne radar systems are designed to perform a large number of functions which range from detection and discrimination of targets to mapping large areas of ground terrain. This mapping can be performed by the Synthetic Aperture Radar (SAR). Through illuminating the ground with coherent radiation and measuring the echo signals, SAR can produce high resolution two-dimensional (and in some cases three-dimensional) imagery of the ground surface. The quality of ground maps generated by SAR is determined by the size of the resolution cell. A resolution cell is specified by both range and azimuth resolutions of the system. Other factors affecting the size of the resolution cells are (1) size of the processed map and the amount of signal processing involved; (2) cost consideration; and (3) size of the objects that need to be resolved in the map. For example, mapping gross features of cities and coastlines does not require as much resolution when compared to resolving houses, vehicles, and streets.

---

1. Dr. Brian J. Smith is with the US Army Aviation and Missile Command (AMCOM), Redstone Arsenal, Alabama.

SAR systems can produce maps of reflectivity versus range and Doppler (cross range). Range resolution is accomplished through range gating. Fine range resolution can be accomplished by using pulse compression techniques. The azimuth resolution depends on antenna size and radar wavelength. Fine azimuth resolution is enhanced by taking advantage of the radar motion in order to synthesize a larger antenna aperture. Let N_r denote the number of range bins and let N_a denote the number of azimuth cells. It follows that the total number of resolution cells in the map is $N_r N_a$. SAR systems that are generally concerned with improving azimuth resolution are often referred to as Doppler Beam-Sharpening (DBS) SARs. In this case, each range bin is processed to resolve targets in Doppler which correspond to azimuth. This chapter is presented in the context of DBS.

Due to the large amount of signal processing required in SAR imagery, the early SAR designs implemented optical processing techniques. Although such optical processors can produce high quality radar images, they have several shortcomings. They can be very costly and are, in general, limited to making strip maps. Motion compensation is not easy to implement for radars that utilize optical processors. With the recent advances in solid state electronics and Very Large Scale Integration (VLSI) technologies, digital signal processing in real time has been made possible in SAR systems.

12.2. Side Looking SAR Geometry

Fig. 12.1 shows the geometry of the standard side looking SAR. We will assume that the platform carrying the radar maintains both fixed altitude h and velocity v. The antenna $3dB$ beamwidth is θ, and the elevation angle (measured from the z-axis to the antenna axis) is β. The intersection of the antenna beam with the ground defines a footprint. As the platform moves, the footprint scans a swath on the ground.

The radar position with respect to the absolute origin $\vec{O} = (0, 0, 0)$, at any time, is the vector $\vec{a}(t)$. The velocity vector $\vec{a}'(t)$ is

$$\vec{a}'(t) = 0 \times \hat{a}_x + v \times \hat{a}_y + 0 \times \hat{a}_z \tag{12.1}$$

The Line of Sight (LOS) for the current footprint centered at $\vec{q}(t_c)$ is defined by the vector $\vec{R}(t_c)$, where t_c denotes the central time of the observation interval T_{ob} (coherent integration interval). More precisely,

$$(t = t_a + t_c) \; ; \; -\frac{T_{ob}}{2} \le t \le \frac{T_{ob}}{2} \tag{12.2}$$

Side Looking SAR Geometry

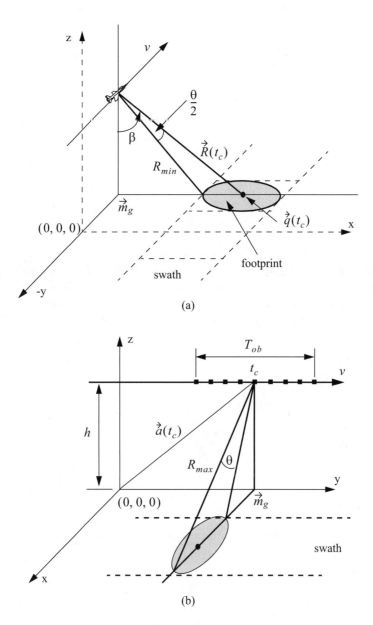

Figure 12.1. Side looking SAR geometry.

where t_a and t are the absolute and relative times, respectively. The vector $\vec{m}_g$ defines the ground projection of the antenna at central time. The minimum slant range to the swath is R_{min}, and the maximum range is denoted R_{max}, as illustrated by Fig. 12.2. It follows that

$$R_{min} = h/\cos(\beta - \theta/2)$$
$$R_{max} = h/\cos(\beta + \theta/2) \quad (12.3)$$
$$|\vec{R}(t_c)| = h/\cos\beta$$

Notice that the elevation angle β is equal to

$$\beta = 90 - \psi_g \quad (12.4)$$

where ψ_g is the grazing angle. The size of the footprint is a function of the grazing angle and the antenna beamwidth, as illustrated in Fig. 12.3. The SAR geometry described in this section is referred to as SAR "strip mode" of operation. Another SAR mode of operation, which will not be discussed in this chapter, is called "spot-light mode," where the antenna is steered (mechanically or electronically) to continuously illuminate one spot (footprint) on the ground. In this case, one high resolution image of the current footprint is generated during an observation interval.

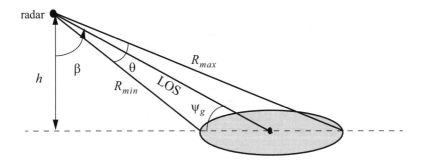

Figure 12.2. Definition of minimum and maximum range.

12.3. SAR Design Considerations

The quality of SAR images is heavily dependent on the size of the map resolution cell shown in Fig. 12.4. The range resolution, ΔR, is computed on the beam LOS, and is given by

$$\Delta R = (c\tau)/2 \quad (12.5)$$

SAR Design Considerations

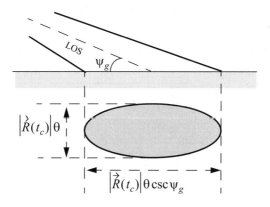

Figure 12.3. Footprint definition.

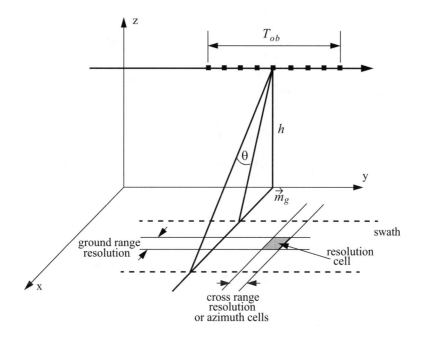

Figure 12.4a. Definition of a resolution cell.

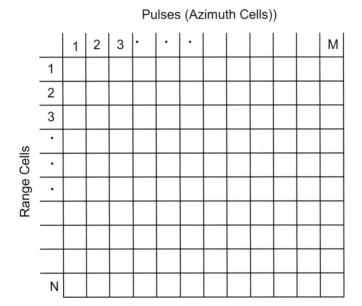

Figure 12.4b. Definition of a resolution cell.

where τ is the pulsewidth. From the geometry in Fig. 12.5 the extent of the range cell ground projection ΔR_g is computed as

$$\Delta R_g = \frac{c\tau}{2} \sec \psi_g \qquad (12.6)$$

The azimuth or cross range resolution for a real antenna with a $3dB$ beamwidth θ (radians) at range R is

$$\Delta A = \theta R \qquad (12.7)$$

However, the antenna beamwidth is proportional to the aperture size,

$$\theta \approx \frac{\lambda}{L} \qquad (12.8)$$

where λ is the wavelength and L is the aperture length. It follows that

$$\Delta A = \frac{\lambda R}{L} \qquad (12.9)$$

And since the effective synthetic aperture size is twice that of a real array, the azimuth resolution for a synthetic array is then given by

SAR Design Considerations

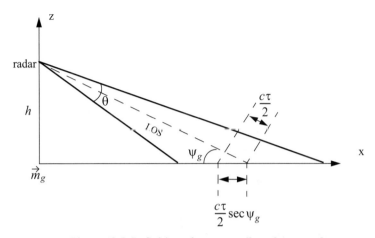

Figure 12.5. Definition of a range cell on the ground.

$$\Delta A = \frac{\lambda R}{2L} \tag{12.10}$$

Furthermore, since the synthetic aperture length L is equal to vT_{ob}, Eq. (12.10) can be rewritten as

$$\Delta A = \frac{\lambda R}{2vT_{ob}} \tag{12.11}$$

The azimuth resolution can be greatly improved by taking advantage of the Doppler variation within a footprint (or a beam). As the radar travels along its flight path the radial velocity to a ground scatterer (point target) within a footprint varies as a function of the radar radial velocity in the direction of that scatterer. The variation of Doppler frequency for a certain scatterer is called the "Doppler history."

Let $R(t)$ denote the range to a scatterer at time t, and v_r be the corresponding radial velocity; thus the Doppler shift is

$$f_d = -\frac{2R'(t)}{\lambda} = \frac{2v_r}{\lambda} \tag{12.12}$$

where $R'(t)$ is the range rate to the scatterer. Let t_1 and t_2 be the times when the scatterer enters and leaves the radar beam, respectively, and t_c be the time that corresponds to minimum range. Fig. 12.6 shows a sketch of the corresponding $R(t)$. Since the radial velocity can be computed as the derivative of $R(t)$ with respect to time, one can clearly see that Doppler frequency is maximum at t_1, zero at t_c, and minimum at t_2, as illustrated in Fig. 12.7.

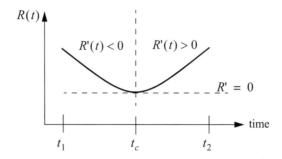

Figure 12.6. Sketch of range versus time for a scatterer.

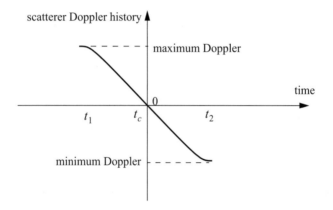

Figure 12.7. Point scatterer Doppler history.

In general, the radar maximum PRF, $f_{r_{max}}$, must be low enough to avoid range ambiguity. Alternatively, the minimum PRF, $f_{r_{min}}$, must be high enough to avoid Doppler ambiguity. SAR unambiguous range must be at least as wide as the extent of a footprint. More precisely, since target returns from maximum range due to the current pulse must be received by the radar before the next pulse is transmitted, it follows that SAR unambiguous range is given by

$$R_u = R_{max} - R_{min} \tag{12.13}$$

An expression for unambiguous range was derived in Chapter 1, and is repeated here as Eq. (12.14),

$$R_u = \frac{c}{2f_r} \tag{12.14}$$

SAR Design Considerations

Combining Eq. (12.14) and Eq. (12.13) yields

$$f_{r_{max}} \leq \frac{c}{2(R_{max} - R_{min})} \tag{12.15}$$

SAR minimum PRF, $f_{r_{min}}$, is selected so that Doppler ambiguity is avoided. In other words, $f_{r_{min}}$ must be greater than the maximum expected Doppler spread within a footprint. From the geometry of Fig. 12.8, the maximum and minimum Doppler frequencies are, respectively, given by

$$f_{d_{max}} = \frac{2v}{\lambda} \sin\left(\frac{\theta}{2}\right) \sin\beta \; ; \; at \; t_1 \tag{12.16}$$

$$f_{d_{min}} = -\frac{2v}{\lambda} \sin\left(\frac{\theta}{2}\right) \sin\beta \; ; \; at \; t_2 \tag{12.17}$$

It follows that the maximum Doppler spread is

$$\Delta f_d = f_{d_{max}} - f_{d_{min}} \tag{12.18}$$

Substituting Eqs. (12.16) and (12.17) into Eq. (12.18) and applying the proper trigonometric identities yield

$$\Delta f_d = \frac{4v}{\lambda} \sin\frac{\theta}{2} \sin\beta \tag{12.19}$$

Finally, by using the small angle approximation we get

$$\Delta f_d \approx \frac{4v}{\lambda} \frac{\theta}{2} \sin\beta = \frac{2v}{\lambda} \theta \sin\beta \tag{12.20}$$

Therefore, the minimum PRF is

$$f_{r_{min}} \geq \frac{2v}{\lambda} \theta \sin\beta \tag{12.21}$$

Combining Eqs. (11.15) and (11.21) we get

$$\frac{c}{2(R_{max} - R_{min})} \geq f_r \geq \frac{2v}{\lambda} \theta \sin\beta \tag{12.22}$$

It is possible to resolve adjacent scatterers at the same range within a footprint based only on the difference of their Doppler histories. For this purpose, assume that the two scatterers are within the *k*th range bin.

598 *MATLAB Simulations for Radar Systems Design*

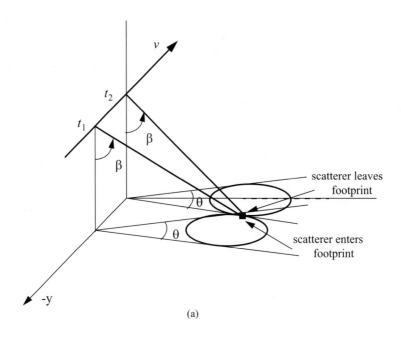

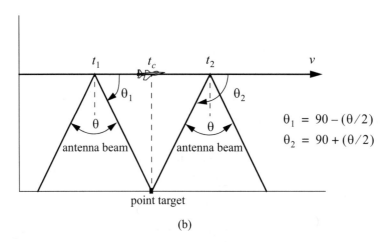

Figure 12.8. Doppler history computation. (a) Full view; (b) top view.

Denote their angular displacement as $\Delta\theta$, and let $\Delta f_{d_{min}}$ be the minimum Doppler spread between the two scatterers such that they will appear in two distinct Doppler filters. Using the same methodology that led to Eq. (12.20) we get

$$\Delta f_{d_{min}} = \frac{2v}{\lambda} \Delta\theta \sin\beta_k \qquad (12.23)$$

where β_k is the elevation angle corresponding to the kth range bin.

The bandwidth of the individual Doppler filters must be equal to the inverse of the coherent integration interval T_{ob} (i.e., $\Delta f_{d_{min}} = 1/T_{ob}$). It follows that

$$\Delta\theta = \frac{\lambda}{2vT_{ob}\sin\beta_k} \qquad (12.24)$$

Substituting L for vT_{ob} yields

$$\Delta\theta = \frac{\lambda}{2L\sin\beta_k} \qquad (12.25)$$

Therefore, the SAR azimuth resolution (within the kth range bin) is

$$\Delta A_g = \Delta\theta R_k = R_k \frac{\lambda}{2L\sin\beta_k} \qquad (12.26)$$

Note that when $\beta_k = 90°$, Eq. (12.26) is identical to Eq. (12.10).

12.4. SAR Radar Equation

The single pulse radar equation was derived in Chapter 1, and is repeated here as Eq. (12.27),

$$SNR = \frac{P_t G^2 \lambda^2 \sigma}{(4\pi)^3 R_k^4 kT_0 BL_{Loss}} \qquad (12.27)$$

where: P_t is peak power; G is antenna gain; λ is wavelength; σ is radar cross section; R_k is radar slant range to the kth range bin; k is Boltzman's constant; T_0 is receiver noise temperature; B is receiver bandwidth; and L_{Loss} is radar losses. The radar cross section is a function of the radar resolution cell and terrain reflectivity. More precisely,

$$\sigma = \sigma^0 \Delta R_g \Delta A_g = \sigma^0 \Delta A_g \frac{c\tau}{2} \sec\psi_g \qquad (12.28)$$

where σ^0 is the clutter scattering coefficient, ΔA_g is the azimuth resolution, and Eq. (12.6) was used to replace the ground range resolution. The number of coherently integrated pulses within an observation interval is

$$n = f_r T_{ob} = \frac{f_r L}{v} \tag{12.29}$$

where L is the synthetic aperture size. Using Eq. (12.26) in Eq. (12.29) and rearranging terms yield

$$n = \frac{\lambda R f_r}{2 \Delta A_g v} \csc \beta_k \tag{12.30}$$

The radar average power over the observation interval is

$$P_{av} = (P_t / B) f_r \tag{12.31}$$

The SNR for n coherently integrated pulses is then

$$(SNR)_n = nSNR = n \frac{P_t G^2 \lambda^2 \sigma}{(4\pi)^3 R_k^4 k T_0 B L_{Loss}} \tag{12.32}$$

Substituting Eqs. (11.31), (11.30), and (11.28) into Eq. (12.32) and performing some algebraic manipulations give the SAR radar equation,

$$(SNR)_n = \frac{P_{av} G^2 \lambda^3 \sigma^0}{(4\pi)^3 R_k^3 k T_0 L_{Loss}} \frac{\Delta R_g}{2v} \csc \beta_k \tag{12.33}$$

Eq. (12.33) leads to the conclusion that in SAR systems the SNR is (1) inversely proportional to the third power of range; (2) independent of azimuth resolution; (3) function of the ground range resolution; (4) inversely proportional to the velocity v; and (5) proportional to the third power of wavelength.

12.5. SAR Signal Processing

There are two signal processing techniques to sequentially produce a SAR map or image; they are line-by-line processing and Doppler processing. The concept of SAR line-by-line processing is as follows: Through the radar linear motion a synthetic array is formed, where the elements of the current synthetic array correspond to the position of the antenna transmissions during the last observation interval. Azimuth resolution is obtained by forming narrow synthetic beams through combinations of the last observation interval returns. Fine range resolution is accomplished in real time by utilizing range gating and pulse compression. For each range bin and each of the transmitted pulses during the last observation interval, the returns are recorded in a two-dimensional array of data that is updated for every pulse. Denote the two-dimensional array of data as MAP.

To further illustrate the concept of line-by-line processing, consider the case where a map of size $N_a \times N_r$ is to be produced, where N_a is the number of azimuth cells and N_r is the number of range bins. Hence, MAP is of size $N_a \times N_r$, where the columns refer to range bins, and the rows refer to azimuth cells. For each transmitted pulse, the echoes from consecutive range bins are recorded sequentially in the first row of MAP. Once the first row is completely filled (i.e., returns from all range bins have been received), all data (in all rows) are shifted downward one row before the next pulse is transmitted. Thus, one row of MAP is generated for every transmitted pulse. Consequently, for the current observation interval, returns from the first transmitted pulse will be located in the bottom row of MAP, and returns from the last transmitted pulse will be in the first row of MAP.

In SAR Doppler processing, the array MAP is updated once every N pulses so that a block of N columns is generated simultaneously. In this case, N refers to the number of transmissions during an observation interval (i.e., size of the synthetic array). From an antenna point of view, this is equivalent to having N adjacent synthetic beams formed in parallel through electronic steering.

12.6. Side Looking SAR Doppler Processing

Consider the geometry shown in Fig. 12.9, and assume that the scatterer C_i is located within the kth range bin. The scatterer azimuth and elevation angles are μ_i and β_i, respectively. The scatterer elevation angle β_i is assumed to be equal to β_k, the range bin elevation angle. This assumption is true if the ground range resolution, ΔR_g, is small; otherwise, $\beta_i = \beta_k + \varepsilon_i$ for some small ε_i; in this chapter $\varepsilon_i = 0$.

The normalized transmitted signal can be represented by

$$s(t) = \cos(2\pi f_0 t - \xi_0) \tag{12.34}$$

where f_0 is the radar operating frequency, and ξ_0 denotes the transmitter phase. The returned radar signal from C_i is then equal to

$$s_i(t, \mu_i) = A_i \cos[2\pi f_0(t - \tau_i(t, \mu_i)) - \xi_0] \tag{12.35}$$

where $\tau_i(t, \mu_i)$ is the round-trip delay to the scatterer, and A_i includes scatterer strength, range attenuation, and antenna gain. The round-trip delay is

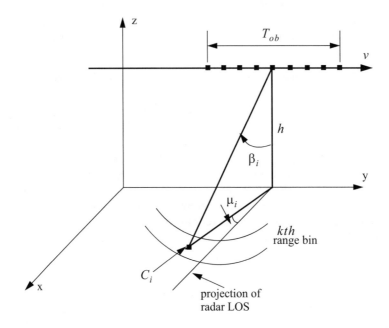

Figure 12.9. A scatterer C_i within the kth range bin.

$$\tau_i(t, \mu_i) = \frac{2r_i(t, \mu_i)}{c} \quad (12.36)$$

where c is the speed of light and $r_i(t, \mu_i)$ is the scatterer slant range. From the geometry in Fig. 12.9, one can write the expression for the slant range to the ith scatterer within the kth range bin as

$$r_i(t, \mu_i) = \frac{h}{\cos\beta_i}\sqrt{1 - \frac{2vt}{h}\cos\beta_i\cos\mu_i\sin\beta_i + \left(\frac{vt}{h}\cos\beta_i\right)^2} \quad (12.37)$$

And by using Eq. (12.36) the round-trip delay can be written as

$$\tau_i(t, \mu_i) = \frac{2}{c}\frac{h}{\cos\beta_i}\sqrt{1 - \frac{2vt}{h}\cos\beta_i\cos\mu_i\sin\beta_i + \left(\frac{vt}{h}\cos\beta_i\right)^2} \quad (12.38)$$

The round-trip delay can be approximated using a two-dimensional second order Taylor series expansion about the reference state $(t, \mu) = (0, 0)$. Performing this Taylor series expansion yields

$$\tau_i(t, \mu_i) \approx \bar{\tau} + \bar{\tau}_{t\mu} \mu_i t + \bar{\tau}_{tt} \frac{t^2}{2} \tag{12.39}$$

where the over-bar indicates evaluation at the state $(0, 0)$, and the subscripts denote partial derivatives. For example, $\bar{\tau}_{t\mu}$ means

$$\bar{\tau}_{t\mu} = \frac{\partial^2}{\partial t \partial \mu} \tau_i(t, \mu_i) \bigg|_{(t, \mu) = (0, 0)} \tag{12.40}$$

The Taylor series coefficients are

$$\bar{\tau} = \left(\frac{2h}{c}\right) \frac{1}{\cos \beta_i} \tag{12.41}$$

$$\bar{\tau}_{t\mu} = \left(\frac{2v}{c}\right) \sin \beta_i \tag{12.42}$$

$$\bar{\tau}_{tt} = \left(\frac{2v^2}{hc}\right) \cos \beta_i \tag{12.43}$$

Note that other Taylor series coefficients are either zeros or very small. Hence, they are neglected. Finally, we can rewrite the returned radar signal as

$$s_i(t, \mu_i) = A_i \cos[\hat{\psi}_i(t, \mu_i) - \xi_0]$$
$$\hat{\psi}_i(t, \mu_i) = 2\pi f_0 \left[(1 - \bar{\tau}_{t\mu} \mu_i) t - \bar{\tau} - \bar{\tau}_{tt} \frac{t^2}{2} \right] \tag{12.44}$$

Observation of Eq. (12.44) indicates that the instantaneous frequency for the ith scatterer varies as a linear function of time due to the second order phase term $2\pi f_0(\bar{\tau}_{tt} t^2 / 2)$ (this confirms the result we concluded about a scatterer Doppler history). Furthermore, since this phase term is range-bin dependent and not scatterer dependent, all scatterers within the same range bin produce this exact second order phase term. It follows that scatterers within a range bin have identical Doppler histories. These Doppler histories are separated by the time delay required to fly between them, as illustrated in Fig. 12.10.

Suppose that there are I scatterers within the kth range bin. In this case, the combined returns for this cell are the sum of the individual returns due to each scatterer as defined by Eq. (12.44). In other words, superposition holds, and the overall echo signal is

$$s_r(t) = \sum_{i=1}^{I} s_i(t, \mu_i) \tag{12.45}$$

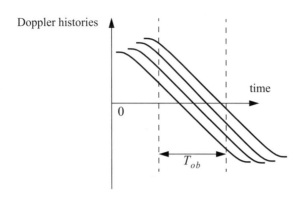

Figure 12.10. Doppler histories for several scatterers within the same range bin.

A signal processing block diagram for the *kth* range bin is illustrated in Fig. 12.11. It consists of the following steps. First, heterodyning with the carrier frequency is performed to extract the quadrature components.

This is followed by LP filtering and A/D conversion. Next, deramping or focusing to remove the second order phase term of the quadrature components is carried out using a phase rotation matrix. The last stage of the processing includes windowing, performing an FFT on the windowed quadrature components, and scaling the amplitude spectrum to account for range attenuation and antenna gain.

The discrete quadrature components are

$$\tilde{x}_I(t_n) = \tilde{x}_I(n) = A_i \cos[\tilde{\psi}_i(t_n, \mu_i) - \xi_0]$$
$$\tilde{x}_Q(t_n) = \tilde{x}_Q(n) = A_i \sin[\tilde{\psi}_i(t_n, \mu_i) - \xi_0]$$
(12.46)

$$\tilde{\psi}_i(t_n, \mu_i) = \hat{\psi}_i(t_n, \mu_i) - 2\pi f_0 t_n$$
(12.47)

and t_n denotes the *nth* sampling time (remember that $-T_{ob}/2 \leq t_n \leq T_{ob}/2$). The quadrature components after deramping (i.e., removal of the phase $\psi = -\pi f_0 \bar{\tau}_{tt} t_n^2$) are given by

$$\begin{bmatrix} x_I(n) \\ x_Q(n) \end{bmatrix} = \begin{bmatrix} \cos\psi & -\sin\psi \\ \sin\psi & \cos\psi \end{bmatrix} \begin{bmatrix} \tilde{x}_I(n) \\ \tilde{x}_Q(n) \end{bmatrix}$$
(12.48)

Side Looking SAR Doppler Processing

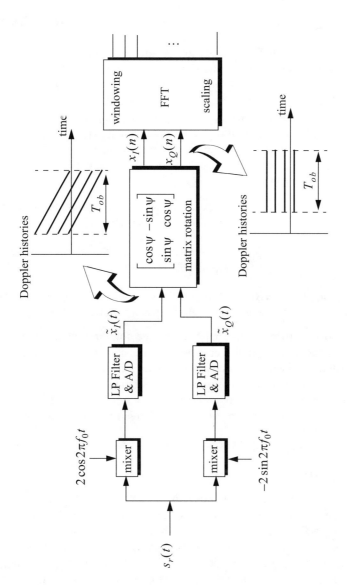

Figure 12.11. Signal processing block diagram for the kth range bin.

12.7. SAR Imaging Using Doppler Processing

It was mentioned earlier that SAR imaging is performed using two orthogonal dimensions (range and azimuth). Range resolution is controlled by the receiver bandwidth and pulse compression. Azimuth resolution is limited by the antenna beamwidth. A one-to-one correspondence between the FFT bins and the azimuth resolution cells can be established by utilizing the signal model described in the previous section. Therefore, the problem of target detection is transformed into a spectral analysis problem, where detection is based on the amplitude spectrum of the returned signal. The FFT frequency resolution Δf is equal to the inverse of the observation interval T_{ob}. It follows that a peak in the amplitude spectrum at $k_1 \Delta f$ indicates the presence of a scatterer at frequency $f_{d1} = k_1 \Delta f$.

For an example, consider the scatterer C_i within the $k\text{th}$ range bin. The instantaneous frequency f_{di} corresponding to this scatterer is

$$f_{di} = \frac{1}{2\pi} \frac{d\psi}{dt} = f_0 \bar{\tau}_{t\mu} \mu_i = \frac{2v}{\lambda} \sin\beta_i \mu_i \qquad (12.49)$$

This is the same result derived in Eq. (12.23), with $\mu_i = \Delta\theta$. Therefore, the scatterers separated in Doppler by more than Δf can then be resolved.

Fig. 12.12 shows a two-dimensional SAR image for three point scatterers located 10 Km down-range. In this case, the azimuth and range resolutions are equal to 1 m and the operating frequency is 35GHz. Fig. 12.13 is similar to Fig. 12.12, except in this case the resolution cell is equal to 6 inches. One can clearly see the blurring that occurs in the image. Figs. 12.12 and 12.13 can be reproduced using the program *"fig12_12_13.m"* given in Listing 12.1 in Section 12.10.

12.8. Range Walk

As shown earlier, SAR Doppler processing is achieved in two steps: first, range gating and second, azimuth compression within each bin at the end of the observation interval. For this purpose, azimuth compression assumes that each scatterer remains within the same range bin during the observation interval. However, since the range gates are defined with respect to a radar that is moving, the range gate grid is also moving relative to the ground. As a result a scatterer appears to be moving within its range bin. This phenomenon is known as range walk. A small amount of range walk does not bother Doppler processing as long as the scatterer remains within the same range bin. However, range walk over several range bins can constitute serious problems, where in this case Doppler processing is meaningless.

Range Walk

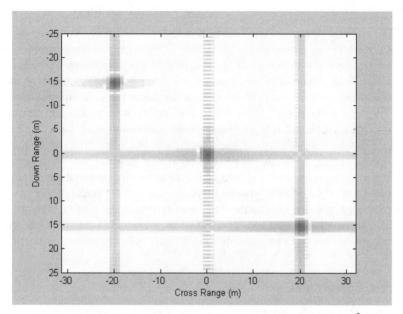

Figure 12.12. Three point scatterer image. Resolution cell is 1m².

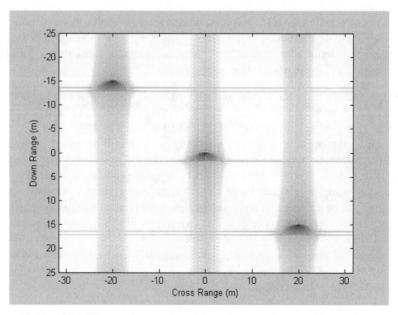

Figure 12.13. Three point scatterer image. Resolution cell is squared inches.

12.9. A Three-Dimensional SAR Imaging Technique

This section presents a new three-dimensional (3-D) Synthetic Aperture Radar (SAR) imaging technique.[1] It utilizes a linear array in transverse motion to synthesize a two-dimensional (2-D) synthetic array. Elements of the linear array are fired sequentially (one element at a time), while all elements receive in parallel. A 2-D information sequence is computed from the equiphase two-way signal returns. A signal model based on a third-order Taylor series expansion about incremental relative time, azimuth, elevation, and target height is used. Scatterers are detected as peaks in the amplitude spectrum of the information sequence. Detection is performed in two stages. First, all scatterers within a footprint are detected using an incomplete signal model where target height is set to zero. Then, processing using the complete signal model is performed only on range bins containing significant scatterer returns. The difference between the two images is used to measure target height. Computer simulation shows that this technique is accurate and virtually impulse invariant.

12.9.1. Background

Standard Synthetic Aperture Radar (SAR) imaging systems are generally used to generate high resolution two-dimensional (2-D) images of ground terrain. Range gating determines resolution along the first dimension. Pulse compression techniques are usually used to achieve fine range resolution. Such techniques require the use of wide band receiver and display devices in order to resolve the time structure in the returned signals. The width of azimuth cells provides resolution along the other dimension. Azimuth resolution is limited by the duration of the observation interval.

This section presents a three-dimensional (3-D) SAR imaging technique based on Discrete Fourier Transform (DFT) processing of equiphase data collected in sequential mode (DFTSQM). It uses a linear array in transverse motion to synthesize a 2-D synthetic array. A 2-D information sequence is computed from the equiphase two-way signal returns. To this end, a new signal model based on a third-order Taylor series expansion about incremental relative time, azimuth, elevation, and target height is introduced. Standard SAR imaging can be achieved using an incomplete signal model where target height is set to zero. Detection is performed in two stages. First, all scatterers within a footprint are detected using an incomplete signal model, where target height is set to zero. Then, processing using the complete signal model is performed

---

1. This section is extracted from: Mahafza, B. R. and Sajjadi, M., Three-Dimensional SAR Imaging Using a Linear Array in Transverse Motion, *IEEE - AES Trans.*, Vol. 32, No. 1, January 1996, pp. 499-510.

only on range bins containing significant scatterer returns. The difference between the two images is used as an indication of target height. Computer simulation shows that this technique is accurate and virtually impulse invariant.

12.9.2. DFTSQM Operation and Signal Processing

Linear Arrays

Consider a linear array of size N, uniform element spacing d, and wavelength λ. Assume a far field scatterer P located at direction-sine $\sin\beta_l$. DFTSQM operation for this array can be described as follows. The elements are fired sequentially, one at a time, while all elements receive in parallel. The echoes are collected and integrated coherently on the basis of equal phase to compute a complex information sequence $\{b(m); m = 0, 2N-1\}$. The x-coordinates, in d-units, of the x_n^{th} element with respect to the center of the array is

$$x_n = \left(-\frac{N-1}{2} + n\right); n = 0, \ldots N-1 \quad (12.50)$$

The electric field received by the x_2^{th} element due to the firing of the x_1^{th}, and reflection by the lth far field scatterer P, is

$$E(x_1, x_2; s_l) = G^2(s_l)\left(\frac{R_0}{R}\right)^4 \sqrt{\sigma_l}\, exp(j\phi(x_1, x_2; s_l)) \quad (12.51)$$

$$\phi(x_1, x_2; s_l) = \frac{2\pi}{\lambda}(x_1 + x_2)(s_l) \quad (12.52)$$

$$s_l = \sin\beta_l \quad (12.53)$$

where $\sqrt{\sigma_l}$ is the target cross section, $G^2(s_l)$ is the two-way element gain, and $(R_0/R)^4$ is the range attenuation with respect to reference range R_0. The scatterer phase is assumed to be zero; however it could be easily included. Assuming multiple scatterers in the array's FOV, the cumulative electric field in the path $x_1 \Rightarrow x_2$ due to reflections from all scatterers is

$$E(x_1, x_2) = \sum_{\text{all } l} [E_I(x_1, x_2; s_l) + jE_Q(x_1, x_2; s_l)] \quad (12.54)$$

where the subscripts (I, Q) denote the quadrature components. Note that the variable part of the phase given in Eq. (12.52) is proportional to the integers resulting from the sums $\{(x_{n1} + x_{n2}); (n1, n2) = 0, \ldots N-1\}$. In the far field operation there are a total of $(2N-1)$ distinct $(x_{n1} + x_{n2})$ sums. Therefore, the electric fields with paths of the same $(x_{n1} + x_{n2})$ sums can be collected coherently. In this manner the information sequence $\{b(m); m = 0, 2N-1\}$ is computed, where $b(2N-1)$ is set to zero. At the same time one forms the sequence $\{c(m); m = 0, \ldots 2N-2\}$ which keeps track of the number of returns that have the same $(x_{n1} + x_{n2})$ sum. More precisely, for $m = n1 + n2$; $(n1, n2) = \ldots 0, N-1$

$$b(m) = b(m) + E(x_{n1}, x_{n2}) \tag{12.55}$$

$$c(m) = c(m) + 1 \tag{12.56}$$

It follows that

$$\{c(m); m = 0, \ldots 2N-2\} = \begin{cases} m+1 \ ; \ m = 0, \ldots N-2 \\ N \ ; \ m = N-1 \\ 2N-1-m \ \ m = N, \ldots 2N-2 \end{cases} \tag{12.57}$$

which is a triangular shape sequence.

The processing of the sequence $\{b(m)\}$ is performed as follows: (1) the weighting takes the sequence $\{c(m)\}$ into account; (2) the complex sequence $\{b(m)\}$ is extended to size N_F, a power integer of two, by zero padding; (3) the DFT of the extended sequence $\{b'(m); m = 0, N_F - 1\}$ is computed,

$$B(q) = \sum_{m=0}^{N_F-1} b'(m) \cdot exp\left(-j\frac{2\pi qm}{N_F}\right); q = 0, \ldots N_F - 1 \tag{12.58}$$

and, (4) after compensation for antenna gain and range attenuation, scatterers are detected as peaks in the amplitude spectrum $|B(q)|$. Note that step (4) is true only when

$$\sin\beta_q = \frac{\lambda q}{2Nd}; q = 0, \ldots 2N-1 \tag{12.59}$$

where $\sin\beta_q$ denotes the direction-sine of the qth scatterer, and $N_F = 2N$ is implied in Eq. (12.59).

The classical approach to multiple target detection is to use a phased array antenna with phase shifting and tapering hardware. The array beamwidth is proportional to (λ/Nd), and the first sidelobe is at about -13 dB. On the other hand, multiple target detection using DFTSQM provides a beamwidth proportional to $(\lambda/2Nd)$ as indicated by (Eq. (12.59), which has the effect of doubling the array's resolution. The first sidelobe is at about -27 dB due to the triangular sequence $\{c(m)\}$. Additionally, no phase shifting hardware is required for detection of targets within a single element's field of view.

Rectangular Arrays

DFTSQM operation and signal processing for 2-D arrays can be described as follows. Consider an $N_x \times N_y$ rectangular array. All $N_x N_y$ elements are fired sequentially, one at a time. After each firing, all the $N_x N_y$ array elements receive in parallel. Thus, $N_x N_y$ samples of the quadrature components are collected after each firing, and a total of $(N_x N_y)^2$ samples will be collected. However, in the far field operation, there are only $(2N_x - 1) \times (2N_y - 1)$ distinct equiphase returns. Therefore, the collected data can be added coherently to form a 2-D information array of size $(2N_x - 1) \times (2N_y - 1)$. The two-way radiation pattern is computed as the modulus of the 2-D amplitude spectrum of the information array. The processing includes 2-D windowing, 2-D Discrete Fourier Transformation, antenna gain, and range attenuation compensation. The field of view of the 2-D array is determined by the 3 dB pattern of a single element. All the scatterers within this field will be detected simultaneously as peaks in the amplitude spectrum.

Consider a rectangular array of size $N \times N$, with uniform element spacing $d_x = d_y = d$, and wavelength λ. The coordinates of the nth element, in d-units, are

$$x_n = \left(-\frac{N-1}{2} + n\right) \quad ; n = 0, \ldots N-1 \quad (12.60)$$

$$y_n = \left(-\frac{N-1}{2} + n\right) \quad ; n = 0, \ldots N-1 \quad (12.61)$$

Assume a far field point P defined by the azimuth and elevation angles (α, β). In this case, the one-way geometric phase for an element is

$$\varphi'(x, y) = \frac{2\pi}{\lambda}[x \sin\beta \cos\alpha + y \sin\beta \sin\alpha] \quad (12.62)$$

Therefore, the two-way geometric phase between the (x_1, y_1) and (x_2, y_2) elements is

$$\varphi(x_1, y_1, x_2, y_2) = \frac{2\pi}{\lambda} \sin\beta[(x_1 + x_2)\cos\alpha + (y_1 + y_2)\sin\alpha] \quad (12.63)$$

The two-way electric field for the lth scatterer at (α_l, β_l) is

$$E(x_1, x_2, y_1, y_2; \alpha_l, \beta_l) = G^2(\beta_l)\left(\frac{R_0}{R}\right)^4 \sqrt{\sigma_l} \, exp[j(\varphi(x_1, y_1, x_2, y_2))] \quad (12.64)$$

Assuming multiple scatterers within the array's FOV, then the cumulative electric field for the two-way path $(x_1, y_1) \Rightarrow (x_2, y_2)$ is given by

$$E(x_1, x_2, y_1, y_2) = \sum_{\text{all scatterers}} E(x_1, x_2, y_1, y_2; \alpha_l, \beta_l) \tag{12.65}$$

All formulas for the 2-D case reduce to those of a linear array case by setting $N_y = 1$ and $\alpha = 0$.

The variable part of the phase given in Eq. (12.63) is proportional to the integers $(x_1 + x_2)$ and $(y_1 + y_2)$. Therefore, after completion of the sequential firing, electric fields with paths of the same (i, j) sums, where

$$\{i = x_{n1} + x_{n2}; i = -(N-1), \ldots(N-1)\} \tag{12.66}$$

$$\{j = y_{n1} + y_{n2}; j = -(N-1), \ldots(N-1)\} \tag{12.67}$$

can be collected coherently. In this manner the 2-D information array $\{b(m_x, m_y); (m_x, m_y) = 0, \ldots 2N-1\}$ is computed. The coefficient sequence $\{c(m_x, m_y); (m_x, m_y) = 0, \ldots 2N-2\}$ is also computed. More precisely,

$$\begin{aligned} \text{for } m_x &= n1 + n2 \text{ and } m_y = n1 + n2 \\ n1 &= 0, \ldots N-1, \text{ and } n2 = 0, \ldots N-1 \end{aligned} \tag{12.68}$$

$$b(m_x, m_y) = b(m_x, m_y) + E(x_{n1}, y_{n1}, x_{n2}, y_{n2}) \tag{12.69}$$

It follows that

$$c(m_x, m_y) = (N_x - |m_x - (N_x - 1)|) \times (N_y - |m_y - (N_y - 1)|) \tag{12.70}$$

The processing of the complex 2-D information array $\{b(m_x, m_y)\}$ is similar to that of the linear case with the exception that one should use a 2-D DFT. After antenna gain and range attenuation compensation, scatterers are detected as peaks in the 2-D amplitude spectrum of the information array. A scatterer located at angles (α_l, β_l) will produce a peak in the amplitude spectrum at DFT indexes (p_l, q_l), where

$$\alpha_l = \operatorname{atan}\left(\frac{q_l}{p_l}\right) \tag{12.71}$$

$$\sin\beta_l = \frac{\lambda p_l}{2Nd\cos\alpha_l} = \frac{\lambda q_l}{2Nd\sin\alpha_l} \tag{12.72}$$

Derivation of Eq. (12.71) is in Section 12.9.7.

12.9.3. Geometry for DFTSQM SAR Imaging

Fig. 12.14 shows the geometry of the DFTSQM SAR imaging system. In this case, t_c denotes the central time of the observation interval, D_{ob}. The aircraft maintains both constant velocity v and height h. The origin for the relative system of coordinates is denoted as $\vec{O}$. The vector $\vec{OM}$ defines the radar location at time t_c. The transmitting antenna consists of a linear real array operating in the sequential mode. The real array is of size N, element spacing d, and the radiators are circular dishes of diameter $D = d$. Assuming that the aircraft scans M transmitting locations along the flight path, then a rectangular array of size $N \times M$ is synthesized, as illustrated in Fig. 12.15.

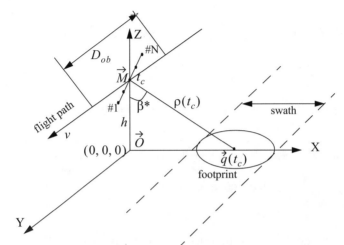

Figure 12.14. Geometry for DFTSQM imaging system.

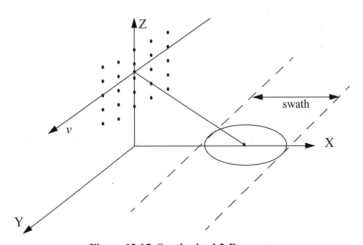

Figure 12.15. Synthesized 2-D array.

The vector $\vec{q}(t_c)$ defines the center of the 3 dB footprint at time t_c. The center of the array coincides with the flight path, and it is assumed to be perpendicular to both the flight path and the line of sight $\rho(t_c)$. The unit vector $\vec{a}$ along the real array is

$$\vec{a} = \cos\beta^* \vec{a}_x + \sin\beta^* \vec{a}_z \quad (12.73)$$

where β^* is the elevation angle, or the complement of the depression angle, for the center of the footprint at central time t_c.

12.9.4. Slant Range Equation

Consider the geometry shown in Fig. 12.16 and assume that there is a scatterer $\vec{C}_i$ within the kth range cell. This scatterer is defined by

$$\{amplitude, phase, elevation, azimuth, height\} = \quad (12.74)$$
$$\{a_i, \phi_i, \beta_i, \mu_i, \tilde{h}_i\}$$

The scatterer $\vec{C}_i$ (assuming rectangular coordinates) is given by

$$\vec{C}_i = h\tan\beta_i \cos\mu_i \vec{a}_x + h\tan\beta_i \sin\mu_i \vec{a}_y + \tilde{h}_i \vec{a}_z \quad (12.75)$$

$$\beta_i = \beta_k + \varepsilon \quad (12.76)$$

where β_k denotes the elevation angle for the kth range cell at the center of the observation interval and ε is an incremental angle. Let $\vec{O}e_n$ refer to the vector between the nth array element and the point $\vec{O}$, then

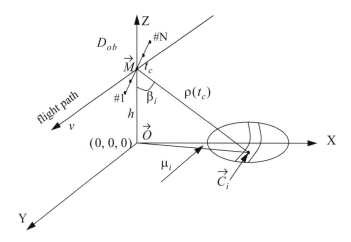

Figure 12.16. Scatterer $\vec{C}_i$ within a range cell.

$$\vec{O}e_n = D_n\cos\beta^*\vec{a}_x + vt\vec{a}_y + (D_n\sin\beta^* + h)\vec{a}_z \qquad (12.77)$$

$$D_n = \left(\frac{1-N}{2}+n\right)d \; ;n = 0,...N-1 \qquad (12.78)$$

The range between a scatterer $\vec{C}$ within the kth range cell and the nth element of the real array is

$$r_n^2(t,\varepsilon,\mu,\tilde{h};D_n) = D_n^2 + v^2t^2 + (h-\tilde{h})^2 + 2D_n\sin\beta^*(h-\tilde{h}) + \qquad (12.79)$$
$$h\tan(\beta_k+\varepsilon)[h\tan(\beta_k+\varepsilon) - 2D_n\cos\beta^*\cos\mu - 2vt\sin\mu]$$

It is more practical to use the scatterer's elevation and azimuth direction-sines rather than the corresponding increments. Therefore, define the scatterer's azimuth and elevation direction-sines as

$$s = \sin\mu \qquad (12.80)$$

$$u = \sin\varepsilon \qquad (12.81)$$

Then, one can rewrite Eq. (12.79) as

$$r_n^2(t,s,u,\tilde{h};D_n) = D_n^2 + v^2t^2 + (h-\tilde{h})^2 + h^2f^2(u) + \qquad (12.82)$$
$$2D_n\sin\beta^*(h-\tilde{h}) - (2D_nh\cos\beta^*f(u))\sqrt{1-s^2} - 2vhtf(u)s$$

$$f(u) = \tan(\beta_k + a\sin u) \qquad (12.83)$$

Expanding r_n as a third order Taylor series expansion about incremental $(t,s,u,\tilde{h})$ yields

$$r(t,s,u,\tilde{h};D_n) = \bar{r} + \bar{r}_{\tilde{h}}\tilde{h} + \bar{r}_u u + \bar{r}_{\tilde{h}\tilde{h}}\frac{\tilde{h}^2}{2} + \bar{r}_{hu}\tilde{h}u + \bar{r}_{ss}\frac{s^2}{2} + \bar{r}_{st}st + \qquad (12.84)$$
$$\bar{r}_{tt}\frac{t^2}{2} + \bar{r}_{uu}\frac{u^2}{2} + \bar{r}_{\tilde{h}\tilde{h}\tilde{h}}\frac{\tilde{h}^3}{6} + \bar{r}_{\tilde{h}\tilde{h}u}\frac{\tilde{h}^2u}{2} + \bar{r}_{hst}hst + \bar{r}_{huu}\frac{hu^2}{2} +$$
$$\bar{r}_{hss}\frac{hs^2}{2} + \bar{r}_{uss}\frac{us^2}{2} + \bar{r}_{stu}stu + \bar{r}_{suu}\frac{su^2}{2} + \bar{r}_{\tilde{t}\tilde{h}\tilde{h}}\frac{t\tilde{h}^2}{2} + \bar{r}_{utt}\frac{ut^2}{2} + \bar{r}_{uuu}\frac{u^3}{6}$$

where subscripts denote partial derivations, and the over-bar indicates evaluation at the state $(t,s,u,\tilde{h}) = (0,0,0,0)$. Note that

$$\{\bar{r}_s = \bar{r}_t = \bar{r}_{\tilde{h}s} = \bar{r}_{\tilde{h}t} = \bar{r}_{su} = \bar{r}_{tu} = \bar{r}_{\tilde{h}\tilde{h}s} = \bar{r}_{\tilde{h}\tilde{h}t} = \bar{r}_{\tilde{h}su} = \bar{r}_{\tilde{h}tu} = \qquad (12.85)$$
$$\bar{r}_{sss} = \bar{r}_{sst} = \bar{r}_{stt} = \bar{r}_{ttt} = \bar{r}_{tsu} = 0\}$$

Section 12.9.8 has detailed expressions of all non-zero Taylor series coefficients for the kth range cell.

Even at the maximum increments $t_{mx}, s_{mx}, u_{mx}, \tilde{h}_{mx}$, the terms:

$$\left\{ \tilde{r}_{\tilde{h}\tilde{h}\tilde{h}}\frac{\tilde{h}^3}{6}, \tilde{r}_{\tilde{h}\tilde{h}u}\frac{\tilde{h}^2 u}{2}, \tilde{r}_{\tilde{h}uu}\frac{\tilde{h}u^2}{2}, \tilde{r}_{\tilde{h}ss}\frac{\tilde{h}s^2}{2}, \right.$$ (12.86)

$$\left. \tilde{r}_{uss}\frac{us^2}{2}, \tilde{r}_{stu}stu, \tilde{r}_{suu}\frac{su^2}{2}, \tilde{r}_{t\tilde{h}\tilde{h}}\frac{t\tilde{h}^2}{2}, \tilde{r}_{utt}\frac{ut^2}{2}, \tilde{r}_{uuu}\frac{u^3}{6} \right\}$$

are small and can be neglected. Thus, the range r_n is approximated by

$$r(t, s, u, \tilde{h}; D_n) = \tilde{r} + \tilde{r}_{\tilde{h}}\tilde{h} + \tilde{r}_u u + \tilde{r}_{\tilde{h}\tilde{h}}\frac{\tilde{h}^2}{2} + \tilde{r}_{\tilde{h}u}\tilde{h}u +$$ (12.87)

$$\tilde{r}_{ss}\frac{s^2}{2} + \tilde{r}_{st}st + \tilde{r}_{tt}\frac{t^2}{2} + \tilde{r}_{uu}\frac{u^2}{2} + \tilde{r}_{hst}\tilde{h}st$$

Consider the following two-way path: the n_1^{th} element transmitting, scatterer $\vec{C}_i$ reflecting, and the n_2^{th} element receiving. It follows that the round trip delay corresponding to this two-way path is

$$\tau_{n_1 n_2} = \frac{1}{c}(r_{n_1}(t, s, u, \tilde{h}; D_{n_1}) + r_{n_2}(t, s, u, \tilde{h}; D_{n_2}))$$ (12.88)

where c is the speed of light.

12.9.5. Signal Synthesis

The observation interval is divided into M subintervals of width $\Delta t = (D_{ob} \div M)$. During each subinterval, the real array is operated in sequential mode, and an array length of $2N$ is synthesized. The number of subintervals M is computed such that Δt is large enough to allow sequential transmission for the real array without causing range ambiguities. In other words, if the maximum range is denoted as R_{mx} then

$$\Delta t > N\frac{2R_{mx}}{c}$$ (12.89)

Each subinterval is then partitioned into N sampling subintervals of width $2R_{mx}/c$. The location t_{mn} represents the sampling time at which the nth element is transmitting during the mth subinterval.

The normalized transmitted signal during the mth subinterval for the nth element is defined as

$$s_n(t_{mn}) = \cos(2\pi f_o t_{mn} + \zeta)$$ (12.90)

where ζ denotes the transmitter phase, and f_o is the system operating frequency. Assume that there is only one scatterer, $\vec{C}_i$, within the kth range cell

A Three-Dimensional SAR Imaging Technique

defined by $(a_i, \phi_i, s_i, u_i, \tilde{h}_i)$. The returned signal at the n_2^{th} element due to firing from the n_1^{th} element and reflection from the $\vec{C}_i$ scatterer is

$$s_i(n_1, n_2; t_{mn_1}) = a_i G^2 (\sin\beta_i)(\rho_k(t_c)/\rho(t_c))^4 \cos[2\pi f_o(t_{mn_1} - \tau_{n_1 n_2}) + \zeta - \phi_i] \quad (12.91)$$

where G^2 represents the two-way antenna gain, and the term $(\rho_k(t_c)/\rho(t_c))^4$ denotes the range attenuation at the kth range cell. The analysis in this paper will assume hereon that ζ and ϕ_i are both equal to zeroes.

Suppose that there are N_o scatterers within the kth range cell, with angular locations given by

$$\{(a_i, \phi_i, s_i, u_i, \tilde{h}_i); i = 1, \ldots N_o\} \quad (12.92)$$

The composite returned signal at time t_{mn_1} within this range cell due to the path $(n_1 \Rightarrow all \ \vec{C}_i \Rightarrow n_2)$ is

$$s(n_1, n_2; t_{mn_1}) = \sum_{i=1}^{N_o} s_i(n_1, n_2; t_{mn_1}) \quad (12.93)$$

The platform motion synthesizes a rectangular array of size $N \times M$, where only one column of N elements exists at a time. However, if $M = 2N$ and the real array is operated in the sequential mode, a square planar array of size $2N \times 2N$ is synthesized. The element spacing along the flight path is $d_y = vD_{ob}/M$.

Consider the kth range bin. The corresponding two-dimensional information sequence $\{b_k(n, m); (n, m) = 0, \ldots 2N-2\}$ consists of $2N$ similar vectors. The mth vector represents the returns due to the sequential firing of all N elements during the mth subinterval. Each vector has $(2N-1)$ rows, and it is extended, by adding zeroes, to the next power of two. For example, consider the mth subinterval, and let $M = 2N = 4$. Then, the elements of the extended column $\{b_k(n, m)\}$ are

$$\{b_k(0, m), b_k(1, m), b_k(2, m), b_k(3, m), b_k(4, m), b_k(5, m), \quad (12.94)$$
$$b_k(6, m), b_k(7, m)\} = \{s(0, 0; t_{mn_0}), s(0, 1; t_{mn_0}) + s(1, 0; t_{mn_1}),$$
$$s(0, 2; t_{mn_0}) + s(1, 1; t_{mn_1}) + s(2, 0; t_{mn_2}), s(0, 3; t_{mn_0}) + s(1, 2; t_{mn_1}) +$$
$$s(2, 1; t_{mn_2}) + s(3, 0; t_{mn_3}), s(1, 3; t_{mn_1}) + s(2, 2; t_{mn_2}) +$$
$$s(3, 1; t_{mn_3}), s(2, 3; t_{mn_2}) + s(3, 2; t_{mn_3}), s(3, 3; t_{mn_3}), 0\}$$

12.9.6. Electronic Processing

Consider again the kth range cell during the mth subinterval, and the two-way path: n_1^{th} element transmitting and n_2^{th} element receiving. The analog quadrature components corresponding to this two-way path are

$$s_I^\perp(n_1, n_2; t) = B\cos\psi^\perp \tag{12.95}$$

$$s_Q^\perp(n_1, n_2; t) = B\sin\psi^\perp \tag{12.96}$$

$$\psi^\perp = 2\pi f_0 \Bigg\{ t - \frac{1}{c}\bigg[2\tilde{r} + (\tilde{r}_{\tilde{h}}(D_{n_1}) + \tilde{r}_{\tilde{h}}(D_{n_2}))\tilde{h} + (\tilde{r}_u(D_{n_1}) + \tilde{r}_u(D_{n_2}))u + \tag{12.97}$$

$$(\tilde{r}_{\tilde{h}\tilde{h}}(D_{n_1}) + \tilde{r}_{\tilde{h}\tilde{h}}(D_{n_2}))\frac{\tilde{h}^2}{2} + (\tilde{r}_{\tilde{h}u}(D_{n_1}) + \tilde{r}_{\tilde{h}u}(D_{n_2}))\tilde{h}u +$$

$$(\tilde{r}_{ss}(D_{n_1}) + \tilde{r}_{ss}(D_{n_2}))\frac{s^2}{2} + 2\tilde{r}_{st}st + 2\tilde{r}_{tt}\frac{t^2}{2} +$$

$$(\tilde{r}_{uu}(D_{n_1}) + \tilde{r}_{uu}(D_{n_2}))\frac{u^2}{2} + (\tilde{r}_{\tilde{h}st}(D_{n_1}) + \tilde{r}_{\tilde{h}st}(D_{n_2}))\tilde{h}st] \Bigg\}$$

where B denotes antenna gain, range attenuation, and scatterers' strengths. The subscripts for t have been dropped for notation simplicity. Rearranging Eq. (12.97) and collecting terms yields

$$\psi^\perp = \frac{2\pi f_0}{c} \Bigg\{ \{tc - [2\tilde{r}_{st}s + (\tilde{r}_{\tilde{h}st}(D_{n_1}) + \tilde{r}_{\tilde{h}st}(D_{n_2}))\tilde{h}s]t - \tilde{r}_{tt}t^2\} - \tag{12.98}$$

$$\bigg[2\tilde{r} + (\tilde{r}_{\tilde{h}}(D_{n_1}) + \tilde{r}_{\tilde{h}}(D_{n_2}))\tilde{h} + (\tilde{r}_u(D_{n_1}) + \tilde{r}_u(D_{n_2}))u +$$

$$(\tilde{r}_{uu}(D_{n_1}) + \tilde{r}_{uu}(D_{n_2}))\frac{u^2}{2} + (\tilde{r}_{\tilde{h}\tilde{h}}(D_{n_1}) + \tilde{r}_{\tilde{h}\tilde{h}}(D_{n_2}))\frac{\tilde{h}^2}{2} +$$

$$(\tilde{r}_{\tilde{h}u}(D_{n_1}) + \tilde{r}_{\tilde{h}u}(D_{n_2}))\tilde{h}u + (\tilde{r}_{ss}(D_{n_1}) + \tilde{r}_{ss}(D_{n_2}))\frac{s^2}{2} \bigg] \Bigg\}$$

After analog to digital (A/D) conversion, deramping of the quadrature components to cancel the quadratic phase $(-2\pi f_0 \tilde{r}_{tt} t^2/c)$ is performed. Then, the digital quadrature components are

$$s_I(n_1, n_2; t) = B\cos\psi \tag{12.99}$$

$$s_Q(n_1, n_2; t) = B\sin\psi \tag{12.100}$$

A Three-Dimensional SAR Imaging Technique

$$\psi = \psi^{\perp} - 2\pi f_0 t + 2\pi f_0 \ddot{r}_{tt} \frac{t^2}{c} \tag{12.101}$$

The instantaneous frequency for the ith scatterer within the kth range cell is computed as

$$f_{di} = \frac{1}{2\pi} \frac{d\psi}{dt} = \frac{f_0}{c}[2\dot{r}_{st}s + (\ddot{r}_{hst}(D_{n_1}) + \ddot{r}_{hst}(D_{n_2}))\tilde{hs}] \tag{12.102}$$

Substituting the actual values for $\ddot{r}_{st}$, $\ddot{r}_{hst}(D_{n_1})$, $\ddot{r}_{hst}(D_{n_2})$ and collecting terms yields

$$f_{di} = -\left(\frac{2v\sin\beta_k}{\lambda}\right)\left(\frac{\tilde{hs}}{\rho_k^2(t_c)}(h + (D_{n_1} + D_{n_2})\sin\beta^*) - s\right) \tag{12.103}$$

Note that if $\tilde{h} = 0$, then

$$f_{di} = \frac{2v}{\lambda} \sin\beta_k \sin\mu \tag{12.104}$$

which is the Doppler value corresponding to a ground patch (see Eq. (12.49)).

The last stage of the processing consists of three steps: (1) two-dimensional windowing; (2) performing a two-dimensional DFT on the windowed quadrature components; and (3) scaling to compensate for antenna gain and range attenuation.

12.9.7. Derivation of Eq. (12.71)

Consider a rectangular array of size $N \times N$, with uniform element spacing $d_x = d_y = d$, and wavelength λ. Assume sequential mode operation where elements are fired sequentially, one at a time, while all elements receive in parallel. Assume far field observation defined by azimuth and elevation angles (α, β). The unit vector $\vec{u}$ on the line of sight, with respect to $\vec{O}$, is given by

$$\vec{u} = \sin\beta\cos\alpha \ \vec{a}_x + \sin\beta\sin\alpha \ \vec{a}_y + \cos\beta \ \vec{a}_z \tag{12.105}$$

The $(n_x, n_y)^{th}$ element of the array can be defined by the vector

$$\vec{e}(n_x, n_y) = \left(n_x - \frac{N-1}{2}\right)d \ \vec{a}_x + \left(n_y - \frac{N-1}{2}\right)d \ \vec{a}_y \tag{12.106}$$

where $(n_x, n_y = 0, \ldots N-1)$. The one-way geometric phase for this element is

$$\varphi'(n_x, n_y) = k(\vec{u} \bullet \vec{e}(n_x, n_y)) \tag{12.107}$$

where $k = 2\pi/\lambda$ is the wave-number, and the operator ($\bullet$) indicates dot product. Therefore, the two-way geometric phase between the (n_{x1}, n_{y1}) and (n_{x2}, n_{y2}) elements is

$$\varphi(n_{x1}, n_{y1}, n_{x2}, n_{y2}) = k[\hat{u} \bullet \{\hat{e}(n_{x1}, n_{y1}) + \hat{e}(n_{x2}, n_{y2})\}] \tag{12.108}$$

The cumulative two-way normalized electric field due to all transmissions is

$$E(\hat{u}) = E_t(\hat{u})E_r(\hat{u}) \tag{12.109}$$

where the subscripts t and r, respectively, refer to the transmitted and received electric fields. More precisely,

$$E_t(\hat{u}) = \sum_{n_{xt}=0}^{N-1} \sum_{n_{yt}=0}^{N-1} w(n_{xt}, n_{yt}) \exp[jk\{\hat{u} \bullet \hat{e}(n_{xt}, n_{yt})\}] \tag{12.110}$$

$$E_r(\hat{u}) = \sum_{n_{xr}=0}^{N-1} \sum_{n_{yr}=0}^{N-1} w(n_{xr}, n_{yr}) \exp[jk\{\hat{u} \bullet \hat{e}(n_{xr}, n_{yr})\}] \tag{12.111}$$

In this case, $w(n_x, n_y)$ denotes the tapering sequence. Substituting Eqs. (12.108), (12.110), and (12.111) into Eq. (12.109) and grouping all fields with the same two-way geometric phase yields

$$E(\hat{u}) = e^{j\delta} \sum_{m=0}^{N_a-1} \sum_{n=0}^{N_a-1} w'(m,n) \exp[jkd\sin\beta(m\cos\alpha + n\sin\alpha)] \tag{12.112}$$

$$N_a = 2N - 1 \tag{12.113}$$

$$m = n_{xt} + n_{xr}; m = 0, \ldots 2N-2 \tag{12.114}$$

$$n = n_{yt} + n_{yr}; n = 0, \ldots 2N-2 \tag{12.115}$$

$$\delta = \left(\frac{-d\sin\beta}{2}\right)(N-1)(\cos\alpha + \sin\alpha) \tag{12.116}$$

The two-way array pattern is then computed as

$$|E(\hat{u})| = \left|\sum_{m=0}^{N_a-1} \sum_{n=0}^{N_a-1} w'(m,n) \exp[jkd\sin\beta(m\cos\alpha + n\sin\alpha)]\right| \tag{12.117}$$

Consider the two-dimensional DFT transform, $W'(p, q)$, of the array $w'(n_x, n_y)$.

$$W'(p, q) = \qquad (12.118)$$

$$\sum_{m=0}^{N_a-1}\sum_{n=0}^{N_a-1} w'(m,n)\exp\left(-j\frac{2\pi}{N_a}(pm+qn)\right); p,q = 0, \ldots N_a-1$$

Comparison of Eqs. (12.117) and Eq. (12.118) indicates that $|E(\tilde{u})|$ is equal to $|W'(p,q)|$ if

$$-\left(\frac{2\pi}{N_a}\right)p = \frac{2\pi}{\lambda}d\sin\beta\cos\alpha \qquad (12.119)$$

$$-\left(\frac{2\pi}{N_a}\right)q = \frac{2\pi}{\lambda}d\sin\beta\sin\alpha \qquad (12.120)$$

It follows that

$$\alpha = \tan^{-1}\left(\frac{q}{p}\right) \qquad (12.121)$$

12.9.8. Non-Zero Taylor Series Coefficients for the kth Range Cell

$$\bar{r} = \sqrt{D_n^2 + h^2(1+\tan\beta_k) + 2hD_n\sin\beta^* - 2hD_n\cos\beta^*\tan\beta_k} = \rho_k(t_c) \quad (12.122)$$

$$\bar{r}_{\tilde{h}} = \left(\frac{-1}{\bar{r}}\right)(h + D_n\sin\beta^*) \qquad (12.123)$$

$$\bar{r}_u = \left(\frac{h}{\bar{r}\cos^2\beta_k}\right)(h\tan\beta_k - D_n\cos\beta^*) \qquad (12.124)$$

$$\bar{r}_{\tilde{h}\tilde{h}} = \left(\frac{1}{\bar{r}}\right) - \left(\frac{1}{\bar{r}^3}\right)(h + D_n\sin\beta^*) \qquad (12.125)$$

$$\bar{r}_{\tilde{h}u} = \left(\frac{1}{\bar{r}^3}\right)\left(\frac{h}{\cos^2\beta_k}\right)(h + D_n\tan\beta^*)(h\tan\beta_k - D_n\cos\beta^*) \qquad (12.126)$$

$$\bar{r}_{ss} = \left(\frac{-1}{4\bar{r}^3}\right) + \left(\frac{1}{\bar{r}}\right)(h\tan\beta_k - D_n\cos\beta^*) \qquad (12.127)$$

$$\bar{r}_{st} = \left(\frac{-1}{\bar{r}}\right)hv\tan\beta_k \qquad (12.128)$$

$$\bar{r}_{tt} = \frac{v^2}{\bar{r}} \qquad (12.129)$$

$$\bar{r}_{uu} = \left(\frac{h}{\bar{r}\cos^3\beta_k}\right)\left\{\left(\frac{h}{\bar{r}^2\cos\beta_k}\right)(h\tan\beta_k - D_n\cos\beta^*) + \right.$$
$$\left. h\left(\left(\frac{1}{\cos\beta_k}\right) + 2\tan\beta_k\sin\beta_k\right) - 2\sin\beta_k D_n\cos\beta^*\right\} \tag{12.130}$$

$$\bar{r}_{\tilde{h}\tilde{h}\tilde{h}} = \left(\frac{3}{\bar{r}^3}\right)(h + D_n\sin\beta^*)\left[\left(\frac{1}{\bar{r}^2}\right)(h + D_n\sin\beta^*)^2 - 1\right] \tag{12.131}$$

$$\bar{r}_{\tilde{h}\tilde{h}u} = \left(\frac{h}{\bar{r}^3\cos^2\beta_k}\right)(h\tan\beta_k - D_n\cos\beta^*)\left[\left(\frac{-3}{\bar{r}^2}\right)(h + D_n\sin\beta^*)^2 + 1\right] \tag{12.132}$$

$$\bar{r}_{\tilde{h}st} = \left(\frac{hv\tan\beta_k}{\bar{r}^3}\right)(h + D_n\sin\beta^*) \tag{12.133}$$

$$\bar{r}_{\tilde{h}uu} = \left(\frac{-3}{\bar{r}^5}\right)\left(\frac{h^2}{\cos^4\beta_k}\right)(h + D_n\sin\beta^*)(h\tan\beta_k - D_n\cos\beta^*) \tag{12.134}$$

$$\bar{r}_{\tilde{h}ss} = \left(\frac{-1}{\bar{r}^3}\right)(h\tan\beta_k - D_n\cos\beta^*)(h + D_n\sin\beta^*) \tag{12.135}$$

$$\bar{r}_{uss} = \left(\frac{h}{\bar{r}\cos^2\beta_k}\right)(D_n\cos\beta^*)\left[\left(\frac{1}{\bar{r}^2}\right)(h\tan\beta_k - D_n\cos\beta^*)(h\tan\beta_k) + 1\right] \tag{12.136}$$

$$\bar{r}_{stu} = \left(\frac{-h\tan\beta_k}{\bar{r}^3\cos^2\beta_k}\right)(h\tan\beta_k - D_n\cos\beta^*) \tag{12.137}$$

$$\bar{r}_{suu} = \left(\frac{hD_n\cos\beta^*}{\bar{r}\cos^2\beta_k}\right)\left[\left(\frac{h\tan\beta_k}{\bar{r}^2}\right)(h\tan\beta_k - D_n\cos\beta^*) + 1\right] \tag{12.138}$$

$$\bar{r}_{\tilde{h}tt} = \left(\frac{v^2 h}{\bar{r}^3\cos^2\beta_k}\right)(h\tan\beta_k - D_n\cos\beta^*) \tag{12.139}$$

$$\bar{r}_{uuu} = \tag{12.140}$$
$$\left(\frac{h}{\bar{r}\cos^4\beta_k}\right)[8h\tan\beta_k + \sin^2\beta_k(h - D_n\cos\beta^*) - 2D_n\cos\beta^*] +$$
$$\left(\frac{3h^2}{\bar{r}^3\cos^5\beta_k}\right)(h\tan\beta_k - D_n\cos\beta^*) + \left[\left(\frac{3h^2}{\bar{r}^3\cos^5\beta_k}\right)(h\tan\beta_k - D_n\cos\beta^*)\right.$$
$$\left.\left(\frac{1}{2\cos\beta_k} + (h\tan\beta_k - D_n\cos\beta^*)\right)\right] + \left(\frac{3h^3}{\bar{r}^5\cos^6\beta_k}\right)(h\tan\beta_k - D_n\cos\beta^*)$$

12.10. MATLAB Programs and Functions

Listing 12.1. MATLAB Program "fig12_12-13.m"

```
%               Figures 12.12 and 12.13
%   Program to do Spotlight SAR using the rectangular format and
%   HRR for range compression.
%               13 June 2003
%               Dr. Brian J. Smith
clear all;
%%%%%%%%%% SAR Image Resolution %%%%
dr = .50;
da = .10;
% dr = 6*2.54/100;
% da = 6*2.54/100;
%%%%%%%%%% Scatter Locations %%%%%%%%
xn = [10000 10015 9985];   % Scatter Location, x-axis
yn = [0 -20 20];           % Scatter Location, y-axis
Num_Scatter = 3;           % Number of Scatters
Rnom = 10000;
%%%%%%%%%% Radar Parameters %%%%%%%%
f_0 =  35.0e9;   % Lowest Freq. in the HRR Waveform
df =   3.0e6;    % Freq. step size for HRR, Hz
c =    3e8;      % Speed of light, m/s
Kr = 1.33;
Num_Pulse = 2^(round(log2(Kr*c/(2*dr*df))));
Lambda = c/(f_0 + Num_Pulse*df/2);
%%%%%%%%%% Synthetic Array Parameters %%%%%%%
du = 0.2;
L = round(Kr*Lambda*Rnom/(2*da));
U = -(L/2):du:(L/2);
Num_du = length(U);
%%%%%%%%%% This section generates the target returns %%%%%%%
Num_U = round(L/du);
I_Temp = 0;
Q_Temp = 0;
for I = 1:Num_U
   for J = 1:Num_Pulse
      for K = 1:Num_Scatter
         Yr = yn(K) - ((I-1)*du - (L/2));
         Rt = sqrt(xn(K)^2 + Yr^2);
         F_ci = f_0 + (J-1)*df;
         PHI = -4*pi*Rt*F_ci/c;
         I_Temp = cos(PHI) + I_Temp;
```

```
      Q_Temp = sin(PHI) + Q_Temp;
   end;
   IQ_Raw(J,I) = I_Temp + i*Q_Temp;
   I_Temp = 0.0;
   Q_Temp = 0.0;
  end;
end;
%%%%%%%%%% End target return section %%%%%%
%%%%%%%%%% Range Compression %%%%%%%%%%%%%%
Num_RB = 2*Num_Pulse;
WR = hamming(Num_Pulse);
for I = 1:Num_U
   Range_Compressed(:,I) = fftshift(ifft(IQ_Raw(:,I).*WR,Num_RB));
end;
%%%%%%%%%% Focus Range Compressed Data %%%%%
dn = (1:Num_U)*du - L/2;
PHI_Focus = -2*pi*(dn.^2)/(Lambda*xn(1));
for I = 1:Num_RB
   Temp = angle(Range_Compressed(I,:)) - PHI_Focus;
   Focused(I,:) = abs(Range_Compressed(I,:)).*exp(i*Temp);
end;
%Focused = Range_Compressed;
%%%%%%%%%% Azimuth Compression %%%%%%%%%%%%%
WA = hamming(Num_U);
for I = 1:Num_RB
   AZ_Compressed(I,:) = fftshift(ifft(Focused(I,:).*WA'));
end;
 SAR_Map = 10*log10(abs(AZ_Compressed));
Y_Temp = (1:Num_RB)*(c/(2*Num_RB*df));
Y = Y_Temp - max(Y_Temp)/2;
X_Temp = (1:length(IQ_Raw))*(Lambda*xn(1)/(2*L));
X = X_Temp - max(X_Temp)/2;
image(X,Y,20-SAR_Map);  %
%image(X,Y,5-SAR_Map);  %
axis([-25 25 -25 25]); axis equal; colormap(gray(64));
xlabel('Cross Range (m)'); ylabel('Down Range (m)');
grid
%print -djpeg .jpg
```

Chapter 13 Signal Processing

13.1. Signal and System Classifications

In general, electrical signals can represent either current or voltage, and may be classified into two main categories: energy signals and power signals. Energy signals can be deterministic or random, while power signals can be periodic or random. A signal is said to be random if it is a function of a random parameter (such as random phase or random amplitude). Additionally, signals may be divided into low pass or band pass signals. Signals that contain very low frequencies (close to DC) are called low pass signals; otherwise they are referred to as band pass signals. Through modulation, low pass signals can be mapped into band pass signals.

The average power P for the current or voltage signal $x(t)$ over the interval (t_1, t_2) across a 1Ω resistor is

$$P = \frac{1}{t_2 - t_1} \int_{t_1}^{t_2} |x(t)|^2 \, dt \qquad (13.1)$$

The signal $x(t)$ is said to be a power signal over a very large interval $T = t_2 - t_1$, if and only if it has finite power; it must satisfy the following relation:

$$0 < \lim_{T \to \infty} \frac{1}{T} \int_{-T/2}^{T/2} |x(t)|^2 \, dt < \infty \qquad (13.2)$$

Using Parseval's theorem, the energy E dissipated by the current or voltage signal $x(t)$ across a 1Ω resistor, over the interval (t_1, t_2), is

$$E = \int_{t_1}^{t_2} |x(t)|^2 \, dt \qquad (13.3)$$

The signal $x(t)$ is said to be an energy signal if and only if it has finite energy,

$$E = \int_{-\infty}^{\infty} |x(t)|^2 \, dt < \infty \qquad (13.4)$$

A signal $x(t)$ is said to be periodic with period T if and only if

$$x(t) = x(t + nT) \qquad \text{for all } t \qquad (13.5)$$

where n is an integer.

Example:

Classify each of the following signals as an energy signal, as a power signal, or as neither. All signals are defined over the interval $(-\infty < t < \infty)$: $x_1(t) = \cos t + \cos 2t$, $x_2(t) = \exp(-\alpha^2 t^2)$.

Solution:

$$P_{x_1} = \frac{1}{T} \int_{-T/2}^{T/2} (\cos t + \cos 2t)^2 \, dt = 1 \Rightarrow \text{power signal}$$

Note that since the cosine function is periodic, the limit is not necessary.

$$E_{x_2} = \int_{-\infty}^{\infty} (e^{-\alpha^2 t^2})^2 \, dt = 2 \int_0^{\infty} e^{-2\alpha^2 t^2} \, dt = 2 \frac{\sqrt{\pi}}{2\sqrt{2}\alpha} = \frac{1}{\alpha}\sqrt{\frac{\pi}{2}} \Rightarrow \text{energy signal}$$

Electrical systems can be linear or nonlinear. Furthermore, linear systems may be divided into continuous or discrete. A system is linear if the input signal $x_1(t)$ produces $y_1(t)$ and $x_2(t)$ produces $y_2(t)$; then for some arbitrary constants a_1 and a_2 the input signal $a_1 x_1(t) + a_2 x_2(t)$ produces the output $a_1 y_1(t) + a_2 y_2(t)$. A linear system is said to be shift invariant (or time invariant) if a time shift at its input produces the same shift at its output. More precisely, if the input signal $x(t)$ produces $y(t)$ then the delayed signal $x(t - t_0)$ produces the output $y(t - t_0)$. The impulse response of a Linear Time Invariant (LTI) system, $h(t)$, is defined to be the system's output when the input is an impulse (delta function).

13.2. The Fourier Transform

The Fourier Transform (FT) of the signal $x(t)$ is

$$F\{x(t)\} = X(\omega) = \int_{-\infty}^{\infty} x(t) e^{-j\omega t} \, dt \tag{13.6}$$

or

$$F\{x(t)\} = X(f) = \int_{-\infty}^{\infty} x(t) e^{-j2\pi f t} \, dt \tag{13.7}$$

and the Inverse Fourier Transform (IFT) is

$$F^{-1}\{X(\omega)\} = x(t) = \frac{1}{2\pi} \int_{-\infty}^{\infty} X(\omega) e^{j\omega t} \, d\omega \tag{13.8}$$

or

$$F^{-1}\{X(f)\} = x(t) = \int_{-\infty}^{\infty} X(f) e^{j2\pi f t} \, df \tag{13.9}$$

where, in general, t represents time, while $\omega = 2\pi f$ and f represent frequency in radians per second and Hertz, respectively. In this book we will use both notations for the transform, as appropriate (i.e., $X(\omega)$ and $X(f)$).

A detailed table of the FT pairs is listed in Appendix 13A. The FT properties are (the proofs are left as an exercise):

1. Linearity:

$$F\{a_1 x_1(t) + a_2 x_2(t)\} = a_1 X_1(\omega) + a_2 X_2(\omega) \tag{13.10}$$

2. Symmetry: If $F\{x(t)\} = X(\omega)$ then

$$2\pi X(-\omega) = \int_{-\infty}^{\infty} X(t) e^{-j\omega t} dt \tag{13.11}$$

3. Shifting: For any real time t_0

$$F\{x(t \pm t_0)\} = e^{\pm j\omega t_0} X(\omega) \tag{13.12}$$

4. Scaling: If $F\{x(t)\} = X(\omega)$ then

$$F\{x(at)\} = \frac{1}{|a|} X\left(\frac{\omega}{a}\right) \tag{13.13}$$

5. Central Ordinate:

$$X(0) = \int_{-\infty}^{\infty} x(t) dt \tag{13.14}$$

$$x(0) = \frac{1}{2\pi} \int_{-\infty}^{\infty} X(\omega) d\omega \tag{13.15}$$

6. Frequency Shift: If $F\{x(t)\} = X(\omega)$ then

$$F\{e^{\pm \omega_0 t} x(t)\} = X(\omega \mp \omega_0) \tag{13.16}$$

7. Modulation: If $F\{x(t)\} = X(\omega)$ then

$$F\{x(t)\cos\omega_0 t\} = \frac{1}{2}[X(\omega+\omega_0) + X(\omega-\omega_0)] \tag{13.17}$$

$$F\{x(t)\sin(\omega_0 t)\} = \frac{1}{2j}[X(\omega-\omega_0) - X(\omega+\omega_0)] \tag{13.18}$$

8. Derivatives:

$$F\left\{\frac{d^n}{dt^n}(x(t))\right\} = (j\omega)^n X(\omega) \tag{13.19}$$

9. Time Convolution: if $x(t)$ and $h(t)$ have Fourier transforms $X(\omega)$ and $H(\omega)$, respectively, then

$$F\left\{\int_{-\infty}^{\infty} x(\tau) h(t-\tau) d\tau\right\} = X(\omega) H(\omega) \tag{13.20}$$

10. Frequency Convolution:

$$F\{x(t) h(t)\} = \frac{1}{2\pi} \int_{-\infty}^{\infty} X(\tau) H(\omega-\tau) d\tau \tag{13.21}$$

11. *Autocorrelation:*

$$F\left\{\int_{-\infty}^{\infty} x(\tau)x^*(\tau-t)d\tau\right\} = X(\omega)X^*(\omega) = |X(\omega)|^2 \quad (13.22)$$

12. *Parseval's Theorem:* The energy associated with the signal $x(t)$ is

$$E = \int_{-\infty}^{\infty} |x(t)|^2 dt = \int_{-\infty}^{\infty} |X(\omega)|^2 d\omega \quad (13.23)$$

13. *Moments:* The nth moment is

$$m_n = \int_0^{\infty} t^n x(t) dt = \frac{d^n}{d\omega^n} X(\omega)\bigg|_{\omega=0} \quad (13.24)$$

13.3. The Fourier Series

A set of functions $S = \{\varphi_n(t) \; ; \; n = 1, \ldots, N\}$ is said to be orthogonal over the interval (t_1, t_2) if and only if

$$\int_{t_1}^{t_2} \varphi_i^*(t)\varphi_j(t)dt = \int_{t_1}^{t_2} \varphi_i(t)\varphi_j^*(t)dt = \begin{Bmatrix} 0 & i \neq j \\ \lambda_i & i = j \end{Bmatrix} \quad (13.25)$$

where the asterisk indicates complex conjugate, and λ_i are constants. If $\lambda_i = 1$ for all i, then the set S is said to be an orthonormal set.

An electrical signal $x(t)$ can be expressed over the interval (t_1, t_2) as a weighted sum of a set of orthogonal functions as

$$x(t) \approx \sum_{n=1}^{N} X_n \varphi_n(t) \quad (13.26)$$

where X_n are, in general, complex constants, and the orthogonal functions $\varphi_n(t)$ are called basis functions. If the integral-square error over the interval (t_1, t_2) is equal to zero as N approaches infinity, i.e.,

$$\lim_{N \to \infty} \int_{t_1}^{t_2} \left| x(t) - \sum_{n=1}^{N} X_n \varphi_n(t) \right|^2 dt = 0 \quad (13.27)$$

then the set $S = \{\varphi_n(t)\}$ is said to be complete, and Eq. (13.26) becomes an equality. The constants X_n are computed as

$$X_n = \frac{\int_{t_1}^{t_2} x(t)\varphi_n^*(t)dt}{\int_{t_1}^{t_2} |\varphi_n(t)|^2 dt} \tag{13.28}$$

Let the signal $x(t)$ be periodic with period T, and let the complete orthogonal set S be

$$S = \left\{ e^{\frac{j2\pi nt}{T}} \; ; \; n = -\infty, \infty \right\} \tag{13.29}$$

Then the complex exponential Fourier series of $x(t)$ is

$$x(t) = \sum_{n=-\infty}^{\infty} X_n e^{\frac{j2\pi nt}{T}} \tag{13.30}$$

Using Eq. (13.28) yields

$$X_n = \frac{1}{T} \int_{-T/2}^{T/2} x(t) e^{\frac{-j2\pi nt}{T}} dt \tag{13.31}$$

The FT of Eq. (13.30) is given by

$$X(\omega) = 2\pi \sum_{n=-\infty}^{\infty} X_n \delta\left(\omega - \frac{2\pi n}{T}\right) \tag{13.32}$$

where $\delta(\cdot)$ is delta function. When the signal $x(t)$ is real we can compute its trigonometric Fourier series from Eq. (13.30) as

$$x(t) = a_0 + \sum_{n=1}^{\omega} a_n \cos\left(\frac{2\pi nt}{T}\right) + \sum_{n=1}^{\omega} b_n \sin\left(\frac{2\pi nt}{T}\right) \tag{13.33}$$

$$a_0 = X_0$$

$$a_n = \frac{1}{T}\int_{-T/2}^{T/2} x(t)\cos\left(\frac{2\pi nt}{T}\right)dt \quad (13.34)$$

$$b_n = \frac{1}{T}\int_{-T/2}^{T/2} x(t)\sin\left(\frac{2\pi nt}{T}\right)dt$$

The coefficients a_n are all zeros when the signal $x(t)$ is an odd function of time. Alternatively, when the signal is an even function of time, then all b_n are equal to zero.

Consider the periodic energy signal defined in Eq. (13.33). The total energy associated with this signal is then given by

$$E = \frac{1}{T}\int_{t_0}^{t_0+T} |x(t)|^2 dt = \frac{a_0^2}{4} + \sum_{n=1}^{\infty}\left(\frac{a_n^2}{2} + \frac{b_n^2}{2}\right) \quad (13.35)$$

13.4. Convolution and Correlation Integrals

The convolution $\phi_{xh}(t)$ between the signals $x(t)$ and $h(t)$ is defined by

$$\phi_{xh}(t) = x(t) \bullet h(t) = \int_{-\infty}^{\infty} x(\tau)h(t-\tau)d\tau \quad (13.36)$$

where τ is a dummy variable, and the operator $\bullet$ is used to symbolically describe the convolution integral. Convolution is commutative, associative, and distributive. More precisely,

$$x(t) \bullet h(t) = h(t) \bullet x(t)$$
$$x(t) \bullet h(t) \bullet g(t) = (x(t) \bullet h(t)) \bullet g(t) = x(t) \bullet (h(t) \bullet g(t)) \quad (13.37)$$

For the convolution integral to be finite at least one of the two signals must be an energy signal. The convolution between two signals can be computed using the FT

$$\phi_{xh}(t) = F^{-1}\{X(\omega)H(\omega)\} \quad (13.38)$$

Consider an LTI system with impulse response $h(t)$ and input signal $x(t)$. It follows that the output signal $y(t)$ is equal to the convolution between the input signal and the system impulse response,

$$y(t) = \int_{-\infty}^{\infty} x(\tau)h(t-\tau)d\tau = \int_{-\infty}^{\infty} h(\tau)x(t-\tau)d\tau \qquad (13.39)$$

The cross-correlation function between the signals $x(t)$ and $g(t)$ is defined as

$$R_{xg}(t) = \int_{-\infty}^{\infty} x^*(\tau)g(t+\tau)d\tau \qquad (13.40)$$

Again, at least one of the two signals should be an energy signal for the correlation integral to be finite. The cross-correlation function measures the similarity between the two signals. The peak value of $R_{xg}(t)$ and its spread around this peak are an indication of how good this similarity is. The cross-correlation integral can be computed as

$$R_{xg}(t) = F^{-1}\{X^*(\omega)G(\omega)\} \qquad (13.41)$$

When $x(t) = g(t)$ we get the autocorrelation integral,

$$R_x(t) = \int_{-\infty}^{\infty} x^*(\tau)x(t+\tau)d\tau \qquad (13.42)$$

Note that the autocorrelation function is denoted by $R_x(t)$ rather than $R_{xx}(t)$. When the signals $x(t)$ and $g(t)$ are power signals, the correlation integral becomes infinite and, thus, time averaging must be included. More precisely,

$$\bar{R}_{xg}(t) = \lim_{T\to\infty} \frac{1}{T} \int_{-T/2}^{T/2} x^*(\tau)g(t+\tau)d\tau \qquad (13.43)$$

13.5. Energy and Power Spectrum Densities

Consider an energy signal $x(t)$. From Parseval's theorem, the total energy associated with this signal is

$$E = \int_{-\infty}^{\infty} |x(t)|^2 dt = \frac{1}{2\pi} \int_{-\infty}^{\infty} |X(\omega)|^2 d\omega \qquad (13.44)$$

When $x(t)$ is a voltage signal, the amount of energy dissipated by this signal when applied across a network of resistance R is

Energy and Power Spectrum Densities

$$E = \frac{1}{R}\int_{-\infty}^{\infty}|x(t)|^2 dt = \frac{1}{2\pi R}\int_{-\infty}^{\infty}|X(\omega)|^2 d\omega \tag{13.45}$$

Alternatively, when $x(t)$ is a current signal we get

$$E = R\int_{-\infty}^{\infty}|x(t)|^2 dt = \frac{R}{2\pi}\int_{-\infty}^{\infty}|X(\omega)|^2 d\omega \tag{13.46}$$

The quantity $\int|X(\omega)|^2 d\omega$ represents the amount of energy spread per unit frequency across a 1Ω resistor; therefore, the Energy Spectrum Density (ESD) function for the energy signal $x(t)$ is defined as

$$ESD = |X(\omega)|^2 \tag{13.47}$$

The ESD at the output of an LTI system when $x(t)$ is at its input is

$$|Y(\omega)|^2 = |X(\omega)|^2 |H(\omega)|^2 \tag{13.48}$$

where $H(\omega)$ is the FT of the system impulse response, $h(t)$. It follows that the energy present at the output of the system is

$$E_y = \frac{1}{2\pi}\int_{-\infty}^{\infty}|X(\omega)|^2 |H(\omega)|^2 d\omega \tag{13.49}$$

Example:

The voltage signal $x(t) = e^{-5t}$; $t \geq 0$ is applied to the input of a low pass LTI system. The system bandwidth is $5 Hz$, and its input resistance is 5Ω. If $H(\omega) = 1$ over the interval $(-10\pi < \omega < 10\pi)$ and zero elsewhere, compute the energy at the output.

Solution:

From Eqs. (13.45) and (13.49) we get

$$E_y = \frac{1}{2\pi R}\int_{\omega = -10\pi}^{10\pi}|X(\omega)|^2 |H(\omega)|^2 d\omega$$

Using Fourier transform tables and substituting $R = 5$ yield

$$E_y = \frac{1}{5\pi}\int_0^{10\pi}\frac{1}{\omega^2 + 25} d\omega$$

Completing the integration yields

$$E_y = \frac{1}{25\pi}[\operatorname{atanh}(2\pi) - \operatorname{atanh}(0)] = 0.01799 \; Joules$$

Note that an infinite bandwidth would give $E_y = 0.02$, only 11% larger.

The total power associated with a power signal $g(t)$ is

$$P = \lim_{T \to \infty} \frac{1}{T} \int_{-T/2}^{T/2} |g(t)|^2 dt \qquad (13.50)$$

Define the Power Spectrum Density (PSD) function for the signal $g(t)$ as $S_g(\omega)$, where

$$P = \lim_{T \to \infty} \frac{1}{T} \int_{-T/2}^{T/2} |g(t)|^2 dt = \frac{1}{2\pi} \int_{-\infty}^{\infty} S_g(\omega) d\omega \qquad (13.51)$$

It can be shown that (see Problem 1.13)

$$S_g(\omega) = \lim_{T \to \infty} \frac{|G(\omega)|^2}{T} \qquad (13.52)$$

Let the signals $x(t)$ and $g(t)$ be two periodic signals with period T. The complex exponential Fourier series expansions for those signals are, respectively, given by

$$x(t) = \sum_{n=-\infty}^{\infty} X_n e^{\frac{j2\pi nt}{T}} \qquad (13.53)$$

$$g(t) = \sum_{m=-\infty}^{\infty} G_m e^{\frac{j2\pi mt}{T}} \qquad (13.54)$$

The power cross-correlation function $\bar{R}_{gx}(t)$ was given in Eq. (13.43), and is repeated here as Eq. (13.55),

$$\bar{R}_{gx}(t) = \frac{1}{T} \int_{-T/2}^{T/2} g^*(\tau) x(t+\tau) d\tau \qquad (13.55)$$

Note that because both signals are periodic the limit is no longer necessary. Substituting Eqs. (13.53) and (13.54) into Eq. (13.55), collecting terms, and using the definition of orthogonality, we get

$$\bar{R}_{gx}(t) = \sum_{n=-\infty}^{\infty} G_n^* X_n e^{\frac{j2n\pi t}{T}} \tag{13.56}$$

When $x(t) = g(t)$, Eq. (13.56) becomes the power autocorrelation function,

$$\bar{R}_x(t) = \sum_{n=-\infty}^{\infty} |X_n|^2 e^{\frac{j2n\pi t}{T}} = |X_0|^2 + 2\sum_{n=1}^{\infty} |X_n|^2 e^{\frac{j2n\pi t}{T}} \tag{13.57}$$

The power spectrum and cross-power spectrum density functions are then computed as the FT of Eqs. (13.57) and (13.56), respectively. More precisely,

$$\bar{S}_x(\omega) = 2\pi \sum_{n=-\infty}^{\infty} |X_n|^2 \delta\left(\omega - \frac{2n\pi}{T}\right)$$

$$\bar{S}_{gx}(\omega) = 2\pi \sum_{n=-\infty}^{\infty} G_n^* X_n \delta\left(\omega - \frac{2n\pi}{T}\right) \tag{13.58}$$

The line (or discrete) power spectrum is defined as the plot of $|X_n|^2$ versus n, where the lines are $\Delta f = 1/T$ apart. The DC power is $|X_0|^2$, and the total power is $\sum_{n=-\infty}^{\infty} |X_n|^2$.

13.6. Random Variables

Consider an experiment with outcomes defined by a certain sample space. The rule or functional relationship that maps each point in this sample space into a real number is called "random variable." Random variables are designated by capital letters (e.g., $X, Y, \ldots$), and a particular value of a random variable is denoted by a lowercase letter (e.g., $x, y, \ldots$).

The Cumulative Distribution Function (*cdf*) associated with the random variable X is denoted as $F_X(x)$, and is interpreted as the total probability that the random variable X is less or equal to the value x. More precisely,

$$F_X(x) = Pr\{X \le x\} \tag{13.59}$$

The probability that the random variable X is in the interval (x_1, x_2) is then given by

$$F_X(x_2) - F_X(x_1) = Pr\{x_1 \leq X \leq x_2\} \tag{13.60}$$

The *cdf* has the following properties:

$$\begin{aligned} 0 &\leq F_X(x) \leq 1 \\ F_X(-\infty) &= 0 \\ F_X(\infty) &= 1 \\ F_X(x_1) &\leq F_X(x_2) \Leftrightarrow x_1 \leq x_2 \end{aligned} \tag{13.61}$$

It is often practical to describe a random variable by the derivative of its *cdf*, which is called the Probability Density Function *(pdf)*. The *pdf* of the random variable X is

$$f_X(x) = \frac{d}{dx} F_X(x) \tag{13.62}$$

or, equivalently,

$$F_X(x) = Pr\{X \leq x\} = \int_{-\infty}^{x} f_X(\lambda) d\lambda \tag{13.63}$$

The probability that a random variable X has values in the interval (x_1, x_2) is

$$F_X(x_2) - F_X(x_1) = Pr\{x_1 \leq X \leq x_2\} = \int_{x_1}^{x_2} f_X(x) dx \tag{13.64}$$

Define the *nth* moment for the random variable X as

$$E[X^n] = \overline{X^n} = \int_{-\infty}^{\infty} x^n f_X(x) dx \tag{13.65}$$

The first moment, $E[X]$, is called the mean value, while the second moment, $E[X^2]$, is called the mean squared value. When the random variable X represents an electrical signal across a 1Ω resistor, then $E[X]$ is the DC component, and $E[X^2]$ is the total average power.

The *nth* central moment is defined as

$$E[(X - \bar{X})^n] = \overline{(X - \bar{X})^n} = \int_{-\infty}^{\infty} (x - \bar{x})^n f_X(x) dx \tag{13.66}$$

Random Variables

and, thus, the first central moment is zero. The second central moment is called the variance and is denoted by the symbol σ_X^2,

$$\sigma_X^2 = \overline{(X-\bar{X})^2} \tag{13.67}$$

Appendix 13B has some common *pdf*s and their means and variances.

In practice, the random nature of an electrical signal may need to be described by more than one random variable. In this case, the joint *cdf* and *pdf* functions need to be considered. The joint *cdf* and *pdf* for the two random variables X and Y are, respectively, defined by

$$F_{XY}(x, y) = Pr\{X \le x; Y \le y\} \tag{13.68}$$

$$f_{XY}(x, y) = \frac{\partial^2}{\partial x \partial y} F_{XY}(x, y) \tag{13.69}$$

The marginal *cdf*s are obtained as follows:

$$F_X(x) = \int_{-\infty}^{\infty} \int_{-\infty}^{x} f_{UV}(u, v) du\, dv = F_{XY}(x, \infty)$$

$$F_Y(y) = \int_{-\infty}^{\infty} \int_{-\infty}^{y} f_{UV}(u, v) dv\, du = F_{XY}(\infty, y) \tag{13.70}$$

If the two random variables are statistically independent, then the joint *cdf*s and *pdf*s are, respectively, given by

$$F_{XY}(x, y) = F_X(x) F_Y(y) \tag{13.71}$$

$$f_{XY}(x, y) = f_X(x) f_Y(y) \tag{13.72}$$

Let us now consider a case when the two random variables X and Y are mapped into two new variables U and V through some transformations T_1 and T_2 defined by

$$U = T_1(X, Y)$$
$$V = T_2(X, Y) \tag{13.73}$$

The joint *pdf*, $f_{UV}(u, v)$, may be computed based on the invariance of probability under the transformation. One must first compute the matrix of derivatives; then the new joint *pdf* is computed as

$$f_{UV}(u, v) = f_{XY}(x, y) |J| \tag{13.74}$$

$$|J| = \begin{vmatrix} \dfrac{\partial x}{\partial u} & \dfrac{\partial x}{\partial v} \\ \dfrac{\partial y}{\partial u} & \dfrac{\partial y}{\partial v} \end{vmatrix} \qquad (13.75)$$

where the determinant of the matrix of derivatives $|J|$ is called the Jacobian.

The characteristic function for the random variable X is defined as

$$C_X(\omega) = E[e^{j\omega X}] = \int_{-\infty}^{\infty} f_X(x) e^{j\omega x} dx \qquad (13.76)$$

The characteristic function can be used to compute the *pdf* for a sum of independent random variables. More precisely, let the random variable Y be equal to

$$Y = X_1 + X_2 + \ldots + X_N \qquad (13.77)$$

where $\{X_i \; ; \; i = 1, \ldots N\}$ is a set of independent random variables. It can be shown that

$$C_Y(\omega) = C_{X_1}(\omega) C_{X_2}(\omega) \ldots C_{X_N}(\omega) \qquad (13.78)$$

and the *pdf* $f_Y(y)$ is computed as the inverse Fourier transform of $C_Y(\omega)$ (with the sign of y reversed),

$$f_Y(y) = \frac{1}{2\pi} \int_{-\infty}^{\infty} C_Y(\omega) e^{-j\omega y} d\omega \qquad (13.79)$$

The characteristic function may also be used to compute the *nth* moment for the random variable X as

$$E[X^n] = (-j)^n \frac{d^n}{d\omega^n} C_X(\omega) \bigg|_{\omega = 0} \qquad (13.80)$$

13.7. Multivariate Gaussian Distribution

Consider a joint probability for m random variables, $X_1, X_2, \ldots, X_m$. These variables can be represented as components of an $m \times 1$ random column vector, $\underline{X}$. More precisely,

$$\underline{X}^t = \begin{bmatrix} X_1 & X_2 & \dots & X_m \end{bmatrix} \quad (13.81)$$

where the superscript indicates the transpose operation. The joint *pdf* for the vector $\underline{X}$ is

$$f_{\underline{x}}(\underline{x}) = f_{x_1, x_2, \dots, x_m}(x_1, x_2, \dots, x_m) \quad (13.82)$$

The mean vector is defined as

$$\mu_x = \begin{bmatrix} E[X_1] & E[X_2] & \dots & E[X_m] \end{bmatrix}^t \quad (13.83)$$

and the covariance is an $m \times m$ matrix given by

$$C_x = E[\underline{X}\,\underline{X}^t] - \mu_x\, \mu_x^t \quad (13.84)$$

Note that if the elements of the vector $\underline{X}$ are independent, then the covariance matrix is a diagonal matrix.

By definition a random vector $\underline{X}$ is multivariate Gaussian if its *pdf* has the form

$$f_{\underline{x}}(\underline{x}) = [(2\pi)^{m/2} |C_x|^{1/2}]^{-1} \exp\left(-\frac{1}{2}(\underline{x} - \mu_x)^t C_x^{-1}(\underline{x} - \mu_x)\right) \quad (13.85)$$

where μ_x is the mean vector, C_x is the covariance matrix, C_x^{-1} is inverse of the covariance matrix and $|C_x|$ is its determinant, and $\underline{X}$ is of dimension m. If $\underline{A}$ is a $k \times m$ matrix of rank k, then the random vector $\underline{Y} = \underline{A}\underline{X}$ is a k-variate Gaussian vector with

$$\mu_y = \underline{A}\mu_x \quad (13.86)$$

$$C_y = \underline{A} C_x \underline{A}^t \quad (13.87)$$

The characteristic function for a multivariate Gaussian *pdf* is defined by

$$C_{\underline{X}} = E[\exp\{j(\omega_1 X_1 + \omega_2 X_2 + \dots + \omega_m X_m)\}] = \quad (13.88)$$

$$\exp\left\{j\mu_x^t \underline{\omega} - \frac{1}{2}\underline{\omega}^t C_x \underline{\omega}\right\}$$

Then the moments for the joint distribution can be obtained by partial differentiation. For example,

$$E[X_1 X_2 X_3] = \frac{\partial^3}{\partial \omega_1 \partial \omega_2 \partial \omega_3} C_{\underline{X}}(\omega_1, \omega_2, \omega_3) \quad \text{at} \quad \underline{\omega} = \underline{0} \quad (13.89)$$

Example:

The vector $\underline{X}$ is a 4-variate Gaussian with

$$\mu_x = \begin{bmatrix} 2 & 1 & 1 & 0 \end{bmatrix}^t$$

$$C_x = \begin{bmatrix} 6 & 3 & 2 & 1 \\ 3 & 4 & 3 & 2 \\ 2 & 3 & 4 & 3 \\ 1 & 2 & 3 & 3 \end{bmatrix}$$

Define

$$\underline{X}_1 = \begin{bmatrix} X_1 \\ X_2 \end{bmatrix} \quad \underline{X}_2 = \begin{bmatrix} X_3 \\ X_4 \end{bmatrix}$$

Find the distribution of $\underline{X}_1$ and the distribution of

$$\underline{Y} = \begin{bmatrix} 2X_1 \\ X_1 + 2X_2 \\ X_3 + X_4 \end{bmatrix}$$

Solution:

$\underline{X}_1$ has a bivariate Gaussian distribution with

$$\mu_{x_1} = \begin{bmatrix} 2 \\ 1 \end{bmatrix} \quad C_{x_1} = \begin{bmatrix} 6 & 3 \\ 3 & 4 \end{bmatrix}$$

The vector $\underline{Y}$ can be expressed as

$$\underline{Y} = \begin{bmatrix} 2 & 0 & 0 & 0 \\ 1 & 2 & 0 & 0 \\ 0 & 0 & 1 & 1 \end{bmatrix} \begin{bmatrix} X_1 \\ X_2 \\ X_3 \\ X_4 \end{bmatrix} = \underline{A}\underline{X}$$

It follows that

$$\mu_y = \underline{A}\mu_x = \begin{bmatrix} 4 & 4 & 1 \end{bmatrix}^t$$

$$C_y = \underline{A}C_x\underline{A}^t = \begin{bmatrix} 24 & 24 & 6 \\ 24 & 34 & 13 \\ 6 & 13 & 13 \end{bmatrix}$$

13.8. Random Processes

A random variable X is by definition a mapping of all possible outcomes of a random experiment to numbers. When the random variable becomes a function of both the outcomes of the experiment as well as time, it is called a random process and is denoted by $X(t)$. Thus, one can view a random process as an ensemble of time domain functions that are the outcome of a certain random experiment, as compared to single real numbers in the case of a random variable.

Since the *cdf* and *pdf* of a random process are time dependent, we will denote them as $F_X(x;t)$ and $f_X(x;t)$, respectively. The *nth* moment for the random process $X(t)$ is

$$E[X^n(t)] = \int_{-\infty}^{\infty} x^n f_X(x;t)dx \qquad (13.90)$$

A random process $X(t)$ is referred to as stationary to order one if all its statistical properties do not change with time. Consequently, $E[X(t)] = \bar{X}$, where $\bar{X}$ is a constant. A random process $X(t)$ is called stationary to order two (or wide sense stationary) if

$$f_X(x_1, x_2; t_1, t_2) = f_X(x_1, x_2; t_1 + \Delta t, t_2 + \Delta t) \qquad (13.91)$$

for all t_1, t_2 and Δt.

Define the statistical autocorrelation function for the random process $X(t)$ as

$$\Re_X(t_1, t_2) = E[X(t_1)X(t_2)] \qquad (13.92)$$

The correlation $E[X(t_1)X(t_2)]$ is, in general, a function of (t_1, t_2). As a consequence of the wide sense stationary definition, the autocorrelation function depends on the time difference $\tau = t_2 - t_1$, rather than on absolute time; and thus, for a wide sense stationary process we have

$$E[X(t)] = \bar{X}$$
$$\Re_X(\tau) = E[X(t)X(t+\tau)] \qquad (13.93)$$

If the time average and time correlation functions are equal to the statistical average and statistical correlation functions, the random process is referred to as an ergodic random process. The following is true for an ergodic process:

$$\lim_{T \to \infty} \frac{1}{T} \int_{-T/2}^{T/2} x(t)dt = E[X(t)] = \bar{X} \qquad (13.94)$$

$$\lim_{T \to \infty} \frac{1}{T} \int_{-T/2}^{T/2} x^*(t)x(t+\tau)dt = \Re_X(\tau) \qquad (13.95)$$

The covariance of two random processes $X(t)$ and $Y(t)$ is defined by

$$C_{XY}(t, t+\tau) = E[\{X(t) - E[X(t)]\}\{Y(t+\tau) - E[Y(t+\tau)]\}] \qquad (13.96)$$

which can be written as

$$C_{XY}(t, t+\tau) = \Re_{XY}(\tau) - \bar{X}\bar{Y} \qquad (13.97)$$

13.9. Sampling Theorem

Most modern communication and radar systems are designed to process discrete samples of signals bearing information. In general, we would like to determine the necessary condition such that a signal can be fully reconstructed from its samples by filtering, or data processing in general. The answer to this question lies in the sampling theorem which may be stated as follows: let the signal $x(t)$ be real-valued and band-limited with bandwidth B; this signal can be fully reconstructed from its samples if the time interval between samples is no greater than $1/(2B)$.

Fig. 13.1 illustrates the sampling process concept. The sampling signal $p(t)$ is periodic with period T_s, which is called the sampling interval. The Fourier series expansion of $p(t)$ is

$$p(t) = \sum_{n=-\infty}^{\infty} P_n e^{\frac{j2\pi nt}{T_s}} \qquad (13.98)$$

The sampled signal $x_s(t)$ is then given by

Sampling Theorem

$$x_s(t) = \sum_{n=-\infty}^{\infty} x(t) P_n e^{\frac{j2\pi nt}{T_s}} \tag{13.99}$$

Taking the FT of Eq. (13.99) yields

$$X_s(\omega) = \sum_{n=-\infty}^{\infty} P_n X\left(\omega - \frac{2\pi n}{T_s}\right) = P_0 X(\omega) + \sum_{\substack{n=-\infty \\ n \neq 0}}^{\infty} P_n X\left(\omega - \frac{2\pi n}{T_s}\right) \tag{13.100}$$

where $X(\omega)$ is the FT of $x(t)$. Therefore, we conclude that the spectral density, $X_s(\omega)$, consists of replicas of $X(\omega)$ spaced $(2\pi/T_s)$ apart and scaled by the Fourier series coefficients P_n. A Low Pass Filter (LPF) of bandwidth B can then be used to recover the original signal $x(t)$.

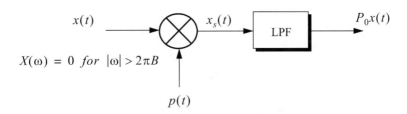

Figure 13.1. Concept of sampling.

When the sampling rate is increased (i.e., T_s decreases), the replicas of $X(\omega)$ move farther apart from each other. Alternatively, when the sampling rate is decreased (i.e., T_s increases), the replicas get closer to one another. The value of T_s such that the replicas are tangent to one another defines the minimum required sampling rate so that $x(t)$ can be recovered from its samples by using an LPF. It follows that

$$\frac{2\pi}{T_s} = 2\pi(2B) \Leftrightarrow T_s = \frac{1}{2B} \tag{13.101}$$

The sampling rate defined by Eq. (13.101) is known as the Nyquist sampling rate. When $T_s > (1/2B)$, the replicas of $X(\omega)$ overlap and, thus, $x(t)$ cannot be recovered cleanly from its samples. This is known as aliasing. In practice, ideal LPF cannot be implemented; hence, practical systems tend to over-sample in order to avoid aliasing.

Example:

Assume that the sampling signal $p(t)$ is given by

$$p(t) = \sum_{n=-\infty}^{\infty} \delta(t - nT_s)$$

Compute an expression for $X_s(\omega)$.

Solution:

The signal $p(t)$ is called the Comb function. Its exponential Fourier series is

$$p(t) = \sum_{n=-\infty}^{\infty} \frac{1}{T_s} e^{\frac{2\pi n t}{T_s}}$$

It follows that

$$x_s(t) = \sum_{n=-\infty}^{\infty} x(t) \frac{1}{T_s} e^{\frac{2\pi n t}{T_s}}$$

Taking the Fourier transform of this equation yields

$$X_s(\omega) = \frac{2\pi}{T_s} \sum_{n=-\infty}^{\infty} X\left(\omega - \frac{2\pi n}{T_s}\right).$$

Before proceeding to the next section, we will establish the following notation: samples of the signal $x(t)$ are denoted by $x(n)$ and referred to as a discrete time domain sequence, or simply a sequence. If the signal $x(t)$ is periodic, we will denote its sample by the periodic sequence $\tilde{x}(n)$.

13.10. The Z-Transform

The Z-transform is a transformation that maps samples of a discrete time domain sequence into a new domain known as the z-domain. It is defined as

$$Z\{x(n)\} = X(z) = \sum_{n=-\infty}^{\infty} x(n) z^{-n} \qquad (13.102)$$

The Z-Transform

where $z = re^{j\omega}$, and for most cases, $r = 1$. It follows that Eq. (13.102) can be rewritten as

$$X(e^{j\omega}) = \sum_{n=-\infty}^{\infty} x(n)e^{-jn\omega} \tag{13.103}$$

In the z-domain, the region over which $X(z)$ is finite is called the Region of Convergence (ROC). Appendix 13C has a list of most common Z-transform pairs. The Z-transform properties are (the proofs are left as an exercise):

1. **Linearity:**

$$Z\{ax_1(n) + bx_2(n)\} = aX_1(z) + bX_2(z) \tag{13.104}$$

2. **Right-Shifting Property:**

$$Z\{x(n-k)\} = z^{-k}X(z) \tag{13.105}$$

3. **Left-Shifting Property:**

$$Z\{x(n+k)\} = z^k X(z) - \sum_{n=0}^{k-1} x(n)z^{k-n} \tag{13.106}$$

4. **Time Scaling:**

$$Z\{a^n x(n)\} = X(a^{-1}z) = \sum_{n=0}^{\infty} (a^{-1}z)^{-n} x(n) \tag{13.107}$$

5. **Periodic Sequences:**

$$Z\{x(n)\} = \frac{z^N}{z^N - 1} Z\{x(n)\} \tag{13.108}$$

where N is the period.

6. **Multiplication by n:**

$$Z\{nx(n)\} = -z\frac{d}{dz}X(z) \tag{13.109}$$

7. **Division by $n + a$; a is a real number:**

$$Z\left\{\frac{x(n)}{n+a}\right\} = \sum_{n=0}^{\infty} x(n) z^a \left(-\int_0^z u^{-k-a-1} du\right) \quad (13.110)$$

8. *Initial Value:*

$$x(n_0) = z^{n_0} X(z)\big|_{z \to \infty} \quad (13.111)$$

9. *Final Value:*

$$\lim_{n \to \infty} x(n) = \lim_{z \to 1} (1 - z^{-1}) X(z) \quad (13.112)$$

10. *Convolution:*

$$Z\left\{\sum_{k=0}^{\infty} h(n-k) x(k)\right\} = H(z) X(z) \quad (13.113)$$

11. *Bilateral Convolution:*

$$Z\left\{\sum_{k=-\infty}^{\infty} h(n-k) x(k)\right\} = H(z) X(z) \quad (13.114)$$

Example:

Prove Eq. (13.109).

Solution:

Starting with the definition of the Z-transform,

$$X(z) = \sum_{n=-\infty}^{\infty} x(n) z^{-n}$$

Taking the derivative, with respect to z, of the above equation yields

$$\frac{d}{dz} X(z) = \sum_{n=-\infty}^{\infty} x(n)(-n) z^{-n-1}$$

$$= (-z^{-1}) \sum_{n=-\infty}^{\infty} nx(n)z^{-n}$$

It follows that

$$Z\{nx(n)\} = (-z)\frac{d}{dz}X(z)$$

In general, a discrete LTI system has a transfer function $H(z)$ which describes how the system operates on its input sequence $x(n)$ in order to produce the output sequence $y(n)$. The output sequence $y(n)$ is computed from the discrete convolution between the sequences $x(n)$ and $h(n)$,

$$y(n) = \sum_{m=-\infty}^{\infty} x(m)h(n-m) \tag{13.115}$$

However, since practical systems require that the sequence $x(n)$ be of finite length, we can rewrite Eq. (13.115) as

$$y(n) = \sum_{m=0}^{N} x(m)h(n-m) \tag{13.116}$$

where N denotes the input sequence length. Taking the Z-transform of Eq. (13.116) yields

$$Y(z) = X(z)H(z) \tag{13.117}$$

and the discrete system transfer function is

$$H(z) = \frac{Y(z)}{X(z)} \tag{13.118}$$

Finally, the transfer function $H(z)$ can be written as

$$H(z)\big|_{z=e^{j\omega}} = |H(e^{j\omega})|e^{\angle H(e^{j\omega})} \tag{13.119}$$

where $|H(e^{j\omega})|$ is the amplitude response, and $\angle H(e^{j\omega})$ is the phase response.

13.11. The Discrete Fourier Transform

The Discrete Fourier Transform (DFT) is a mathematical operation that transforms a discrete sequence, usually from the time domain into the frequency domain, in order to explicitly determine the spectral information for the sequence. The time domain sequence can be real or complex. The DFT has finite length N, and is periodic with period equal to N.

The discrete Fourier transform for the finite sequence $x(n)$ is defined by

$$\tilde{X}(k) = \sum_{n=0}^{N-1} x(n) e^{-\frac{j2\pi nk}{N}} \quad ; \; k = 0, ..., N-1 \quad (13.120)$$

The inverse DFT is given by

$$\tilde{x}(n) = \frac{1}{N} \sum_{k=0}^{N-1} \tilde{X}(k) e^{\frac{j2\pi nk}{N}} \quad ; \; n = 0, ..., N-1 \quad (13.121)$$

The Fast Fourier Transform (FFT) is not a new kind of transform different from the DFT. Instead, it is an algorithm used to compute the DFT more efficiently. There are numerous FFT algorithms that can be found in the literature. In this book we will interchangeably use the DFT and the FFT to mean the same thing. Furthermore, we will assume radix-2 FFT algorithm, where the FFT size is equal to $N = 2^m$ for some integer m.

13.12. Discrete Power Spectrum

Practical discrete systems utilize DFTs of finite length as a means of numerical approximation for the Fourier transform. It follows that input signals must be truncated to a finite duration (denoted by T) before they are sampled. This is necessary so that a finite length sequence is generated prior to signal processing. Unfortunately, this truncation process may cause some serious problems.

To demonstrate this difficulty, consider the time domain signal $x(t) = \sin 2\pi f_0 t$. The spectrum of $x(t)$ consists of two spectral lines at $\pm f_0$. Now, when $x(t)$ is truncated to length T seconds and sampled at a rate $T_s = T/N$, where N is the number of desired samples, we produce the sequence $\{x(n) \; ; \; n = 0, 1, ..., N-1\}$. The spectrum of $x(n)$ would still be composed of the same spectral lines if T is an integer multiple of T_s and if the DFT frequency resolution Δf is an integer multiple of f_0. Unfortunately, those two conditions are rarely met and, as a consequence, the spectrum of $x(n)$

spreads over several lines (normally the spread may extend up to three lines). This is known as spectral leakage. Since f_0 is normally unknown, this discontinuity caused by an arbitrary choice of T cannot be avoided. Windowing techniques can be used to mitigate the effect of this discontinuity by applying smaller weights to samples close to the edges.

A truncated sequence $x(n)$ can be viewed as one period of some periodic sequence $\tilde{x}(n)$ with period N. The discrete Fourier series expansion of $x(n)$ is

$$x(n) = \sum_{k=0}^{N-1} X_k e^{\frac{j2\pi nk}{N}} \qquad (13.122)$$

It can be shown that the coefficients X_k are given by

$$X_k = \frac{1}{N}\sum_{n=0}^{N-1} x(n) e^{\frac{-j2\pi nk}{N}} = \frac{1}{N}X(k) \qquad (13.123)$$

where $X(k)$ is the DFT of $x(n)$. Therefore, the Discrete Power Spectrum (DPS) for the band limited sequence $x(n)$ is the plot of $|X_k|^2$ versus k, where the lines are Δf apart,

$$P_0 = \frac{1}{N^2}|X(0)|^2$$

$$P_k = \frac{1}{N^2}\{|X(k)|^2 + |X(N-k)|^2\} \quad ; k = 1, 2, ..., \frac{N}{2}-1 \qquad (13.124)$$

$$P_{N/2} = \frac{1}{N^2}|X(N/2)|^2$$

Before proceeding to the next section, we will show how to select the FFT parameters. For this purpose, consider a band limited signal $x(t)$ with bandwidth B. If the signal is not band limited, a LPF can be used to eliminate frequencies greater than B. In order to satisfy the sampling theorem, one must choose a sampling frequency $f_s = 1/T_s$, such that

$$f_s \geq 2B \qquad (13.125)$$

The truncated sequence duration T and the total number of samples N are related by

$$T = NT_s \tag{13.126}$$

or equivalently,

$$f_s = \frac{N}{T} \tag{13.127}$$

It follows that

$$f_s = \frac{N}{T} \geq 2B \tag{13.128}$$

and the frequency resolution is

$$\Delta f = \frac{1}{NT_s} = \frac{f_s}{N} = \frac{1}{T} \geq \frac{2B}{N} \tag{13.129}$$

13.13. Windowing Techniques

Truncation of the sequence $x(n)$ can be accomplished by computing the product,

$$x_w(n) = x(n)w(n) \tag{13.130}$$

where

$$w(n) = \begin{cases} f(n) & ; n = 0, 1, \ldots, N-1 \\ 0 & \text{otherwise} \end{cases} \tag{13.131}$$

where $f(n) \leq 1$. The finite sequence $w(n)$ is called a windowing sequence, or simply a window. The windowing process should not impact the phase response of the truncated sequence. Consequently, the sequence $w(n)$ must retain linear phase. This can be accomplished by making the window symmetrical with respect to its central point.

If $f(n) = 1$ for all n we have what is known as the rectangular window. It leads to the Gibbs phenomenon which manifests itself as an overshoot and a ripple before and after a discontinuity. Fig. 13.2 shows the amplitude spectrum of a rectangular window. Note that the first side lobe is at $-13.46 dB$ below the main lobe. Windows that place smaller weights on the samples near the edges will have lesser overshoot at the discontinuity points (lower side lobes); hence, they are more desirable than a rectangular window. However, sidelobes reduction is offset by a widening of the main lobe. Therefore, the proper choice of a windowing sequence is continuous trade-off between side lobe reduction and main lobe widening. Table 13.1 gives a summary of some windows with the corresponding impact on main beam widening and peak reduction.

TABLE 13.1. Common windows.

Window	Null-to-null Beamwidth. Rectangular window is the reference.	Peak Reduction
Rectangular	1	1
Hamming	2	0.73
Hanning	2	0.664
Blackman	6	0.577
Kaiser ($\beta = 6$)	2.76	0.683
Kaiser ($\beta = 3$)	1.75	0.882

The multiplication process defined in Eq. (13.131) is equivalent to cyclic convolution in the frequency domain. It follows that $X_w(k)$ is a smeared (distorted) version of $X(k)$. To minimize this distortion, we would seek windows that have a narrow main lobe and small side lobes. Additionally, using a window other than a rectangular window reduces the power by a factor P_w, where

$$P_w = \frac{1}{N}\sum_{n=0}^{N-1} w^2(n) = \sum_{k=0}^{N-1} |W(k)|^2 \qquad (13.132)$$

It follows that the DPS for the sequence $x_w(n)$ is now given by

$$P_0^w = \frac{1}{P_w N^2}|X(0)|^2$$

$$P_k^w = \frac{1}{P_w N^2}\{|X(k)|^2 + |X(N-k)|^2\} \quad ; k = 1, 2, ..., \frac{N}{2}-1 \qquad (13.133)$$

$$P_{N/2}^w = \frac{1}{P_w N^2}|X(N/2)|^2$$

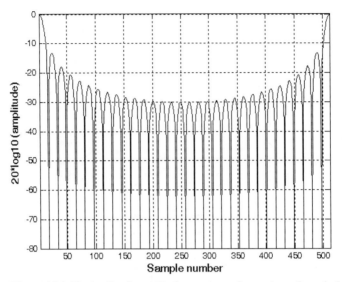

Figure 13.2. Normalized amplitude spectrum for rectangular window.

where P_w is defined in Eq. (13.132). Table 13.2 lists some common windows. Figs. 13.3 through 13.5 show the frequency domain characteristics for these windows. These figures can be reproduced using MATLAB program *"figs13.m"*.

TABLE 13.2. Some common windows. $n = 0, N-1$.

Window	Expression	First side lobe	Main lobe width
rectangular	$w(n) = 1$	$-13.46 dB$	1
Hamming	$w(n) = 0.54 - 0.46\cos\left(\frac{2\pi n}{N-1}\right)$	$-41 dB$	2
Hanning	$w(n) = 0.5\left[1 - \cos\left(\frac{2\pi n}{N-1}\right)\right]$	$-32 dB$	2
Kaiser	$w(n) = \dfrac{I_0[\beta\sqrt{1-(2n/N)^2}]}{I_0(\beta)}$ I_0 is the zero-order modified Bessel function of the first kind	$-46 dB$ for $\beta = 2\pi$	$\sqrt{5}$ for $\beta = 2\pi$

Windowing Techniques

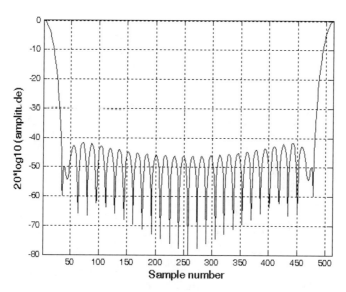

Figure 13.3. Normalized amplitude spectrum for Hamming window.

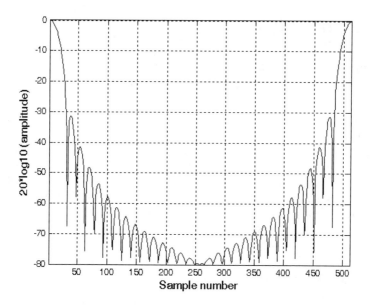

Figure 13.4. Normalized amplitude spectrum for Hanning window.

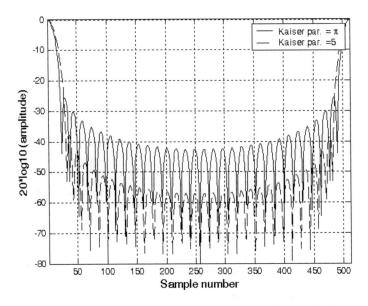

Figure 13.5. Normalized amplitude spectrum for Kaiser window.

13.14. MATLAB Programs

Listing 13.1. MATLAB Program "figs13.m"

```
%Use this program to reproduce figures in Section 13.13.
clear all
close all
eps = 0.0001;
N = 32;
win_rect (1:N) = 1;
win_ham = hamming(N);
win_han = hanning(N);
win_kaiser = kaiser(N, pi);
win_kaiser2 = kaiser(N, 5);
Yrect = abs(fft(win_rect, 512));
Yrectn = Yrect ./ max(Yrect);
Yham = abs(fft(win_ham, 512));
Yhamn = Yham ./ max(Yham);
Yhan = abs(fft(win_han, 512));
Yhann = Yhan ./ max(Yhan);
```

```
YK = abs(fft(win_kaiser, 512));
YKn = YK ./ max(YK);
YK2 = abs(fft(win_kaiser2, 512));
YKn2 = YK2 ./ max(YK2);
figure (1)
plot(20*log10(Yrectn+eps),'k')
xlabel('Sample number')
ylabel('20*log10(amplitude)')
axis tight
grid
figure(2)
plot(20*log10(Yhamn + eps),'k')
xlabel('Sample number')
ylabel('20*log10(amplitude)')
grid
axis tight
figure (3)
plot(20*log10(Yhann+eps),'k')
xlabel('Sample number')
ylabel('20*log10(amplitude)')
grid
axis tight
figure(4)
plot(20*log10(YKn+eps),'k')
grid
hold on
plot(20*log10(YKn2+eps),'k--')
xlabel('Sample number')
ylabel('20*log10(amplitude)')
legend('Kaiser par. = \pi','Kaiser par. =5')
axis tight
hold off
```

Appendix 13A *Fourier Transform Table*

$x(t)$	$X(\omega)$
$A\,Rect(t/\tau)$; rectangular pulse	$A\tau\,Sinc(\omega\tau/2)$
$A\Delta(t/\tau)$; triangular pulse	$A\dfrac{\tau}{2}Sinc^2(\tau\omega/4)$
$\dfrac{1}{\sqrt{2\pi}\sigma}\exp\left(-\dfrac{t^2}{2\sigma^2}\right)$; Gaussian pulse	$\exp\left(-\dfrac{\sigma^2\omega^2}{2}\right)$
$e^{-at}u(t)$	$1/(a+j\omega)$
$e^{-a|t|}$	$\dfrac{2a}{a^2+\omega^2}$
$e^{-at}\sin\omega_0 t\ u(t)$	$\dfrac{\omega_0}{\omega_0^2+(a+j\omega)^2}$
$e^{-at}\cos\omega_0 t\ u(t)$	$\dfrac{a+j\omega}{\omega_0^2+(a+j\omega)^2}$
$\delta(t)$	1
1	$2\pi\delta(\omega)$
$u(t)$	$\pi\delta(\omega)+\dfrac{1}{j\omega}$
$sgn(t)$	$\dfrac{2}{j\omega}$

$x(t)$	$X(\omega)$
$\cos\omega_0 t$	$\pi[\delta(\omega-\omega_0)+\delta(\omega+\omega_0)]$
$\sin\omega_0 t$	$j\pi[\delta(\omega+\omega_0)-\delta(\omega-\omega_0)]$
$u(t)\cos\omega_0 t$	$\dfrac{\pi}{2}[\delta(\omega-\omega_0)+\delta(\omega+\omega_0)]+\dfrac{j\omega}{\omega_0^2-\omega^2}$
$u(t)\sin\omega_0 t$	$\dfrac{\pi}{2j}[\delta(\omega+\omega_0)-\delta(\omega-\omega_0)]+\dfrac{\omega_0}{\omega_0^2-\omega^2}$
$|t|$	$\dfrac{-2}{\omega^2}$

Appendix 13B *Some Common Probability Densities*

Chi-Square with N degrees of freedom

$$f_X(x) = \frac{x^{(N/2)-1}}{2^{N/2}\Gamma(N/2)} \exp\left\{\frac{-x}{2}\right\} \; ; \; x > 0$$

$$\bar{X} = N \; ; \; \sigma_X^2 = 2N$$

$$\text{gamma function} = \Gamma(z) = \int_0^\infty \lambda^{z-1} e^{-\lambda} d\lambda \; ; \; Re\{z\} > 0$$

Exponential

$$(f_X(x) = a\exp\{-ax\}) \; ; \; x > 0$$

$$\bar{X} = \frac{1}{a} \; ; \; \sigma_X^2 = \frac{1}{a^2}$$

Gaussian

$$f_X(x) = \frac{1}{\sqrt{2\pi}\sigma} \exp\left\{-\frac{1}{2}\left(\frac{x-x_m}{\sigma}\right)^2\right\} ; \; \bar{X} = x_m \; ; \; \sigma_X^2 = \sigma^2$$

Laplace

$$f_X(x) = \frac{\sigma}{2} \exp\{-\sigma|x-x_m|\}$$

$$\bar{X} = x_m \; ; \; \sigma_X^2 = \frac{2}{\sigma^2}$$

Log-Normal

$$f_X(x) = \frac{1}{x\sigma\sqrt{2\pi}} \exp\left(-\frac{(\ln x - \ln x_m)^2}{2\sigma^2}\right) \; ; \; x > 0$$

$$\bar{X} = \exp\left\{\ln x_m + \frac{\sigma^2}{2}\right\} \; ; \; \sigma_X^2 = [\exp\{2\ln x_m + \sigma^2\}][\exp\{\sigma^2\} - 1]$$

Rayleigh

$$f_X(x) = \frac{x}{\sigma^2}\exp\left\{\frac{-x^2}{2\sigma^2}\right\} \; ; \; x \geq 0$$

$$\bar{X} = \sqrt{\frac{\pi}{2}}\sigma \; ; \; \sigma_X^2 = \frac{\sigma^2}{2}(4 - \pi)$$

Uniform

$$f_X(x) = \frac{1}{b-a} \; ; \; a < b \quad ; \quad \bar{X} = \frac{a+b}{2} \; ; \; \sigma_X^2 = \frac{(b-a)^2}{12}$$

Weibull

$$f_X(x) = \frac{bx^{b-1}}{\bar{\sigma}_0} \exp\left(-\frac{(x)^b}{\bar{\sigma}_0}\right) \; ; \; (x, b, \bar{\sigma}_0) \geq 0$$

$$\bar{X} = \frac{\Gamma(1+b^{-1})}{1/(\sqrt[b]{\bar{\sigma}_0})} \; ; \; \sigma_X^2 = \frac{\Gamma(1+2b^{-1}) - [\Gamma(1+b^{-1})]^2}{1/[\sqrt[b]{(\bar{\sigma}_0)^2}]}$$

Appendix 13C *Z - Transform Table*

$x(n);\ n \geq 0$	$X(z)$	ROC; $|z| > R$
$\delta(n)$	1	0
1	$\dfrac{z}{z-1}$	1
n	$\dfrac{z}{(z-1)^2}$	1
n^2	$\dfrac{z(z+1)}{(z-1)^3}$	1
a^n	$\dfrac{z}{z-a}$	$|a|$
na^n	$\dfrac{az}{(z-a)^2}$	$|a|$
$\dfrac{a^n}{n!}$	$e^{a/z}$	0
$(n+1)a^n$	$\dfrac{z^2}{(z-a)^2}$	$|a|$
$\sin n\omega T$	$\dfrac{z \sin \omega T}{z^2 - 2z\cos \omega T + 1}$	1
$\cos n\omega T$	$\dfrac{z(z - \cos \omega T)}{z^2 - 2z\cos \omega T + 1}$	1

$x(n); \ n \geq 0$	$X(z)$	ROC; $\|z\| > R$
$a^n \sin n\omega T$	$\dfrac{az \sin \omega T}{z^2 - 2az \cos \omega T + a^2}$	$\dfrac{1}{\|a\|}$
$a^n \cos n\omega T$	$\dfrac{z(z - a^2 \cos \omega T)}{z^2 - 2az \cos \omega T + a^2}$	$\dfrac{1}{\|a\|}$
$\dfrac{n(n-1)}{2!}$	$\dfrac{z}{(z-1)^3}$	1
$\dfrac{n(n-1)(n-2)}{3!}$	$\dfrac{z}{(z-1)^4}$	1
$\dfrac{(n+1)(n+2)a^n}{2!}$	$\dfrac{z^3}{(z-a)^3}$	$\|a\|$
$\dfrac{(n+1)(n+2)\ldots(n+m)a^n}{m!}$	$\dfrac{z^{m+1}}{(z-a)^{m+1}}$	$\|a\|$

Chapter 14   *MATLAB Program and Function Name List*

This chapter provides a summary of all MATLAB program and function names used throughout this book. All these programs and functions can be downloaded from the CRC Press Web site (*www.crcpress.com*). For this purpose, follow this procedure: 1) from your Web browser type *"http://www.crcpress.com"*, 2) click on *"Electronic Products"*, 3) click on *"Download & Updates"*, and finally 4) follow instructions of how to download a certain set of code off that Web page. Furthermore, this MATLAB code can also be downloaded from The MathWorks Web site by following these steps: 1) from the Web browser type: *"http://mathworks.com/matlabcentral/fileexchange/"*, 2) place the curser on *"Companion Software for Books"* and click on *"Communications"*.

Chapter 1: Introduction to Radar Basics

Name	Purpose
radar_eq	Implements radar equation
fig1_12	Reproduces Fig. 1.12
fig1_13	Reproduces Fig. 1.13
ref_snr	Calculates the radar reference range or SNR
power_aperture	Implements the power aperture radar equation
fig1_16	Reproduces Fig. 1.16
casestudy1_1	Program for mini design case study 1.1
fig1_19	Reproduces Fig. 1.19
fig1_21	Reproduces Fig. 1.21
pulse_integration	Performs coherent or non-coherent pulse integration

Name	Purpose
myradarvisit1_1	Program for "MyRadar" design case study - visit 1
fig1_27	Reproduces Fig. 1.27
fig1_28	Reproduces Fig. 1.128

Chapter 2: Radar Detection

Name	Purpose (all functions have associated GUI)
fig2_2	Reproduces Fig. 2.2
que_func	Implements Marcum's Q-function
fig2_3	Reproduces Fig. 2.3
prob_snr1	Calculates single pulse probability of detection
fig2_6a	Reproduces Fig. 2.6a
improv_fac	Calculates the improvement factor
fig2_6b	Reproduces Fig. 2.6b
incomplete_gamma	Calculates the incomplete Gamma function
factor	Calculates the factorial of an integer
fig2_7	Reproduces Fig. 2.7
threshold	Calculates the detection threshold value
fig2_8	Reproduces Fig. 2.8
pd_swerling5	Calculates the Swerling 0 or 5 Prob. of detection
fig2_9	Reproduces Fig. 2.9
pd_swrling1	Calculates the Swerling 1 Prob. of detection
fig2_10	Reproduces Fig. 2.10
pd_swrling2	Calculates the Swerling 2 Prob. of detection
fig2_11ab	Reproduces Fig.s 2.11 a and b
pd_swrling3	Calculates the Swerling 3 Prob. of detection
fig2_12	Reproduces Fig. 2.12
pd_swrling4	Calculates the Swerling 4 Prob. of detection
fig2_13	Reproduces Fig. 2.13
fig2_14	Reproduces Fig. 2.14

Name	Purpose (all functions have associated GUI)
fluct_loss	Calculates the SNR loss due to RCS fluctuation
fig2_15	Reproduces Fig. 2.15
myradar_visit2_1	Program for "MyRadar" design case study visit 2.1
myradar_visit2_2	Program for "MyRadar" design case study visit 2.2
fig2_21	Reproduces Fig. 2.21

Chapter 3: Radar Waveforms

Name	Purpose
fig3_7	Reproduces Fig. 3.7
fig3_8	Reproduces Fig. 3.8
hrr_profile	Computes and plots HRR profile
fig3_17	Reproduces Fig. 3.17

Chapter 4: The Radar Ambiguity Function

Name	Purpose
single_pulse_ambg	Calculate and plot ambiguity function for a single pulse
fig4_2	Reproduces Fig. 4.2
fig4_4	Reproduces Fig. 4.4
lfm_ambig	Calculates and plot LFM ambiguity function
fig4_5	Reproduces Fig. 4.5
fig4_6	Reproduces Fig. 4.6
train_ambg	Calculates and plots ambiguity function for a train of coherent pulses
fig4_8	Reproduces Fig. 4.8
barker_ambg	Calculates and plots ambiguity function corresponding to a Barker code

Name	Purpose
prn_ambig	Calculates and plots ambiguity function corresponding to a PRN code
myradar_visit4	Program for "MyRadar" design case study visit 4

Chapter 5: Pulse Compression

Name	Purpose
fig5_3	Reproduces Fig. 5.3
matched_filter	Performs pulse compression using a matched filter
power_integer_2	Calculates the power integer of 2 for a given positive integer
stretch	Performs pulse compression using stretch processing
fig5_14	Reproduces Fig. 5.14

Chapter 6: Surface and Volume Clutter

Name	Purpose
clutter_rcs	Calculates and plots clutter RCS versus range
myradar_visit6	Program for "MyRadar" design case study visit 6

Chapter 7: Moving Target Indicator (MTI) - Clutter Mitigation

Name	Purpose
single_canceler	Performs single delay line MTI operation
double_canceler	Performs double delay line MTI operation
fig7_9	Reproduces Fig. 7.9
fig7_10	Reproduces Fig. 7.10
fig7_11	Reproduces Fig. 7.11
myradar_visit7	Program for "MyRadar" design case study visit 7

Chapter 8: Phased Arrays

Name	Purpose
fig8_5	Reproduces Fig. 8.5
fig8_7	Reproduces Fig. 8.7
linear_array	Calculates the linear array gain pattern
circular_array	Calculates the array pattern for a circular array
rect_array	Calculates the rectangular array gain pattern
circ_array	Calculates the circular array gain pattern
rec_to_circ	Calculates the boundary for rectangular array with circular boundary
fig8_52	Reproduces Fig. 8.52

Chapter 9: Target Tracking

Name	Purpose
mono_pulse	Calculate the sum and difference antenna patterns
ghk_tracker	implements the GHK filter
fir9_21	Reproduces Fig. 9.21
kalman_filter	Implements a 3-state Kalman filter
fig9_28	Reproduces Fig. 9.28
maketraj	Calculates and generates a trajectory
addnoise	Corrupts a trajectory
kalfilt	Implements a 6-state Kalman filter

Chapter 10: Electronic Countermeasures (ECM)

Name	Purpose
ssj_req	Implements SSJ radar equation
sir	Calculates and plots the $S/(J+N)$ ratio
bun_thru	Calculates the burnthrough range
soj_req	Implements the SOJ radar equation

Name	Purpose
range_red_factor	*Calculates the range reduction factor*
fig10_8	*Reproduces Fig. 10.8*

Chapter 11: Radar Cross Section (RCS)

Name	Purpose (all functions have associated GUI)
rcs_aspect	*compute and plot RCS dependency on aspect angle*
rcs_frequency	*compute and plot RCS dependency on frequency*
example11_1	*Used in solving Example on page*
rcs_sphere	*compute and plot RCS of a sphere*
rcs_ellipsoid	*compute and plot RCS of an ellipsoid*
rcs_circ_plate	*compute and plot RCS of a circular flat plate*
rcs_frustum	*compute and plot RCS of a truncated cone*
rcs_cylinder	*compute and plot RCS of a cylinder*
rcs_rect_plate	*compute and plot RCS of a rectangular flat plate*
rcs_isosceles	*compute and plot RCS of a triangular flat plate*
CappedWedgeTM	*Used to calculate the TM E-field for a capped wedge*
rcs_cylinder_complex	*reproduce Fig. 2.22*
swerlin_models	*reproduce Fig. 2.24*

Chapter 12: High Resolution Tactical Synthetic Aperture Radar (TSAR)

Name	Purpose
fig12_12_13	*Reproduces Figs. 12.12 and 12.13*

Chapter 13: Signal Processing

Name	Purpose
figs13	*Reproduces Fig. 13.2 through Fig. 13.5.*

Bibliography

Abramowitz, M. and Stegun, I. A., Editors, *Handbook of Mathematical Functions, with Formulas, Graphs, and Mathematical Tables*, Dover Publications, 1970.

Balanis, C. A., *Antenna Theory, Analysis and Design*, Harper & Row, New York, 1982.

Barkat, M., *Signal Detection and Estimation*, Artech House, Norwood, MA, 1991.

Barton, D. K., *Modern Radar System Analysis*, Artech House, Norwood, MA, 1988.

Benedict, T. and Bordner, G., Synthesis of an Optimal Set of Radar Track-While-Scan Smoothing Equations, *IRE Transaction on Automatic Control, Ac-7*, July 1962, pp. 27-32.

Berkowitz, R. S., *Modern Radar - Analysis, Evaluation, and System Design*, John Wiley & Sons, Inc, New York, 1965.

Beyer, W. H., *CRC Standard Mathematical Tables*, 26th edition, CRC Press, Boca Raton, FL, 1981.

Billetter, D. R., *Multifunction Array Radar*, Artech House, Norwood, MA, 1989.

Blackman, S. S., *Multiple-Target Tracking with Radar Application*, Artech House, Norwood, MA, 1986.

Blake, L. V., *A Guide to Basic Pulse-Radar Maximum Range Calculation Part-I Equations, Definitions, and Aids to Calculation*, Naval Res. Lab. Report 5868, 1969.

Blake, L. V., *Radar-Range Performance Analysis*, Lexington Books, Lexington, MA, 1980.

Boothe, R. R., *A Digital Computer Program for Determining the Performance of an Acquisition Radar Through Application of Radar Detection Probability Theory*, U.S. Army Missile Command: Report No. RD-TR-64-2. Redstone Arsenal, Alabama, 1964.

Brookner, E., Editor, *Aspects of Modern Radar*, Artech House, Norwood, MA, 1988.

Brookner, E., Editor, *Practical Phased Array Antenna System*, Artech House, Norwood, MA, 1991.

Brookner, E., *Radar Technology*, Lexington Books, Lexington, MA, 1996.

Burdic, W. S., *Radar Signal Analysis*, Prentice-Hall, Englewood Cliffs, NJ, 1968.

Brookner, E., *Tracking and Kalman Filtering Made Easy*, John Wiley & Sons, New York, 1998.

Cadzow, J. A., *Discrete-Time Systems, an Introduction with Interdisciplinary Applications*, Prentice-Hall, Englewood Cliffs, NJ, 1973.

Carlson, A. B., *Communication Systems, An Introduction to Signals and Noise in Electrical Communication*, 3rd edition, McGraw-Hill, New York, 1986.

Carpentier, M. H., *Principles of Modern Radar Systems*, Artech House, Norwood, MA, 1988.

Compton, R. T., *Adaptive Antennas*, Prentice-Hall, Englewood Cliffs, NJ, 1988.

Costas, J. P., A Study of a Class of Detection Waveforms Having Nearly Ideal Range-Doppler Ambiguity Properties, *Proc. IEEE 72*, 1984, pp. 996-1009.

Curry, G. R., *Radar System Performance Modeling*, Artech House, Norwood, 2001.

DiFranco, J. V. and Rubin, W. L., *Radar Detection*. Artech House, Norwood, MA, 1980.

Dillard, R. A. and Dillard, G. M., *Detectability of Spread-Spectrum Signals*, Artech House, Norwood, MA, 1989.

Edde, B., *Radar - Principles, Technology, Applications*, Prentice-Hall, Englewood Cliffs, NJ, 1993.

Elsherbeni, Atef, Inman, M. J., and Riley, C., "Antenna Design and Radiation Pattern Visualization," The 19th Annual Review of Progress in Applied Computational Electromagnetics, ACES'03, Monterey, California, March 2003.

Fehlner, L. F., *Marcum's and Swerling's Data on Target Detection by a Pulsed Radar*, Johns Hopkins University, Applied Physics Lab. Rpt. # TG451, July 2, 1962, and Rpt. # TG451A, Septemeber 1964.

Fielding, J. E. and Reynolds, G. D., *VCCALC: Vertical Coverage Calculation Software and Users Manual*, Artech House, Norwood, MA, 1988.

Gabriel, W. F., Spectral Analysis and Adaptive Array Superresolution Techniques, *Proc. IEEE*, Vol. 68, June 1980, pp. 654-666.

Gelb, A., Editor, *Applied Optimal Estimation*, MIT Press, Cambridge, MA, 1974.

Grewal, M. S. and Andrews, A. P., *Kalman Filtering - Theory and Practice Using MATLAB*, 2nd edition, Wiley & Sons Inc., New York, 2001.

Hamming, R. W., *Digital Filters*, 2nd edition, Prentice-Hall, Englewood Cliffs, NJ, 1983.

Hanselman, D. and Littlefield, B., *Mastering Matlab 5, A Complete Tutorial and Reference,* Malab Curriculum Series, Prentice-Hall, Englewood Cliffs, NJ, 1998.

Hirsch, H. L. and Grove, D. C., *Practical Simulation of Radar Antennas and Radomes*, Artech House, Norwood, MA, 1987.

Hovanessian, S. A., *Radar System Design and Analysis*, Artech House, Norwood, MA, 1984.

James, D. A., *Radar Homing Guidance for Tactical Missiles*, John Wiley & Sons, New York, 1986.

Kanter, I., *Exact Detection Probability for Partially Correlated Rayleigh Targets*, IEEE Trans. AES-22, pp. 184-196. March 1986.

Kay, S. M., *Fundamentals of Statistical Signal Processing - Estimation Theory*, Volume I, Prentice Hall Signal Processing Series, New Jersey, 1993.

Kay, S. M., *Fundamentals of Statistical Signal Processing - Detection Theory*, Volume II, Prentice Hall Signal Processing Series, New Jersey, 1993.

Klauder, J. R., Price, A. C., Darlington, S., and Albershiem, W. J., The Theory and Design of Chirp Radars, *The Bell System Technical Journal*, Vol. 39, No. 4, 1960.

Knott, E. F., Shaeffer, J. F., and Tuley, M. T., *Radar Cross Section*, 2nd edition, Artech House, Norwood, MA, 1993.

Lativa, J., Low-Angle Tracking Using Multifrequency Sampled Aperture Radar, *IEEE - AES Trans.*, Vol. 27, No. 5, September 1991, pp.797-805.

Levanon, N., *Radar Principles*, John Wiley & Sons, New York, 1988.

Lewis, B. L., Kretschmer, Jr., F. F., and Shelton, W. W., *Aspects of Radar Signal Processing*, Artech House, Norwood, MA, 1986.

Long, M. W., *Radar Reflectivity of Land and Sea*, Artech House, Norwood, MA, 1983.

Lothes, R. N., Szymanski, M. B., and Wiley, R. G., *Radar Vulnerability to Jamming*, Artech House, Norwood, MA, 1990.

Mahafza, B. R. and Polge, R. J., Multiple Target Detection Through DFT Processing in a Sequential Mode Operation of Real Two-Dimensional Arrays, *Proc. of the IEEE Southeast Conf. '90*, New Orleans, LA, April 1990, pp. 168-170.

Mahafza, B. R., Heifner, L.A., and Gracchi, V. C., Multitarget Detection Using Synthetic Sampled Aperture Radars (SSAMAR), *IEEE - AES Trans.*, Vol. 31, No. 3, July 1995, pp. 1127-1132.

Mahafza, B. R. and Sajjadi, M., Three-Dimensional SAR Imaging Using a Linear Array in Transverse Motion, *IEEE - AES Trans.*, Vol. 32, No. 1, January 1996, pp. 499-510.

Mahafza, B. R., *Introduction to Radar Analysis*, CRC Press, Boca Raton, FL, 1998.

Mahafza, B. R., *Radar Systems Analysis and Design Using MATLAB*, CRC Press, Boca Raton, FL, 2000.

Marchand, P., *Graphics and GUIs with Matlab*, 2nd edition, CRC Press, Boca Raton, FL, 1999.

Marcum, J. I., A Statistical Theory of Target Detection by Pulsed Radar, Mathematical Appendix, *IRE Trans.*, Vol. IT-6, April 1960, pp. 59-267.

Meeks, M. L., *Radar Propagation at Low Altitudes*, Artech House, Norwood, MA, 1982.

Melsa, J. L. and Cohn, D. L., *Decision and Estimation Theory*, McGraw-Hill, New York, 1978.

Mensa, D. L., *High Resolution Radar Imaging*, Artech House, Norwood, MA, 1984.

Meyer, D. P. and Mayer, H. A., *Radar Target Detection: Handbook of Theory and Practice*, Academic Press, New York, 1973.

Monzingo, R. A. and Miller, T. W., *Introduction to Adaptive Arrays*, John Wiley & Sons, New York, 1980.

Morchin, W., *Radar Engineer's Sourcebook*, Artech House, Norwood, MA, 1993.

Morris, G. V., *Airborne Pulsed Doppler Radar*, Artech House, Norwood, MA, 1988.

Nathanson, F. E., *Radar Design Principles*, 2nd edition, McGraw-Hill, New York, 1991.

Navarro, Jr., A. M., *General Properties of Alpha Beta, and Alpha Beta Gamma Tracking Filters*, Physics Laboratory of the National Defense Research Organization TNO, Report PHL 1977-92, January 1977.

North, D. O., An Analysis of the Factors which Determine Signal/Noise Discrimination in Pulsed Carrier Systems, *Proc. IEEE 51*, No. 7, July 1963, pp. 1015-1027.

Oppenheim, A. V. and Schafer, R. W., *Discrete-Time Signal Processing*, Prentice-Hall, Englewood Cliffs, NJ, 1989.

Oppenheim, A. V., Willsky, A. S., and Young, I. T., *Signals and Systems*, Prentice-Hall, Englewood Cliffs, NJ, 1983.

Orfanidis, S. J., *Optimum Signal Processing, an Introduction*, 2nd edition, McGraw-Hill, New York, 1988.

Papoulis, A., *Probability, Random Variables, and Stochastic Processes*, second edition, McGraw-Hill, New York, 1984.

Parl, S. A., New Method of Calculating the Generalized Q Function, *IEEE Trans. Information Theory*, Vol. IT-26, No. 1, January 1980, pp. 121-124.

Peebles, Jr., P. Z., *Probability, Random Variables, and Random Signal Principles*, McGraw-Hill, New York, 1987.

Peebles, Jr., P. Z., *Radar Principles*, John Wiley & Sons, New York, 1998.

Pettit, R. H., *ECM and ECCM Techniques for Digital Communication Systems*, Lifetime Learning Publications, New York, 1982.

Polge, R. J., Mahafza, B. R., and Kim, J. G., *Extension and Updating of the Computer Simulation of Range Relative Doppler Processing for MM Wave Seekers*, Interim Technical Report, Vol. I, prepared for the U.S. Army Missile Command, Redstone Arsenal, Alabama, January 1989.

Polge, R. J., Mahafza, B. R., and Kim, J. G., Multiple Target Detection Through DFT Processing in a Sequential Mode Operation of Real or Synthetic Arrays, *IEEE 21st Southeastern Symposium on System Theory*, Tallahassee, FL, 1989, pp. 264-267.

Poularikas, A. and Seely, S., *Signals and Systems*, PWS Publishers, Boston, MA, 1984.

Rihaczek, A. W., *Principles of High Resolution Radars*, McGraw-Hill, New York, 1969.

Ross, R. A., Radar Cross Section of Rectangular Flat Plate as a Function of Aspect Angle, *IEEE Trans*. AP-14:320, 1966.

Ruck, G. T., Barrick, D. E., Stuart, W. D., and Krichbaum, C. K., *Radar Cross Section Handbook*, Volume 1, Plenum Press, New York, 1970.

Ruck, G. T., Barrick, D. E., Stuart, W. D., and Krichbaum, C. K., *Radar Cross Section Handbook*, Volume 2, Plenum Press, New York, 1970.

Rulf, B. and Robertshaw, G. A., *Understanding Antennas for Radar, Communications, and Avionics*, Van Nostrand Reinhold, 1987.

Scanlan, M.J., Editor, *Modern Radar Techniques*, Macmillan, New York, 1987.

Scheer, J. A. and Kurtz, J. L., Editors, *Coherent Radar Performance Estimation*, Artech House, Norwood, MA, 1993.

Shanmugan, K. S. and Breipohl, A. M., *Random Signals: Detection, Estimation and Data Analysis*, John Wiley & Sons, New York, 1988.

Sherman, S. M., *Monopulse Principles and Techniques*, Artech House, Norwood, MA.

Singer, R. A., Estimating Optimal Tracking Filter Performance for Manned Maneuvering Targets, *IEEE Transaction on Aerospace and Electronics*, AES-5, July 1970, pp. 473-483.

Skillman, W. A., *DETPROB: Probability of Detection Calculation Software and User's Manual*, Artech House, Norwood, MA, 1991.

Skolnik, M. I., *Introduction to Radar Systems*, McGraw-Hill, New York, 1982.

Skolnik, M. I., Editor, *Radar Handbook*, 2nd edition, McGraw-Hill, New York, 1990.

Stearns, S. D. and David, R. A., *Signal Processing Algorithms*, Prentice-Hall, Englewood Cliffs, NJ, 1988.

Stimson, G. W., *Introduction to Airborne Radar*, Hughes Aircraft Company, El Segundo, CA, 1983.

Stratton, J. A., *Electromagnetic Theory*, McGraw-Hill, New York, 1941.

Stremler, F. G., *Introduction to Communication Systems*, 3rd edition, Addison-Wesley, New York, 1990.

Stutzman, G. E., *Estimating Directivity and Gain of Antennas*, IEEE Antennas and Propagation Magazine 40, August 1998, pp 7-11.

Swerling, P., *Probability of Detection for Fluctuating Targets*, IRE Transaction on Information Theory, Vol IT-6, April 1960, pp. 269-308.

Van Trees, H. L., *Detection, Estimation, and Modeling Theory*, Part I. Wiley & Sons, Inc., New York, 2001.

Van Trees, H. L., *Detection, Estimation, and Modeling Theory*, Part III.Wiley & Sons, Inc., New York, 2001.

Van Trees, H. L., *Optimum Array Processing*, Part IV of Detection, Estimation, and Modeling Theory, Wiley & Sons, Inc., New York, 2002.

Tzannes, N. S., *Communication and Radar Systems*, Prentice-Hall, Englewood Cliffs, NJ, 1985.

Urkowtiz, H., *Signal Theory and Random Processes*, Artech House, Norwood, MA, 1983.

Urkowitz, H., *Decision and Detection Theory*, Unpublished Lecture Notes, Lockheed Martin Co., Moorestown, NJ.

Vaughn, C. R., Birds and Insects as Radar Targets: A Review, *Proc. IEEE*, Vol. 73, No. 2, February 1985, pp. 205-227.

Wehner, D. R., *High Resolution Radar*, Artech House, Norwood, MA, 1987.

White, J. E., Mueller, D. D., and Bate, R. R., *Fundamentals of Astrodynamics*, Dover Publications, 1971.

Ziemer, R. E. and Tranter, W. H., *Principles of Communications, Systems, Modulation, and Noise,* 2nd edition, Houghton Mifflin, Boston, MA, 1985.

Zierler, N., *Several Binary-Sequence Generators*, MIT Technical Report No. 95, Sept. 1955.

Index

A

Active correlation, 247-254
  also see stretch processing and pulse compression
Ambiguity function, 187-224
  Barker code, 212-215, 229
  contour diagrams, 204-206
  ideal, 189
  LFM, 193-198
  PRN, 215-221, 231-233
  pulse train, 198-203
  single pulse, 188-193
  definition, 187
  properties, 188
Analytic signal, 143
Angle error measurement, 456
Angle tracking - *see* tracking
Antenna
  antenna pattern loss, 36-37
  aperture efficiency, 14, 320
  definition, 322, 323
  directivity, 319
  effective aperture, 14
  far field, 321
  gain, 14, 274, 319
  near field, 321
Arrays
  circular arrays, 344-352
  circular array using rectangular grid, 353,54
  concentric grid circular arrays, 352, 353
  general array, 321
  grating lobes, 327
  hexagonal grid, 354
  linear, 325-330
  planar arrays, 341-375
  rectangular arrays, 342-344
  scan loss, 375-378
  tapering, 330

B

Band pass waveforms, 141

Barker code - *see* binary phase codes, 210-214
  ambiguity diagrams, 212-214
  combined Barker codes, 211
Barrage jammers, *see* Jammers
Binary phase codes, 209
Blind Doppler, 295
Blind speeds, 295
Boltzman's constant, 15

C

Cancelers - *see* MTI
Chaff, 280, 485-493
Chirp waveforms
  down chirp, 149
  up chirp, 149
Clutter
  area clutter, 270-279
  chaff- *see* Chaff
  CNR, 275
  cross section, 274
  definition, 267-268
  main beam, 273
  regions, 268
  sidelode clutter, 273
  spectrum, 293-294, 295
  statistical models, 283-284, 294
  surface clutter, 268-270
  surface height irregularity, 269
  volume, 280-283
Code
  Barker, 209-215
  Costas, 206-208
  PRN, 215-222 *also see* Pseudo-Random Number
Coherent integration - *see* pulse integration
Complementary error function, 82
Compressed pulse width, 197
Compression gain, 198
Compression ratio, 198
Conical scan - *see* tracking
Constant false alarm rate (CFAR), 109-112

678 Index

110-111
cell-averaging CFAR (non-coherent integration), 111
cell-averaging CFAR (single pulse)
Convolution integral, 631
Correlation integral, 631
Costas codes - *see* frequency coding
Cross-over range, 474, 481
Cumulative probability of detection, 106-109

D

Delay line cancelers - *see* MTI
Detection in the presence of noise, 75-83
Detection of fluctuating targets - *see* probability of detection
Detection threshold, 79, 92, 93
Digital coded waveforms, 206-222
Directivity gain, 319
Discrete Fourier Transform, 647
Discrete power spectrum, 647-649
Distortion
  in pulse compression, 254-257
Doppler beam sharpening - see compression scaling factor
Doppler frequency, 7-13, 595
  Doppler resolution, 174-177
  Doppler scaling factor, 11
  Doppler shift spectra, 12
  velocity error, 456
Dynamic RCS, 558, 559
Duty factor, 4

E

Effective aperture, 14, 320, 377
Electronic countermeasure (ECM), 471-493, *also see* chaff
Energy spectrum density, 633
Euler's phi function, 218

F

False alarm time, 80
FFT parameters - selection, 648-649
Fixed gain tracking filters
  filter equation, 428
  the $\alpha\beta$ filter, 430-434
  the $\alpha\beta\gamma$ filter, 434-445
  steady state error, 430, 431

variance reduction ratio, 432, 433
Footprint, 590
Fourier series, 629-631
Fourier transform, 627
  properties, 627-628
  Table, 655-656
Frequency coding
  Costas codes, 206-209
Frequency modulation - *see* linear frequency modulation
Fresnel integrals, 150-152
Fresnel spectrum, 151
Frustum, *see* RCS
Fundamental matrix - *see* state transition matrix

G

Geometrical theory of diffraction, 504
gh filter - *see* $\alpha\beta$ filter under fixed gain tracking filters
ghk filter - *see* $\alpha\beta\gamma$ filter under fixed gain tracking filters
Gram-Charlier series coefficients, 95, 96, 100, 103
Grating lobes - *see* arrays
Grazing angle, 269, 270

H

Half cone angle, 524
Hamming - *see* windows
Hanning - *see* windows
High range resolution (HRR), 157-165

I

Incomplete gamma function, 92-93
Integration - *see* pulse integration

J

Jacobian, 77, 637
Jammers
  Barrage, 472
  burn through range, 476
  range reduction factor, 482
  Repeater, 472
  Self screening, 473-476
  Stand off, 480-482

K

Kalman filter, 445-453

Index 679

advantages, 445
equations, 446-447
relation to $\alpha\beta\gamma$ filter, 450-453
Singer-$\alpha\beta\gamma$ Kalman filter, 447-450
structure, 446

L

Linear frequency modulation,148-153
LFM waveforms - *see* waveforms
Line power spectrum, 648
Linear Shift Register, 216
Low pass waveforms - *see* waveforms
Lobe switching - *see* tracking, sequential lobing
Losses - *see* radar losses

M

Marcum's Q-function - *see* que-function
Matched filter
  causal, 168
  impulse response, 168
  non-causal, 168
  processor - *see* pulse compression
  response to LFM waveforms, 170-172
  SNR, 168, 169
Maximum length sequences, 215
Mie
  region, 518
  series, 518
Minimum detectable signal, 16
Monopulse - *see* tracking
Moving target indicator (MTI), 293-309
  delay line with feedback, 300-302
  double delay line, 298-300
  single delay lines, 296-298
MTI improvement factor, 303-309
  definition, 303
  general case, 309
  two-pulse MTI case, 307-308
Multiple MTI Doppler filter, 488-493
MyRadar design case study
  visit 1, 41-48
  visit 2, 113-117
  visit 3, 177-181
  visit 4, 223-224
  visit 5, 257-261
  visit 6, 284-288
  visit 7, 309-313
  visit 8, 378-380
  visit 9, 454-462
Multiple PRFs - *see* PRF staggering

N

Near field - *see* arrays
Newtonian matrix, 427
Noise
  bandwidth, 15, 69
  effective noise temperature, 15, 71
  noise figure, 15, 69-74
  noise power, 15
Non-coherent integration - *see* pulse integration
Number of false alarm, 80
Nyquist sampling rate, 643

O

Orthogonal functions, 629
Orthonormal functions, 629

P

Phi function, 218 - *see* Euler's phi function
Phased arrays - *see* arrays
Plank's constant, 69
Polarization, 510-515
Polynomial
  Characteristic, 218, 219
Power
  average - definition, 4
  peak - definition, 4
Power aperture product, 22
Power spectrum density, 632-633
PRN codes - *see* Pseudo random codes
Probability density functions
  Chi-square, 91, 657
  exponential, 657
  Gaussian, 657
  Laplace, 507
  Log normal, 283, 658
  Rayleigh, 283, 658
  Rice, 144
  uniform, 658
  Weibull, 658
Probability of detection, 81-83, 95
  cumulative, 106-107 - *also see* Cumulative probability of detection

Swerling I model, 97-99
Swerling II model, 100
Swerling III model, 100-103
Swerling IV model, 102-104
Swerling V model, 96-98
Probability of false alarm, 79-81, 111
Pseudo random codes, 215-222
   ambiguity diagrams, 219-222
Pulse compression, 235-247
   active correlation, 247-254
   correlation processor, 240-246
   LFM pulse compression, 237-239
   stretch processor, 247-254
   time bandwidth product, 235-236
Pulse energy, 4
Pulse integration, 28-31, 83-88
   coherent integration, 29, 83
   improvement factor, 86
   loss, 30, 88
   non-coherent integration, 30, 84
   time, 30
Pulse repetition frequency (PRF), 3
   staggering, 62,302
Pulse repetition interval, 4
Pulsed radar, 2, 59-68
   block diagram, 61
   high PRF, 60
   low PRF, 60
   medium PRF, 60
   resolving Doppler ambiguity, 64-67
   resolving range ambiguity, 61-64

Q

Quadrature components, 142, 242, 604, 618
Que-function, 79

R

Radar classifications, 1-2
Radar cross section (RCS), 15, 501-560
Radar equation, 13-19, 104
   area clutter, 270
   pulse compression, 236-238
   SAR - *see* synthetic aperture radar
   self-screening jammer, 473
   standoff-jammers (SOJ), 480
   surveillance (search), 20-25
   volume clutter, 280-283

   waveforms, 141-165
Radar losses
   antenna pattern, 36
   collapsing, 37
   processing losses, 38-40
   scan, 36, 37, 375-378
Random processes, 640-642
Random variables
   cdf, 635
   central moments, 636
   characteristic function, 637
   definition, 473
   joint cdf, 637, 639
   joint pdf, 637
   moments, 636
   pdf definition, 636
Range
   ambiguity, 60-64
   burn-through range, 476-478
   definition, 3-5
   ground projection, 594
   keep out, 44
   measurement error, 954
   profile, 165, 508
   reference, 19, 20
   resolution, 5, 6, 172, 176
   resolution in SFW, 157-159
   unambiguous, 5, 158
Range reduction factor, 482-484
Range rate, 59, 595
Repeater jammers - *see* Jammers
Replica, 169
Resolving Doppler ambiguity, 64-67
Resolving range ambiguity, 61-64
RCS fluctuation, 90-92
RCS of complex objects, 558-559
RCS of simple objects
   circular flat plate RCS, 522
   cylinder, 528
   dielectric capped wedge, 536
   definition, 501-503
   dependency on aspect angle, 504-508
   dependency on frequency, 504-508
   dependency on polarization, 508-515
   ellipsoid RCS, 520
   prediction methods, 503, 504
   rectangular flat plate, 531
   sphere RCS, 518

Index 681

triangular flat plate, 534
truncated cone (frustum), 524
RCS statistical models
  chi-square of degree, 559-560
  Swerling models - *see* probability of detection

S

Sampling theorem, 642-644
Scan loss - *see* array *and* radar losses
Scattering matrix, 515, 516
Schwartz inequality, 167
Scintillation - *see* target scintillation
Search volume, 21, 25, 45, 378
Signal-to-clutter-ratio, 271, 282
Signal-to-interference ratio, 275, 482
Signal-to-noise ratio (SNR)
  matched filter, 169
  minimum SNR, 16
Squint angle, 404
State transition matrix, 423, 424
State variable representation, 422-426
Stepped frequency waveforms (SFW) - *see* waveforms
Stretch processing, 247-254
  see Active correlation
Swath, 591
Swerling targets, 559-560
  also see detection probability of detection
Synthetic Aperture Radar (SAR)
  azimuth resolution, 599
  coherent integration interval, 590
  deramping, 604
  design considerations, 592-599
  Doppler history, 596
  Doppler processing, 601, 602
  maximum range, 592
  minimum range, 592
  PRF selection, 597
  range walk, 606
  SAR radar equation, 599, 600
  side looking, 590-599
  signal processing, 600
  unambiguous range, 596

T

Target fluctuation - *see* probability of detection

Target scintillation, 190-92
Threshold, 92, 93, 109, 111, *also see* minimum detectable signal
Time bandwidth product, 235-236- *also see* pulse compression
Time of false alarm - *see* false alarm time
Time on target, 22
Tracking
  amplitude comparison monopulse, 407-416
  angle tracking, 401
  conical scan, 403-407
  difference beam, 408
  mono pulse error, 414
  phase comparison monopulse, 416-418
  range tracking, 418-419
  sequential lobing, 402-403
  sum beam, 408
  track while scan (TWS), 420

U

Unambiguous range, 5, 158, 160, 596
Unambiguous range window, 158
Uncertainty function, 187
Uncertainty in Doppler, 174
Uncertainty in range, 172
U, V space, 343

V

Velocity distortion
  in SWF, 162-164
  in LFM, 254-257
Velocity error, 456
Velocity rms, 306

W

Waveforms
  CW, 144, 148
  LFM, 148-153
  low pass and high pass, 141-142
  stepped frequency, 153-165
Windowing techniques, 330-332, 649-653
Windows
  Hamming, 332, 650
  Hanning, 332, 650
  Kaiser, 332, 650
  rectangular, 332, 650

Z

Z-transform
  definition, 644
  discrete convolution, 646
  properties, 644, 645
  table, 659, 660